Interaction of Mechanics and Mathematics

Series editor

Lev Truskinovsky, Laboratoire de Mechanique des Solid, Palaiseau, France
e-mail: trusk@lms.polytechnique.fr

About this Series

The Interaction of Mechanics and Mathematics (IMM) series publishes advanced textbooks and introductory scientific monographs devoted to modern research in the wide area of mechanics. The authors are distinguished specialists with international reputation in their field of expertise. The books are intended to serve as modern guides in their fields and anticipated to be accessible to advanced graduate students. IMM books are planned to be comprehensive reviews developed to the cutting edge of their respective field and to list the major references.

Advisory Board

D. Colton, USA
R. Knops, UK
G. DelPiero, Italy
Z. Mroz, Poland
M. Slemrod, USA
S. Seelecke, USA
L. Truskinovsky, France

IMM is promoted under the auspices of ISIMM (International Society for the Interaction of Mechanics and Mathematics).

More information about this series at http://www.springer.com/series/5395

Taeyoung Lee • Melvin Leok
N. Harris McClamroch

Global Formulations of Lagrangian and Hamiltonian Dynamics on Manifolds

A Geometric Approach to Modeling and Analysis

 Springer

Taeyoung Lee [iD]
The George Washington University
Washington, District of Columbia, USA

Melvin Leok [iD]
Department of Mathematics
University of California, San Diego
La Jolla, California, USA

N. Harris McClamroch [iD]
Department of Aerospace Engineering
The University of Michigan
Ann Arbor, Michigan, USA

ISSN 1860-6245   ISSN 1860-6253 (electronic)
Interaction of Mechanics and Mathematics
ISBN 978-3-319-56951-2   ISBN 978-3-319-56953-6 (eBook)
DOI 10.1007/978-3-319-56953-6

Library of Congress Control Number: 2017938585

Mathematics Subject Classification (2010): 70-XX, 70-02, 70Exx, 70Gxx, 70Hxx

Printed on acid-free paper

This Springer imprint is published by Springer Nature
The registered company is Springer International Publishing AG
The registered company address is: Gewerbestrasse 11, 6330 Cham, Switzerland

Preface

This book is a tutorial on foundational geometric principles of Lagrangian and Hamiltonian dynamics and their application in studying important physical systems. As the title indicates, the emphasis is on describing Lagrangian and Hamiltonian dynamics in a form that enables global formulations and, where suitable mathematical tools are available, global analysis of dynamical properties. This emphasis on global descriptions, that is, descriptions that hold everywhere on the configuration manifold, as a means of determining global dynamical properties is in marked contrast to the most common approach in the literature on Lagrangian and Hamiltonian dynamics that makes use of local coordinates on the configuration manifold, thereby resulting in formulations that are typically limited to a small open subset of the configuration manifold. In this sense, the material that we introduce and develop represents a *significant conceptual departure* from the traditional methods of studying Lagrangian and Hamiltonian dynamics.

More specifically, this book differs from most of the traditional studies of analytical mechanics on Euclidean spaces, such as [13, 75]. Moreover, the global formulation of mechanics presented in this book should be distinguished from the geometric treatments that appear in [1, 10, 16, 25, 27, 37, 38, 39, 69, 70], which explicitly make use of local coordinates when illustrating the abstract formulation through specific examples. In contrast, we directly use the representations in the embedding space of the configuration manifold, without resorting to an atlas of coordinate charts. This allows us to obtain equations of motion that are globally valid and do not require changes of coordinates. This is particularly useful in constructing a compact and elegant form of Lagrangian and Hamiltonian mechanics for complex dynamical systems without algebraic constraints or coordinate singularities. This treatment is novel and unique, and it is the most important distinction and contribution of this monograph to the existing literature.

This book is the result of a research collaboration that began in 2005, when the first author initiated his doctoral research at the University of Michigan with the other two authors as his graduate advisers. That research program led to the completion of his doctoral degree and to numerous conference and journal publications.

The research plan, initiated in 2005, was based on our belief that there were advantages to be gained by the formulation, analysis, and computation of Lagrangian or Hamiltonian dynamics by explicitly viewing configurations of the system as elements of a manifold embedded in a finite-dimensional vector space. This viewpoint was not new in 2005, but we believed that the potential of this perspective had not been fully exploited in the research literature available at that time. This led us to embark on a long-term research program that would make use of powerful methods of variational calculus, differential geometry, and Lie groups for studying the dynamics of Lagrangian and Hamiltonian systems. Our subsequent research since 2005 confirms that there are important practical benefits to be gained by this perspective, especially for multi-body and other mechanical systems with dynamics that evolve in three dimensions.

This book arose from our research and the resulting publications in [21], [46, 47], and [49, 50, 51, 52, 53, 54, 55, 56, 57, 58, 59, 61, 62, 63] since 2005, but it goes substantially beyond this earlier work. During the writing of this book, we were motivated to consider many new issues that we had not previously studied; in this sense, all of Chapter 4 is new material. We also had many new insights and obtained new results that have not been previously published. Nevertheless, this book is intended to be a self-contained treatment containing many of the results of those publications plus new tutorial material to provide a unifying framework for Lagrangian and Hamiltonian dynamics on a manifold. As our research has progressed, we have come to realize the practical importance and effectiveness of this geometric perspective.

This book is not a complete treatment of Lagrangian and Hamiltonian dynamics; many important topics, such as geometric reduction, canonical transformations, Hamilton–Jacobi theory, Poisson geometry, and nonholonomic constraints, are not treated. These subjects are nicely covered in many excellent books [10, 37, 38, 39, 70]. All of these developments, as well as the development in this book, treat Lagrangian and Hamiltonian dynamics that are smooth in the sense that they can be described by differentiable vector fields. We note the important literature, summarized in [15], that treats non-smooth Lagrangian and Hamiltonian dynamics. A complete development of these topics, within the global geometric framework proposed in this book, remains to be accomplished.

The following manifolds, which naturally arise as configuration manifolds for Lagrangian and Hamiltonian systems, are of primary importance in our subsequent development. The standard linear vector spaces of two- and three-dimensional vectors are denoted by $\mathbb{R}^2$ and $\mathbb{R}^3$, endowed with the usual dot product operation; the cross product operation is also fundamental in $\mathbb{R}^3$. As usual, $\mathbb{R}^n$ denotes the linear space of ordered real n-tuples. All translations of subspaces in $\mathbb{R}^n$, e.g., lines, planes, and hyperplanes, are examples of embedded manifolds. The unit sphere in two dimensions is denoted by S^1; it is a one-dimensional manifold embedded in $\mathbb{R}^2$; similarly, the unit sphere in three dimensions is denoted by S^2; it is a two-dimensional manifold embedded in $\mathbb{R}^3$. The Lie group of orthogonal transformations in three dimensions is denoted by $SO(3)$. The Lie group of homogeneous transformations in three dimensions is denoted by $SE(3)$. Each of these Lie groups has an additional structure based on a group operation, which in each case corresponds to matrix multiplication. Finally, products of the above manifolds also commonly arise as configuration manifolds.

All of the manifolds that we consider are embedded in a finite-dimensional vector space. Hence, the geometry of these manifolds can be described using mathematical tools and operations in the embedding vector space. Although we are only interested in Lagrangian and Hamiltonian dynamics that evolve on such an embedded manifold, it is sometimes convenient to extend the dynamics to the embedding vector space. In fact, most of the results in the subsequent chapters can be viewed from this perspective.

It is important to justify our geometric assumption that the configurations constitute a manifold for Lagrangian and Hamiltonian systems. First, manifolds can be used to encode certain types of important motion constraints that arise in many mechanical systems; such constraints may arise from restrictions on the allowed motion due to physical restrictions. A formulation in terms of manifolds is a direct encoding of the constraints and does not require the use of additional holonomic constraints and associated Lagrange multipliers. Second, there is a beautiful theory of embedded manifolds, including Lie group manifolds, that can be brought to bear on the development of geometric mechanics in this context. It is important to recognize that configurations, as elements in a manifold, may often be described and analyzed in a globally valid way that does not require the use of local charts, coordinates, or parameters that may lead to singularities or ambiguities in the representation. We make extensive use of Euclidean frames in $\mathbb{R}^3$ and associated Euclidean coordinates in $\mathbb{R}^3$, $\mathbb{R}^n$, and $\mathbb{R}^{n \times n}$, but we do *not* use coordinates to describe the configuration manifolds. In this sense, this geometric formulation is said to be *coordinate-free*. Third, this geometric formulation turns out to be an efficient way to formulate, analyze, and compute the kinematics, dynamics, and their temporal evolution on the configuration manifold. This representational efficiency has a major practical advantage for many complex dynamical systems that has not been widely appreciated by the applied scientific and engineering communities. The associated cost of this efficiency is the requirement to make use of the well-developed mathematical machinery of manifolds, calculus on manifolds, and Lie groups.

We study dynamical systems that can be viewed as Lagrangian or Hamiltonian systems. Under appropriate assumptions, such dynamical systems are conservative in the sense that the Hamiltonian, which oftentimes coincides with the total energy of the system, is conserved. This is an ideal assumption but a very useful one in many applications. Although our main attention is given to dynamical systems that are conservative, many of the results can be extended to dissipative dynamical systems and to dynamical systems with inputs.

There are two basic requirements to make use of the Lagrangian perspective in obtaining the equations of motion. Based on the physical properties of the dynamical system, it is first necessary to select the set of possible configurations of the system and to identify the set of all configurations M as a manifold. The second requirement is to develop a Lagrangian function $L : \mathsf{T}M \to \mathbb{R}^1$ which is a real-valued function defined on the tangent bundle $\mathsf{T}M$ of the configuration manifold and satisfying certain assumptions. The Lagrangian function is the difference of the kinetic energy of the system and the potential energy of the system. It is assumed that the reader has sufficient background to construct the kinetic energy function and the potential energy function; we do not go into detail on the basic physics to construct these energy functions. Rather, numerous specific examples of Lagrangian and Hamiltonian systems are introduced and used to illustrate the concepts.

Hamilton's variational principle is the fundamental basis for the theory of Lagrangian and Hamiltonian dynamics. The action integral is the integral of the Lagrangian function over a fixed time period. Along a motion of the system, a specific value of the action integral is induced. Small variations of the system motion, which are consistent with the configuration manifold but not necessarily possible motions of the system, induce variations in the value of the action integral. Hamilton's variational principle states that these variations in the value of the action integral are necessarily of higher than first order for arbitrarily small variations about any system motion. In other words, the directional or Gateaux derivative of the action integral vanishes for all allowable variations of the system motion. Using methods of variational calculus where variations are introduced in terms of a small scalar parameter, this principle leads to Euler–Lagrange equations which characterize all possible system motions.

Hamilton's equations of motion are obtained by introducing the Legendre transformation that is a mapping from the tangent bundle of the configuration manifold to the cotangent bundle of the configuration manifold. A Hamiltonian function is introduced, and Hamilton's equations are obtained using a phase space version of Hamilton's variational principle. Methods of variational calculus are used to express the dynamics on the cotangent bundle of the configuration manifold.

It is admitted that some of the derivations are lengthy and the details and formulas are sometimes complicated. However, most of the formulations of Lagrangian and Hamiltonian dynamics on specific configuration manifolds, considered in this book, are relatively simple and elegant. Consequently, their application to the formulation of the dynamics of mass particles, rigid bodies, deformable bodies, and multi-body systems follows a relatively straightforward pattern that is, in fact, both more general and simpler than the traditional formulations that make use of local coordinates.

This book presents a unifying framework for this geometric perspective that we intend to be accessible to a wide audience. In concrete terms, the book is intended to achieve the following objectives:

- Study the geometric formulations of dynamical equations of motion for Lagrangian and Hamiltonian systems that evolve on a configuration manifold using variational methods.
- Express theoretical results in a global geometric form that does not require local charts or coordinates for the configuration manifold.
- Demonstrate simple methods for the analysis of solution properties.
- Present numerous illustrations of theory and analysis for the dynamics of multiple interacting particles and of rigid and deformable bodies.
- Identify theoretical and analytical benefits to be gained by the proposed treatment of geometric mechanics.

The book is also intended to set the stage for a treatment of computational issues associated with Lagrangian and Hamiltonian dynamics that evolve on a configuration manifold. In particular, the material in this book can be extended to obtain a framework for computational aspects of Lagrangian and Hamiltonian dynamics that achieve the analogous objectives:

- Study the geometric formulations of discrete-time dynamical equations of motion for Lagrangian and Hamiltonian systems that evolve on an embedded configuration manifold using discrete-time variational methods.
- Develop discrete-time versions of Lagrangian and Hamiltonian dynamics; these are referred to as geometric variational integrators to reflect the configuration manifold for the problems considered.
- Demonstrate the benefits of these discrete-time dynamics as a computational approximation of the continuous-time Lagrangian or Hamiltonian dynamics.
- Express computational dynamics in a global geometric form that does not require local charts.
- Present numerous computational illustrations for the dynamics of multiple interacting particles, and of rigid and deformable bodies.
- Identify computational benefits to be gained by the proposed treatment of geometric mechanics.

Computational developments for Lagrangian and Hamiltonian dynamics, following the above prescription, lead to computational algorithms that are not based on the discretization of differential equations on a manifold, but are based on the discretization of variational principles on a manifold. The above computational approach has been developed in [46, 50, 51, 54]. A symbolic approach to obtaining differential equations on a manifold has been proposed in [9], without addressing computational issues.

This book is written for a general audience of mathematicians, engineers, and physicists who have a basic knowledge of classical Lagrangian and Hamiltonian dynamics. Some background in differential geometry would be helpful to the reader, but it is not essential as arguments in the book make primary use of basic differential geometric concepts that are introduced in the book. Hence, our hope is that the material in this book is accessible to a wide range of readers.

In this book, Chapter 1 provides a summary of mathematical material required for the subsequent development; in particular, manifolds and Lie groups are introduced. Chapter 2 then introduces kinematics relationships for ideal particles, rigid bodies, multi-bodies, and deformable bodies, expressed in terms of differential equations that evolve on a configuration manifold.

Chapter 3 treats the classical approach to variational mechanics where the configurations lie in an open set of a vector space $\mathbb{R}^n$. This is standard material, but the presentation provides a development that is followed in subsequent chapters. Chapters 4 and 5 develop the fundamental results for Lagrangian and Hamiltonian dynamics when the configuration manifold $(\mathsf{S}^1)^n$ is the product of n copies of the one-sphere in $\mathbb{R}^2$ (in Chapter 4) and the configuration manifold $(\mathsf{S}^2)^n$ is the product of n copies of the two-sphere in $\mathbb{R}^3$ (Chapter 5). The geometries of these two configuration manifolds are exploited in the developments, especially the definitions of variations. Chapter 6 introduces the geometric approach for rigid body rotation in three dimensions using configurations in the Lie group $\mathsf{SO}(3)$. The development follows Chapter 3, Chapter 4, and Chapter 5, except that the variations are carefully defined to be consistent with the Lie group structure of $\mathsf{SO}(3)$. Chapter 7 introduces the geometric approach for rigid body rotation and translation in three dimensions using configurations in the Lie group $\mathsf{SE}(3)$. The development reflects the fact that the variations are defined to be consistent with the Lie group structure of $\mathsf{SE}(3)$. The results in Chapters 3–7 are developed using only well-known results from linear algebra and elementary properties of orthogonal matrices and skew-symmetric matrices; minimal knowledge of differential geometry or Lie groups is required, and all of it is introduced in the book.

Chapter 8 makes use of the notation and formalism of differential geometry and Lie groups. This mathematical machinery enables the development of Lagrangian and Hamiltonian dynamics with configurations that lie in an arbitrary differentiable manifold, an arbitrary matrix Lie group, or an arbitrary homogeneous manifold (a manifold that is transitive with respect to a Lie group action). The power of this mathematical formalism is that it allows a relatively straightforward development that follows the variational calculus approach of the previous chapters and it results in a simple abstract statement of the results in each case. The development, however, does require a level of abstraction and some knowledge of differential geometry and Lie groups.

Chapter 9 makes use of the prior results to treat the dynamics of various multi-body systems; Chapter 10 treats the dynamics of various deformable multi-body systems. In each of these example illustrations, the equations of motion are obtained in several different forms. The equations of motion are used to study conservation properties and equilibrium properties in each example illustration. The book concludes with two appendices that provide brief summaries of fundamental lemmas of the calculus of variations and procedures for linearization of a vector field on a manifold in a neighborhood of an equilibrium solution.

Numerous examples of mechanical and multi-body systems are developed in the text and introduced in the end of chapter problems. These examples form a core part of the book, since they illustrate the way in which the developed theory can be applied in practice. Many of these examples are classical, and they are studied in the existing literature using local coordinates; some of the examples are apparently novel. Various multi-body examples, involving pendulums, are introduced, since these provide good illustrations for the theory. The books [6, 29] include many examples developed using local coordinates.

This book could form the basis for a graduate-level course in applied mathematics, classical physics, or engineering. For students with some prior background in differential geometry, a course could begin with the theoretical material in Chapter 8 and then cover applications in Chapters 3–7 and 9–10 as time permits. For students with primary interest in the applications, the course could treat the topics in the order presented in the book, covering the theoretical topics in Chapter 8 as time permits. This book is also intended for self-study; these two paths through the material in the book may aid readers in this category.

In conclusion, the authors are excited to share our perspective on "global formulations of Lagrangian and Hamiltonian dynamics on manifolds" with a wide audience. We welcome feedback about theoretical issues the book introduces, the practical value of the proposed perspective, and indeed any aspect of this book.

TAEYOUNG LEE
Washington, DC

MELVIN LEOK
La Jolla, CA

N. HARRIS MCCLAMROCH
Ann Arbor, MI

January, 2017

Acknowledgments

This book provides our perspective on geometric mechanics, and it depends on the foundations provided by many scholars and researchers. Although this book is somewhat unique in its geometric perspective, we recognize the long history of Lagrangian and Hamiltonian dynamics and the many persons over the last two centuries who have contributed to its current status as a central subject area in mathematics, physics, and engineering.

The three authors would like to acknowledge several persons who have had great influence on our understanding of geometric mechanics. Our professional colleagues Tony Bloch, Dennis Bernstein, Jessy Grizzle, John Baillieul, Naomi Leonard, Francesco Bullo, and P. S. Krishnaprasad have had significant influence through their outstanding research and personal interactions.

TL wishes to thank Harris McClamroch, Melvin Leok, and Youdan Kim for their contributions to his inspiration and knowledge and especially for showing how to enjoy life in academia while being respected by peers and students. He personally dedicates this book to his wife Rachel with love, and he wishes that this book would help his two-year-old son Nathan soon to understand that the trains that he loves most wouldn't have been built without math.

ML dedicates this book to his wife Lina, for her love, support, and patience; to his children Richard and Catherine, for constantly reminding him that the world is full of curious and wonderful things; and to the memory of his adviser Jerry Marsden, for teaching him by example that mathematics at its best is both beautiful and useful.

Finally, NHM would like to acknowledge the contributions that his doctoral students Danwei Wang, Mahmut Reyhanoglu, Hariharan Krishnan, Ilya Kolmanovsky, Chunlei Rui, Sangbum Cho, Jinglai Shen, Amit Sanyal, Nalin Chaturvedi, and Taeyoung Lee have made to his understanding of geometric mechanics. His doctoral adviser, Lyle Clark at the University of Texas in the

1960s, did not know anything of the modern treatment of geometric mechanics, but his insight into and passion for analytical mechanics had a profound influence.

All three of us have been inspired by the seminal contributions of Professor Jerry Marsden through his publications and his personal example. His deep mathematical and physical insights and the clarity of his written publications and his oral presentations have inspired us and motivated us in the writing of this book.

The authors gratefully acknowledge support from the National Science Foundation.

TL

ML

NHM

Contents

List of Symbols

Vectors and Matrices

$\mathbb{R}^n$ Vector space of n-tuples of real numbers over the real field

$\mathbb{R}^{n \times m}$ Vector space of matrices with n rows and m columns over the real field

$I_{n \times n}$ $n \times n$ identity matrix

$e_1, \ldots, e_n$ Standard basis vectors in $\mathbb{R}^n$

$x \cdot y$ Inner or dot product of vectors in $\mathbb{R}^n$

$x \times y$ Cross product of vectors in $\mathbb{R}^3$

$\|x\|$ Norm of vector in $\mathbb{R}^n$

A^T Matrix transpose

A^{-1} Matrix inverse

$\text{trace}(A)$ Trace of a matrix

$\mathcal{N}(A)$ Null space of a matrix A

$S = \begin{bmatrix} 0 & -1 \\ 1 & 0 \end{bmatrix}$ 2×2 skew-symmetric matrix

$S(x)$ Isomorphism between $\mathbb{R}^3$ and $\mathfrak{so}(3)$, or cross product operator on $\mathbb{R}^3$, that is, $S(x)y = x \times y$ and $S(x)^T = -S(x)$

$\mathcal{S}(\omega, v) = \mathcal{S}\left(\begin{bmatrix} \omega \\ v \end{bmatrix}\right)$ Isomorphism between $\mathbb{R}^6$ and $\mathfrak{se}(3)$, defined as

$$\mathcal{S}(\omega, v) = \begin{bmatrix} S(\omega) & v \\ 0 & 0 \end{bmatrix} \in \mathbb{R}^{4 \times 4}$$

Manifolds

$GL(n)$ The set of $n \times n$ real nonsingular matrices

M Differentiable, embedded manifold

S^1 Unit sphere manifold embedded in $\mathbb{R}^2$

S^2	Unit sphere manifold embedded in $\mathbb{R}^3$
$SO(3)$	Special orthogonal group, that is, the set of all 3×3 orthogonal matrices with determinant $+1$
$SE(3)$	Special Euclidean group
$(M)^n$	Manifold, that is, the n-fold product $M \times \cdots \times M$
$T_x M$	Tangent space of the manifold M at $x \in M$
TM	Tangent bundle of the manifold M
$T_x^* M$	Cotangent space of the manifold M at $x \in M$
$T^* M$	Cotangent bundle of the manifold M
$P(x) : \mathbb{R}^n \to T_x M$	Orthogonal projection for $x \in M$
$f : M \to \mathbb{R}^n$	Vector-valued function defined on a manifold M
$\frac{\partial f(x)}{\partial x} = D_x f(x)$	Derivative function
$x(t, t_0, x_0) \in M$	Solution of initial-value problem at t corresponding to initial time t_0 and initial-value $x_0 \in M$ associated with a vector field on a manifold M
$F_{t,t_0} : M \to M$	Flow map associated with a vector field on a manifold M

Mechanics

q	Configuration vector in $\mathbb{R}^n$ or S^1 or S^2
R	Configuration in $SO(3)$, referred to as an attitude or rotation matrix
(R, x)	Configuration in $SE(3)$
$\omega \in T_q S^1$	Scalar angular velocity associated with configuration $q \in S^1$
$\omega \in T_q S^2$	Angular velocity vector associated with configuration $q \in S^2$
$\omega \in T_R SO(3)$	Angular velocity vector associated with configuration $R \in SO(3)$
$\Omega \in T_R SO(3)$	Angular velocity vector associated with configuration $R \in SO(3)$
$\mathfrak{G}$	Action integral
$\delta(\cdot)$	Infinitesimal variation
$T : TM \to \mathbb{R}^1$	Kinetic energy function
$U : M \to \mathbb{R}^1$	Potential energy function
$L : TM \to \mathbb{R}^1$	Lagrangian function, $L(q, \dot{q}) = T(q, \dot{q}) - U(q)$
$E : TM \to \mathbb{R}^1$	Energy function, $E(q, \dot{q}) = T(q, \dot{q}) + U(q)$
$\lambda_1, \ldots, \lambda_m$	Lagrange multipliers
$L^a : TM \times \mathbb{R}^m \to \mathbb{R}^1$	Augmented Lagrangian function
$\tilde{L} : TM \to \mathbb{R}^1$	Modified Lagrangian function
p, π, Π, μ	Conjugate momentum
$H : T^* M \to \mathbb{R}^1$	Hamiltonian function
$H^a : T^* M \times \mathbb{R}^m \to \mathbb{R}^1$	Augmented Hamiltonian function
$\tilde{H} : T^* M \to \mathbb{R}^1$	Modified Hamiltonian function

$R(q, \dot{q})$	Routhian function
Ω_{can}	Symplectic form
$\mathbb{J}$	Symplectic matrix
$\int_{\mathcal{B}}$	Integral over material points that constitute a rigid body $\mathcal{B}$

Lie Groups and Lie Algebras

G	Lie group
$\mathfrak{g}$	Lie algebra
$\mathsf{L}_h : \mathsf{G} \to \mathsf{G}$	Left translation map
$\mathsf{R}_h : \mathsf{G} \to \mathsf{G}$	Right translation map
$\exp : \mathfrak{g} \to \mathsf{G}$	Exponential map
$\mathrm{Ad}_g : \mathfrak{g} \to \mathfrak{g}$	Adjoint operator
$\mathrm{ad}_\xi : \mathfrak{g} \to \mathfrak{g}$	Adjoint operator
$\mathrm{Ad}_g^* : \mathsf{G} \times \mathfrak{g}^* \to \mathfrak{g}^*$	Coadjoint operator
$\mathrm{ad}^* : \mathfrak{g} \times \mathfrak{g}^* \to \mathfrak{g}^*$	Coadjoint operator

Chapter 1
Mathematical Background

Dynamical systems are most naturally described in the language of differential equations. For many, but not all dynamical systems, the variables that describe the motion dynamics can be viewed as elements of a finite-dimensional vector space, e.g., $\mathbb{R}^n$. But there are important situations where the variables that describe the dynamics do not lie in a vector space but rather lie in a set with a different mathematical structure, in particular a manifold structure. In this case, the solution flow of the differential equations that describe the dynamics of the system must evolve on this manifold.

In this chapter, we summarize the mathematical background that is used subsequently. Important results in linear algebra are introduced for finite-dimensional vectors and matrices viewed as linear transformations. A summary is given of manifold concepts and related differential geometric concepts are introduced; a summary of results for vector fields on a manifold is given. Further mathematical background is presented in Chapter 8, where additional details on Lie groups and homogeneous manifolds are provided. That material appears there since it is not required for the prior developments.

1.1 Vectors and Matrices

A vector is an n-tuple of real numbers. Vector addition and scalar multiplication are defined as usual. A matrix is an $n \times m$ ordered array of real numbers. Matrix addition, for compatible matrices, and scalar multiplication are defined as usual.

The transpose of an $n \times m$ matrix A is an $m \times n$ matrix, denoted by A^T, obtained by interchange of the rows and columns. The $n \times n$ identity matrix is denoted by $I_{n \times n}$. The $n \times m$ *zero* matrix composed of zero elements is denoted by $0_{n \times m}$ or more often by 0.

© Springer International Publishing AG 2018
T. Lee et al., *Global Formulations of Lagrangian and Hamiltonian Dynamics on Manifolds*, Interaction of Mechanics and Mathematics,
DOI 10.1007/978-3-319-56953-6_1

Vector spaces in this book should be understood as being defined over the real field; the only exception occurs when we occasionally use eigenvalue and eigenvector concepts, in which case the field is the complex field. An excellent reference on matrix theory is [7], while a comprehensive treatment is given in [8].

1.1.1 Vector Spaces

As usual, $\mathbb{R}^n$ denotes the set of all ordered n-tuples of real numbers, with the usual definition of vector addition and scalar multiplication. Thus, $\mathbb{R}^n$ is a real vector space. Also, $\mathbb{R}^{n \times m}$ denotes the set of all $n \times m$ real matrices consisting of n rows and m columns. With the usual definition of matrix addition and scalar multiplication of a matrix, $\mathbb{R}^{n \times m}$ is a real vector space. Unless indicated otherwise, we view an n-tuple of real numbers as a *column vector* and we view a matrix as an array of real numbers.

The common notions of span, linear independence, basis, and subspace are fundamental. The dimension of a vector space is the number of elements in a basis.

Linear transformations can be defined from the vector space $\mathbb{R}^m$ to the vector space $\mathbb{R}^n$; such a linear transformation can be represented by a real matrix in $\mathbb{R}^{n \times m}$. Linear transformations from a vector space to the vector space $\mathbb{R}^1$ are referred to as linear functionals.

We also make use of common matrix notions of rank, determinant, singular matrix, and eigenvalues and eigenvectors.

The usual Euclidean inner product or dot product of two vectors x, $y \in \mathbb{R}^n$ is the real number

$$x \cdot y = x^T y = y^T x, \tag{1.1}$$

and the Euclidean norm on the real vector space $\mathbb{R}^n$ is the nonnegative real number

$$\|x\| = \sqrt{x^T x}. \tag{1.2}$$

Vectors x, $y \in \mathbb{R}^n$ that satisfy $x \cdot y = 0$ are said to be orthogonal or normal. If $y \in \mathbb{R}^n$ is nonzero, then the set of vectors that are orthogonal to y is referred to as the orthogonal complement of y; the orthogonal complement is an $(n-1)$-dimensional subspace of $\mathbb{R}^n$.

Suppose that a vector $y \in \mathbb{R}^n$ is nonzero, then a vector $x \in \mathbb{R}^n$ can be uniquely decomposed into the linear combination of a vector in the direction of $y \in \mathbb{R}^n$ and a vector orthogonal to $y \in \mathbb{R}^n$; this is given by the expression

$$x = \frac{(y^T x)y}{\|y\|^2} + \left\{ I_{n \times n} - \frac{yy^T}{\|y\|^2} \right\} x. \tag{1.3}$$

The first term on the right is the component of $x \in \mathbb{R}^n$ in the direction of $y \in \mathbb{R}^n$, referred to as the orthogonal projection of x onto y. The second term on the right is the component of $x \in \mathbb{R}^n$ orthogonal to $y \in \mathbb{R}^n$. These properties are easily verified. Since the two terms in the decomposition are orthogonal, this is referred to as an orthogonal decomposition. The above decomposition formula is simplified if $\|y\| = 1$.

The standard basis vectors in $\mathbb{R}^n$ are denoted by $e_1, \ldots, e_n$ where $e_i \in \mathbb{R}^n$ denotes the n-tuple with 1 in the i-th place and zeros elsewhere. Note that each of the standard basis vectors has unit norm and they are mutually orthogonal, that is, they form a set of orthonormal vectors. In particular, in $\mathbb{R}^2$ the standard basis vectors are $e_1 = [1,0]^T$, $e_2 = [0,1]^T$; in $\mathbb{R}^3$ the standard basis vectors are $e_1 = [1,0,0]^T$, $e_2 = [0,1,0]^T$, $e_3 = [0,0,1]^T$.

Important subsets of the real vector space $\mathbb{R}^n$ are the linear subspaces, namely subsets that are closed under the operations of vector addition and scalar multiplication. Examples of important subspaces are the span of a set of vectors and the orthogonal complement of a set of vectors. There are important subsets of $\mathbb{R}^n$, other than subspaces; we introduce subsets which are referred to as manifolds subsequently. All of these concepts can be given an important geometric interpretation.

A linear transformation $A : \mathbb{R}^m \to \mathbb{R}^n$ can be represented, with respect to basis sets for $\mathbb{R}^m$ and $\mathbb{R}^n$, by an $n \times m$ matrix that we also denote by A. We denote the set of all real $n \times m$ matrices by $\mathbb{R}^{n \times m}$. As mentioned previously, $\mathbb{R}^{n \times m}$ is a vector space, and it is isomorphic to $\mathbb{R}^{nm}$. If A is a linear transformation on $\mathbb{R}^n$, or equivalently if $A \in \mathbb{R}^{n \times n}$, then A is singular if there is a nonzero $x \in \mathbb{R}^n$ such that $Ax = 0$. If A is nonsingular, there is an inverse linear transformation, or equivalently a matrix in $\mathbb{R}^{n \times n}$ denoted by A^{-1}, that is nonsingular and satisfies $AA^{-1} = A^{-1}A = I_{n \times n}$.

Important subspaces associated with a matrix $A \in \mathbb{R}^{n \times n}$ are the range of A, the null space of A, denoted by $\mathcal{N}(A)$, and the invariant subspaces of the matrix A; these invariant subspaces include the eigenspaces of A associated with the eigenvalues of A, which are the complex numbers $s \in \mathbb{C}$ for which the matrix $sI_{n \times n} - A$ is singular.

An important linear functional on $\mathbb{R}^{n \times n}$ is the trace, which is the sum of the diagonal terms. In particular, the inner product $A \cdot B$ of two matrices $A, B \in \mathbb{R}^{n \times n}$ is the real number

$$A \cdot B = \mathsf{trace}(A^T B) = \mathsf{trace}(B^T A). \tag{1.4}$$

Thus, the norm on the real vector space $\mathbb{R}^{n \times n}$ is the nonnegative real number

$$\|A\| = \sqrt{\mathsf{trace}(A^T A)}. \tag{1.5}$$

The determinant is not a linear functional on $\mathbb{R}^{n \times n}$, but it can be viewed as a multilinear functional on $\mathbb{R}^n \times \cdots \times \mathbb{R}^n$. It has the important property that $\det(A) \neq 0$ if and only if $A \in \mathbb{R}^{n \times n}$ is nonsingular.

1.1.2 Symmetric and Skew-Symmetric Matrices

A symmetric matrix $A \in \mathbb{R}^{n \times n}$ has the property that $A = A^T$. It is easy to show that the set of all symmetric matrices is a subspace of $\mathbb{R}^{n \times n}$. A basis of $\frac{n(n+1)}{2}$ symmetric matrices can be constructed for the subspace of symmetric matrices; the dimension of the subspace of symmetric matrices is $\frac{n(n+1)}{2}$. For example, the standard basis for all 2×2 real symmetric matrices is

$$\begin{bmatrix} 1 & 0 \\ 0 & 0 \end{bmatrix}, \begin{bmatrix} 0 & 1 \\ 1 & 0 \end{bmatrix}, \begin{bmatrix} 0 & 0 \\ 0 & 1 \end{bmatrix},$$

and the standard basis for all 3×3 real symmetric matrices is

$$\begin{bmatrix} 1 & 0 & 0 \\ 0 & 0 & 0 \\ 0 & 0 & 0 \end{bmatrix}, \begin{bmatrix} 0 & 1 & 0 \\ 1 & 0 & 0 \\ 0 & 0 & 0 \end{bmatrix}, \begin{bmatrix} 0 & 0 & 0 \\ 0 & 1 & 0 \\ 0 & 0 & 0 \end{bmatrix}, \begin{bmatrix} 0 & 0 & 1 \\ 0 & 0 & 0 \\ 1 & 0 & 0 \end{bmatrix}, \begin{bmatrix} 0 & 0 & 0 \\ 0 & 0 & 1 \\ 0 & 1 & 0 \end{bmatrix}, \begin{bmatrix} 0 & 0 & 0 \\ 0 & 0 & 0 \\ 0 & 0 & 1 \end{bmatrix}.$$

A skew-symmetric matrix $A \in \mathbb{R}^{n \times n}$ has the property that $A + A^T = 0$. A basis of $\frac{n(n-1)}{2}$ skew-symmetric matrices can be constructed for the subspace of skew-symmetric matrices; the dimension of the subspace of skew-symmetric matrices is $\frac{n(n-1)}{2}$. For example, the standard basis for all 2×2 real skew-symmetric matrices is

$$\begin{bmatrix} 0 & -1 \\ 1 & 0 \end{bmatrix},$$

and the standard basis for all 3×3 real skew-symmetric matrices is

$$\begin{bmatrix} 0 & -1 & 0 \\ 1 & 0 & 0 \\ 0 & 0 & 0 \end{bmatrix}, \begin{bmatrix} 0 & 0 & 1 \\ 0 & 0 & 0 \\ -1 & 0 & 0 \end{bmatrix}, \begin{bmatrix} 0 & 0 & 0 \\ 0 & 0 & -1 \\ 0 & 1 & 0 \end{bmatrix}.$$

A simple but important result is that each matrix $A \in \mathbb{R}^{n \times n}$ can be written as the sum of a symmetric matrix and a skew-symmetric matrix, namely

$$A = \frac{1}{2}\left(A + A^T\right) + \frac{1}{2}\left(A - A^T\right).$$

Symmetric $n \times n$ matrices are useful in defining quadratic forms on $\mathbb{R}^n$. The eigenvalues of symmetric matrices are necessarily real; the associated quadratic form is positive-definite, negative-definite, positive-semidefinite, or negative-semidefinite if the eigenvalues of the symmetric matrix that defines the quadratic form are all positive, negative, nonnegative, or nonpositive, respectively.

1.1.3 Vector Operations in $\mathbb{R}^2$

The real vector space $\mathbb{R}^2$ has special importance, reflecting our interest in describing objects that have a conceptual existence in two-dimensional space. An important vector operation on pairs of vectors in $\mathbb{R}^2$ is the dot product. The dot product, also known as the inner product, of two vectors x, $y \in \mathbb{R}^2$ is the real number

$$x \cdot y = x^T y.$$

Thus, the norm in $\mathbb{R}^2$ is given by

$$\|x\| = +\sqrt{x^T x}.$$

We will also make use of the 2×2 skew-symmetric matrix

$$S = \begin{bmatrix} 0 & -1 \\ 1 & 0 \end{bmatrix}, \tag{1.6}$$

which rotates a vector by $\frac{\pi}{2}$ counterclockwise. Thus, for any $x \in \mathbb{R}^2$, the vector $Sx \in \mathbb{R}^2$ is orthogonal to $x \in \mathbb{R}^2$ since $x^T S x = 0$.

Although the cross product can be defined on $\mathbb{R}^3$ as shown in the next section, the cross product is not defined on $\mathbb{R}^2$.

Consider two unit vectors x, $y \in \mathbb{R}^2$, which we can think of as direction vectors in $\mathbb{R}^2$. Plane geometry allows us to define the angle θ between these two unit vectors. The angle θ, assumed to lie in the interval $[0, 2\pi)$, can be shown to satisfy

$$\cos \theta = x^T y,$$
$$\sin \theta = x^T S^T y.$$

Consequently, if $\theta = 0$, it follows that $x^T y = 1$ and $x^T S^T y = 0$; that is, the two vectors are collinear. If $\theta = \frac{\pi}{2}$, it follows that $x^T y = 0$ and $x^T S^T y = 1$; that is, the two vectors are orthogonal.

1.1.4 Vector Operations in $\mathbb{R}^3$

The real vector space $\mathbb{R}^3$ has special importance, reflecting our interest in describing physical objects that exist in three-dimensional space. Two important vector operations on pairs of vectors in $\mathbb{R}^3$ are the dot product and the cross product. The dot product, also known as the inner product, of two vectors x, $y \in \mathbb{R}^3$ is the real number

$$x \cdot y = x^T y.$$

Then, as before, the norm in $\mathbb{R}^3$ is given by

$$\|x\| = +\sqrt{x^T x}.$$

The cross product of two vectors x, $y \in \mathbb{R}^3$ is the real vector in $\mathbb{R}^3$

$$x \times y = \begin{bmatrix} x_2 y_3 - x_3 y_2 \\ x_3 y_1 - x_1 y_3 \\ x_1 y_2 - x_2 y_1 \end{bmatrix}. \tag{1.7}$$

The cross product, viewed as an operator on $\mathbb{R}^3$, can also be written in terms of a linear transformation and a matrix multiplication as

$$x \times y = S(x)y,$$

where $S : \mathbb{R}^3 \to \mathbb{R}^{3 \times 3}$ is

$$S(x) = \begin{bmatrix} 0 & -x_3 & x_2 \\ x_3 & 0 & -x_1 \\ -x_2 & x_1 & 0 \end{bmatrix}, \tag{1.8}$$

the skew-symmetric matrix-valued function on $\mathbb{R}^3$.

Thus, for any x, $y \in \mathbb{R}^3$, the vector $S(x)y \in \mathbb{R}^3$ is orthogonal to $y \in \mathbb{R}^3$ since $S(x)y \cdot y = y^T S(x)^T y = 0$. Note that $\det(S(x)) = 0$, for all $x \in \mathbb{R}^3$, so that each 3×3 skew-symmetric matrix is singular and does not have a matrix inverse. Alternatively, any 3×3 skew-symmetric matrix can be viewed as the image of a linear transformation on $\mathbb{R}^3$ that represents the cross product operation, $y \mapsto x \times y$, for some $x \in \mathbb{R}^3$.

The shorthand notation

$$\hat{x} = S(x)$$

is sometimes convenient, where we view $S : \mathbb{R}^3 \to \mathfrak{so}(3)$, where $\mathfrak{so}(3)$ denotes the subspace of 3×3 skew-symmetric matrices. Thus, the linear transformation _hat_ map $\hat{\ } : \mathbb{R}^3 \to \mathfrak{so}(3)$ transforms a vector x in $\mathbb{R}^3$ to a 3×3 skew-symmetric matrix $\hat{x}$ such that $\hat{x}y = x \times y$ for any $y \in \mathbb{R}^3$. This linear transformation has an inverse and the inverse of the hat map is denoted by the _vee_ map $\cdot^\vee : \mathfrak{so}(3) \to \mathbb{R}^3$. This demonstrates an isomorphism between the vector spaces $\mathbb{R}^3$ and $\mathfrak{so}(3)$.

Consider two unit vectors x, y in $\mathbb{R}^3$, which we can think of as direction vectors in $\mathbb{R}^3$. The two unit vectors define a plane in $\mathbb{R}^3$ wherein the angle θ between these two unit vectors can be defined. The angle θ, assumed to lie in the interval $[0, 2\pi)$, can be shown to satisfy

$$\cos\theta = x^T y,$$
$$e \sin\theta = S(x)y,$$

for a unit vector $e \in \mathbb{R}^3$ that is orthogonal to the plane defined by the two unit vectors; the vector $e \in \mathbb{R}^3$ is defined according to the *right-hand rule*, that is, $e = \frac{x \times y}{\|x \times y\|}$. Note that $\sin \theta$ also satisfies

$$\sin \theta = \sqrt{1 - (x^T y)^2}.$$

Consequently, if the two unit vectors are collinear, then either $\theta = 0$ and it follows that $x^T y = 1$ and $S(x)y = 0$, or $\theta = \pi$ and it follows that $x^T y = -1$ and $S(x)y = 0$. If the two unit vectors are orthogonal, then either $\theta = \frac{\pi}{2}$ and it follows that $x^T y = 0$ and $S(x)y = 1$, or $\theta = \frac{3\pi}{2}$ and it follows that $x^T y = 0$ and $S(x)y = -1$.

1.1.5 Orthogonal Matrices on $\mathbb{R}^3$

We use the notation $\mathsf{GL}(n)$ for the set of all $n \times n$ real nonsingular matrices. Since $\mathsf{GL}(n)$ is a subset of $\mathbb{R}^{n \times n}$, the inner product and the norm for matrices in $\mathsf{GL}(n)$ are defined. It can be shown that $\mathsf{GL}(n)$ has the properties of a group, formally introduced shortly, where matrix multiplication is the group operation.

An important subset of $\mathsf{GL}(3)$ is the set of real 3×3 orthogonal matrices, that is, matrices whose inverses are equal to their transposes. In other words, a matrix $R \in \mathbb{R}^{3 \times 3}$ is orthogonal if

$$RR^T = I_{3 \times 3}, \qquad R^T R = I_{3 \times 3}. \tag{1.9}$$

Orthogonal matrices have the property that their columns, as vectors in $\mathbb{R}^3$, are orthonormal and their rows, as vectors in $\mathbb{R}^3$, are orthonormal. An orthogonal matrix can be viewed as an invertible linear transformation on $\mathbb{R}^3$. The set of 3×3 orthogonal matrices, with determinant $+1$, is denoted subsequently as $\mathsf{SO}(3)$, referred to as the special orthogonal group or the group of rotations.

There is a close connection between skew-symmetric matrices in $\mathfrak{so}(3)$ and orthogonal matrices in $\mathsf{SO}(3)$ in the sense that an orthogonal matrix can be expressed in terms of a skew-symmetric matrix. We now show two different representations that illustrate this connection.

For each orthogonal matrix $R \in \mathsf{SO}(3)$, there is a skew-symmetric matrix $\xi \in \mathfrak{so}(3)$ such that $I_{3 \times 3} + \xi$ is nonsingular and the orthogonal matrix can be expressed as:

$$R = (I_{3 \times 3} - \xi)(I_{3 \times 3} + \xi)^{-1}. \tag{1.10}$$

This provides an expression for an orthogonal matrix in terms of a skew-symmetric matrix. This relationship is often referred to as the Cayley transformation.

The following is also true. For each orthogonal matrix $R \in \mathsf{SO}(3)$, there is a skew-symmetric matrix $\xi \in \mathfrak{so}(3)$ such that the orthogonal matrix can be expressed as:

$$R = e^{\xi} = \sum_{n=0}^{\infty} \frac{(\xi)^n}{n!}, \tag{1.11}$$

where the right-hand side gives the definition of the matrix exponential. This provides another expression for an orthogonal matrix in terms of a skew-symmetric matrix, using the exponential map.

Since the vee map provides an isomorphism between skew-symmetric matrices and vectors in $\mathbb{R}^3$, the above properties can also be expressed in the following ways. For each orthogonal matrix $R \in \mathsf{SO}(3)$, there is a vector $x \in \mathbb{R}^3$, such that the orthogonal matrix can be expressed in terms of the Cayley transformation as

$$R = (I_{3 \times 3} - S(x))(I_{3 \times 3} + S(x))^{-1}. \tag{1.12}$$

For each orthogonal matrix $R \in \mathsf{SO}(3)$, there is a vector $x \in \mathbb{R}^3$ such that the orthogonal matrix can be expressed in terms of the matrix exponential as

$$R = e^{S(x)} = \sum_{n=0}^{\infty} \frac{(S(x))^n}{n!}. \tag{1.13}$$

The Cayley and exponential representations for elements in $\mathsf{SO}(3)$ have important theoretical and computational implications. In particular, they suggest that various operations and computations involving elements in $\mathsf{SO}(3)$ correspond to vector space operations and computations on the associated vector spaces $\mathfrak{so}(3)$ or equivalently on $\mathbb{R}^3$. However, these representations in terms of skew-symmetric matrices $\mathfrak{so}(3)$ are invertible only in a neighborhood of the identity transformation in $\mathsf{SO}(3)$, so the identification does not hold globally on $\mathsf{SO}(3)$.

1.1.6 Homogeneous Matrices as Actions on $\mathbb{R}^3$

An important subset of $\mathsf{GL}(4)$ is the set of real 4×4 matrices with the following partitioned matrix structure

$$\begin{bmatrix} R & x \\ 0 & 1 \end{bmatrix},$$

where $R \in \mathsf{SO}(3)$ is a 3×3 orthogonal matrix with determinant $+1$ and $x \in \mathbb{R}^3$ is a column vector. Here, the 0 is a row vector in $\mathbb{R}^3$ and '1 is a real number in $\mathbb{R}^1$. Such matrices are said to be homogeneous matrices. A homogeneous matrix can also be viewed as a linear transformation on $\mathbb{R}^4$. The set of 4×4 homogeneous matrices is denoted subsequently as $\mathsf{SE}(3)$, and the homogeneous matrix above is sometimes denoted by $(R, x) \in \mathsf{SE}(3)$. It has important properties that we subsequently describe.

The set of all real 4×4 homogeneous matrices is closed under matrix multiplication. To illustrate, the matrix product of two homogeneous matrices is a homogeneous matrix since

$$\begin{bmatrix} R_2 & x_2 \\ 0 & 1 \end{bmatrix} \begin{bmatrix} R_1 & x_1 \\ 0 & 1 \end{bmatrix} = \begin{bmatrix} R_2 R_1 & x_2 + R_2 x_1 \\ 0 & 1 \end{bmatrix}.$$

Each homogeneous matrix has an inverse given by

$$\begin{bmatrix} R & x \\ 0 & 1 \end{bmatrix}^{-1} = \begin{bmatrix} R^T & -R^T x \\ 0 & 1 \end{bmatrix},$$

which is also a homogeneous matrix. The identity matrix $I_{4 \times 4}$ is a homogeneous matrix. Consequently, the set of all homogeneous matrices is closed under matrix multiplication and is a group.

The matrix product represents the composition of the two linear transformations represented by the individual homogeneous matrices. As a set, $\mathsf{SE}(3)$ can be identified with $\mathsf{SO}(3) \times \mathbb{R}^3$; however the calculation above indicates that the group composition on $\mathsf{SE}(3)$ is given by $(R_2, x_2)(R_1, x_1) = (R_2 R_1, x_2 + R_2 x_1)$, as opposed to the natural composition on $\mathsf{SO}(3) \times \mathbb{R}^3$, $(R_2, x_2)(R_1, x_1) = (R_2 R_1, x_2 + x_1)$. If we endow the set $\mathsf{SO}(3) \times \mathbb{R}^3$ with the operation defined by the second composition, it is referred to as the direct product of $\mathsf{SO}(3)$ and $\mathbb{R}^3$. Since the product is endowed with the first group composition, $\mathsf{SE}(3)$ is referred to as the semidirect product of $\mathsf{SO}(3)$ and $\mathbb{R}^3$. The homogeneous matrix representation of $\mathsf{SE}(3)$ provides a convenient way of encoding the semidirect product structure in terms of matrix multiplication on $\mathsf{GL}(4)$.

We can identify a vector $x \in \mathbb{R}^3$ with the vector $(x, 1) \in \mathbb{R}^4$. The group action of a homogeneous matrix in $\mathsf{SE}(3)$ acting on $\mathbb{R}^3$ can be expressed as

$$\begin{bmatrix} R & x_2 \\ 0 & 1 \end{bmatrix} \begin{bmatrix} x_1 \\ 1 \end{bmatrix} = \begin{bmatrix} x_2 + R x_1 \\ 1 \end{bmatrix},$$

that represents the action $x_1 \mapsto x_2 + R x_1$ of a homogeneous matrix in $\mathsf{GL}(4)$ on a vector in $\mathbb{R}^3$. In geometric terms, the element $(R, x_2) \in \mathsf{SE}(3)$ acts on $x_1 \in \mathbb{R}^3$ by first rotating the vector x_1 by R, followed by a translation by x_2.

1.1.7 Identities Involving Vectors, Orthogonal Matrices, and Skew-Symmetric Matrices

There are a number of matrix identities that involve vectors, orthogonal matrices, and skew-symmetric matrices. A few of these identities are summarized.

For any vectors x, $y \in \mathbb{R}^n$:

$$x^T y = \text{trace}[xy^T]. \tag{1.14}$$

For any $m \times n$ matrix A and any $n \times m$ matrix B:

$$\text{trace}[AB] = \text{trace}[BA]. \tag{1.15}$$

For any $x \in \mathbb{R}^3$:

$$S^T(x) = -S(x), \tag{1.16}$$

$$x \times x = S(x)x = 0, \tag{1.17}$$

$$S(x)^2 = xx^T - \|x\|^2 I_{3 \times 3}, \tag{1.18}$$

$$S(x)^3 = -\|x\|^2 S(x). \tag{1.19}$$

For any $R \in \text{SO}(3)$ and any x, $y \in \mathbb{R}^3$:

$$R(x \times y) = (Rx) \times (Ry), \tag{1.20}$$

$$RS(x)R^T = S(Rx). \tag{1.21}$$

For any x, y, $z \in \mathbb{R}^3$:

$$S(x \times y) = yx^T - xy^T, \tag{1.22}$$

$$S(x \times y) = S(x)S(y) - S(y)S(x), \tag{1.23}$$

$$S(x)S(y) = -x^T y I_{3 \times 3} + yx^T, \tag{1.24}$$

$$(x \times y) \cdot z = x \cdot (y \times z), \tag{1.25}$$

$$x \times (y \times z) = (x \cdot z)y - (x \cdot y)z, \tag{1.26}$$

$$y \times (x \times z) + x \times (z \times y) + z \times (y \times x) = 0, \tag{1.27}$$

$$\|x \times y\|^2 = \|x\|^2 \|y\|^2 - (x^T y)^2. \tag{1.28}$$

The proofs of these identities are not given here but the proofs depend only on the definitions and properties previously introduced.

1.1.8 Derivative Functions

Here, we review the notation and terminology for derivatives of scalar and vector-valued functions.

Suppose that $D \subset \mathbb{R}^n$ is an open set and the scalar-valued function $f : D \to \mathbb{R}^1$ is a differentiable function on D. Then we use the notation $\frac{\partial f(x)}{\partial x} \in \mathbb{R}^n$ to describe the derivative function, also referred to as the gradient; we usually view the gradient as a column vector in $\mathbb{R}^n$. The directional derivative of $f : D \to \mathbb{R}^1$ at $x \in \mathbb{R}^n$ in the direction $\xi \in \mathbb{R}^n$ is given by

$$\left(\frac{\partial f(x)}{\partial x} \cdot \xi \right) = \left(\frac{\partial f(x)}{\partial x} \right)^T \xi \in \mathbb{R}^1,$$

where we use the inner product on $\mathbb{R}^n$.

Suppose that $D \subset \mathbb{R}^n$ is an open set and the vector-valued function $f : D \to \mathbb{R}^m$ is a differentiable function on D. We use the notation $\frac{\partial f(x)}{\partial x} \in \mathbb{R}^{m \times n}$ to describe the derivative function; we usually view this as an $m \times n$ matrix of partial derivatives of the scalar component functions with respect to the scalar components of the argument vector. This is often referred to as a Jacobian matrix.

For either scalar or vector-valued functions that are differentiable, we subsequently use the notation $\frac{\partial f(x_0)}{\partial x}$ to denote the derivative function $\frac{\partial f(x)}{\partial x}$ evaluated at $x_0 \in D$.

Suppose that $D \subset \mathbb{R}^{n \times n}$ is an open set and the scalar-valued function $f : D \to \mathbb{R}^1$ is a differentiable function on D. Then we use the notation $\frac{\partial f(A)}{\partial A} \in \mathbb{R}^{n \times n}$ to describe the derivative function; we usually view this derivative as a matrix-valued function with values in $\mathbb{R}^{n \times n}$. The directional derivative of $f : D \to \mathbb{R}^1$ at $A \in \mathbb{R}^{n \times n}$ in the direction $B \in \mathbb{R}^{n \times n}$ is given by

$$\left(\frac{\partial f(A)}{\partial A} \cdot B \right) = \operatorname{trace}\left(\left(\frac{\partial f(A)}{\partial A} \right)^T B \right) \in \mathbb{R}^1,$$

where we use the inner product on $\mathbb{R}^{n \times n}$.

1.2 Manifold Concepts

Finite-dimensional differentiable manifolds are sets with a mathematical structure, referred to as a differential geometric structure, that supports calculus. Locally, all m-dimensional, differentiable manifolds are equivalent, or diffeomorphic to $\mathbb{R}^m$, in the sense that there is a local diffeomorphism between the manifold and $\mathbb{R}^m$; that is, in a neighborhood of each point on the manifold, the points in the neighborhood can be labeled using m local coordinates, or equivalently, each point in the neighborhood can be identified with a point in $\mathbb{R}^m$. This important fact is valid only locally in a possibly small neighborhood of each point on the manifold. The global geometry of the manifold may be quite different from the linear geometry of $\mathbb{R}^m$. These concepts are formalized in the subsequent sections.

Differentiable manifolds are used to describe the sets on which dynamical systems evolve. This motivates us to use mathematical developments that characterize the global properties of the manifolds and are *not* based only on local manifold representations. We summarize the most important differential geometric features of differentiable manifolds that are required. There are many good references on the mathematical theory of differential geometry and manifolds; references that emphasize connections with mechanics include gentle introductions in [81, 89] and the important references [5, 42, 70].

1.2.1 Manifolds

A point in a differentiable manifold M is most often described in terms of local coordinates, called charts, and a collection of charts, called an atlas. Each vector in M is specified by real-valued coordinates. These coordinates are referred to as *local coordinates* on M since they are defined only on an open set in M, that is the coordinates, in general, are not (and typically cannot be) globally defined everywhere on M.

We do not make extensive use of local coordinates in our subsequent development of geometric mechanics, since such a development is necessarily restricted by the inherent limitations of local coordinates. Rather, we take a different point of view: viewing a manifold as a submersion or embedded manifold. In particular, an embedded differentiable manifold is a subset of a finite-dimensional vector space defined as the zero set of scalar differentiable functions, where the derivatives (gradients) of the functions are assumed to be linearly independent on the manifold.

A differentiable manifold, as a submersion or embedded manifold in $\mathbb{R}^n$, is described by

$$M = \{x \in \mathbb{R}^n : f_i(x) = 0, \ i = 1, \ldots, l\}, \tag{1.29}$$

where $f_i : \mathbb{R}^n \to \mathbb{R}^1$, $i = 1, \ldots, l$ are scalar differentiable functions with the property that the vectors $\frac{\partial f_i(x)}{\partial x}$, $i = 1, \ldots, l$ are linearly independent vectors in $\mathbb{R}^n$ for each $x \in M$. Thus, necessarily $1 \leq l \leq n$. We say that the manifold M has dimension $n - l$ and codimension l. Consequently, we can represent a vector in an embedded manifold $M \subset \mathbb{R}^n$ as a vector in $\mathbb{R}^n$ so long as the equality conditions defining the embedded manifold are satisfied. Such representations for a point in M have the geometric advantage that they are *global* in the sense that they are defined everywhere that the manifold M is defined. In the subsequent analysis, a differentiable manifold is assumed to have a constant dimension everywhere on the manifold.

Manifolds are fundamental to our subsequent development, and there is an extensive theory described in the mathematical literature. Subsequent references to a smooth manifold or simply a manifold imply that the manifold is differentiable. Although more abstract notions of a manifold can be introduced, it is sufficient for our subsequent development to view manifolds as embedded manifolds.

A differentiable manifold M is also assumed to have an inner product, typically the inner product that arises from the finite-dimensional vector space within which it is embedded.

1.2.2 Tangent Vectors, Tangent Spaces and Tangent Bundles

Let $\gamma : [-1, 1] \to M$ denote a differentiable curve on M with $\gamma(0) = x \in M$. Then $\frac{d\gamma(s)}{ds}$, evaluated at $s = 0$, is a tangent vector to M at $x \in M$. For each $x \in M$, the set of all such tangent vectors to M at $x \in M$, denoted by $\mathsf{T}_x M$, is a subspace of $\mathbb{R}^n$, referred to as the tangent space of M at $x \in M$, and hence it is a linear manifold. We refer to $\xi \in \mathsf{T}_x M$ as a tangent to M at $x \in M$. For a manifold M as given above, it can be shown that the tangent space

$$\mathsf{T}_x M = \left\{ \xi \in \mathbb{R}^n : \left(\frac{\partial f_i(x)}{\partial x} \cdot \xi \right) = 0, \ i = 1, \ldots, m \right\}, \tag{1.30}$$

so that the tangent space consists of the set of vectors in $\mathbb{R}^n$ that are orthogonal to all of the gradients of the functions that define the manifold. The dimension of the tangent space is $n - m$.

The tangent bundle of a manifold M, denoted by $\mathsf{T}M$, is the set of pairs $(x, \xi) \in M \times \mathsf{T}_x M$, and it has its own manifold structure. The dimension of the tangent bundle is $2(n - m)$.

We are also interested in functions (maps, or transformations) from one manifold to another manifold and calculus associated with such functions. A map is said to be a diffeomorphism if it is continuously differentiable and it has an inverse map that is continuously differentiable. We distinguish between a local diffeomorphism that is defined on open subsets of the manifolds and a global diffeomorphism that is defined everywhere on the two manifolds.

Two manifolds are said to be diffeomorphic if there exists a (global) diffeomorphism from one of the manifolds to the other. Two diffeomorphic manifolds necessarily have the same dimension. Two manifolds embedded in the same vector space are diffeomorphic if one manifold can be smoothly deformed into the other manifold. Two diffeomorphic manifolds are, from a differential geometry perspective, equivalent.

1.2.3 Cotangent Vectors, Cotangent Spaces, and Cotangent Bundles

In the following, if D is a subspace of $\mathbb{R}^n$ then D^* denotes the set of all linear functionals from D to $\mathbb{R}^1$. It can be shown that D^* is a subspace of $(\mathbb{R}^n)^*$ and it is referred to as the dual of D.

This notation is used to introduce cotangent vectors (or covectors). If $x \in M$, then a cotangent vector is a linear functional defined on the tangent space $\mathsf{T}_x M$. The set of all such cotangents, denoted $\mathsf{T}_x^* M$, is the set of all linear functionals on the tangent space at $x \in M$, namely

$$\mathsf{T}_x^* M = (\mathsf{T}_x M)^*, \tag{1.31}$$

and it is a subspace, referred to as the cotangent space of M at $x \in M$, and hence it is a linear manifold. The dimension of the cotangent space is $n - m$.

The cotangent bundle of a manifold M, denoted by $\mathsf{T}^* M$, is the set of pairs $(x, \zeta) \in M \times \mathsf{T}_x^* M$, and it has its own manifold structure. The dimension of the cotangent bundle is $2(n - m)$.

Both $\mathsf{T}M$ and $\mathsf{T}^* M$ are manifolds in their own right, so it makes sense to consider tangent and cotangent spaces to these objects. In particular, the tangent space of $\mathsf{T}M$ at $(x, \xi) \in \mathsf{T}M$ can be defined and it is denoted by $\mathsf{T}_{(x,\xi)} \mathsf{T}M$. Further, the cotangent space of $\mathsf{T}M$ at $(x, \xi) \in \mathsf{T}M$ can be defined and it is denoted by $\mathsf{T}_{(x,\xi)}^* \mathsf{T}M$. The tangent bundle of the manifold $\mathsf{T}M$ is denoted by $\mathsf{T}\mathsf{T}M$, and the cotangent bundle of the manifold $\mathsf{T}M$ is denoted by $\mathsf{T}^* \mathsf{T}M$. Similarly, there are tangent spaces and cotangent spaces for $\mathsf{T}^* M$; further, the tangent bundle of $\mathsf{T}^* M$ is denoted by $\mathsf{T}\mathsf{T}^* M$ and the cotangent bundle of $\mathsf{T}^* M$ is denoted by $\mathsf{T}^* \mathsf{T}^* M$.

Typically, one thinks of the tangent vectors as column vectors, and the covectors as row vectors, which can be viewed as the transpose of a column vector. This is because the matrix product of a row vector with a column vector yields a scalar, which provides a natural way of representing the action of a linear functional (or covector) on a vector to yield a scalar. Given a covector represented by α^T, with $\alpha \in \mathbb{R}^n$, and a vector $v \in \mathbb{R}^n$, the natural pairing of $\alpha \in \mathbb{R}^n$ and $v \in \mathbb{R}^n$ is given by the matrix product or inner product $\alpha^T v = \alpha \cdot v \in \mathbb{R}^1$. Thus, it is possible to identify the cotangent space with the tangent space, by associating each (row) covector $\alpha^T \in \mathbb{R}^n$ with the (column) vector $\alpha \in \mathbb{R}^n$.

With this identification in mind, we represent covectors as vectors in $\mathbb{R}^n$, but we emphasize the dual nature of covectors by expressing them in terms of the inner product with an element of a tangent space. Since a covector is defined by its action on a tangent vector, there is an ambiguity in the representation since we can add a vector that is orthogonal to M to the vector proxy for the covector and not change its inner product (or action) on tangent vectors in $\mathsf{T}_x M$. To resolve this ambiguity, we identify the cotangent space of M at each point $x \in M$ with the corresponding tangent space.

Consider an orthogonal projection operator $P(x) : \mathbb{R}^n \to \mathsf{T}_x M$. As in the prior orthogonal decomposition result for vectors in $\mathbb{R}^n$ (1.3), we can express any vector $v \in \mathbb{R}^n$ uniquely as

$$v = P(x)v + (I_{n \times n} - P(x))v, \tag{1.32}$$

where the first term on the right is the projection of $v \in \mathbb{R}^n$ onto $\mathsf{T}_x M$ and the second term on the right is orthogonal to $\mathsf{T}_x M$.

This orthogonal decomposition implies the following. A covector $\alpha^T \in \mathbb{R}^n$ is defined by its action $\alpha \cdot (P(x)v) = \alpha^T (P(x)v) = (P(x)^T \alpha)^T v$ on all $v \in \mathsf{T}_x M$. The ambiguity in defining a covector $x \in \mathsf{T}_x^* M$ can be resolved by projecting α to $P(x)^T \alpha$. This resolution of the ambiguity in defining covectors is extensively used in the subsequent chapters.

1.2.4 Intersections and Products of Manifolds

As previously in (1.29), assume

$$M = \{x \in \mathbb{R}^n : f_i(x) = 0, \ i = 1, \dots, m\}, \tag{1.33}$$

where $f_i : \mathbb{R}^n \to \mathbb{R}^1$, $i = 1, \dots, m$ are scalar differentiable functions with the property that the vectors $\frac{\partial f_i(x)}{\partial x}$, $i = 1, \dots, m$ are linearly independent vectors in $\mathbb{R}^n$ on M. Thus, M is an $(n - m)$-dimensional differentiable manifold embedded in $\mathbb{R}^n$. Now define

$$M_i = \{x \in \mathbb{R}^n : f_i(x) = 0\}, \tag{1.34}$$

for $i = 1, \dots, m$. Thus, M_i is an $(n - 1)$-dimensional differentiable manifold embedded in $\mathbb{R}^n$ for each $i = 1, \dots, m$. It can be shown that M is the intersection of these m manifolds; that is

$$M = \bigcap_{i=1}^{m} M_i. \tag{1.35}$$

Further, it can be shown that for each $x \in M$, the tangent and cotangent spaces of the intersection manifold M satisfy

$$\mathsf{T}_x M = \bigcap_{i=1}^{m} \mathsf{T}_x M_i, \tag{1.36}$$

$$\mathsf{T}_x^* M = \bigcap_{i=1}^{m} \mathsf{T}_x^* M_i, \tag{1.37}$$

and the tangent and cotangent bundles of the intersection manifold M are

$$\mathsf{T}M = \bigcap_{i=1}^{m} \mathsf{T}M_i, \tag{1.38}$$

$$\mathsf{T}^*M = \bigcap_{i=1}^{m} \mathsf{T}^*M_i. \tag{1.39}$$

Now suppose that M_i is a differentiable manifold embedded in $\mathbb{R}^{n_i}$ for $i = 1, \ldots, m$. It can be shown that the product set defined by

$$M = \{(x_1, \ldots, x_m) : x_i \in M_i, \, i = 1, \ldots, m\} \tag{1.40}$$
$$= M_1 \times \ldots \times M_m, \tag{1.41}$$

is a differentiable manifold embedded in $\mathbb{R}^n$, where $n = \sum_{i=1}^{m} n_i$. Further, it can be shown that for each $x = (x_1, \ldots, x_m) \in M$, the tangent and cotangent spaces of the product manifold M are

$$\mathsf{T}_x M = \mathsf{T}_{x_1} M_1 \times \cdots \times \mathsf{T}_{x_m} M_m, \tag{1.42}$$
$$\mathsf{T}_x^* M = \mathsf{T}_{x_1}^* M_1 \times \cdots \times \mathsf{T}_{x_m}^* M_m, \tag{1.43}$$

and the tangent and cotangent bundles of the product manifold M are

$$\mathsf{T}M = \mathsf{T}M_1 \times \cdots \times \mathsf{T}M_m, \tag{1.44}$$
$$\mathsf{T}^*M = \mathsf{T}^*M_1 \times \cdots \times \mathsf{T}^*M_m. \tag{1.45}$$

1.2.5 Examples of Manifolds, Tangent Bundles, and Cotangent Bundles

In this section, several illustrations of manifolds, their tangent bundles, and their cotangent bundles are described. In particular, we provide examples of manifolds which are subsets of $\mathbb{R}^2$, $\mathbb{R}^3$, $\mathbb{R}^n$, or $\mathsf{GL}(n)$.

1.2.5.1 Manifolds Embedded in $\mathbb{R}^2$

The interesting nontrivial manifolds embedded in $\mathbb{R}^2$ are one-dimensional. The one-dimensional subspaces of $\mathbb{R}^2$, namely straight lines through the origin, are examples of one-dimensional manifolds. Each subspace can be translated by addition of a fixed vector; the resulting sets are also linear manifolds. For any element in a linear manifold defined by the translation of a subspace, the tangent space at that element coincides with the linear subspace. The tangent bundle of such a linear manifold can be identified with the product of the linear manifold and the subspace. The cotangent space can be identified with the dual of the subspace and the cotangent bundle can be identified with the product of the linear manifold and the dual of the subspace.

Assume $f : \mathbb{R}^2 \to \mathbb{R}^1$ is continuously differentiable and $f'(x) = \frac{\partial f(x)}{\partial x} \neq 0$ when $f(x) = 0$; then

$$M = \left\{ x \in \mathbb{R}^2 : f(x) = 0 \right\}$$

is a one-dimensional manifold embedded in $\mathbb{R}^2$. The tangent space of M at $x \in M$ is

$$\mathsf{T}_x M = \left\{ \xi \in \mathbb{R}^2 : (f'(x) \cdot \xi) = 0 \right\},$$

which can also be expressed as the range of a 2×2 projection matrix as

$$\mathsf{T}_x M = \left\{ \xi \in \mathbb{R}^2 : \xi = \left(I_{2 \times 2} - \frac{f'(x) f'^T(x)}{\| f'(x) \|^2} \right) y, \ y \in \mathbb{R}^2 \right\},$$

or in terms of skew-symmetric matrices as

$$\mathsf{T}_x M = \left\{ \xi \in \mathbb{R}^2 : \xi = S f'(x), \ S \in \mathbb{R}^{2 \times 2}, \ S + S^T = 0 \right\}.$$

The tangent bundle of M is given by

$$\mathsf{T} M = \left\{ (x, \xi) \in \mathbb{R}^2 \times \mathbb{R}^2 : x \in M, \ \xi \in \mathsf{T}_x M \right\}.$$

The dimension of the tangent bundle is two.

The one-dimensional cotangent space of M at $x \in M$ is

$$\mathsf{T}_x^* M = (\mathsf{T}_x M)^*,$$

the space of linear functionals on the tangent space $\mathsf{T}_x M$. The cotangent bundle of M is

$$\mathsf{T}^* M = \left\{ (x, \zeta) \in \mathbb{R}^2 \times (\mathbb{R}^2)^* : x \in M, \ \zeta \in \mathsf{T}_x^* M \right\}.$$

The dimension of the cotangent bundle is two.

An important example of a one-dimensional manifold is the unit sphere in $\mathbb{R}^2$:

$$\mathsf{S}^1 = \left\{ q \in \mathbb{R}^2 : \| q \|^2 = 1 \right\}. \tag{1.46}$$

For any $q \in \mathsf{S}^1$, it can be shown that the tangent space to S^1 at $q \in \mathsf{S}^1$ is the one-dimensional subspace

$$\mathsf{T}_q \mathsf{S}^1 = \left\{ \xi \in \mathbb{R}^2 : q \cdot \xi = 0 \right\},$$

which can also be expressed as the range of a 2×2 projection matrix as

$$T_q S^1 = \left\{ \xi \in \mathbb{R}^2 : \xi = (I_{2\times 2} - qq^T)y, \ y \in \mathbb{R}^2 \right\},$$

or in terms of skew-symmetric matrices as

$$T_q S^1 = \left\{ \xi \in \mathbb{R}^2 : \xi = Sq, \ S \in \mathbb{R}^{2\times 2}, \ +S^T = 0 \right\}.$$

The tangent bundle of S^1 is given by

$$TS^1 = \left\{ (q, \xi) \in \mathbb{R}^2 \times \mathbb{R}^2 : q \in S^1, \ \xi \in T_q S^1 \right\}.$$

The dimension of the tangent bundle is two.

The cotangent space of S^1 at $q \in S^1$ is

$$T_q^* S^1 = (T_q S^1)^*.$$

The dimension of the cotangent space $T_q^* S^1$ is one. The cotangent bundle of S^1 is

$$T^* S^1 = \left\{ (q, \zeta) \in \mathbb{R}^2 \times (\mathbb{R}^2)^* : q \in S^1, \ \zeta \in T_q^* S^1 \right\}.$$

The dimension of the cotangent bundle of S^1 is two.

1.2.5.2 Manifolds Embedded in $\mathbb{R}^3$

The interesting nontrivial manifolds embedded in $\mathbb{R}^3$ have dimension one or dimension two. The subspaces of $\mathbb{R}^3$ are examples of linear manifolds. In geometric terms, the one-dimensional subspaces can be viewed as straight lines in $\mathbb{R}^3$ that contain the origin; the two-dimensional subspaces can be viewed as planes in $\mathbb{R}^3$ that contain the origin. Each subspace can be translated by addition of a fixed vector; the resulting sets are linear manifolds. The one-dimensional linear manifolds are translations of straight lines; the two-dimensional linear manifolds are translations of planes. For any vector in such a linear manifold, the tangent space at that vector coincides with the subspace. The tangent bundle of a linear manifold can be identified with the product of the linear manifold and the subspace. The cotangent space can be identified with the dual of the subspace and the cotangent bundle can be identified with the product of the linear manifold and the dual of the subspace.

Assume $f : \mathbb{R}^3 \to \mathbb{R}^1$ is continuously differentiable and $f'(x) = \frac{\partial f(x)}{\partial x} \neq 0$ when $f(x) = 0$; then

$$M = \left\{ x \in \mathbb{R}^3 : f(x) = 0 \right\}$$

is a two-dimensional manifold embedded in $\mathbb{R}^3$. The tangent space of M at $x \in M$ is

$$T_x M = \left\{ \xi \in \mathbb{R}^3 : (f'(x) \cdot \xi) = 0 \right\},$$

which can also be expressed as the range of a 3×3 projection matrix as

$$T_x M = \left\{ \xi \in \mathbb{R}^3 : \xi = \left(I_{3\times3} - \frac{f'(x) f'^T(x)}{\| f'(x) \|^2} \right) y, \ y \in \mathbb{R}^3 \right\},$$

or in terms of skew-symmetric matrices as

$$T_x M = \left\{ \xi \in \mathbb{R}^3 : \xi = S f'(x), \ S \in \mathbb{R}^{3\times3}, \ S + S^T = 0 \right\}.$$

The tangent bundle of M is given by

$$TM = \left\{ (x, \xi) \in \mathbb{R}^3 \times \mathbb{R}^3 : x \in M, \ \xi \in T_x M \right\}.$$

The dimension of the tangent bundle is four.

The two-dimensional cotangent space of M at $x \in M$ is

$$T_x^* M = (T_x M)^*.$$

The cotangent bundle of M is

$$T^* M = \left\{ (x, \zeta) \in \mathbb{R}^3 \times (\mathbb{R}^3)^* : x \in M, \ \zeta \in T_x^* M \right\}.$$

The dimension of the cotangent bundle is four.

An important example of a two-dimensional manifold is the unit sphere in $\mathbb{R}^3$:

$$S^2 = \left\{ q \in \mathbb{R}^3 : \| q \|^2 = 1 \right\}. \tag{1.47}$$

For any $q \in S^2$, the tangent space to S^2 at $q \in S^2$ is a two-dimensional vector space

$$T_q S^2 = \left\{ \xi \in \mathbb{R}^3 : q \cdot \xi = 0 \right\},$$

which can also be expressed as the range of a 3×3 projection matrix as

$$T_q S^2 = \left\{ \xi \in \mathbb{R}^3 : \xi = (I_{3\times3} - q q^T) y, \ y \in \mathbb{R}^3 \right\},$$

or in terms of skew-symmetric matrices as

$$T_q S^2 = \left\{ \xi \in \mathbb{R}^3 : \xi = S q, \ S \in \mathbb{R}^{3\times3}, \ S + S^T = 0 \right\}.$$

The tangent bundle of S^2 is given by

$$TS^2 = \left\{ (q, \xi) \in \mathbb{R}^3 \times \mathbb{R}^3 : q \in S^2, \ \xi \in T_q S^2 \right\}.$$

The dimension of the tangent bundle is four.

The cotangent space $T_q^* S^2$ of S^2 at $q \in S^2$ is two dimensional. The cotangent bundle of S^2 is

$$T^* S^2 = \left\{ (q, \zeta) \in \mathbb{R}^3 \times (\mathbb{R}^3)^* : q \in S^2, \zeta \in T_q^* S^2 \right\}.$$

The dimension of the cotangent bundle is four.

Assume $f_i : \mathbb{R}^3 \to \mathbb{R}^1$, $i = 1, 2$ are continuously differentiable and $f_i'(x) = \frac{\partial f_i(x)}{\partial x}$, $i = 1, 2$, are linearly independent; then

$$M = \left\{ x \in \mathbb{R}^3 : f_i(x) = 0, i = 1, 2 \right\}$$

is a one-dimensional manifold embedded in $\mathbb{R}^3$. The tangent space of M at $x \in M$ is

$$T_x M = \left\{ \xi \in \mathbb{R}^3 : (f_i'(x) \cdot \xi) = 0, \, i = 1, 2 \right\},$$

which can also be expressed in terms of 3×3 projection matrices

$$T_x M = \left\{ \xi \in \mathbb{R}^3 : \xi = \left(I_{3 \times 3} - \frac{f_i'(x) f_i'(x)^T}{\|f_i'(x)\|^2} \right) y_i, \, y_i \in \mathbb{R}^3, \, i = 1, 2 \right\}.$$

The tangent bundle of M is given by

$$TM = \left\{ (x, \xi) \in \mathbb{R}^3 \times \mathbb{R}^3 : x \in M, \, \xi \in T_x M \right\}.$$

The dimension of the tangent bundle is two.

The two-dimensional cotangent space of M at $x \in M$ is

$$T_x^* M = (T_x M)^*.$$

The cotangent bundle of M is

$$T^* M = \left\{ (x, \zeta) \in \mathbb{R}^3 \times (\mathbb{R}^3)^* : x \in M, \, \zeta \in T_x^* M \right\}.$$

The dimension of the cotangent bundle is two.

Let $a \in \mathbb{R}^3$ with $\|a\| = 1$. An example of a one-dimensional manifold in $\mathbb{R}^3$ is given by

$$M = \left\{ x \in \mathbb{R}^3 : \|x\|^2 = 1, \, a \cdot x = 0 \right\}.$$

For any $x \in M$, it can be shown that the tangent space to M at $x \in M$ is the one-dimensional vector space

$$T_x M = \left\{ \xi \in \mathbb{R}^3 : x \cdot \xi = 0, \, a \cdot \xi = 0 \right\},$$

which can also be expressed in terms of 3×3 projection matrices as

$$\mathsf{T}_x M \;=\; \left\{ \xi \in \mathbb{R}^3 : \xi = (I_{3\times 3} - xx^T)y_1 = (I_{3\times 3} - aa^T)y_2, \; y_1, y_2 \in \mathbb{R}^3 \right\}.$$

The tangent bundle of M is given by

$$\mathsf{T}M = \left\{ (x, \xi) \in \mathbb{R}^3 \times \mathbb{R}^3 : x \in M, \; \xi \in \mathsf{T}_x M \right\}.$$

The dimension of the tangent bundle is two.

The cotangent space $\mathsf{T}_x^* M = (\mathsf{T}_x M)^*$ is one dimensional. The cotangent bundle of M is

$$\mathsf{T}^* M = \left\{ (x, \zeta) \in \mathbb{R}^3 \times (\mathbb{R}^3)^* : x \in M, \; \zeta \in \mathsf{T}_x^* M \right\}.$$

The dimension of the cotangent bundle is two.

1.2.5.3 Manifolds Embedded in $\mathbb{R}^n$

The subspaces of $\mathbb{R}^n$ are examples of linear manifolds. In geometric terms, the one-dimensional subspaces can be viewed as straight lines in $\mathbb{R}^n$ that contain the origin; the m-dimensional subspaces can be viewed as m-dimensional hyperplanes in $\mathbb{R}^n$ that contain the origin. Each subspace can be translated by addition of a fixed vector; the resulting sets are linear manifolds. The one-dimensional linear manifolds are translations of one-dimensional subspaces; the m-dimensional linear manifolds are translations of m-dimensional subspaces. For any vector in such a linear manifold, the tangent space at that vector coincides with the subspace. The tangent bundle of such a linear manifold can be viewed as the product of the linear manifold and the subspace. Similarly, the cotangent space is the dual of the subspace and the cotangent bundle can be viewed as the product of the linear manifold and the dual of the subspace.

The $(n-1)$-dimensional unit sphere is an important example of a manifold in $\mathbb{R}^n$:

$$\mathsf{S}^{n-1} = \left\{ q \in \mathbb{R}^n : \|q\|^2 = 1 \right\}.$$

For any $q \in \mathsf{S}^{n-1}$, the tangent space to S^{n-1} at q is

$$\mathsf{T}_q \mathsf{S}^{n-1} = \left\{ \xi \in \mathbb{R}^n : q \cdot \xi = 0 \right\},$$

which can also be expressed as the range of an $n \times n$ projection matrix as

$$\mathsf{T}_q \mathsf{S}^{n-1} = \left\{ \xi \in \mathbb{R}^n : \xi = (I_{n\times n} - qq^T)y, \; y \in \mathbb{R}^n \right\},$$

or in terms of skew-symmetric matrices as

$$\mathsf{T}_q \mathsf{S}^{n-1} = \left\{ \xi \in \mathbb{R}^n : \xi = Sq, \; S \in \mathbb{R}^{n\times n}, \; S + S^T = 0 \right\}.$$

The tangent bundle of the $(n-1)$-dimensional unit sphere is given by

$$\mathsf{T}\mathsf{S}^{n-1} = \left\{ (q, \xi) \in \mathbb{R}^n \times \mathbb{R}^n : q \in \mathsf{S}^{n-1}, \xi \in \mathsf{T}_q \mathsf{S}^{n-1} \right\},$$

and can be viewed as a $2(n-1)$-dimensional manifold embedded in $\mathbb{R}^{2n}$. The cotangent space to S^{n-1} at $q \in \mathsf{S}^{n-1}$ is the $(n-1)$-dimensional dual space

$$\mathsf{T}_q^* \mathsf{S}^{n-1} = (\mathsf{T}_q \mathsf{S}^{n-1})^*.$$

The cotangent bundle of S^{n-1} is

$$\mathsf{T}^* \mathsf{S}^{n-1} = \left\{ (q, \zeta) \in \mathbb{R}^3 \times (\mathbb{R}^3)^* : q \in \mathsf{S}^{n-1}, \zeta \in \mathsf{T}_q^* \mathsf{S}^{n-1} \right\}.$$

The dimension of the cotangent bundle is $2(n-1)$.

We now consider manifolds that have a product structure. Introduce

$$(\mathsf{S}^1)^n = \left\{ (q_1, \ldots, q_n) \in \mathbb{R}^{2n} : q_i \in \mathbb{R}^2, \|q_i\|^2 = 1, i = 1, \ldots, n \right\},$$

which an n-dimensional manifold embedded in $\mathbb{R}^{2n}$. We also write $(\mathsf{S}^1)^n = \mathsf{S}^1 \times \cdots \times \mathsf{S}^1$ as the product of n copies of the one-dimensional unit sphere in $\mathbb{R}^2$. For any $q \in (\mathsf{S}^1)^n$, the n-dimensional tangent space to $(\mathsf{S}^1)^n$ at $q \in (\mathsf{S}^1)^n$ is

$$\mathsf{T}_q (\mathsf{S}^1)^n = \left\{ (\xi_1, \ldots, \xi_n) \in \mathbb{R}^{2n} : (q_i \cdot \xi_i) = 0, i = 1, \ldots, n \right\}.$$

This tangent space can also be expressed in terms of the range of 2×2 projection matrices as

$$\mathsf{T}_q (\mathsf{S}^1)^n = \left\{ (\xi_1, \ldots, \xi_n) \in \mathbb{R}^{2n} : \xi_i = (I_{2 \times 2} - q_i q_i^T) y_i, y_i \in \mathbb{R}^2, i = 1, \ldots, n \right\},$$

or in terms of skew-symmetric matrices as

$$\mathsf{T}_q (\mathsf{S}^1)^n = \left\{ (\xi_1, \ldots, \xi_n) \in \mathbb{R}^{2n} : \xi_i = S_i q_i, S_i \in \mathbb{R}^{2 \times 2}, S_i^T = -S_i, i = 1, \ldots, n \right\}.$$

The tangent bundle of $(\mathsf{S}^1)^n$ is

$$\mathsf{T}(\mathsf{S}^1)^n = \left\{ (q, \xi) \in \mathbb{R}^{2n} \times \mathbb{R}^{2n} : q \in (\mathsf{S}^1)^n, \xi \in \mathsf{T}_q (\mathsf{S}^1)^n \right\},$$

and can be viewed as a $2n$-dimensional manifold embedded in $\mathbb{R}^{4n}$.

The cotangent space to $(\mathsf{S}^1)^n$ at $q \in (\mathsf{S}^1)^n$ is the n-dimensional dual of $\mathsf{T}_q (\mathsf{S}^1)^n$

$$\mathsf{T}_q^* (\mathsf{S}^1)^n = (\mathsf{T}_q (\mathsf{S}^1)^n)^*.$$

The cotangent bundle of $(\mathsf{S}^1)^n$ is

$$\mathsf{T}^* (\mathsf{S}^1)^n = \left\{ (q, \zeta) \in \mathbb{R}^{2n} \times (\mathbb{R}^{2n})^* : q \in (\mathsf{S}^1)^n, \zeta \in \mathsf{T}_q^* (\mathsf{S}^1)^n \right\}.$$

The dimension of the cotangent bundle is $2n$.

Similarly, we introduce a manifold that is the product of n copies of the two-dimensional unit sphere in $\mathbb{R}^3$. In set-theoretic notation this manifold is

$$(\mathsf{S}^2)^n = \left\{(q_1, \ldots, q_n) \in \mathbb{R}^{3n} : q_i \in \mathbb{R}^3, \|q_i\|^2 = 1, \, i = 1, \ldots, n\right\},$$

which is a $2n$-dimensional manifold embedded in $\mathbb{R}^{3n}$. We also write $(\mathsf{S}^2)^n = \mathsf{S}^2 \times \cdots \times \mathsf{S}^2$. For any $q \in (\mathsf{S}^2)^n$, the $2n$-dimensional tangent space to $(\mathsf{S}^2)^n$ at q is

$$T_q(\mathsf{S}^2)^n = \left\{\xi = (\xi_1, \ldots, \xi_n) \in \mathbb{R}^{3n} : (q_i \cdot \xi_i) = 0, \, i = 1, \ldots, n\right\}.$$

This tangent space can also be expressed in terms of the range of 3×3 projection matrices as

$$T_q(\mathsf{S}^2)^n = \left\{(\xi_1, \ldots, \xi_n) \in \mathbb{R}^{3n} : \xi_i = (I_{3 \times 3} - q_i q_i^T) y_i, \, y_i \in \mathbb{R}^3, \, i = 1, \ldots, n\right\},$$

or in terms of skew-symmetric matrices as

$$T_q(\mathsf{S}^2)^n = \left\{(\xi_1, \ldots, \xi_n) \in \mathbb{R}^{3n} : \xi_i = S_i q_i, S_i \in \mathbb{R}^{3 \times 3}, S_i^T = -S_i, \, i = 1, \ldots, n\right\}.$$

The tangent bundle of $(\mathsf{S}^2)^n$ is

$$T(\mathsf{S}^2)^n = \left\{(q, \xi) \in \mathbb{R}^{3n} \times \mathbb{R}^{3n} : q \in (\mathsf{S}^2)^n, \xi \in T_q(\mathsf{S}^2)^n\right\},$$

which is a $4n$-dimensional manifold embedded in $\mathbb{R}^{6n}$.

The cotangent space to $(\mathsf{S}^2)^n$ at $q \in (\mathsf{S}^2)^n$ is

$$T_q^*(\mathsf{S}^2)^n = (T_q(\mathsf{S}^2)^n)^*,$$

which has dimension $2n$. The cotangent bundle of $(\mathsf{S}^2)^n$ is

$$T^*(\mathsf{S}^2)^n = \left\{(q, \zeta) \in \mathbb{R}^{3n} \times (\mathbb{R}^{3n})^* : q \in (\mathsf{S}^2)^n, \zeta \in T_q^*(\mathsf{S}^2)^n\right\}.$$

The dimension of the cotangent bundle is $4n$.

1.2.5.4 Manifolds Embedded in $\mathbb{R}^{n \times n}$

Here we consider differentiable manifolds that are subsets of the vector space $\mathbb{R}^{n \times n}$ of $n \times n$ matrices. The matrix inner product $A \cdot B = \text{trace}(A^T B)$.

All subspaces and translations of subspaces of $\mathbb{R}^{n \times n}$ are examples of linear manifolds. Consequently, the tangent space of a linear manifold can be viewed as the associated subspace; the tangent bundle of a linear manifold can be viewed as the product of the linear manifold and the associated subspace. Similarly, the cotangent space of a linear manifold in $\mathbb{R}^{n \times n}$ can be viewed as the dual of the associated subspace; the cotangent bundle of a linear manifold in $\mathbb{R}^{n \times n}$ can be viewed as the product of the linear manifold and the dual of the associated subspace.

The set of $n \times n$ real, nonsingular matrices is denoted by $\mathsf{GL}(n)$. It is both a group under matrix multiplication and a manifold.

As previously introduced, the rotation group $\mathsf{SO}(3)$ has a group structure with matrix multiplication as the group operation, and it can be viewed as a subgroup of $\mathsf{GL}(3)$ consisting of orthogonal 3×3 matrices with determinant $+1$. But $\mathsf{SO}(3)$ is also a manifold and here we describe some of its differential geometric features.

It can be shown that $\mathsf{SO}(3)$ is a three-dimensional manifold embedded in the Lie group $\mathsf{GL}(3)$ (or embedded in the vector space $\mathbb{R}^{3\times3}$). For any $R \in \mathsf{SO}(3)$, the tangent space, denoted by $\mathsf{T}_R\mathsf{SO}(3)$, consists of all tangent matrices to $\mathsf{SO}(3)$ at $R \in \mathsf{SO}(3)$. It can be shown, using the exponential representation of orthogonal matrices, that the tangent space is

$$\mathsf{T}_R\mathsf{SO}(3) = \left\{ R\xi \in \mathbb{R}^{3\times3} : \xi \in \mathfrak{so}(3) \right\}.$$

That is, a tangent of the manifold $\mathsf{SO}(3)$ at $R \in \mathsf{SO}(3)$ is the product of R and a 3×3 skew-symmetric matrix. Thus, we can associate the tangent space $\mathsf{T}_R\mathsf{SO}(3)$ with the subspace $\mathfrak{so}(3)$ of all real skew-symmetric matrices in $\mathbb{R}^{3\times3}$. This shows that $\mathsf{T}_R\mathsf{SO}(3)$ is a three-dimensional subspace. Using the inner product introduced on $\mathbb{R}^{n\times n}$, it can be seen that the inner product of $R \in \mathsf{SO}(3)$ and of $R\xi \in \mathsf{T}_R\mathsf{SO}(3)$, denoted by the inner product pairing $(R \cdot R\xi)$ on $\mathbb{R}^{3\times3}$, satisfies

$$(R \cdot R\xi) = \mathsf{trace}(R^T \xi R) = 0.$$

Thus, each tangent to $\mathsf{SO}(3)$ at $R \in \mathsf{SO}(3)$ is orthogonal to $R \in \mathsf{SO}(3)$. The tangent bundle of $\mathsf{SO}(3)$ is

$$\mathsf{TSO}(3) = \left\{ (R, R\xi) \in \mathsf{SO}(3) \times \mathbb{R}^{3\times3} : \xi \in \mathfrak{so}(3) \right\},$$

and is a six-dimensional manifold. The cotangent space to $\mathsf{SO}(3)$ at $R \in \mathsf{SO}(3)$, denoted by $\mathsf{T}_R^*\mathsf{SO}(3)$, is the dual of the tangent space $\mathsf{T}_R\mathsf{SO}(3)$ of $\mathsf{SO}(3)$ at $R \in \mathsf{SO}(3)$. The cotangent space $\mathsf{T}_R^*\mathsf{SO}(3)$ is three dimensional. Thus, the cotangent bundle of $\mathsf{SO}(3)$ is the six-dimensional manifold

$$\mathsf{T}^*\mathsf{SO}(3) = \left\{ (R, \zeta) \in \mathsf{SO}(3) \times (\mathbb{R}^{3\times3})^* : \zeta \in \mathsf{T}_R^*\mathsf{SO}(3) \right\}.$$

We have seen that $\mathsf{SE}(3)$ can be viewed as a subset of $\mathsf{GL}(4)$ consisting of homogeneous 4×4 matrices with matrix multiplication as the group operation. But $\mathsf{SE}(3)$ is also a manifold and here we describe some of its differential geometric features.

It can be shown that $\mathsf{SE}(3)$ is a six-dimensional manifold embedded in the Lie group $\mathsf{GL}(4)$ (or embedded in the vector space $\mathbb{R}^{4\times4}$). For any $(R, x) \in \mathsf{SE}(3)$, the tangent space, denoted by $\mathsf{T}_{(R,x)}\mathsf{SE}(3)$, consists of all matrix and vector pairs that are tangent to $\mathsf{SE}(3)$ at $(R, x) \in \mathsf{SE}(3)$. The tangent space can be shown to be

$$\mathsf{T}_{(R,x)}\mathsf{SE}(3) = \left\{ (R\xi, y) \in \mathbb{R}^{3\times3} \times \mathbb{R}^3 : \xi \in \mathfrak{so}(3) \right\}.$$

That is, tangents to the manifold $\mathsf{SE}(3)$ at $(R, x) \in \mathsf{SE}(3)$ are ordered pairs of the form $(R\xi, y)$ where $\xi \in \mathfrak{so}(3)$ and $y \in \mathbb{R}^3$. This shows that $\mathsf{T}_{(R,x)}\mathsf{SE}(3)$ is a six-dimensional subspace. The tangent bundle of $\mathsf{SE}(3)$ is

$$\mathsf{TSE}(3) = \left\{ (R, x, R\xi, y) \in \mathsf{SO}(3) \times \mathbb{R}^3 \times \mathbb{R}^{3\times3} \times \mathbb{R}^3 : \xi \in \mathfrak{so}(3) \right\},$$

which is a twelve-dimensional manifold. The cotangent space to $\mathsf{SE}(3)$ at $(R, x) \in \mathsf{SE}(3)$, denoted by $\mathsf{T}_{(R,x)}^*\mathsf{SE}(3)$, is the dual of the tangent space; it is a six-dimensional subspace. The cotangent bundle of $\mathsf{SE}(3)$ is the twelve-dimensional manifold

$$\mathsf{T}^*\mathsf{SE}(3) = \left\{ (R, x, \zeta, y) \in \mathsf{SO}(3) \times \mathbb{R}^3 \times (\mathbb{R}^{3\times3})^* \times (\mathbb{R}^3)^* : \zeta \in \mathsf{T}_R^*\mathsf{SO}(3) \right\}.$$

1.2.6 Lie Groups and Lie Algebras

It is convenient to make use of certain aspects of the mathematical theory of Lie groups and Lie algebras. In fact, we use only concepts of matrix Lie groups and Lie algebras. There are many good references on this subject; references that make use of these concepts in the context of mechanics include [5, 23, 41, 70, 77].

A matrix Lie group, denoted by G, is a subset of $\mathsf{GL}(\mathsf{n})$, the set of all $n \times n$ real invertible matrices, that has both group properties and manifold properties. That is, G satisfies the group properties:

- Closure: for any two matrices A, $B \in \mathsf{G}$, the matrix product $AB \in \mathsf{G}$;
- Associativity: for any three matrices A, B, $C \in \mathsf{G}$, $(AB)C = A(BC)$;
- Identity: The $n \times n$ matrix $I_{n \times n} \in \mathsf{G}$ is the group identity with the property that for each matrix $A \in \mathsf{G}$, $I_{n \times n}A = AI_{n \times n} = A$;
- Inverse: For each matrix $A \in \mathsf{G}$, there is a matrix denoted by $A^{-1} \in \mathsf{G}$ such that $A^{-1}A = AA^{-1} = I_{n \times n}$.

In particular, $\mathsf{GL}(\mathsf{n})$ is a matrix Lie group. Furthermore, any subset of $\mathsf{GL}(\mathsf{n})$ that is closed under matrix multiplication and matrix inverse must contain the identity matrix, and since it inherits the associativity property, it is also a matrix group and it is referred to as a matrix subgroup of $\mathsf{GL}(\mathsf{n})$. We identify two important matrix subgroups, namely $\mathsf{SO}(3)$ and $\mathsf{SE}(3)$, as matrix Lie groups since they each have both group properties and differential geometric or manifold properties. Lie algebras, associated with these matrix Lie groups, are also introduced.

The set of all orthogonal matrices with determinant $+1$ is a matrix group and also a manifold; as such it is a Lie group and is denoted by $\mathsf{SO}(3)$ and is referred to as the special orthogonal group or the group of rotations. The Lie group $\mathsf{SO}(3)$ is important in our subsequent development since it is used to characterize rigid body rotation in three dimensions.

We have already seen that $R \in \mathsf{SO}(3)$ can be represented in terms of the matrix exponential map as $R = e^{\xi}$, where $\xi \in \mathfrak{so}(3)$. This implies a close association between the Lie group $\mathsf{SO}(3)$ and the vector space $\mathfrak{so}(3)$; this close association is formalized by referring to $\mathfrak{so}(3)$ as the Lie algebra associated with the Lie group $\mathsf{SO}(3)$. Recall that the tangent space of $\mathsf{SO}(3)$ at the identity, namely $T_{I_{3 \times 3}}\mathsf{SO}(3)$, is in fact the Lie algebra $\mathfrak{so}(3)$. More generally, given a Lie group G, we refer to the tangent space $T_e\mathsf{G}$ to the identity $e \in \mathsf{G}$ as the Lie algebra $\mathfrak{g}$. This association suggests that analysis on the manifold $\mathsf{SO}(3)$ can be translated into analysis on the Lie algebra $\mathfrak{so}(3)$, which is a vector space, at least locally on $\mathsf{SO}(3)$ near the identity. This concept is central to the variational calculus on $\mathsf{SO}(3)$ that is at the core of the development in Chapter 6.

The set of all homogeneous matrices is a matrix group and also a manifold; as such it is a Lie group and is denoted by $\mathsf{SE}(3)$ and is referred to as the special Euclidean group. The Lie group $\mathsf{SE}(3)$ is important in our subsequent development since it is used to characterize rigid body Euclidean motion, that is combined rotation and translation in three dimensions.

We know that elements in the Lie group $\mathsf{SE}(3)$ can be represented by $(R, x) \in \mathsf{SE}(3)$, with $R \in \mathsf{SO}(3)$ and $x \in \mathbb{R}^3$, where R can be represented in terms of the matrix exponential map as $R = e^{\xi}$, where $\xi \in \mathfrak{so}(3)$. This implies a close association between the Lie group $\mathsf{SE}(3)$ and the vector space $\mathfrak{so}(3) \times \mathbb{R}^3$ which we denote by $\mathfrak{se}(3)$; this close association is formalized by referring to $\mathfrak{se}(3)$ as the Lie algebra associated with the Lie group $\mathsf{SE}(3)$. In particular, the tangent space of $\mathsf{SE}(3)$ at $I_{4 \times 4} \in \mathsf{SE}(3)$ is in fact the Lie algebra $\mathfrak{se}(3)$. This association suggests that analysis on the manifold $\mathsf{SE}(3)$ can be translated into analysis on the Lie algebra $\mathfrak{se}(3)$, which is a vector space, at least locally on $\mathsf{SE}(3)$ near the identity. This concept is central to the variational calculus on $\mathsf{SE}(3)$ that is at the core of the development in Chapter 7.

1.2.7 Homogeneous Manifolds

A homogeneous manifold is a manifold that is associated with a Lie group. This association can be exploited in characterizing the differential geometry of homogeneous manifolds.

Consider a manifold M; then the Lie group G is said to act on M if there is a smooth map $A : \mathsf{G} \to \mathrm{Diff}(M)$, which maps each element $g \in \mathsf{G}$ into a diffeomorphism $A(g) : M \to M$. Alternatively, this can be expressed in terms of the smooth map $A : \mathsf{G} \times M \to M$, and we write $g \cdot x = (A(g))$ $(x) = A(g, x)$ for a left action, and $x \cdot g = (x)(A(g)) = A(g, x)$ for a right action. Furthermore, we have the property that $h \cdot (g \cdot x) = (hg) \cdot x$ for a left action, and $(x \cdot g) \cdot h = x \cdot (gh)$ for a right action. For the rest of the discussion, we assume that the group acts on the left, but the extension to right group actions is straightforward.

The group orbit $\mathsf{G} \cdot x$ of a point $x \in M$ is given by

$$\mathsf{G} \cdot x = \{g \cdot x \in M : g \in \mathsf{G}\}.$$

A manifold M is a homogeneous manifold if it has a transitive Lie group action, which is to say that $M = \mathsf{G} \cdot x$ for some $x \in M$. It is easy to check using the properties of groups and group actions that if this is true for one $x \in M$, then it is true for all $x \in M$. This is equivalent to saying that there is only one group orbit, or that given two points $x, y \in M$, it is always possible to find $g \in \mathsf{G}$ such that $y = g \cdot x$.

It is easy to see that S^1 is a homogeneous manifold with the associated Lie group of planar rotations. Furthermore, S^2 is a homogeneous manifold with the associated Lie group $\mathsf{SO}(3)$ of three-dimensional rotations.

1.3 Vector Fields on a Manifold

Vector fields and differential equations are closely related. Although much of the classical literature on vector fields treats the case of vector fields defined on an open subset of the Euclidean vector space $\mathbb{R}^n$, it is natural to introduce the concept of a vector field defined on a manifold.

Let M be a differentiable manifold embedded in $\mathbb{R}^n$. A vector field on M is defined by a mapping from M to $\mathbb{R}^n$ with the property that for each $x \in M$ there is a unique $y \in \mathsf{T}_x M$ and this correspondence $x \to y$ is continuous. The interpretation is that a vector field associates with each point $x \in M$ in the manifold a unique tangent vector $y \in \mathsf{T}_x M$ in the tangent space of the manifold. The vector field perspective is important in the subsequent development since it emphasizes the geometry of the manifold.

Calculus associated with vector fields is used in the subsequent variational developments: Euler–Lagrange equations are defined on the tangent bundle of the configuration manifold and the Lagrangian vector fields on the tangent bundle are defined by the Euler–Lagrange equations. We now review a few basic results for vector fields.

1.3.1 Vector Fields on a Manifold that Arise from Differential Equations

Differential equations are fundamental to representing vector fields in general and, in particular, to the representation of the dynamics of Lagrangian and Hamiltonian systems. In the classical situation, differential equations are defined on an open subset of $\mathbb{R}^n$; a vector field on a manifold can always be represented, at least locally in a neighborhood of a point on a manifold, by vector differential equations on an open set of $\mathbb{R}^n$. Consequently, many references on (nonlinear) differential equations only treat this case; see [44, 85]. However, we are subsequently interested in global dynamics of a vector field on a manifold; thus, such a local approach is not appropriate. References that provide a geometric perspective of vector fields on a manifold, suitable for studying global properties of Lagrangian and Hamiltonian dynamics, are [5, 33, 34, 83].

Let M be an $(n - m)$-dimensional differentiable manifold embedded in $\mathbb{R}^n$. A vector field on the manifold M associates with each point in the manifold a unique vector in the tangent space of the manifold at that point. Differential equations on a manifold can represent a vector field on a manifold by expressing the rate of change of a configuration point as a tangent vector at the configuration point on the manifold. If this dependence is sufficiently smooth, the classical theory of differential equations guarantees that there exists, at least locally in time, a unique solution of the differential equation satisfying a specified initial-value on the manifold. In particular, let $f : M \times \mathbb{R}^1 \to \mathbb{R}^n$ be a differentiable time-dependent vector-valued function that satisfies $f(x, t) \in \mathsf{T}_x M$ for each $x \in M$ and $t \in \mathbb{R}^1$. If the vector field $f(x, t)$ is time-independent, then it is said to be autonomous, otherwise it is nonautonomous. The differential equation

$$\dot{x}(t) = f(x(t), t) \tag{1.48}$$

on the manifold M is well-posed in the sense that for each initial condition in the manifold M, the initial-value problem has a unique solution. Suppose that a solution at time instant t, corresponding to an initial-value $x_0 \in M$ at time instant t_0, is given by $x(t, t_0, x_0) \in M$. Then $x(t, t_0, x_0) = F_{t,t_0}(x_0) \in M$ denotes the one-parameter time evolution or motion on M. The operator $F_{t,t_0} : M \to M$ defines the flow map. The initial time t_0 and final time t in the flow map F_{t,t_0} needs to be explicitly specified when the vector field is nonautonomous (or time-dependent), but if the vector field is autonomous (or time-independent), then we can without loss of generality assume that the initial conditions are given at time $t_0 = 0$, and we can suppress the initial time t_0 in the flow map notation by writing $F_t : M \to M$.

For the rest of the book, we will restrict ourselves to the case of time-independent vector fields $f(x)$, and their associated flow maps F_t. The manifold M is an invariant manifold in the sense that the flow necessarily remains on M. Throughout the subsequent chapters, we assume that each solution of an initial-value problem for a vector field on a manifold is defined for all time t.

It is important to recognize that the vector differential equations (1.48) are most often expressed in terms of vectors in the embedding space, namely $\mathbb{R}^n$, but the vector field of interest is only defined on the $(n-m)$-dimensional manifold M. It is, of course, possible to introduce a chart and local coordinates on M so that the differential equations (1.48) can be described locally in terms of the $n-m$ coordinates. This is a traditional approach, but the limitations imposed by the use of local coordinates means that we generally avoid this approach in this book.

Solutions of differential equations (1.48) on a manifold M can have a number of interesting properties. For example, solutions that are constant in time, $x = F_t(x)$ for all t, are equilibrium solutions of the differential equations or vector field and necessarily satisfy $f(x) = 0 \in T_x M$.

A function $\phi : M \to \mathbb{R}^1$ is an invariant function of the flow on the manifold M if for each $x \in M$, $\phi(F_t(x))$ is constant for all t. Such a function is referred to as an integral function or a conserved quantity of the differential equations or the vector field. If $\phi : M \to \mathbb{R}^1$ is differentiable, then we can compute the time derivative

$$\frac{d}{dt}\phi(x(t; x_0)),$$

where $x(t; x_0)$ denotes the fact that x is a function of time t, and it is parametrized by the initial data x_0. The chain rule yields the expression

$$\dot{\phi}(x) = \frac{\partial \phi(x)}{\partial x} \cdot f(x),$$

which is the total derivative of $\phi(x)$ along the vector field defined by $f(x)$ on M. Note that $\dot{\phi}(x) = 0$, $x \in M$ implies that $\phi : M \to \mathbb{R}^1$ is a conserved quantity on the flow defined by the vector field on M.

The theory of nonlinear differential equations, including differential equations and vector fields on manifolds, is well developed. Much of this theory makes use of tools from differential geometry; see [33, 100]. This material is not required for the subsequent development in this volume, but it is important for analytic studies of the geometric properties of vector fields on manifolds.

1.3.2 Vector Fields on a Manifold that Arise from Differential-Algebraic Equations

As shown in subsequent sections, vector fields on a manifold can also be associated with sets of differential-algebraic equations. Differential-algebraic equations arise in many applications, including constrained mechanical and electrical systems. Differential-algebraic equations are sometimes referred to as generalized, singular, or descriptor differential equations, or differential and algebraic equations, and they appear in several different forms. Overviews are given in [12, 79, 82, 83]. Here we summarize a few results for two categories of differential-algebraic equations.

Under appropriate assumptions, differential-algebraic equations are consistent in the sense that they define unique solutions of initial-value problems. Without going into detail in the general case, one of the key assumptions is the index. The index is essentially the number of times that the algebraic equations need to be differentiated with respect to time so that the resulting differential equations, together with the given differential equations, have unique solutions for the initial-value problem. The index is discussed in substantial detail in the above references.

We begin with an examination of a category of differential-algebraic equations that satisfy an index one assumption; this material is presented for completeness and background. Then we present results for a category of differential-algebraic equations that satisfy an index two assumption. This material is important for the subsequent treatment of Lagrangian dynamics that evolve on a manifold.

1.3.2.1 Differential-Algebraic Equations: Index One

Let $g : \mathbb{R}^n \to \mathbb{R}^n$ and $f : \mathbb{R}^n \to \mathbb{R}^m$ be continuously differentiable functions and $0 < m < n$. Semi-explicit differential-algebraic equations with index one are given by

$$\dot{x} = g(x) + \left(\frac{\partial f(x)}{\partial x}\right)^T \lambda, \tag{1.49}$$

$$f(x) = 0, \tag{1.50}$$

where $x \in \mathbb{R}^n$, $\lambda \in \mathbb{R}^m$. The set $M = \{x \in \mathbb{R}^n : f(x) = 0\}$ is assumed to be an $(n-m)$-dimensional differentiable manifold embedded in $\mathbb{R}^n$. Further we assume that the $m \times m$ matrix-valued function

$$\left(\frac{\partial f(x)}{\partial x}\right)\left(\frac{\partial f(x)}{\partial x}\right)^T$$

is full rank m for all $x \in M$. This assumption is referred to as a differential index one assumption [12] for the differential-algebraic equations (1.49) and (1.50).

A simple calculation shows that the index one differential-algebraic equations (1.49) and (1.50) are equivalent to a vector field on the manifold M. Multiplying (1.49) by the Jacobian of the constraint functions and using the assumptions allows the multipliers to be determined; substituting this expression for the multipliers into (1.49) and using (1.50) leads to the first-order vector differential equations

$$\dot{x} = g(x) - \left(\frac{\partial f(x)}{\partial x}\right)^T \left(\frac{\partial f(x)}{\partial x}\left(\frac{\partial f(x)}{\partial x}\right)^T\right)^{-1} \left(\frac{\partial f(x)}{\partial x}\right) g(x) \qquad (1.51)$$

on $\mathbb{R}^n$. Since the right-hand side of (1.51) projects $g(x)$ onto $T_x M$, it follows that $\dot{x} \in T_x M$; in other words M is an invariant manifold of the flow defined by (1.51). Consequently, the differential-algebraic equations (1.49) and (1.50) define a continuous vector field on M and for each initial condition in M there exists a unique solution of the initial-value problem defined on M.

The above index one assumptions are sufficient conditions for existence and uniqueness of initial-value solutions of differential-algebraic equations (1.49) and (1.50).

1.3.2.2 Differential-Algebraic Equations: Index Two

Let $g : \mathbb{R}^{2n} \to \mathbb{R}^n$ and $f : \mathbb{R}^n \to \mathbb{R}^m$ be continuously differentiable functions and $0 < m < n$. Semi-explicit differential-algebraic equations with index two are given by

$$\ddot{x} = g(x, \dot{x}) + \left(\frac{\partial f(x)}{\partial x}\right)^T \lambda, \qquad (1.52)$$

$$f(x) = 0, \qquad (1.53)$$

where $x \in \mathbb{R}^n$, $\lambda \in \mathbb{R}^m$. The set $M = \{x \in \mathbb{R}^n : f(x) = 0\}$ is assumed to be an $(n - m)$-dimensional differentiable manifold embedded in $\mathbb{R}^n$. Further we assume that the $m \times m$ matrix-valued function

$$\left(\frac{\partial f(x)}{\partial x}\right)\left(\frac{\partial f(x)}{\partial x}\right)^T$$

is full rank m for all $x \in M$. This assumption is referred to as a differential index two assumption [12] for the differential-algebraic equations (1.52) and (1.53).

A simple calculation shows that the index two differential-algebraic equations (1.52) and (1.53) are equivalent to a vector field on the tangent bundle TM. Multiplying (1.52) by the Jacobian of the constraint function and using the assumptions allows the multipliers to be determined; substituting this expression for the multipliers into (1.52) and using (1.53) leads to the second-order vector differential equations

$$\ddot{x} = g(x, \dot{x}) - \left(\frac{\partial f(x)}{\partial x}\right)^T \left(\frac{\partial f(x)}{\partial x} \left(\frac{\partial f(x)}{\partial x}\right)^T\right)^{-1}$$
$$\left(\left(\frac{\partial f(x)}{\partial x}\right) g(x, \dot{x}) + \left(\frac{d}{dt} \frac{\partial f(x)}{\partial x}\right)^T \dot{x}\right) \qquad (1.54)$$

on $\mathbb{R}^{2n}$. The expression for $\ddot{x}$ guarantees that $(x, \dot{x}) \in \mathsf{T}M$ implies $(x, \dot{x}, \dot{x}, \ddot{x}) \in \mathsf{T}_{(x,\dot{x})} \mathsf{T}M$; that is $\mathsf{T}M$ is an invariant manifold of the flow defined by (1.54). Consequently, the differential-algebraic equations (1.52) and (1.53) define a continuous vector field on $\mathsf{T}M$ and for each initial condition in $\mathsf{T}M$ there exists a unique solution of the initial-value problem defined on $\mathsf{T}M$.

The above index two assumptions are sufficient conditions for existence and uniqueness of initial-value solutions of differential-algebraic equations (1.52) and (1.53).

1.3.3 Linearized Vector Fields

We now return to the prior notation, assuming M is a differentiable $(n-m)$-dimensional manifold embedded in $\mathbb{R}^n$ and $f : M \to \mathbb{R}^n$ satisfies $f(x) \in \mathsf{T}_x M$ for each $x \in M$, thus defining a differentiable vector field on M.

Suppose that $x_0 \in M$ satisfies $f(x_0) = 0$ so that it is an equilibrium solution. Initial-value problems are well-posed for all initial-values in a small neighborhood of $x_0 \in M$. If the neighborhood is chosen sufficiently small then the manifold can be approximated near $x_0 \in M$ by the tangent space $\mathsf{T}_{x_0} M$. Furthermore, the function $f : M \to \mathbb{R}^n$ can be approximated in this neighborhood of $x_0 \in M$ by the linear vector field $\xi \to \left(\frac{\partial f(x_0)}{\partial x}\right) \xi$ where $\xi \in \mathsf{T}_{x_0} M$ are viewed as *perturbations from equilibrium*. The linearized vector field on $\mathbb{R}^n$ can be described by the linear vector differential equation

$$\dot{\xi} = \left(\frac{\partial f(x_0)}{\partial x}\right) \xi. \qquad (1.55)$$

This linear vector field, restricted to the invariant subspace $\mathsf{T}_{x_0} M$, can be viewed as a linearization of the original vector field on M at $x_0 \in M$.

We emphasize that the linear vector differential equations (1.55) are expressed in terms of vectors in the embedding space, namely $\mathbb{R}^n$, but the linearized vector field is only of interest on the $(n-m)$-dimensional invariant subspace $\mathsf{T}_{x_0} M$.

Local coordinates can be introduced on the subspace $\mathsf{T}_{x_0} M$ by selecting $(n-m)$ orthonormal basis vectors $\{\xi_1, \ldots, \xi_{n-m}\}$ in $\mathbb{R}^n$ for $\mathsf{T}_{x_0} M$. Thus, for any $\xi \in \mathsf{T}_{x_0} M$ there exist local coordinates $[\sigma_1, \ldots, \sigma_{n-m}]^T \in \mathbb{R}^{n-m}$ such that $\xi = \sum_{i=1}^{n-m} \sigma_i \xi_i$. Equivalently, it can be expressed in terms of a matrix-vector product $\xi = [\xi_1| \cdots |\xi_{n-m}][\sigma_1, \ldots, \sigma_{n-m}]^T$. Consequently, the linear differential equations can be expressed in terms of the local coordinates as

$$
\begin{bmatrix} \dot{\sigma}_1 \\ \vdots \\ \dot{\sigma}_{n-m} \end{bmatrix} = \begin{bmatrix} \xi_1 & \cdots & \xi_{n-m} \end{bmatrix}^T \left(\frac{\partial f(x_0)}{\partial x} \right) \begin{bmatrix} \xi_1 & \cdots & \xi_{n-m} \end{bmatrix} \begin{bmatrix} \sigma_1 \\ \vdots \\ \sigma_{n-m} \end{bmatrix} . \quad (1.56)
$$

This approach using local coordinates provides only a local approximation, since the linearized vector field is only a suitable approximation of the original vector field on the manifold in a small neighborhood of $x_0 \in M$. At least locally in a small neighborhood of an equilibrium solution, we can view each solution of an initial-value problem for the linearized vector field (1.55) or (1.56) as an approximation of the solution of the corresponding initial-value problem for (1.48). More precisely, the Hartman–Grobman theorem [33] states that if the equilibrium point x_0 is hyperbolic, which is to say that the eigenvalues of the matrix $\begin{bmatrix} \xi_1 & \cdots & \xi_{n-m} \end{bmatrix}^T \left(\frac{\partial f(x_0)}{\partial x} \right) \begin{bmatrix} \xi_1 & \cdots & \xi_{n-m} \end{bmatrix}$ all have nonzero real part, then there exists a local homeomorphism that takes trajectories of the original system to trajectories of the linearization. As such, the linearization process, in one form or another, is commonly employed in many engineering analyses. It should be emphasized that this approximation is valid only over time periods for which the approximating solution remains in the specified neighborhood of $x_0 \in M$.

This procedure for describing a linearized vector field linearizes first about an equilibrium solution and then expresses the resulting differential equations in local coordinates. It is easy to show that these steps commute: one can introduce local coordinates first in a neighborhood of an equilibrium solution and then linearize the resulting vector field. This latter approach is most common in applications. Further details on linearization of a vector field on a manifold are given in Appendix B.

1.3.4 Stability of an Equilibrium

Stability is an important qualitative property of an equilibrium of a vector field $f : M \to \mathbb{R}^n$ defined on a manifold M embedded in $\mathbb{R}^n$. This property is described in many references [33, 44, 76, 85, 93]. Here we give only a brief description.

An equilibrium x_0 of a vector field on a manifold is stable if given any arbitrarily small neighborhood $N_{x_0} \subset M$ of the equilibrium, there always exists a neighborhood B_{x_0} of the equilibrium such that all the trajectories with initial conditions in $B_{x_0} \subset M$ remain inside the neighborhood N_{x_0}. Otherwise, the equilibrium is said to be unstable. A stable equilibrium is asymptotically stable if there is a neighborhood of the equilibrium on the manifold such that for all initial conditions inside this neighborhood, the resulting flow asymptotically converges to the equilibrium.

There are several different approaches to stability analysis. Some approaches are based on the linearized vector field, while other approaches make use of *energy-like* Lyapunov functions.

The linearized vector field defined by (1.55) necessarily has an invariant $(n - m)$-dimensional subspace, namely the tangent space $\mathsf{T}_{x_0}M$ at the equilibrium, and there are $(n - m)$ eigenvalues associated with this invariant subspace. These $(n - m)$ eigenvalues are the spectrum of the linearized vector field at an equilibrium and they provide insight into the stability property of that equilibrium solution. The stable manifold theorem [33] guarantees that the eigenvalues of the linearized equations determine the stability of the equilibrium if the equilibrium is hyperbolic, that is all eigenvalues of the linearized equations have nonzero real part. In particular, if the real parts of all $(n - m)$ eigenvalues are negative, the equilibrium can be shown to be asymptotically stable. If there is at least one eigenvalue with real part that is positive, the equilibrium can be shown to be unstable. For all other spectral patterns, no definite statement can be made on the basis of the linearized vector field on the tangent space at the equilibrium.

Lyapunov stability methods [44, 85] provide powerful tools for stability analysis of an equilibrium of a vector field. The most common Lyapunov result provides sufficient conditions for stability of an equilibrium if there exists an energy-like Lyapunov function that is zero at the equilibrium and is positive elsewhere, and whose total derivative along the vector field is negative-semidefinite in a neighborhood of the equilibrium. Details on Lyapunov stability methods are given in the cited references.

1.3.5 Examples of Vector Fields

To make the prior ideas more concrete, examples of vector fields are described for several different manifolds of the form that are studied in more detail in subsequent chapters.

1.3.5.1 Example of a Vector Field Defined on S^1

We consider differential equations that define a vector field on the one-dimensional manifold S^1 embedded in $\mathbb{R}^2$. These differential equations are expressed in terms of $q = [q_1, q_2]^T \in \mathbb{R}^2$ as

$$\dot{q} = S(q)q,$$

where

$$S(q) = \begin{bmatrix} 0 & \sin q_2 \\ -\sin q_2 & 0 \end{bmatrix}.$$

Note that the skew-symmetric matrix-valued function $S : \mathsf{S}^1 \to \mathbb{R}^{2 \times 2}$ is continuously differentiable. The differential equations can be expressed in scalar form as

$$\dot{q}_1 = q_2 \sin q_2,$$
$$\dot{q}_2 = -q_1 \sin q_2.$$

It is easy to show that $T_q\mathsf{S}^1 = \{\xi \in \mathbb{R}^2 : \xi^T q = 0\}$ and that $\dot{q} \in T_q\mathsf{S}^1$. Consequently, these differential equations define a vector field on S^1.

There are two equilibrium solutions on S^1, namely $[1,0]^T$, $[-1,0]^T$. We first linearize at the equilibrium $[1,0]^T \in \mathsf{S}^1$. The linearized vector field on $\mathbb{R}^2$ at $[1,0]^T \in \mathbb{R}^2$ can be shown to be

$$\dot{\xi} = \begin{bmatrix} 0 & 0 \\ 0 & -1 \end{bmatrix} \xi,$$

where $\xi \in \mathbb{R}^2$ can be viewed as first-order perturbations from the equilibrium. We now restrict this vector field to $T_{[1,0]^T}\mathsf{S}^1$ by selecting the unit basis vector $e_2 = [0,1]^T \in T_{[1,0]^T}\mathsf{S}^1$. Thus, we let $\xi = \sigma e_2$, where $\sigma \in \mathbb{R}^1$, so that the restricted vector field on S^1 is represented, in a neighborhood of the equilibrium $[1,0]^T \in \mathsf{S}^1$, by the scalar differential equation

$$\dot{\sigma} = -\sigma.$$

The eigenvalue associated with this linear vector field is -1 so that the equilibrium $[1,0]^T \in \mathsf{S}^1$ of the original vector field is asymptotically stable.

The linearized vector field on $\mathbb{R}^2$ at $[-1,0]^T \in \mathbb{R}^2$ can be shown to be

$$\dot{\xi} = \begin{bmatrix} 0 & 0 \\ 0 & 1 \end{bmatrix} \xi,$$

where $\xi \in \mathbb{R}^2$ can be viewed as first-order perturbations from the equilibrium. We now restrict this vector field to $T_{[-1,0]^T}\mathsf{S}^1$ by selecting the unit basis vector $e_2 = [0,1]^T \in T_{[1,0]^T}\mathsf{S}^1$. Thus, we let $\xi = \sigma e_2$, where $\sigma \in \mathbb{R}^1$, so that the restricted vector field on S^1 is represented, in a neighborhood of the equilibrium $[-1,0]^T \in \mathsf{S}^1$, by

$$\dot{\sigma} = \sigma.$$

The eigenvalue associated with this linear vector field is 1 so that the equilibrium $[-1,0]^T \in \mathsf{S}^1$ of the original vector field is unstable.

1.3.5.2 Example of a Vector Field Defined on $\mathbf{S}^2$

We consider differential equations that define a vector field on the two-dimensional manifold S^2 embedded in $\mathbb{R}^3$. These differential equations are expressed in terms of $q = [q_1, q_2, q_3]^T \in \mathbb{R}^3$ as

$$\dot{q} = S(q)q,$$

where

$$S(q) = \begin{bmatrix} 0 & \sin q_2 & \sin q_3 \\ -\sin q_2 & 0 & 0 \\ -\sin q_3 & 0 & 0 \end{bmatrix}.$$

Note that the skew-symmetric matrix-valued function $S : \mathsf{S}^2 \to \mathbb{R}^{3\times3}$ is continuously differentiable. The differential equations can be expressed in scalar form as

$$\dot{q}_1 = q_2 \sin q_2 + q_3 \sin q_3,$$
$$\dot{q}_2 = -q_1 \sin q_2,$$
$$\dot{q}_3 = -q_1 \sin q_3.$$

It is easy to show that $\mathsf{T}_q\mathsf{S}^2 = \{\xi \in \mathbb{R}^3 : \xi^T q = 0\}$ and that $\dot{q} \in \mathsf{T}_q\mathsf{S}^2$. Consequently, the differential equations define a vector field on S^2.

There are two equilibrium solutions on S^2, namely $[1, 0, 0]^T$, $[-1, 0, 0]^T$. We first linearize at the equilibrium $[1, 0, 0]^T \in \mathsf{S}^2$. The linearized vector field on $\mathbb{R}^3$ at $[1, 0, 0]^T \in \mathbb{R}^3$ can be shown to be

$$\dot{\xi} = \begin{bmatrix} 0 & 0 & 0 \\ 0 & -1 & 0 \\ 0 & 0 & -1 \end{bmatrix} \xi,$$

where $\xi \in \mathbb{R}^3$ can be viewed as first-order perturbations from the equilibrium. We now restrict this vector field to $\mathsf{T}_{[1,0,0]^T}\mathsf{S}^2$ by selecting the basis vectors $\{e_2 = [0, 1, 0]^T, e_3 = [0, 0, 1]^T\} \in \mathsf{T}_{[1,0,0]^T}\mathsf{S}^2$. Thus, we let $\xi = \sigma_2 e_2 + \sigma_3 e_3$, where $\sigma = [\sigma_2, \sigma_3]^T \in \mathbb{R}^2$, so that the restricted vector field on S^2 is represented, in a neighborhood of the equilibrium $[1, 0, 0]^T \in \mathsf{S}^2$, by the second-order differential equation

$$\dot{\sigma} = \begin{bmatrix} -1 & 0 \\ 0 & -1 \end{bmatrix} \sigma.$$

The eigenvalues associated with this linear vector field are $-1, -1$ so that the equilibrium $[1, 0, 0]^T \in \mathsf{S}^2$ of the original vector field is asymptotically stable.

We next linearize at the equilibrium $[-1, 0, 0]^T \in \mathsf{S}^2$. The linearized vector field on $\mathbb{R}^3$ at $[-1, 0, 0]^T \in \mathbb{R}^3$ can be shown to be

$$\dot{\xi} = \begin{bmatrix} 0 & 0 & 0 \\ 0 & 1 & 0 \\ 0 & 0 & 1 \end{bmatrix} \xi,$$

where $\xi \in D \subset \mathbb{R}^3$ can be viewed as first-order perturbations from the equilibrium. We now restrict this vector field to the tangent space $\mathsf{T}_{[-1,0,0]^T}\mathsf{S}^2$ by selecting the basis vectors $\{e_2 = [0,1,0]^T, \ e_3 = [0,0,1]^T\}$ for the tangent space $\mathsf{T}_{[1,0,0]^T}\mathsf{S}^2$. Thus, let $\xi = \sigma_2 e_2 + \sigma_3 e_3$, where $\sigma = [\sigma_2, \sigma_3]^T \in \mathbb{R}^2$, so that the restricted vector field on S^2 is represented, in a neighborhood of the equilibrium $[1,0,0]^T \in \mathsf{S}^2$, by the second-order differential equation

$$\dot{\sigma} = \begin{bmatrix} 1 & 0 \\ 0 & 1 \end{bmatrix} \sigma.$$

The eigenvalues associated with this linear vector field are $1,1$ so that the equilibrium $[-1,0,0]^T \in \mathsf{S}^2$ of the original vector field is unstable.

1.3.5.3 Example of a Vector Field Defined on SO(3)

We consider matrix differential equations that evolve on the three-dimensional manifold $\mathsf{SO}(3)$ embedded in $\mathbb{R}^{3\times3}$. These differential equations are expressed in terms of matrices $R \in \mathbb{R}^{3\times3}$ as

$$\dot{R} = RS\left(\sum_{i=1}^{3} e_i \times R^T e_i\right),$$

where $e_1, e_2, e_3 \in \mathbb{R}^3$ denote the standard unit basis vectors and $S : \mathbb{R}^3 \to \mathbb{R}^{3\times3}$ is the skew-symmetric matrix-valued function

$$S(x) = \begin{bmatrix} 0 & -x_3 & x_2 \\ x_3 & 0 & -x_1 \\ -x_2 & x_1 & 0 \end{bmatrix}.$$

Note that the skew-symmetric matrix-valued function $S : \mathbb{R}^3 \to \mathbb{R}^{3\times3}$ is continuously differentiable. Since $\mathsf{T}_R\mathsf{SO}(3) = \{R\xi \in \mathbb{R}^{3\times3} : \xi \in \mathfrak{so}(3)\}$, it follows that $\dot{R} \in \mathsf{T}_R\mathsf{SO}(3)$. Consequently, the matrix differential equation defines a vector field on $\mathsf{SO}(3)$.

The equilibrium solutions on $\mathsf{SO}(3)$ satisfy

$$\sum_{i=1}^{3} e_i \times R^T e_i = 0.$$

It can be shown that there are exactly four equilibrium solutions on $\mathsf{SO}(3)$, namely the diagonal matrices

$$I_{3\times 3} = \begin{bmatrix} 1 & 0 & 0 \\ 0 & 1 & 0 \\ 0 & 0 & 1 \end{bmatrix}, \; E_2 = \begin{bmatrix} -1 & 0 & 0 \\ 0 & -1 & 0 \\ 0 & 0 & 1 \end{bmatrix}, \; E_3 = \begin{bmatrix} -1 & 0 & 0 \\ 0 & 1 & 0 \\ 0 & 0 & -1 \end{bmatrix}, \; E_4 = \begin{bmatrix} 1 & 0 & 0 \\ 0 & -1 & 0 \\ 0 & 0 & -1 \end{bmatrix}.$$

We first determine the linearized vector field at the identity matrix $I_{3\times 3} \in$ SO(3). It is convenient to use the exponential representation

$$R = e^\xi,$$

where $\xi \in \mathfrak{so}(3)$ so that the vector field can also be described by

$$\dot\xi = e^{-\xi} S \left(\sum_{i=1}^{3} e_i \times e^{-\xi} e_i \right).$$

For $\xi \in \mathfrak{so}(3)$ sufficiently small, this vector field is approximated by the linear vector field

$$\dot\xi = -S \left(\sum_{i=1}^{3} e_i \times \xi e_i \right).$$

Using skew-symmetric matrix identities, the linear vector field can be described by

$$\dot\xi = \xi^T - \xi.$$

This linear vector field is restricted to the subspace $\mathfrak{so}(3) \subset \mathbb{R}^{3\times 3}$ by using the basis

$$\xi_1 = \begin{bmatrix} 0 & -1 & 0 \\ 1 & 0 & 0 \\ 0 & 0 & 0 \end{bmatrix}, \quad \xi_2 = \begin{bmatrix} 0 & 0 & 1 \\ 0 & 0 & 0 \\ -1 & 0 & 0 \end{bmatrix}, \quad \xi_3 = \begin{bmatrix} 0 & 0 & 0 \\ 0 & 0 & -1 \\ 0 & 1 & 0 \end{bmatrix},$$

for $\mathfrak{so}(3)$. Thus, we introduce the representation

$$\xi = \sigma_1 \xi_1 + \sigma_2 \xi_2 + \sigma_3 \xi_3,$$

where $[\sigma_1, \sigma_2, \sigma_3]^T \in D \subset \mathbb{R}^3$, where D is an open set containing the origin. It follows that the linear vector field on $\mathfrak{so}(3)$ is described by

$$\dot\sigma_i = -2\sigma_i, \quad i = 1, 2, 3.$$

This is a linear approximation for the original vector field on SO(3) in a neighborhood of $I_{3\times 3} \in$ SO(3). The eigenvalues are $-2, -2, -2$ so that the equilibrium at the identity of the original vector field on SO(3) is asymptotically stable.

The linearized differential equations at the equilibrium $E_i \in$ SO(3), $i = 2, 3, 4$, can be shown to be

$$\dot{\xi} = E_i(\xi^T - \xi),$$

which defines a linear vector field on the invariant subspace of $\mathbb{R}^{3\times 3}$ spanned by the basis

$$E_i \begin{bmatrix} 0 & -1 & 0 \\ 1 & 0 & 0 \\ 0 & 0 & 0 \end{bmatrix}, \quad E_i \begin{bmatrix} 0 & 0 & 1 \\ 0 & 0 & 0 \\ -1 & 0 & 0 \end{bmatrix}, \quad E_i \begin{bmatrix} 0 & 0 & 0 \\ 0 & 0 & -1 \\ 0 & 1 & 0 \end{bmatrix}.$$

The eigenvalues associated with this invariant subspace can be shown to be $2, 2, -2$ so that the equilibrium at $E_i \in \mathsf{SO}(3)$, $i = 2, 3, 4$ of the original vector field on $\mathsf{SO}(3)$ is unstable.

In summary, the equilibrium solution at the identity is asymptotically stable while the other three equilibrium solutions are unstable. This analysis demonstrates local flow properties of this vector field on $\mathsf{SO}(3)$, but it is challenging to determine global flow properties of the vector field.

1.3.5.4 Example of a Vector Field on S^1 Defined by Differential-Algebraic Equations with Index One

We consider differential-algebraic equations that define a vector field on the one-dimensional manifold S^1 embedded in $\mathbb{R}^2$. These differential-algebraic equations are expressed in terms of $x \in \mathbb{R}^2$, $\lambda \in \mathbb{R}^1$ as

$$\dot{x} = e_1 + 2x\lambda,$$
$$\|x\|^2 = 1,$$

where $e_1 = [1, 0]^T$. The second equation constrains x to lie on the manifold S^1.

We differentiate the algebraic constraint once to obtain $x^T \dot{x} = 0$ and thus

$$x^T(e_1 + 2x\lambda) = 0,$$

so that

$$\lambda = -\frac{1}{2}x^T e_1.$$

Consequently, the index is one and the flow defined by the above differential-algebraic equations can also be described by the differential equation

$$\dot{x} = e_1 - xx^T e_1,$$

which defines a differentiable vector field on the manifold S^1.

There are two equilibrium solutions of the vector field on the manifold S^1, namely e_1, corresponding to $\lambda = -\frac{1}{2}$, and $-e_1$, corresponding to $\lambda = \frac{1}{2}$.

The linearized differential-algebraic equations at e_1 can be shown to be

$$\dot{\xi} = -(I_{2 \times 2} + e_1 e_1^T)\xi,$$

which defines a linear vector field on the tangent space $\mathsf{T}_{e_1} \mathsf{S}^1 = \mathrm{span}(e_2)$. Thus, we can write $\xi = \sigma e_2$ so that the linear vector field is described on an open subset of R^1 containing the origin by

$$\dot{\sigma} = -\sigma.$$

The single eigenvalue is -1; consequently this equilibrium solution of the given differential-algebraic equations is asymptotically stable.

The linearized differential-algebraic equations at $-e_1$ can be shown to be

$$\dot{\xi} = (I_{2 \times 2} + e_1 e_1^T)\xi.$$

Thus, we can write $\xi = \sigma e_2$ so that the linear vector field is described on an open subset of R^1 containing the origin by

$$\dot{\sigma} = \sigma.$$

The single eigenvalue is $+1$; consequently this equilibrium of the given differential-algebraic equations is unstable.

This analysis demonstrates a few of the important dynamical properties of the solutions of the given index one differential-algebraic equations.

1.3.5.5 Example of a Vector Field on TS^1 Defined by Differential-Algebraic Equations with Index Two

We consider differential-algebraic equations that define a vector field on the two-dimensional manifold TS^1 embedded in $\mathbb{R}^4$. These differential-algebraic equations are expressed in terms of $x \in \mathbb{R}^2$, $\lambda \in \mathbb{R}^1$ as

$$\ddot{x} = e_1 + 2x\lambda,$$
$$\|x\|^2 = 1,$$

where $e_1 = [1, 0]^T$. As before, the second equation constrains x to lie on the manifold S^1.

We differentiate the algebraic constraint twice and substitute into the differential equation to obtain

$$\|\dot{x}\|^2 + x^T(e_1 + 2x\lambda) = 0,$$

so that

$$\lambda = -\frac{1}{2}\|\dot{x}\|^2 - \frac{1}{2}x^T e_1.$$

Consequently, the index is two and the flow defined by the above differential-algebraic equations can also be described by the differential equation

$$\ddot{x} = e_1 - x\|\dot{x}\|^2 - xx^T e_1.$$

Since $(x,\dot{x}) \in \mathsf{TS}^1$, this differential equation defines a differentiable vector field on the tangent bundle manifold TS^1.

There are two equilibrium solutions of the vector field on the tangent bundle TS^1, namely $[e_1,0]^T$, corresponding to $\lambda = -\frac{1}{2}$, and $[-e_1,0]^T$, corresponding to $\lambda = \frac{1}{2}$.

The linearized differential-algebraic equations at $[e_1,0]^T$ can be shown to be

$$\ddot{\xi} = -(I_{2\times 2} + e_1 e_1^T)\xi,$$

which defines a linear vector field on the tangent space $\mathsf{T}_{(e_1,0)}\mathsf{TS}^1$. Thus, we can write $\xi = \sigma e_2$ so that the linear vector field is described on an open subset of $\mathbb{R}^2$ containing the origin by

$$\ddot{\sigma} = -\sigma.$$

The two eigenvalues are purely imaginary; consequently the equilibrium point is non-hyperbolic and we cannot conclude anything about its nonlinear behavior.

The linearized differential-algebraic equations at $[-e_1,0]^T$ can be shown to be

$$\ddot{\xi} = (I_{2\times 2} + e_1 e_1^T)\xi,$$

which defines a linear vector field on the tangent space $\mathsf{T}_{(-e_1,0)}\mathsf{TS}^1$. Thus, we can write $\xi = \sigma e_2$ so that the linear vector field is described on an open subset of $\mathbb{R}^2$ containing the origin by

$$\ddot{\sigma} = \sigma.$$

The two eigenvalues are $+1, -1$; consequently the dynamics defined by the given differential-algebraic equations are unstable in a neighborhood of this equilibrium.

This analysis demonstrates a few of the important dynamical properties of the solutions of the given index two differential-algebraic equations.

1.3.6 Geometric Integrators

Most standard numerical integration algorithms are developed for initial-value problems that evolve on a vector space. For differential equations or differential-algebraic equations whose solutions evolve on $\mathbb{R}^n$, these algorithms are suitable for studying the local dynamics on a manifold, but they do not typically perform well when applied to the global dynamics of a vector field that evolves on a manifold that is not a vector space.

In our subsequent development for Lagrangian dynamics that evolve on the tangent bundle of a configuration manifold, the differential equations are expressed on a vector space within which the tangent bundle of the configuration manifold is embedded. This means that standard numerical integration algorithms can, in principle, be applied directly to the dynamics on the embedding vector space. However, standard integration algorithms do not guarantee that computed solutions of an initial-value problem that begin on the configuration manifold remain on the configuration manifold. In practice, the numerical solutions tend to drift off the configuration manifold, thus introducing significant errors in the computational results. Although methods can be introduced that project the computed solution back onto the configuration manifold, such steps may be computationally intensive and the projection process typically interferes with the conservation of other physical invariants, such as the energy or momentum.

Our experience, and that of many others [12, 34, 79, 82, 83], is that it is best to make use of special purpose numerical integration algorithms that are constructed to produce solution approximations of an initial-value problem that remain on the manifold if the initial-value is on the manifold. Such algorithms are referred to as geometric integration algorithms and there is an extensive literature on geometric integration and associated numerical computations. Naturally, the details of a geometric integration algorithm depend on the specific differential equations and they also depend on the specific geometry of the manifold.

There is now a substantial body of literature that treats geometric integrators for a vector field on a manifold. This literature is not as well known and utilized as it should be, in our opinion. Important references that provide good insight into this literature include [12, 34, 65, 71].

Computational dynamics are not treated in this text, but many of our publications [46, 48, 49, 50, 51, 52, 53, 54, 56, 57, 58, 59, 61] do treat computational issues for Lagrangian and Hamiltonian dynamics on a manifold following the mathematical and geometric framework introduced here. A few remarks illustrate this approach. The formulation of Lagrangian and Hamiltonian dynamics on a manifold in continuous-time presented in this book can be followed to obtain discrete-time descriptions for Lagrangian and Hamiltonian dynamics that evolve on a manifold. This approach is especially attractive since it makes use of variational calculus on manifolds that closely follows the continuous-time development. These discrete-time descriptions are the basis for constructing geometric and variational integration algorithms. The development is most natural when the manifold is a Lie group or the tangent bundle of a Lie group, in which case the geometric integrators are referred to as Lie group variational integrators. The development can also be adapted for Lagrangian and Hamiltonian dynamics that evolve on a homogeneous manifold. This computational approach is especially suited to computations of multi-body dynamics; many of the references cited above treat multi-body examples that are subsequently studied, in continuous-time, in this book.

1.4 Covector Fields on a Manifold

Let M denote a differentiable manifold embedded in $\mathbb{R}^n$. The cotangent space at $x \in M$, denoted by $\mathsf{T}_x^* M$, is the space of linear functionals on the tangent space $\mathsf{T}_x M$. Since a covector $y \in \mathsf{T}_x^* M$ in the cotangent space is always paired with a vector $z \in \mathsf{T}_x M$ in the tangent space through the inner product $y \cdot z = y^T z$, we can formally identify the cotangent space with the tangent space using this inner product. In this sense, a covector can be formally associated with a unique tangent vector except that we view it as a linear functional that acts on another vector via the inner product.

Recall, a smooth vector field on a manifold M has the property: for each $x \in M$ there is a unique vector $y \in \mathsf{T}_x M$ and this correspondence is continuous. Consequently, the concept of a smooth vector field on a manifold M is naturally extended to the concept of a smooth covector field on a manifold M: for each $x \in M$ there is a unique covector $y \in \mathsf{T}_x^* M$ and this correspondence is continuous. The interpretation is that a covector field associates with each $x \in M$ in the manifold a unique covector (or cotangent vector) $y \in \mathsf{T}_x^* M$ in the cotangent space of the manifold.

Covector fields on a manifold can arise from vector differential equations or from vector differential-algebraic equations as discussed in the prior section of this chapter. The difference is that we view solutions of such differential equations or differential-algebraic equations as linear functionals on the appropriate tangent spaces. The concepts of linearized covector fields, equilibria of a covector field, and stability of an equilibrium follow from the corresponding concepts for a vector field.

1.5 Problems

1.1. Let $A \in \mathbb{R}^{m \times n}$, with $m < n$, have rank m and let $b \in \mathbb{R}^m$. Consider the linear manifold $M = \{ y \in \mathbb{R}^n : Ay = b \}$. Let $x \in \mathbb{R}^n$ and obtain an orthogonal decomposition of x into the sum of a vector in the linear manifold M and a vector orthogonal to M.

1.2. Suppose that $y \in S^2$. Let $x \in \mathbb{R}^3$ and obtain an orthogonal decomposition of x into the sum of a vector in the direction of $y \in S^2$ and a vector orthogonal to $y \in S^2$.

1.3. Suppose that $y = (y_1, \ldots, y_n) \in (S^2)^n$. Let $x \in \mathbb{R}^{3n}$ and obtain an orthogonal decomposition of x into the sum of a vector in the direction of $y \in (S^2)^n$ and a vector orthogonal to $y \in (S^2)^n$.

1.4. Suppose that $y \in \mathbb{R}^3$ is nonzero, and $x \in \mathbb{R}^3$.

(a) Show that $x \in \mathbb{R}^3$ can be uniquely decomposed as

$$x = (I_{3\times 3} - S(y))x + S(y)x,$$

   where $S(y)$ is the 3×3 skew-symmetric matrix-valued function introduced in (1.8).
(b) Show that the second term on the right is orthogonal to $y \in \mathbb{R}^3$.
(c) Show that the first term on the right is not necessarily in the direction $y \in \mathbb{R}^3$, so that this decomposition differs from the orthogonal decomposition given in the text.

1.5. Prove the following:

(a) The set of all $n \times n$ symmetric matrices is an $\frac{n(n+1)}{2}$-dimensional subspace of $\mathbb{R}^{n \times n}$.
(b) The set of all $n \times n$ skew-symmetric matrices is an $\frac{n(n-1)}{2}$-dimensional subspace of $\mathbb{R}^{n \times n}$.
(c) The vector space $\mathbb{R}^{n \times n}$ is the direct sum of the subspace of symmetric matrices and the subspace of skew-symmetric matrices.

1.6. Show that for each $R \in SO(3)$ near the identity, the Cayley transformation $\mathfrak{so}(3) \to SO(3)$ given by (1.10) can be inverted. Give an expression for the inversion formula.

1.7. Show that for each $R \in SO(3)$ near the identity, the exponential transformation $\mathfrak{so}(3) \to SO(3)$ given by (1.11) can be inverted. Give an expression for the inversion formula.

1.8. Prove the following matrix identities.

(a) For any vectors $x, y \in \mathbb{R}^n$:

$$x^T y = \mathsf{trace}[xy^T].$$

(b) For any $m \times n$ matrix A and any $n \times m$ matrix B:

$$\mathsf{trace}[AB] = \mathsf{trace}[BA].$$

(c) For any $x \in \mathbb{R}^3$:

$$S^T(x) = -S(x),$$
$$x \times x = S(x)x = 0,$$
$$S(x)^2 = xx^T - \|x\|^2 I_{3\times 3},$$
$$S(x)^3 = -\|x\|^2 S(x).$$

(d) For any $R \in \mathsf{SO}(3)$ and any $x,\ y \in \mathbb{R}^3$:

$$R(x \times y) = (Rx) \times (Ry),$$
$$RS(x)R^T = S(Rx).$$

(e) For any $x,\ y,\ z \in \mathbb{R}^3$:

$$S(x \times y) = yx^T - xy^T,$$
$$S(x \times y) = S(x)S(y) - S(y)S(x),$$
$$S(x)S(y) = -x^T y I_{3\times 3} + yx^T,$$
$$(x \times y) \cdot z = x \cdot (y \times z),$$
$$x \times (y \times z) = (x \cdot z)y - (x \cdot y)z,$$
$$y \times (x \times z) + x \times (z \times y) + z \times (y \times x) = 0,$$
$$\|x \times y\|^2 = \|x\|^2 \|y\|^2 - (x^T y)^2.$$

1.9. The manifold of 2×2 matrices in $\mathsf{SO}(2)$ consists of all matrices that satisfy $R^T R = RR^T = I_{2\times 2}$ and have determinant $+1$.

(a) Show that $\mathsf{SO}(2)$ is a one-dimensional manifold.
(b) Show that $\mathsf{SO}(2)$ is a Lie group with matrix multiplication as the group operation.
(c) Show that each $R \in \mathsf{SO}(2)$ is nonsingular.
(d) Show that the two rows of each $R \in \mathsf{SO}(2)$ are orthonormal; show that the two columns of each $R \in \mathsf{SO}(2)$ are orthonormal.
(e) Show that each $R \in \mathsf{SO}(2)$ in a neighborhood of the identity can be represented as an exponential matrix, that is there is a skew-symmetric matrix $\xi \in \mathbb{R}^{2\times 2}$, that is $\xi + \xi^T = 0$, such that $R = e^\xi$.

1.10. The set of 2×2 real skew-symmetric matrices S, that satisfy $S + S^T = 0$, is denoted by $\mathfrak{so}(2)$.

(a) Show that each real 2×2 skew-symmetric matrix $S \in \mathfrak{so}(2)$ can be expressed as

$$S = \omega \begin{bmatrix} 0 & -1 \\ 1 & 0 \end{bmatrix},$$

for some $\omega \in \mathbb{R}^1$.

(b) Show that $\mathfrak{so}(2)$ is a one-dimensional vector space.
(c) Show that any $R \in SO(2)$ in a neighborhood of the identity can be expressed as $R = e^S$ for some $S \in \mathfrak{so}(2)$; thus $\mathfrak{so}(2)$ can be viewed as the Lie algebra of $SO(2)$.
(d) Show that the tangent space of $SO(2)$ at the identity matrix $I_{2\times2}$ is in fact $\mathfrak{so}(2)$.

1.11. Let $A \in \mathbb{R}^{m\times n}$ be full rank with $0 < m < n$. Consider the linear manifold $M = \{x \in \mathbb{R}^n : Ax = b\}$.

(a) Let $x \in M$. Describe the tangent space $\mathsf{T}_x M$.
(b) Describe the tangent bundle $\mathsf{T}M$.
(c) Let $x \in M$. Describe the cotangent space $\mathsf{T}_x^* M$.
(d) Describe the cotangent bundle $\mathsf{T}^* M$.

1.12. Consider $q = \left[\frac{4}{5}, \frac{3}{5}\right]^T \in \mathsf{S}^1$.

(a) Describe the tangent space $\mathsf{T}_q \mathsf{S}^1$.
(b) Describe the cotangent space $\mathsf{T}_q^* \mathsf{S}^1$.
(c) Describe the tangent bundle $\mathsf{T}\mathsf{S}^1$.
(d) Describe the cotangent bundle $\mathsf{T}^*\mathsf{S}^1$.
(e) Describe the tangent space $\mathsf{T}_{(q,0)} \mathsf{T}\mathsf{S}^1$.
(f) Describe the tangent bundle $\mathsf{T}\mathsf{T}\mathsf{S}^1$.

1.13. Consider $q = \left[\frac{1}{2}, \frac{1}{2}, \frac{\sqrt{2}}{2}\right]^T \in \mathsf{S}^2$.

(a) Describe the tangent space $\mathsf{T}_q \mathsf{S}^2$.
(b) Describe the cotangent space $\mathsf{T}_q^* \mathsf{S}^2$.
(c) Describe the tangent bundle $\mathsf{T}\mathsf{S}^1$.
(d) Describe the cotangent bundle $\mathsf{T}^*\mathsf{S}^2$.
(e) Describe the tangent space $\mathsf{T}_{(q,0)} \mathsf{T}\mathsf{S}^2$.
(f) Describe the tangent bundle $\mathsf{T}\mathsf{T}\mathsf{S}^2$.

1.14. Consider the manifold $M = \{x \in \mathbb{R}^n : f(x) = 0\}$, where $f : \mathbb{R}^n \to \mathbb{R}^1$ is continuously differentiable and $f'(x) = \frac{\partial f(x)}{\partial x} \neq 0$ when $f(x) = 0$. Show that

$$\mathsf{T}_x M = \left\{\xi \in \mathbb{R}^n : \xi = Sf'(x),\ S \in \mathbb{R}^{n\times n},\ S + S^T = 0\right\}.$$

1.15. Show that $\mathbb{R}^{(n\times n)}$ and $\mathbb{R}^{n^2}$ are diffeomorphic.

1.16. Show that $SO(2)$ and S^1 are diffeomorphic.

1.17. Show that $SO(3)$ and S^2 are not diffeomorphic.

1.18. Let $a > 0$ and $b > 0$. The set $\mathsf{S}_{ab}^1 = \{q \in \mathbb{R}^2 : (\frac{q_1}{a})^2 + (\frac{q_2}{b})^2 = 1\}$ can be viewed as an elliptical manifold embedded in $\mathbb{R}^2$ that is a smooth deformation of the embedded manifold S^1.

(a) Describe the tangent space of S_{ab}^1 at $q \in S_{ab}^1$ and describe the tangent bundle of S_{ab}^1.
(b) Show that this elliptical manifold and the sphere S^1 are diffeomorphic and $\phi : S^1 \to S_{ab}^1$ given by $\phi(q) = [\frac{q_1}{a}, \frac{q_2}{b}]^T$ is a diffeomorphism.
(c) Describe the tangent space of S_{ab}^1 at $q \in S_{ab}^1$ and the tangent bundle of S_{ab}^1 using this diffeomorphism.

1.19. Let $a > 0$, $b > 0$, and $c > 0$. The set $S_{abc}^2 = \{q \in \mathbb{R}^3 : (\frac{q_1}{a})^2 + (\frac{q_2}{b})^2 + (\frac{q_3}{c})^2 = 1\}$ can be viewed as an ellipsoidal manifold embedded in $\mathbb{R}^3$ that is a smooth deformation of the embedded manifold S^2.

(a) Describe the tangent space of S_{abc}^2 at $q \in S_{abc}^2$ and describe the tangent bundle of S_{abc}^2.
(b) Show that this ellipsoidal manifold and the sphere S^2 are diffeomorphic and $\phi : S^2 \to S_{abc}^2$ given by $\phi(q) = [\frac{q_1}{a}, \frac{q_2}{b}, \frac{q_3}{c}]^T$ is a diffeomorphism.
(c) Describe the tangent space of S_{abc}^2 at $q \in S_{abc}^2$ and the tangent bundle of S_{abc}^2 using this diffeomorphism.

1.20. Consider the *parabola* manifold

$$M = \left\{ x \in \mathbb{R}^2 : x_2 = x_1^2 \right\}$$

embedded in $\mathbb{R}^2$. Let $x \in M$.

(a) Determine an analytical expression for the orthogonal projection operator $P(x) : \mathbb{R}^2 \to T_x M$ satisfying (1.32) as a 2×2 matrix-valued function.
(b) Show that the orthogonal projection operator is well defined for all $x \in M$ and it is differentiable.

1.21. Consider the *helix* manifold

$$M = \left\{ x \in \mathbb{R}^3 : x_1 = R \cos\left(\frac{2\pi x_3}{L}\right), \ x_2 = R \sin\left(\frac{2\pi x_3}{L}\right) \right\}$$

embedded in $\mathbb{R}^3$. Let $x \in M$.

(a) Determine an analytical expression for the orthogonal projection operator $P(x) : \mathbb{R}^3 \to T_x M$ satisfying (1.32) as a 3×3 matrix-valued function.
(b) Show that the orthogonal projection operator is well defined for all $x \in M$ and it is differentiable.

1.22. A vector field $F : \mathbb{R}^3 \to T\mathbb{R}^3$ is described by:

$$F(x) = \begin{cases} -k\frac{x}{\|x\|}, & x \neq 0, \\ 0, & x = 0, \end{cases}$$

for a constant $k > 0$.

(a) Describe the geometry of the vector field on $\mathbb{R}^3$. Show that it is continuous everywhere except at the origin.

(b) Describe the flow of the vector field on $\mathbb{R}^3$. What are equilibrium solutions of the flow?

1.23. A vector field $F : \mathbb{R}^3 \to T\mathbb{R}^3$ is described by:

$$F(x) = \begin{cases} k\frac{x}{\|x\|}, & x \neq 0, \\ 0, & x = 0, \end{cases}$$

for a constant $k > 0$.

(a) Describe the geometry of the vector field on $\mathbb{R}^3$. Show that it is continuous everywhere except at the origin.
(b) Describe the flow of the vector field on $\mathbb{R}^3$. What are equilibrium solutions of the flow?

1.24. A vector field $F : \mathbb{R}^3 \to T\mathbb{R}^3$ is described by:

$$F(x) = a \times x,$$

for a constant nonzero vector $a \in \mathbb{R}^3$.

(a) Describe the geometry of the vector field on $\mathbb{R}^3$.
(b) Describe the flow of the vector field on $\mathbb{R}^3$. What are equilibrium solutions of the flow?

1.25. Consider the linear differential equations on $\mathbb{R}^3$:

$$\dot{x}_1 = -1,$$
$$\dot{x}_2 = 2,$$
$$\dot{x}_3 = -1.$$

(a) Show that these differential equations define a continuous vector field on the linear manifold $M = \{x \in \mathbb{R}^3 : x_1 + x_2 + x_3 = 1\}$.
(b) Determine the unique solution for the initial-value problem defined by $x(0) = [1,0,0]^T \in M$ and verify that this solution lies in M.
(c) Are there any equilibrium solutions on this linear manifold? If so, what are the stability properties of these equilibrium solutions?

1.26. Consider the differential equations on $\mathbb{R}^2$:

$$\dot{q}_1 = -q_2,$$
$$\dot{q}_2 = q_1.$$

(a) Show that these differential equations define a continuous vector field on the manifold S^1.
(b) Determine the unique solution for the initial-value problem defined by $q(0) = [1,0]^T \in S^1$ and verify that this solution lies in S^1.

(c) Are there any equilibrium solutions of the vector field on the manifold S^1? If so, what are the stability properties of these equilibrium solutions?

1.27. Consider the differential equations on $\mathbb{R}^3$:

$$\dot{q}_1 = -q_2 + q_3,$$
$$\dot{q}_2 = q_1 - q_3,$$
$$\dot{q}_3 = -q_1 + q_2.$$

(a) Show that these differential equations define a continuous vector field on the manifold S^2.
(b) Determine the unique solution for the initial-value problem defined by $q(0) = [1, 0, 0]^T \in S^2$ and verify that this solution lies in S^2.
(c) Are there any equilibrium solutions of the vector field on the manifold S^2? If so, what are the stability properties of these equilibrium solutions?

1.28. Consider the matrix differential equation

$$\dot{R} = RS$$

defined on $\mathbb{R}^{3 \times 3}$, where

$$S = \begin{bmatrix} 0 & 1 & -1 \\ -1 & 0 & 0 \\ 1 & 0 & 0 \end{bmatrix}.$$

(a) Show that this differential equation defines a continuous vector field on SO(3).
(b) Determine the unique solution for the initial-value problem defined by $R(0) = I_{3 \times 3} \in$ SO(3) and verify that this solution lies in SO(3).
(c) Are there any equilibrium solutions of the vector field on the manifold SO(3)? If so, what are the stability properties of these equilibrium solutions?

1.29. Consider the matrix differential equation

$$\dot{R} = RS$$

defined on $\mathbb{R}^{2 \times 2}$, where

$$S = \begin{bmatrix} 0 & -1 \\ 1 & 0 \end{bmatrix}.$$

(a) Show that this differential equation defines a continuous vector field on SO(2).
(b) Determine the unique solution for the initial-value problem defined by $R(0) = I_{2 \times 2} \in$ SO(2) and verify that this solution lies in SO(2).

(c) Are there any equilibrium solutions of the vector field on the manifold SO(2)? If so, what are the stability properties of these equilibrium solutions?

1.30. Consider the index one differential-algebraic equations (1.49) and (1.50).

(a) Show that the multipliers necessarily satisfy

$$\lambda = -\left(\frac{\partial f(x)}{\partial x}\left(\frac{\partial f(x)}{\partial x}\right)^T\right)^{-1}\left(\frac{\partial f(x)}{\partial x}\right)g(x).$$

(b) Show that the vector field defined by the differential equation (1.51) is invariant on M, that is it defines a continuous vector field restricted to M.

(c) What conditions on the functions $f(x)$ and $g(x)$ guarantee that the multipliers are identically zero?

1.31. Consider the index two differential-algebraic equations (1.52) and (1.53).

(a) Show that the multipliers necessarily satisfy

$$\lambda = -\left(\frac{\partial f(x)}{\partial x}\left(\frac{\partial f(x)}{\partial x}\right)^T\right)^{-1}\left(\left(\frac{\partial f(x)}{\partial x}\right)g(x,\dot{x}) + \left(\frac{d}{dt}\frac{\partial f(x)}{\partial x}\right)^T\dot{x}\right).$$

(b) Show that the vector field defined by the differential equation (1.54) is invariant on $\mathsf{T}M$, that is it defines a continuous vector field on the manifold $\mathsf{T}M$.

(c) What conditions on the functions $f(x)$ and $g(x,\dot{x})$ guarantee that the multipliers are identically zero?

1.32. Consider the differential-algebraic equations

$$\dot{x} + Sx + 2x^T\lambda = 0,$$
$$0 = \|x\|^2 - 1,$$

where $x \in \mathbb{R}^3$, $\lambda \in \mathbb{R}^1$ and

$$S = \begin{bmatrix} 0 & 1 & -1 \\ -1 & 0 & 1 \\ 1 & -1 & 0 \end{bmatrix}.$$

(a) Show that this is a well-posed differential-algebraic system with index one.

(b) Show that these differential-algebraic equations define a vector field on the manifold S^2.

(c) What are equilibrium solutions of this vector field on S^2? What is the corresponding Lagrange multiplier?

(d) What are linearized equations that approximate the dynamics in a neighborhood of each equilibrium solution?

1.33. Consider the differential-algebraic equations

$$\ddot{x} + \dot{x} + Sx + 2x^T\lambda = 0,$$
$$0 = \|x\|^2 - 1,$$

where $x \in \mathbb{R}^3$, $\lambda \in \mathbb{R}^1$ and

$$S = \begin{bmatrix} 0 & 0 & 0 \\ 0 & 0 & 1 \\ 0 & -1 & 0 \end{bmatrix}.$$

(a) Show that this is a well-posed differential-algebraic system with index
two.
(b) Show that these differential-algebraic equations define a vector field on
the tangent bundle $\mathsf{T}S^2$.
(c) What are equilibrium solutions of this vector field on the tangent bundle
$\mathsf{T}S^2$? What is the corresponding Lagrange multiplier at each equilibrium
solution?
(d) What are linearized equations that approximate the dynamics in a neighborhood of each equilibrium solution?

Chapter 2
Kinematics

This chapter first introduces multi-body systems in conceptual terms. It then describes the concept of a Euclidean frame in the material world, following the concept of a *Euclidean structure* introduced in [5]. The Euclidean frame is used to define the set of all possible configurations. This set of configurations is assumed to have the mathematical structure of an embedded finite-dimensional manifold. Differential equations or vector fields on a configuration manifold describe the kinematics or velocity relationships; especially in the case of rotational kinematics, these kinematics relationships are sometimes referred to as *Poisson's* equations. Kinematics equations are obtained for several interesting connections of mass particles and rigid bodies. This chapter provides important background for the subsequent results on the dynamics of Lagrangian and Hamiltonian systems. A classical treatment of kinematics of particles and rigid bodies is given in [99].

2.1 Multi-Body Systems

The concept of a multi-body system is a familiar one in physics and engineering. It consists of a collection of rigid and deformable bodies; these bodies may be physically connected and/or they may interact through forces that arise from a potential. Multi-body systems, as interpreted here and throughout the physics and engineering literature, represent idealizations of real mechanical systems.

Throughout this book, several different categories of multi-body systems are studied. In some cases, an individual body is idealized as consisting of a rigid straight line, referred to as a link, with mass concentrated at one or more points on the link. Such an idealization is a convenient approximation, especially for cases where the physical body is slender and rotational motion

© Springer International Publishing AG 2018
T. Lee et al., *Global Formulations of Lagrangian and Hamiltonian Dynamics on Manifolds*, Interaction of Mechanics and Mathematics,
DOI 10.1007/978-3-319-56953-6_2

about its slender axis and the associated kinetic energy can be ignored. Such approximations are sometimes referred to as lumped mass bodies or bodies defined by mass particles. On the other hand, the concept of a rigid body assumes a full three-dimensional body with spatially distributed mass. Rigidity implies that distances between material points in the body remain constant. Possible interconnection constraints between two bodies include: rotational joints that allow constrained relative rotation between the bodies and prismatic joints that allow constrained relative translation between the bodies. This also allows elastic or gravitational connections that arise from a mutual potential field between the bodies. We do not attempt to provide a theoretical framework for multi-body systems; rather the subsequent development provides numerous instances of multi-body systems that arise as approximations in physics and engineering.

Our interest in multi-body systems is to understand their possible motions, not the structural or design features of multi-body systems. It is convenient to distinguish between multi-body kinematics and multi-body dynamics. Kinematics describe relationships between configuration variables and velocity variables; dynamics describe acceleration relationships. In this chapter, we consider kinematics issues; the remainder of this book presents results on multi-body dynamics, from a Lagrangian and Hamiltonian perspective that make use of the methods of variational calculus.

2.2 Euclidean Frames

In order to describe, in mathematical terms, the mechanics of objects such as ideal mass particles, rigid bodies, and deformable bodies, it is convenient to introduce the concepts of spatial vectors and Euclidean frames that are used to define the motion of objects in the material world or forces that act on those objects.

A spatial vector in mechanics has a direction and magnitude in the material world. A Euclidean frame can be viewed as a construction in the three-dimensional material world consisting of three mutually orthogonal direction vectors that we associate with the standard basis vectors of $\mathbb{R}^3$. We think of the three orthogonal directions as defining (positively) directed axes or spatial vectors in the Euclidean frame, thereby inducing Euclidean coordinates for spatial vectors in the material world. Thus, any spatial vector can be expressed as a linear combination of the three basis vectors of the Euclidean frame, so that a spatial vector is uniquely associated with a vector in $\mathbb{R}^3$. Any direction in the material world is defined by a nonzero vector, typically scaled to be of unit length, in $\mathbb{R}^3$. If we also specify the location of the origin of the Euclidean frame in the material world then the location of any point in the material world is represented by a spatial vector and by a corresponding vector in $\mathbb{R}^3$, expressed with respect to the selected Euclidean frame. It is convenient to order the directed axes of the Euclidean frame according to the usual *right-hand rule*: if the first directed axis is rotated in the direction of the second directed axis according to the fingers of the right hand, then the thumb points in the direction of the third directed axis.

In summary, we have the following. For any given Euclidean frame, we associate a spatial vector in the material world with a vector in $\mathbb{R}^3$; if the Euclidean frame has a specified origin, we can associate the location of a point in the material world by a vector in $\mathbb{R}^3$. These associations imply that the geometry of the material world can be described mathematically in terms of $\mathbb{R}^3$, viewed as a linear vector space with an inner product. Consequently, the following developments do not emphasize the spatial vectors of mechanics but rather the developments are built upon their representations with respect to one or more Euclidean frames.

In the case that the motion of points of a physical object can be characterized to lie within a two-dimensional plane fixed in a three-dimensional Euclidean frame, we often select the three-dimensional frame so that the objects can be easily described with respect to a two-dimensional Euclidean frame consisting of two orthogonal direction vectors; these ideas are natural and are used in examples that are subsequently introduced.

We make use of several categories of Euclidean frames. A Euclidean frame may be fixed or stationary with respect to a background in the material world; such frames are said to be inertial. We refer to such frames as fixed frames or inertial frames. In some cases, we introduce Euclidean frames that are attached to a rigid body, that is the frame translates and rotates with the body; such frames are said to be body-fixed frames. In some cases, it is convenient to introduce a reference Euclidean frame that is neither stationary nor body-fixed but is physically meaningful as a reference. In situations where several Euclidean frames are introduced, it is important to maintain their distinction. We do not introduce any special notation that identifies a specific frame or frames, but rather we hope that this is always clear from the context.

2.3 Kinematics of Ideal Mass Particles

The motion of an ideal mass element or mass particle, viewed as an abstract point in the material world at which mass is concentrated, is naturally characterized with respect to an inertial Euclidean frame. The position of the mass particle in the Euclidean frame, at one instant, is represented by a vector in $\mathbb{R}^3$ where the components in the vector are defined with respect to the standard basis vectors for the Euclidean frame.

If the particle is in motion, then its position vector changes with time t and is represented by a vector-valued function of time $t \to x(t) \in \mathbb{R}^3$. We refer to the position vector $x \in \mathbb{R}^3$ as the configuration of the particle; the space of configurations is represented by the vector space $\mathbb{R}^3$. We can differentiate the position vector once to obtain the velocity vector $v(t) = \frac{dx(t)}{dt}$ of the moving particle, and we observe that the velocity is an element of the tangent space

$\mathsf{T}_x\mathbb{R}^3$, which we can identify in this case with $\mathbb{R}^3$ itself. The tangent bundle $\mathsf{T}\mathbb{R}^3$ consists of all possible pairs of position vectors and velocity vectors, which in this case is identified with $\mathbb{R}^3 \times \mathbb{R}^3$. This characterization defines the kinematics of an ideal mass particle. In fact, this characterization can be used to describe the motion of any fixed point on a body, whether or not this point corresponds to a concentrated mass.

These concepts can easily be extended to a finite number n of interacting particles that are in motion in the material world. Suppose that no constraints are imposed on the possible motions of the n particles; for example, we do not prohibit two or more particles from occupying the same location in the material world. We introduce a Euclidean frame in the material world to describe the motion of these interacting particles. The motion of n interacting particles can be described by an n-tuple, consisting of the position vectors of the n particles, that is functions of time $t \to x_i \in \mathbb{R}^3$, $i = 1, \ldots, n$. Thus, the configuration of the n interacting particles is described by the vector $x = (x_1, \ldots, x_n) \in \mathbb{R}^{3n}$. We can differentiate the configuration vector once to obtain the velocity vector $\dot{x} = (\dot{x}_1, \ldots, \dot{x}_n) \in \mathsf{T}_x\mathbb{R}^{3n}$ of the n particles. The time derivative of the configuration vector, or the velocity vector, is an element of the tangent space $\mathsf{T}_x\mathbb{R}^{3n}$, which we identify in this case with $\mathbb{R}^{3n}$ itself. The tangent bundle consists of all possible pairs of position vectors and velocity vectors, which in this case is identified with $\mathbb{R}^{3n} \times \mathbb{R}^{3n}$.

On the other hand, suppose n interacting particles in motion are subject to algebraic constraints, which can be represented by an embedded manifold M in $\mathbb{R}^{3n}$. The configuration of the n interacting particles is the vector $x = (x_1, \ldots, x_n) \in M$. We can differentiate the configuration vector once to obtain the velocity vector $\dot{x} = (\dot{x}_1, \ldots, \dot{x}_n) \in \mathsf{T}_x M$ of the n particles. That is, the time derivative of the configuration vector, also referred to as the velocity vector, is an element of the tangent space $\mathsf{T}_x M$. The tangent bundle consists of all possible pairs of position vectors and velocity vectors, which is identified with the tangent bundle $\mathsf{T}M$.

2.4 Rigid Body Kinematics

The concept of a rigid body is an idealization of real bodies, but it is a useful approximation that we adopt in the subsequent developments. A rigid body is defined as a collection of material particles, located in the three-dimensional material world. The material particles of a rigid body have the property that the relative distance between any two particles in the body does not change. That is, the body, no matter what forces act on the rigid body or what motion it undergoes, does not deform.

We sometimes consider rigid bodies consisting of an interconnection of a finite number of ideal particles, where the particles are connected by rigid links. Such rigid bodies may be a good approximation to real rigid bodies in the material world, and they have simplified inertial properties. More commonly, we think of a rigid body in the material world as consisting of a spatially distributed mass continuum.

The key in defining rigid body kinematics is the definition of the configuration of the rigid body. The choice of the configuration of a rigid body, and its associated configuration manifold, depends on the perspective and the assumptions or constraints imposed on the rigid body motion. Kinematic relationships describe the rate of change of the configuration as it depends on translational and rotational velocity variables and the configuration. The role of the geometry of the configuration manifold is emphasized in the subsequent development.

As shown subsequently, several configuration manifolds are commonly employed. If the rigid body is constrained to rotate so that each of its material points moves on a circle within a fixed plane, then the attitude configuration of the rigid body can be represented by a point on the configuration manifold S^1. If the rigid body is constrained to rotate so that each of its material points moves on the surface of a sphere in $\mathbb{R}^3$, then the attitude configuration of the rigid body can be represented by a point on the configuration manifold S^2. If the rigid body can rotate arbitrarily in $\mathbb{R}^3$, then the attitude configuration of the rigid body can be represented by a point on the configuration manifold $SO(3)$.

In addition to rotation of rigid bodies, we also consider translation of a rigid body, often characterized by the motion of a selected point in the body, such as its center of mass. If the rigid body is constrained to translate so that each of its material points moves within a fixed plane, then the translational configuration of the rigid body can be selected to lie in the configuration manifold $\mathbb{R}^2$. If the rigid body can translate arbitrarily in $\mathbb{R}^3$, then the translational configuration of the rigid body can be selected to lie in the configuration manifold $\mathbb{R}^3$.

Finally, general rigid body motion, or *Euclidean motion*, can be described by a combination of rotation and translation. For example, a rigid body may be constrained to translate and rotate so that each of its material points lies in a fixed plane or a rigid body may rotate and translate arbitrarily in $\mathbb{R}^3$. We consider each of these situations subsequently, in each case identifying the configuration manifold and the corresponding rigid body kinematics.

2.5 Kinematics of Deformable Bodies

It is challenging to study the kinematics of deformable bodies. The configuration variables and the configuration manifold of a deformable body in motion must be carefully selected to characterize the translational, rotational, and spatial deformation features. This choice is strongly influenced by the assumptions about how the body is spatially distributed and how it may deform in three dimensions. A full treatment of the kinematics and dynamics of deformable bodies requires the use of infinite-dimensional configuration manifolds and is beyond the scope of this book. However, we do treat several challenging examples of finite-dimensional deformable bodies in Chapter 10 that can be viewed as finite element approximations of infinite-dimensional deformable bodies.

2.6 Kinematics on a Manifold

As seen above, the motion of particles, rigid bodies, and deformable bodies is naturally described in terms of the time evolution of configuration variables within a configuration manifold. Kinematics relationships are the mathematical representations that describe this evolution. The time derivative of the configuration variables is necessarily an element of the tangent space at each instant. Thus, the kinematics are the differential equations, and possibly associated algebraic equations, that describe the time derivative of the configuration as it depends on the configuration.

The kinematics relations are typically viewed as describing the time evolution of the configuration or the flow in the configuration manifold. In this chapter, variables that generate the flow within the configuration manifold are often left unspecified. In the subsequent chapters where issues of dynamics are included, the variables that describe the flow within the configuration manifold are specified as part of the dynamics. Alternatively, these flow variables can be specified as having constant values or even as functions of the configuration; mathematical models of this latter type are sometimes referred to as *closed loop kinematics* or kinematics with *feedback*.

Additional structure arises when the configuration manifold is (a) a Lie group manifold, (b) a homogeneous manifold, or (c) a product of Lie groups and homogeneous manifolds. In these cases, the kinematics can be expressed in terms of the Lie algebra of the Lie group manifold or of the Lie group associated with the homogeneous manifold. All of the kinematics examples studied subsequently arise from configuration manifolds of this or a closely related form.

The motion of a particular physical system may be described in different ways, for example by selecting one of several possible configuration manifolds. We demonstrate this possibility in some of the examples to follow. Typically, we would like to select configuration variables and configuration manifolds that are parsimonious in describing the physical features, while providing physical insight without excessive analytical or computational baggage. One of the important features of this book is demonstration of the important role of the geometry of the configuration manifold.

We emphasize an important point made in the Preface: configuration manifolds are fundamental in describing kinematics (and dynamics); although we make use of a Euclidean frame and associated coordinates for the material world, we do *not* use local coordinates to describe the configuration manifolds. In this sense, the subsequent development is said to be *coordinate-free*. This approach is important in obtaining globally valid descriptions of kinematics (and dynamics).

2.7 Kinematics as Descriptions of Velocity Relationships

Several examples of mechanical systems are introduced. In each case, the physical description and assumptions are made clear. The configurations are selected and a configuration manifold is identified. Kinematics relationships are obtained by describing the time derivative of the configuration variables using differential equations, and possibly algebraic equations, on the configuration manifold.

It is important to emphasize that the formulation in each of the following examples leads to representations for the global kinematics; that is, there are no singularities or ambiguities in the expressions for the kinematics. Consequently, this formulation is suitable to describe extreme motions, without requiring an ad-hoc fix or adjustment as is necessary when using local coordinates.

2.7.1 Translational Kinematics of a Particle on an Inclined Plane

A particle is constrained to move in an inclined plane in $\mathbb{R}^3$ with respect to an inertial Euclidean frame. The plane is given by the linear manifold $M = \{(x_1, x_2, x_3) \in \mathbb{R}^3 : x_1 + x_2 + x_3 - 1 = 0\}$. A schematic of the particle on an inclined plane is shown in Figure 2.1.

The manifold M is viewed as the configuration manifold. Since the dimension of the configuration manifold is two, the translational motion of a particle in the plane is said to have two degrees of freedom. We first construct a basis for the tangent space $T_x M$; note that M is the zero level set of the constraint function $f(x) = x_1 + x_2 + x_3 - 1$ and the gradient of this constraint function is $[1, 1, 1]^T \in \mathbb{R}^3$, which is normal to M. Thus, the tangent space $T_x M = \{(y_1, y_2, y_3) \in \mathbb{R}^3 : y_1 + y_2 + y_2 = 0\}$ is a subspace of $\mathbb{R}^3$ that does not depend on $x \in \mathbb{R}^3$. A basis for the tangent space is easily selected, for example as $\{[1, -1, 0]^T, [0, 1, -1]^T\}$. Suppose that a function of time $t \rightarrow x \in M$ represents a translational motion of the particle in the inclined plane. Since $x \in M$, it follows that the time derivative $\dot{x} \in T_x M = \text{span}\{[1, -1, 0]^T, [0, 1, -1]^T\}$. This implies there is a vector-valued function of time $t \rightarrow (v_1, v_2) \in \mathbb{R}^2$ such that

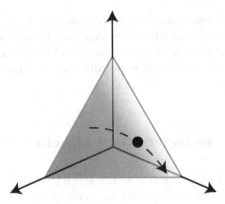

Fig. 2.1 Translational kinematics of a particle on an inclined plane

$$\dot{x} = v_1 \begin{bmatrix} 1 \\ -1 \\ 0 \end{bmatrix} + v_2 \begin{bmatrix} 0 \\ 1 \\ -1 \end{bmatrix}. \tag{2.1}$$

This vector differential equation (2.1) is referred to as the translational kinematics for a particle on the inclined plane. This describes the rate of change of the configuration, namely $\dot{x} \in \mathsf{T}_x M$, in terms of $v = [v_1, v_2]^T \in \mathbb{R}^2$. The vector $v \in \mathbb{R}^2$ is referred to as the translational velocity vector of the particle in the linear manifold. Thus, the translational kinematics of a particle on an inclined plane can be viewed through the evolution of $(x, \dot{x}) \in \mathsf{T}M$ in the tangent bundle or through the evolution of $(x, v) \in M \times \mathbb{R}^2$.

Now, suppose that the translational velocity vector $v \in \mathbb{R}^2$ is a smooth function of the configuration. The translational kinematics equations (2.1) can be viewed as defining a smooth vector field on the linear manifold M. Initial-value problems can be associated with the translational kinematics. The following result can be shown to hold: for any initial-value $x(t_0) = x_0 \in M$, there exists a unique solution of (2.1) satisfying the specified initial-value and this unique solution satisfies $x(t) \in M$ for all t.

2.7.2 Translational Kinematics of a Particle on a Hyperbolic Paraboloid

A particle is constrained to move in a smooth surface in $\mathbb{R}^3$, namely a hyperbolic paraboloid. Its motion is described with respect to an inertial Euclidean frame. The surface is described by the manifold

$$M = \left\{ x \in \mathbb{R}^3 : x_3 = -(x_1)^2 + (x_2)^2 \right\},$$

which we select as the configuration manifold. A schematic of the particle on the surface is shown in Figure 2.2.

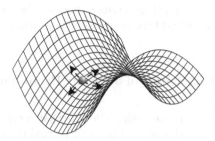

Fig. 2.2 Translational kinematics of a particle on a hyperbolic paraboloid

We determine the tangent space and the tangent bundle of the configuration manifold M. This is done indirectly by first focusing on the kinematics of $[x_1, x_2]^T \in \mathbb{R}^2$ which is the projected position vector of the particle, with respect to the inertial Euclidean frame. Since the dimension of the configuration manifold is two, the translational motion of a particle on the surface is said to have two degrees of freedom.

Suppose that a function of time $t \to [x_1, x_2, x_3]^T \in M$ represents a translational motion of the particle on the surface. It follows that $t \to [x_1, x_2]^T \in \mathbb{R}^2$ and the time derivative $[\dot{x}_1, \dot{x}_2]^T \in \mathsf{T}_{[x_1,x_2]^T}\mathbb{R}^2 = \mathrm{span}\left\{[1, 0]^T, [0, 1]^T\right\}$ using the standard basis for $\mathbb{R}^2$. This implies there is a vector-valued function of time $t \to [v_1, v_2]^T \in \mathbb{R}^2$ such that

$$\begin{bmatrix} \dot{x}_1 \\ \dot{x}_2 \end{bmatrix} = v_1 \begin{bmatrix} 1 \\ 0 \end{bmatrix} + v_2 \begin{bmatrix} 0 \\ 1 \end{bmatrix}.$$

Since $\dot{x}_3 = -2x_1\dot{x}_1 + 2x_2\dot{x}_2$, it follows that

$$\dot{x} = v_1 \begin{bmatrix} 1 \\ 0 \\ -2x_1 \end{bmatrix} + v_2 \begin{bmatrix} 0 \\ 1 \\ 2x_2 \end{bmatrix}. \tag{2.2}$$

This vector differential equation (2.2) describes the translational kinematics for a particle on a hyperbolic paraboloid in $\mathbb{R}^3$. This describes the rate of change of the configuration $\dot{x} \in \mathsf{T}_x M$ in terms of $v = [v_1, v_2]^T \in \mathbb{R}^2$. The vector $v \in \mathbb{R}^2$ is the translational velocity vector of the particle projected onto the $x_1 x_2$-plane. Thus, the translational kinematics of a particle on the surface can be viewed through the evolution of $(x, \dot{x}) \in \mathsf{T}M$ in the tangent bundle or through the evolution of $(x, v) \in M \times \mathbb{R}^2$.

Now, suppose that the translational velocity vector is a smooth function of the configuration. The translational kinematics equations (2.2) can be viewed as defining a smooth vector field on the manifold M. The following result can be shown to hold for an initial-value problem associated with (2.2): for any initial-value $x(t_0) = x_0 \in M$, there exists a unique solution of (2.2) satisfying the specified initial-value and this unique solution satisfies $x(t) \in M$ for all t.

2.7.3 Rotational Kinematics of a Planar Pendulum

A planar pendulum is an inextensible, rigid link that rotates about a fixed axis of rotation, referred to as the pivot. The pivot is fixed in an inertial Euclidean frame. Each material point in the pendulum is constrained to move along a circular path within an inertial two-dimensional plane; thus, it is referred to as a planar pendulum.

The configuration of the planar pendulum is the direction vector of the pendulum link $q \in S^1$ defined in a two-dimensional Euclidean frame, where the two axes define the plane of rotation and the origin of the frame is located at the pivot. For simplicity, the plane of rotation is viewed as defined by the first two axes of a three-dimensional Euclidean frame. The configuration manifold is S^1. Since the dimension of the configuration manifold is one, planar rotations are said to have one degree of freedom. A schematic of the planar pendulum is shown in Figure 2.3.

Suppose that a function of time $t \to q \in S^1$ represents a rotational motion of the planar pendulum. Since $q \in S^1$, it follows that the time derivative $\dot{q} \in T_q S^1$ is necessarily a tangent vector of S^1 at q. Thus, $\dot{q}$ is orthogonal to q, that is $(\dot{q} \cdot q) = 0$. This implies that there is a scalar-valued function of time $t \to \omega \in \mathbb{R}^1$ such that

$$\dot{q} = \omega S q, \tag{2.3}$$

where S is the 2×2 constant skew-symmetric matrix

$$S = \begin{bmatrix} 0 & -1 \\ 1 & 0 \end{bmatrix}$$

introduced in (1.6). Note that when S acts on a vector by multiplication, it rotates the vector by $\frac{\pi}{2}$ counterclockwise. It is easy to verify that equation (2.3) is consistent with the familiar expression in three dimensions for velocity of a point in terms of the cross product of the angular velocity vector and the position vector if we embed $\mathbb{R}^2 \hookrightarrow \mathbb{R}^2 \times \{0\} \subset \mathbb{R}^3$. This embedding is given in terms of the matrix

$$Q = \begin{bmatrix} 1 & 0 \\ 0 & 1 \\ 0 & 0 \end{bmatrix}.$$

Then equation (2.3) is equivalent to

$$Q\dot{q} = \begin{bmatrix} 0 \\ 0 \\ \omega \end{bmatrix} \times (Qq).$$

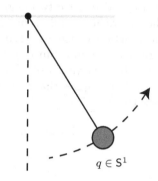

Fig. 2.3 Planar pendulum

This vector differential equation (2.3) is referred to as the rotational kinematics for a planar pendulum. It describes the rate of change of the configuration $\dot{q} \in T_q S^1$ in terms of $\omega \in \mathbb{R}^1$. The scalar ω is referred to as the angular velocity of the planar pendulum. Thus, the rotational kinematics of a planar pendulum can be viewed through the evolution of $(q, \dot{q}) \in TS^1$ in the tangent bundle or through the evolution of $(q, \omega) \in S^1 \times \mathbb{R}^1$.

Suppose the link length of the planar pendulum is L, so that the end of the link is located at $LQq \in \mathbb{R}^3$. Then the velocity vector of the mass particle is $LQ\dot{q} \in \mathbb{R}^3$. It is easy to see that the velocity vector of any material point on the link is proportional to $Q\dot{q} \in \mathbb{R}^3$. Thus, the rotational kinematics of the planar pendulum can be used to characterize the velocity of any point on the planar pendulum.

Now, suppose that the angular velocity is a smooth function of the configuration. The rotational kinematics equation (2.3) can be viewed as defining a smooth vector field on the manifold S^1. We are interested in initial-value problems associated with the rotational kinematics equation (2.3). The following result can be shown to hold: for any initial-value $q(t_0) = q_0 \in S^1$, there exists a unique solution of (2.3) satisfying the specified initial-value and this unique solution satisfies $q(t) \in S^1$ for all t.

2.7.4 Rotational Kinematics of a Spherical Pendulum

A spherical pendulum is an inextensible, rigid link that can rotate about an inertially fixed point on the link, referred to as the pivot. The origin of the inertial Euclidean frame is chosen to be the location of the fixed pivot point. Each material point on the pendulum link is therefore constrained to move on a sphere centered about the pivot, which is why it is referred to as a spherical pendulum. The direction vector $q \in \mathsf{S}^2$ of the link specifies the configuration of the spherical pendulum. The configuration manifold is S^2. Since the dimension of the configuration manifold is two, rotations of the spherical pendulum are said to have two degrees of freedom. This is shown in a schematic of the spherical pendulum in Figure 2.4.

Suppose that the function of time $t \to q \in \mathsf{S}^2$ represents a rotational motion of the spherical pendulum. Since $q \in \mathsf{S}^2$, it follows that the time derivative $\dot{q} \in \mathsf{T}_q\mathsf{S}^2$ is a tangent vector of S^2 at $q \in \mathsf{S}^2$. Thus, $\dot{q}$ is orthogonal to q, that is $(\dot{q} \cdot q) = 0$. This implies that there is a vector-valued function of time $t \to \omega \in \mathbb{R}^3$ such that

$$\dot{q} = \omega \times q,$$

or equivalently

$$\dot{q} = S(\omega)q, \tag{2.4}$$

where

$$S(\omega) = \begin{bmatrix} 0 & -\omega_3 & \omega_2 \\ \omega_3 & 0 & -\omega_1 \\ -\omega_2 & \omega_1 & 0 \end{bmatrix}$$

is a 3×3 skew-symmetric matrix function introduced in (1.8). There is no loss of generality in requiring that $(\omega \cdot q) = 0$. This observation is true since $\omega \in \mathbb{R}^3$ can be decomposed into the sum of a part in the direction of q and a part that is orthogonal to q and the former part does not contribute to the value of $\dot{q}$ in equation (2.4).

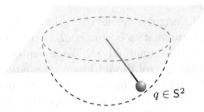

Fig. 2.4 Spherical pendulum

This vector differential equation (2.4) is referred to as the rotational kinematics for a spherical pendulum. It describes the rate of change of the configuration $\dot{q} \in T_q S^2$ in terms of the vector-valued function $\omega \in \mathbb{R}^3$. The vector ω is referred to as the angular velocity vector of the spherical pendulum. Thus, the rotational kinematics of a spherical pendulum can be viewed through the evolution of $(q, \dot{q}) \in TS^2$ in the tangent bundle or through the evolution of $(q, \omega) \in S^2 \times \mathbb{R}^3$. Notice that the dimensions of TS^2 and $S^2 \times \mathbb{R}^3$ are different, which is because of the ambiguity in ω mentioned above. As before, this is resolved by requiring that $(\omega \cdot q) = 0$ for each $(q, \omega) \in S^2 \times \mathbb{R}^3$.

Suppose the link length of the spherical pendulum is L, so that the end of the link is located at $Lq \in \mathbb{R}^3$ and the velocity vector of the point at the end of the link is $L\dot{q} \in \mathbb{R}^3$. It is easy to see that the velocity vector of any point along the massless link of the spherical pendulum is proportional to $\dot{q} \in \mathbb{R}^3$. Thus, the rotational kinematics of the spherical pendulum can be used to characterize the velocity of any material point on the spherical pendulum.

Suppose that the angular velocity vector is a smooth function of the configuration. The rotational kinematics equation (2.4) can be viewed as defining a smooth vector field on the manifold S^2. We are interested in initial-value problems associated with the rotational kinematics equation (2.4). The following result can be shown to hold: for any initial-value $q(t_0) = q_0 \in S^2$, there exists a unique solution of (2.4) satisfying the specified initial-value and this unique solution satisfies $q(t) \in S^2$ for all t.

2.7.5 Rotational Kinematics of a Double Planar Pendulum

The double planar pendulum is an interconnection of two planar pendulum links, with each link constrained to rotate in a common fixed vertical plane. The attitude of each link is defined by a direction vector in the fixed vertical plane. The first link rotates about a fixed one degree of freedom rotational joint or pivot while the second link is connected by a one degree of freedom rotational joint to the end of the first link. A schematic of the double planar pendulum is shown in Figure 2.5.

Here we introduce the attitude of a double planar pendulum as a pair of direction vectors of the two links, with each direction vector in the unit circle S^1. Thus, the configuration of the double planar pendulum is $q = (q_1, q_2) \in (S^1)^2$. These two direction vectors are defined with respect to a fixed two-dimensional Euclidean frame, where the two axes define the plane of rotation of the two pendulums. The origin of the frame is located at the pivot. For simplicity, the plane of rotation is viewed as defined by the first two axes of a three-dimensional Euclidean frame. It follows that $(S^1)^2$ is the two-dimensional configuration manifold of the double planar pendulum, which has two degrees of freedom.

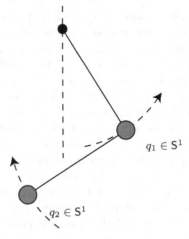

Fig. 2.5 Double planar pendulum

Suppose that a function of time $t \to q = (q_1, q_2) \in (S^1)^2$ represents a rotational motion of the double planar pendulum. Since $q_i \in S^1$, $i = 1, 2$, it follows that the time derivative $\dot{q}_i \in T_{q_i}S^1$ is a tangent vector of S^1 at q_i. Thus, $\dot{q}_i$ is orthogonal to q_i, that is $(\dot{q}_i \cdot q_i) = 0$, $i = 1, 2$. This implies that there is a function of time $t \to \omega = (\omega_1, \omega_2) \in \mathbb{R}^2$ such that

$$\dot{q}_1 = \omega_1 S q_1, \tag{2.5}$$
$$\dot{q}_2 = \omega_2 S q_2. \tag{2.6}$$

As previously, S denotes the 2×2 skew-symmetric matrix given by (1.6).

These vector differential equations (2.5) and (2.6) are referred to as the rotational kinematics for a double planar pendulum. It describes the rate of change of the configuration $\dot{q} \in T_q(S^1)^2$ in terms of $\omega \in \mathbb{R}^2$. The vector $\omega = (\omega_1, \omega_2) \in \mathbb{R}^2$ is referred to as the angular velocity vector of the double planar pendulum. Thus, the rotational kinematics of a double planar pendulum can be viewed through the evolution of $(q, \dot{q}) \in T(S^1)^2$ in the tangent bundle or through the evolution of $(q, \omega) \in (S^1)^2 \times \mathbb{R}^2$.

Suppose the link lengths are L_1 and L_2 for the two planar pendulums. The end of the second planar pendulum, in the fixed Euclidean frame, is located at $L_1 Q q_1 + L_2 Q q_2 \in \mathbb{R}^3$, where

$$Q = \begin{bmatrix} 1 & 0 \\ 0 & 1 \\ 0 & 0 \end{bmatrix},$$

so that the velocity vector of the end of the second spherical pendulum is $L_1 Q \dot{q}_1 + L_2 Q \dot{q}_2 \in \mathbb{R}^3$. It is easy to see that the velocity of any selected body-fixed point on either link of the double planar pendulum is a linear combination of $Q\dot{q}_1 \in \mathbb{R}^3$ and $Q\dot{q}_2 \in \mathbb{R}^3$. Thus, the rotational kinematics of the double planar pendulum can be used to characterize the velocity vector of any material point on either link.

Now, suppose that the angular velocity vector is a smooth function of the configuration. The rotational kinematics equations (2.5) and (2.6) can be viewed as defining a smooth vector field on the manifold $(\mathsf{S}^1)^2$. We are interested in initial-value problems associated with the rotational kinematics equation (2.5) and (2.6). The following result can be shown to hold: for any initial-value $q(t_0) = q_0 \in (\mathsf{S}^1)^2$, there exists a unique solution of (2.5) and (2.6) satisfying the specified initial-value and this unique solution satisfies $q(t) \in (\mathsf{S}^1)^2$ for all t.

2.7.6 Rotational Kinematics of a Double Spherical Pendulum

The double spherical pendulum is an interconnection of two spherical pendulum links, with each link able to rotate in three dimensions. The first link rotates about a fixed spherical pivot while the second link is connected by a spherical pivot located at some point on the first link. A schematic of the double spherical pendulum is shown in Figure 2.6.

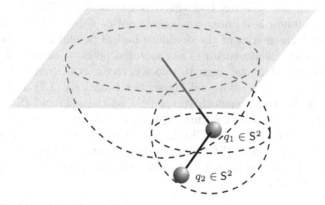

Fig. 2.6 Double spherical pendulum

Here we introduce the attitude of a double spherical pendulum as a pair of direction vectors for the two links, with each direction vector in the unit sphere S^2. Thus, the configuration of the double spherical pendulum is $q = (q_1, q_2) \in (\mathsf{S}^2)^2$. These two direction vectors are defined with respect to an inertially fixed three-dimensional Euclidean frame. The origin of the frame is located at the inertially fixed pivot. It follows that $(\mathsf{S}^2)^2$ is the four-dimensional configuration manifold of the double spherical pendulum, which has four degrees of freedom.

Suppose that a function of time $t \rightarrow q = (q_1, q_2) \in (S^2)^2$ represents a rotational motion of the double spherical pendulum. Since $q_i \in S^2$, $i = 1, 2$, it follows that the time derivative $\dot{q}_i \in T_{q_i}S^2$ is a tangent vector of S^2 at q_i. Thus, $\dot{q}_i$ is orthogonal to q_i, that is $(\dot{q}_i \cdot q_i) = 0$, $i = 1, 2$. This implies that there is a function of time $t \rightarrow \omega = (\omega_1, \omega_2) \in (\mathbb{R}^3)^2$ satisfying $(\omega_i \cdot q_i) = 0$, $i = 1, 2$, such that

$$\dot{q}_1 = S(\omega_1)q_1, \tag{2.7}$$
$$\dot{q}_2 = S(\omega_2)q_2, \tag{2.8}$$

where $S(\omega)$ is the 3×3 skew-symmetric matrix function introduced in (1.8).

These vector differential equations (2.7) and (2.8) are referred to as the rotational kinematics for a double spherical pendulum. It describes the rate of change of the configuration $\dot{q} \in T_q(S^2)^2$ in terms of $\omega \in (\mathbb{R}^3)^2$. The vector $\omega = (\omega_1, \omega_2) \in (\mathbb{R}^3)^2$ is referred to as the angular velocity vector of the double spherical pendulum. Thus, the rotational kinematics of a double spherical pendulum can be viewed through the evolution of $(q, \dot{q}) \in T(S^2)^2$ in the tangent bundle or through the evolution of $(q, \omega) \in (S^2)^2 \times (\mathbb{R}^3)^2$.

Suppose the link lengths are L_1 and L_2 for the two spherical pendulums. The end of the second spherical pendulum, in the fixed Euclidean frame, is located at $L_1 q_1 + L_2 q_2 \in \mathbb{R}^3$, so that the velocity vector of the end of the second spherical pendulum is $L_1 \dot{q}_1 + L_2 \dot{q}_2 \in \mathbb{R}^3$. It is easy to see that the velocity of any selected body-fixed point on either link of the double spherical pendulum is a linear combination of $\dot{q}_1 \in \mathbb{R}^3$ and $\dot{q}_2 \in \mathbb{R}^3$. Thus, the rotational kinematics of the double spherical pendulum can be used to characterize the velocity vector of any material point on either link.

Now, suppose that the angular velocity vector is a smooth function of the configuration. The rotational kinematics equations (2.7) and (2.8) can be viewed as defining a smooth vector field on the manifold $(S^2)^2$. We are interested in initial-value problems associated with the rotational kinematics equation (2.7) and (2.8). The following result can be shown to hold: for any initial-value $q(t_0) = q_0 \in (S^2)^2$, there exists a unique solution of (2.7) and (2.8) satisfying the specified initial-value and this unique solution satisfies $q(t) \in (S^2)^2$ for all t.

2.7.7 Rotational Kinematics of a Planar Pendulum Connected to a Spherical Pendulum

We consider a planar pendulum that can rotate in a fixed plane about a one degree of freedom rotational pivot whose axis is perpendicular to the plane of rotation. One end of a spherical pendulum is connected to the end of the planar pendulum by a two degree of freedom rotational pivot; the spherical pendulum can rotate in three dimensions. An inertially fixed Euclidean frame in three dimensions is constructed so that its first two axes define the plane of rotation of the planar pendulum and its third axis is orthogonal. A schematic of the connection of a planar pendulum and a spherical pendulum is shown in Figure 2.7.

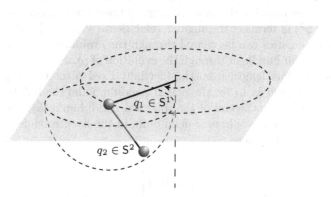

Fig. 2.7 Connection of a planar pendulum and a spherical pendulum

The configuration of the planar pendulum $q_1 \in S^1$ is defined by the direction vector of the planar link, with respect to the first two axes of the Euclidean frame that define the plane of rotation of the link. Thus, S^1 is the configuration manifold of the planar pendulum. The configuration of the spherical pendulum $q_2 \in S^2$ is defined as the direction vector of the spherical link, with respect to the Euclidean frame. Thus, S^2 is the configuration manifold of the spherical pendulum. It follows that $S^1 \times S^2$ is the three-dimensional configuration manifold of the connection of the planar pendulum and the spherical pendulum. The connection has three degrees of freedom.

Suppose the function of time $t \to (q_1, q_2) \in S^1 \times S^2$ represents a motion of the connection of a planar pendulum and a spherical pendulum. As seen previously for a planar pendulum, the time derivative $\dot{q}_1 \in T_{q_1}S^1$ is a tangent vector of S^1 at q_1. Thus, $\dot{q}_1$ is orthogonal to q_1. This implies that there is a scalar-valued function of time $t \to \omega_1 \in \mathbb{R}^1$, referred to as the scalar angular velocity of the planar pendulum about its joint axis, such that

$$\dot{q}_1 = \omega_1 S q_1. \tag{2.9}$$

Here, S denotes the 2×2 skew-symmetric matrix given by (1.6) that rotates a vector by $\frac{\pi}{2}$ counterclockwise.

As seen previously for the spherical pendulum, the time derivative $\dot{q}_2 \in T_{q_2}S^2$ is a tangent vector of S^2 at q_2. Thus, $\dot{q}_2$ is orthogonal to q_2. This implies that there is a vector-valued function of time $t \to \omega_2 \in \mathbb{R}^3$, referred to as the angular velocity vector of the spherical pendulum, such that

$$\dot{q}_2 = S(\omega_2)q_2. \tag{2.10}$$

The skew-symmetric matrix function is defined by (1.8).

These vector differential equations (2.9) and (2.10) are referred to as the rotational kinematics for a connection of a planar pendulum and a spherical pendulum. They describe the rate of change of the configuration $\dot{q} = (\dot{q}_1, \dot{q}_2) \in \mathsf{T}_{(q_1,q_2)}(\mathsf{S}^1 \times \mathsf{S}^2)$ in terms of the angular velocities $\omega = (\omega_1, \omega_2) \in \mathbb{R}^1 \times \mathbb{R}^3$. The rotational kinematics can be viewed through the evolution of $(q, \dot{q}) \in \mathsf{T}(\mathsf{S}^1 \times \mathsf{S}^2)$ in the tangent bundle or through the evolution of $(q, \omega) \in \mathsf{S}^1 \times \mathsf{S}^2 \times \mathbb{R}^1 \times \mathbb{R}^3$.

Suppose the link length from the inertially fixed pivot to the pivot connecting the two links is L_1 for the planar pendulum and the link length from the spherical pivot to the end of the spherical pendulum is L_2. The position vector of the end of the spherical pendulum, in the fixed Euclidean frame, is $L_1 Q q_1 + L_2 q_2 \in \mathbb{R}^3$, where

$$Q = \begin{bmatrix} 1 & 0 \\ 0 & 1 \\ 0 & 0 \end{bmatrix},$$

which defines an embedding of $\mathbb{R}^2$ into $\mathbb{R}^3$. Then the velocity vector of the end of the spherical pendulum is $L_1 Q \dot{q}_1 + L_2 \dot{q}_2 \in \mathbb{R}^3$. It is easy to see that the velocity of any body-fixed point on either link of the connection of a planar pendulum and a spherical pendulum is a linear combination of $Q\dot{q}_1 \in \mathbb{R}^3$ and $\dot{q}_2 \in \mathbb{R}^3$. Thus, the rotational kinematics of the connection of a planar pendulum and a spherical pendulum can be used to characterize the velocity vector of any material point in either link.

Suppose that the angular velocities are smooth functions of the configuration. The rotational kinematics equations can be viewed as defining a smooth vector field on the manifold $\mathsf{S}^1 \times \mathsf{S}^2$. The following result for the initial-value problem holds: for any initial-value $q(t_0) = q_0 \in \mathsf{S}^1 \times \mathsf{S}^2$, there exists a unique solution of (2.9) and (2.10) satisfying the specified initial-value and this unique solution satisfies $q(t) \in \mathsf{S}^1 \times \mathsf{S}^2$ for all t.

2.7.8 Kinematics of a Particle on a Torus

Consider an ideal particle that is constrained to move on the surface of a torus or *doughnut* in $\mathbb{R}^3$, where the torus is the surface of revolution generated by revolving a circle about an axis, coplanar with the circle, that does not touch the circle. Without loss of generality, it is assumed that the torus has major radius $R > 0$ which is the distance from the axis of the torus to the center of the circle and minor radius $0 < r < R$ which is the radius of the revolved circle. A schematic of the particle on a torus is shown in Figure 2.8.

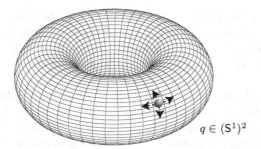

$q \in (S^1)^2$

Fig. 2.8 Particle constrained to a torus

An inertially fixed Euclidean frame is constructed so that the center of the revolved circle is located at $(R, 0, 0) \in \mathbb{R}^3$, the circle lies in the plane defined by the first and third axes and the axis of the torus is the third axis of the Euclidean frame.

The position vector of the particle on the torus is denoted by $x = (x_1, x_2, x_3) \in \mathbb{R}^3$. The configuration of the particle on the torus can be selected as $q = (q_1, q_2) \in (S^1)^2$, and we show that this uniquely determines the position vector of the particle on the torus. Thus, the configuration manifold is $(S^1)^2$ and the particle has two degrees of freedom. We describe the kinematics of the particle on the torus by expressing these kinematics in terms of the evolution on the configuration manifold.

This example should be contrasted with the double planar pendulum example which has the same configuration manifold. It is common to visualize the flow of the double planar pendulum on a torus in $\mathbb{R}^3$, but the values of the major radius and the minor radius of the torus are irrelevant to this visualization. In contrast, the kinematics of the particle on a torus depend on the major radius and the minor radius of the torus, as is subsequently shown. The formulation of the kinematics of a particle on a torus in $\mathbb{R}^3$, as developed here, seems not to have appeared previously in the published literature.

The torus in $\mathbb{R}^3$ can be defined parametrically in terms of two angles

$$x_1 = (R + r \cos \phi) \cos \theta,$$
$$x_2 = (R + r \cos \phi) \sin \theta,$$
$$x_3 = r \sin \phi,$$

but this description leads to an ambiguity in the description of the kinematics.

Alternatively, the torus in $\mathbb{R}^3$ can be defined implicitly by the constraint equation

$$\left(R - \sqrt{x_1^2 + x_2^2} \right)^2 + x_3^2 - r^2 = 0.$$

This formulation can be the basis for describing the kinematics of a particle moving on a torus in terms of a constraint manifold embedded in $\mathbb{R}^3$. This leads to kinematics equations described in terms of differential-algebraic equations on this constraint manifold. We do not develop this formulation any further.

Now, we describe a geometric approach in terms of the configuration manifold $(S^1)^2$. We first express the position of the particle on the torus $x \in \mathbb{R}^3$ in terms of the configuration $q = (q_1, q_2) \in (S^1)^2$. The geometry of the torus implies that an arbitrary vector $x \in \mathbb{R}^3$ on the torus can be uniquely decomposed, in the Euclidean frame, into the sum of a vector from the origin to the center of the embedded circle on which x lies and a vector from the center of this embedded circle to x. This decomposition can be expressed as

$$x = \begin{bmatrix} (R + r(e_1^T q_2))(e_1^T q_1) \\ (R + r(e_1^T q_2))(e_2^T q_1) \\ r(e_2^T q_2) \end{bmatrix}, \tag{2.11}$$

where e_1, e_2 denote the standard basis vectors in $\mathbb{R}^2$. It is easy to see that this is consistent with the parametric representation in terms of two angles given above, but avoids the ambiguity of the angular representation associated with angles that differ by multiples of 2π. This decomposition demonstrates that the position vector of the particle on the torus depends on the configuration of the particle and on the values of the major radius and minor radius of the torus.

The kinematics for the motion of a particle on a torus in $\mathbb{R}^3$ are easily obtained. The velocity vector of the particle on the torus is described by

$$\dot{x} = \begin{bmatrix} (R + r(e_1^T q_2))e_1^T \\ (R + r(e_1^T q_2))e_2^T \\ 0 \end{bmatrix} \dot{q}_1 + \begin{bmatrix} r(e_1^T q_1)e_1^T \\ r(e_2^T q_1)e_1^T \\ re_2^T \end{bmatrix} \dot{q}_2. \tag{2.12}$$

As we have seen previously, there exists an angular velocity vector that is a function of time $t \to \omega = (\omega_1, \omega_2) \in \mathbb{R}^2$ such that the configuration kinematics for $q = (q_1, q_2) \in (S^1)^2$ are given by

$$\dot{q}_1 = \omega_1 S q_1, \tag{2.13}$$

$$\dot{q}_2 = \omega_2 S q_2, \tag{2.14}$$

where S is the 2×2 skew-symmetric matrix that rotates a vector by $\frac{\pi}{2}$ counterclockwise. Thus, the velocity vector of the particle on the torus can also be described by

$$\dot{x} = \begin{bmatrix} -(R + r(e_1^T q_2))e_2^T q_1 \\ (R + r(e_1^T q_2))e_1^T q_1 \\ 0 \end{bmatrix} \omega_1 + \begin{bmatrix} -r(e_1^T q_2)e_2^T q_2 \\ -r(e_1^T q_2)e_2^T q_2 \\ r(e_1^T q_2) \end{bmatrix} \omega_2. \tag{2.15}$$

The vector differential equation (2.12) or equivalently the vector differential equation (2.15), together with (2.13) and (2.14), are referred to as the kinematics of a particle on a torus. They describe the rates of change of the configuration $\dot{q} \in T_q(S^1)^2$ and the particle velocity vector $\dot{x} \in \mathbb{R}^3$ on the torus. The scalars ω_1 and ω_2 are referred to as the angular velocities of the particle on the torus. Thus, the translational kinematics of the particle on a torus can be viewed through the evolution of $(q, \dot{q}) \in T(S^1)^2$ in the tangent bundle or through the evolution of $(q, \omega) \in (S^1)^2 \times (\mathbb{R}^1)^2$.

Suppose that the angular velocities are a smooth function of the configuration. The rotational kinematics (2.13) and (2.14) can be viewed as defining a smooth vector field on the manifold $(S^1)^2$. We are interested in initial-value problems associated with these kinematics equations. The following results can be shown to hold: for any initial-value $q(t_0) = q_0 \in (S^1)^2$, there exists a unique solution of (2.13) and (2.14) satisfying the specified initial-value and this unique solution satisfies $q(t) \in (S^1)^2$ for all t. Each such solution results in a unique solution of (2.12) or (2.15) which has the property: if $x(t_0) \in \mathbb{R}^3$ is on the torus, then $x(t) \in \mathbb{R}^3$ remains on the torus for all t.

2.7.9 Rotational Kinematics of a Free Rigid Body

A rigid body is free to rotate in three dimensions without constraint. In addition to an inertially fixed Euclidean frame, it is convenient to define another Euclidean frame that is fixed to the rigid body. That is, the body-fixed Euclidean frame rotates with the body; in fact, we describe the rotation of the body through the rotation of the body-fixed Euclidean frame. In formal terms, the configuration of a rotating rigid body in three dimensions is the linear transformation that relates the representation of a vector in the body-fixed Euclidean frame to the representation of that vector in the inertially fixed Euclidean frame.

The matrix representing this linear transformation can be constructed by computing the direction cosines between the unit vectors that define the axes of the inertially fixed Euclidean frame and the unit vectors that define the axes of the body-fixed Euclidean frame. This attitude configuration is referred to as a rotation matrix, attitude matrix, or direction cosine matrix of the rigid body. It can be shown that such matrices are orthogonal with determinant $+1$, so that each rotation or attitude matrix $R \in SO(3)$. Hence, the configuration manifold is $SO(3)$, which, as seen previously, is a Lie group. Since the dimension of the configuration manifold is three, rigid body rotations are said to have three degrees of freedom. A schematic of a rotating free rigid body is shown in Figure 2.9.

Suppose that the function of time $t \to R \in SO(3)$ represents a rotational motion of a free rigid body. Since R is an orthogonal matrix, it follows that $R^T R = I_{3 \times 3}$. Differentiating, we obtain $\dot{R}^T R = -R^T \dot{R}$ which implies that

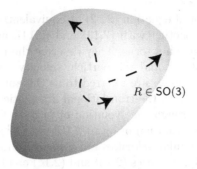

Fig. 2.9 Rotating free rigid body

$R^T \dot{R}$ is always skew-symmetric. This implies that there is a vector-valued function of time $t \to \omega \in \mathbb{R}^3$ such that $R^T \dot{R} = S(\omega)$ where we have used (1.8). This can be written as

$$\dot{R} = RS(\omega), \tag{2.16}$$

which shows that $\dot{R} \in \mathsf{T}_R \mathsf{SO}(3)$, that is $\dot{R}$ is in the tangent space of $\mathsf{SO}(3)$ at $R \in \mathsf{SO}(3)$. This should come as no surprise, since the set of tangent vectors to a curve that takes values on a manifold are the very definition of the tangent space to a manifold. Recall that $S(\omega) \in \mathfrak{so}(3)$, where $\mathfrak{so}(3)$ is the space of skew-symmetric matrices, which is the Lie algebra associated with the Lie group $\mathsf{SO}(3)$. Thus, the time derivative $\dot{R}$ can be expressed as shown in (2.16) so that $S(\omega) = R^T \dot{R} \in \mathfrak{so}(3)$.

Alternatively, we can use the matrix identity $S(R\omega) = RS(\omega)R^T$ to obtain the rotational kinematics for a rotating rigid body given by

$$\dot{R} = S(R\omega)R. \tag{2.17}$$

This alternate form of the rotational kinematics describes the rate of change of the configuration in terms of $R\omega$ which is the angular velocity vector of the rigid body represented in the inertially fixed Euclidean frame, sometimes referred to as the spatial angular velocity of the rigid body. Although (2.16) and (2.17) are equivalent, the form (2.16) using the body-fixed angular velocity is most convenient; we will most often use the rotational kinematics expressed in terms of the body-fixed angular velocity.

Yet another perspective is to view the attitude configuration $R \in \mathsf{SO}(3)$ by partitioning into three rows or equivalently by partitioning $R^T \in \mathsf{SO}(3)$ into three columns, that is

$$R = \begin{bmatrix} r_1^T \\ r_2^T \\ r_3^T \end{bmatrix}, \quad R^T = \begin{bmatrix} r_1 & r_2 & r_3 \end{bmatrix}.$$

The attitude or rotational kinematics of a rotating rigid body can also be described by the three vector differential equations

$$\dot{r}_i = S(r_i)\omega, \quad i = 1, 2, 3. \tag{2.18}$$

The matrix differential equation (2.16) or (2.17), or equivalently the vector differential equations (2.18), are referred to as the rotational kinematics of a free rigid body. They describe the rates of change of the configuration $\dot{R} \in T_R SO(3)$ in terms of the angular velocity $\omega \in \mathbb{R}^3$ represented in the body-fixed frame. Thus, the rotational kinematics of a free rigid body can be viewed through the evolution of $(R, \dot{R}) \in TSO(3)$ in the tangent bundle of $SO(3)$ or through the evolution of $(R, \omega) \in SO(3) \times \mathbb{R}^3$.

Suppose that the angular velocity is a smooth function of the configuration. The rotational kinematics can be viewed as defining a smooth vector field on the Lie group manifold $SO(3)$. We are interested in initial-value problems associated with these rotational kinematics equations. The following results can be shown to hold: for any initial-value $R(t_0) = R_0 \in SO(3)$, there exists a unique solution of the kinematics differential equations satisfying the specified initial-value and this unique solution satisfies $R(t) \in SO(3)$ for all t.

2.7.10 Rotational and Translational Kinematics of a Rigid Body Constrained to a Fixed Plane

Planar Euclidean motion of a rigid body occurs if each point in the body is constrained to move in a fixed two-dimensional plane. Select a point fixed in the rigid body and define a two-dimensional inertial Euclidean frame for this fixed plane within which the selected point moves. Additionally, define a body-fixed Euclidean frame centered at the selected point and spanned by unit vectors e_1 and e_2.

Here, the configuration is taken as $(q, x) \in S^1 \times \mathbb{R}^2$, where $q \in S^1$ denotes the unit vector e_1 with respect to the inertially fixed Euclidean frame and $x \in \mathbb{R}^2$ denotes the position vector of the selected point in the Euclidean frame. In this case, the configuration manifold has a product form. A schematic of a rotating and translating rigid body constrained to a plane is shown in Figure 2.10.

Suppose the function of time $t \to (q, x) \in S^1 \times \mathbb{R}^2$ represents a rotational and translational motion of the constrained rigid body. Since $q \in S^1$ there is a scalar function $t \to \omega \in \mathbb{R}^1$, referred to as the scalar angular velocity of the body and there is a translational velocity vector for the body-fixed point that is a function of time $t \to v \in \mathbb{R}^2$ such that

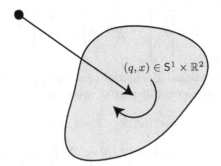

Fig. 2.10 Rotating and translating rigid body in planar motion

$$\dot{q} = \omega S q, \qquad\qquad\qquad\qquad (2.19)$$

$$\dot{x} = [q, Sq]\, v, \qquad\qquad\qquad\qquad (2.20)$$

where S denotes the 2×2 skew-symmetric matrix (1.6) that rotates a vector by $\frac{\pi}{2}$ counterclockwise and the 2×2 partitioned matrix in (2.20) consists of columns $q \in S^1$ and $Sq \in S^1$. Note that $v \in \mathbb{R}^2$ represents the translational velocity of the selected body-fixed point. The rotational and translational kinematics (2.19) and (2.20) can be viewed through the evolution of $(q, x, \dot{q}, \dot{x}) \in T(S^1 \times \mathbb{R}^2)$ in the tangent bundle or through the evolution of $(q, x, \omega, v) \in S^1 \times \mathbb{R}^2 \times \mathbb{R}^1 \times \mathbb{R}^2$.

Now suppose that the angular velocity scalar ω and the translational velocity vector v are smooth functions of the configurations. The kinematics equations can be viewed as defining a smooth vector field on the configuration manifold $S^1 \times \mathbb{R}^2$. We are interested in initial-value problems associated with the kinematics equations (2.19) and (2.20). The following result can be shown to hold: for any initial-values $(q(t_0), x(t_0)) = (q_0, x_0) \in S^1 \times \mathbb{R}^2$, there exists a unique solution of (2.19) and (2.20) satisfying the specified initial-values and this unique solution satisfies $(q(t), x(t)) \in S^1 \times \mathbb{R}^2$ for all t.

2.7.11 Rotational and Translational Kinematics of a Free Rigid Body

A rigid body is free to translate and rotate in three dimensions. As previously, the configuration is defined in terms of an inertially fixed Euclidean frame and a body-fixed Euclidean frame. The configuration of a translating and rotating rigid body in three dimensions consists of an ordered pair $(R, x) \in SE(3)$, where $R \in SO(3)$ is a rotation matrix describing the attitude of the rigid body and $x \in \mathbb{R}^3$ is the vector describing the position of a point in the body, typically the origin of the body-fixed frame, in the inertial frame. The pair $(R, x) \in SE(3)$ can be viewed as a homogenous matrix in $GL(4)$. Hence, the configuration manifold is SE(3), which is a Lie group. Since the dimension of the configuration manifold is six, rigid body rotations and translations are said to have six degrees of freedom. A schematic of a rotating and translating free rigid body is shown in Figure 2.11.

Suppose that the function of time $t \to R(t) \in$ SO(3) is a rotational motion and the position vector of the origin of the body-fixed Euclidean frame with respect to the inertial Euclidean frame is described by the function of time $t \to x(t) \in \mathbb{R}^3$, which defines a translational motion of the rigid body. This implies that there are vector-valued functions of time $t \to \omega \in \mathbb{R}^3$ and $t \to v \in \mathbb{R}^3$ such that the kinematics for Euclidean motion of a rigid body are given by

$$\dot{R} = RS(\omega), \tag{2.21}$$

$$\dot{x} = Rv, \tag{2.22}$$

which shows that $(\dot{R}, \dot{x}) \in$ $\mathsf{T}_{(R,x)}$SE(3). We have used the 3×3 skew-symmetric matrix function given by (1.8).

As in the prior section, it can be shown that if $r_i = R^T e_i$, $i = 1, 2, 3$ then $r_i \in$ S^2, $i = 1, 2, 3$ and the kinematics can be expressed as

$$\dot{r}_i = S(r_i)\omega, \quad i = 1, 2, 3, \tag{2.23}$$

$$\dot{x} = \begin{bmatrix} r_1^T v \\ r_2^T v \\ r_3^T v \end{bmatrix}, \tag{2.24}$$

which also implies that $(\dot{R}, \dot{x}) \in$ $\mathsf{T}_{(R,x)}$SE(3).

Equations (2.21) and (2.22), or equivalently equations (2.23) and (2.24), reflect the fact that the Lie algebra $\mathfrak{se}(3)$ associated with the Lie group SE(3) can be identified with $\mathfrak{so}(3) \times \mathbb{R}^3$.

These differential equations are referred to as the Euclidean kinematics or the rotational and translational kinematics for a rigid body. The matrix differential equation (2.21) describes the rate of change of the rotational configuration in terms of the angular velocity vector $\omega \in \mathbb{R}^3$, represented in the body-fixed frame. The vector differential equation (2.22) is referred to as the translational kinematics for a rigid body. It relates the translational velocity vector $\dot{x} \in \mathbb{R}^3$ represented in the inertial Euclidean frame to the translational velocity vector v represented in the body-fixed Euclidean frame. Thus, the rotational and translational kinematics of a free rigid body can be viewed through the evolution of $(R, x, \dot{R}, \dot{x}) \in$ TSE(3) in the tangent bundle or through the evolution of $(R, x, \omega, v) \in$ SE(3) $\times \mathbb{R}^3 \times \mathbb{R}^3$.

Suppose that the rotational and translational velocity vectors $\omega \in \mathbb{R}^3$, $v \in \mathbb{R}^3$ are smooth functions of the configuration. The translational and rotational kinematics equations (2.21) and (2.22) can be viewed as defining a smooth vector field on the Lie group SE(3). We are interested in initial-value problems. The following result can be shown to hold: for any initial-value

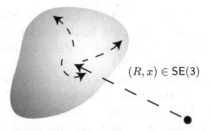

Fig. 2.11 Rotating and translating free rigid body

$(R(t_0), x(t_0)) = (R_0, x_0) \in \mathsf{SE}(3)$, there exists a unique solution of (2.21) and (2.22) satisfying the specified initial-values and this unique solution satisfies $(R(t), x(t)) \in \mathsf{SE}(3)$ for all t.

2.7.12 Translational Kinematics of a Rigid Link with Ends Constrained to Slide Along a Straight Line and a Circle in a Fixed Plane

One end of a rigid link is constrained to slide along a straight line; the other end of the link is constrained to slide along a circle of radius r. For simplicity, assume that the straight line and the circle lie in a common plane, with the straight line passing through the center of the circle. The length of the rigid link is L and assume $r < L$.

Since the ends of the link are constrained, this physical system is referred to as a slider-crank mechanism that can be used to transform circular motion of one end of the link to translational motion of the other end of the link, or vice versa. We now study the kinematics of the slider-crank mechanism. A schematic of this mechanism is shown in Figure 2.12.

A two-dimensional inertially fixed Euclidean frame is constructed with the first axis along the straight line and the second axis orthogonal to the first axis; the origin of the frame is located at the center of the circle.

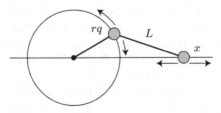

Fig. 2.12 Rigid link with ends constrained to slide along a straight line and a circle

Let $q \in S^1$ denote the direction vector of the end of the link that moves on the circle in the fixed frame; thus, rq is the position vector of the other end of the link in the fixed frame. Let $x \in \mathbb{R}^1$ denote the position of the end of the link that moves on the straight line in the fixed frame. Thus, $(q, x) \in S^1 \times \mathbb{R}^1$, but this is constrained by the fixed length of the rigid link. This constraint is given by

$$\|rq - xe_1\|^2 = L^2.$$

It can be shown that

$$M = \left\{ (q, x) \in S^1 \times \mathbb{R}^1 : \|rq - xe_1\|^2 - L^2 = 0 \right\}$$

is a manifold that characterizes all possible configurations of the physical mechanism. This configuration manifold has dimension one; thus, the mechanism has one degree of freedom.

Suppose that the function of time $t \to (q, x) \in M$ defines a motion for the mechanism. Since one end of the link is constrained to the circle, that is $q \in S^1$, it follows that there is a scalar angular velocity $t \to \omega \in \mathbb{R}^1$ such that

$$\dot{q} = \omega S q,$$

where S is the 2×2 skew-symmetric matrix given by (1.6) that rotates a vector by $\frac{\pi}{2}$ counterclockwise.

Further, since $(\dot{q}, \dot{x}) \in T_{(q,x)}M$, the time derivative of the configuration must satisfy

$$(rq - xe_1)^T (r\omega S q - e_1 \dot{x}) = 0.$$

Since $\dot{x}$ is a scalar, some algebra shows that the time derivative of the configuration can be expressed as

$$\begin{bmatrix} \dot{q} \\ \dot{x} \end{bmatrix} = \omega \begin{bmatrix} I_{2 \times 2} \\ \frac{-rxe_1^T}{(re_1^T q - x)} \end{bmatrix} S q. \tag{2.25}$$

This defines the kinematics for the mechanism by describing the time derivative of the configuration as a tangent vector of the configuration manifold M. Note that the assumption that $r < L$ guarantees that $re_1^T q - x \neq 0$ on M.

Thus, (2.25) guarantees that $(\dot{q}, \dot{x}) \in T_{(q,x)}M$. The kinematics can be viewed through the evolution of $(q, x, \dot{q}, \dot{x}) \in TM$ in the tangent bundle of M or through the evolution of $(q, x, \omega, \dot{x}) \in M \times \mathbb{R}^2$.

If the angular velocity $\omega \in \mathbb{R}^1$ is a smooth function of the configuration, the differential equations (2.25) define a smooth vector field on the configuration manifold M. It follows that for each initial condition $(q(t_0), x(t_0)) = (q_0, x_0) \in M$ there exists a unique solution $(q(t), x(t)) \in M$ for all t.

2.7.13 Rotational and Translational Kinematics of a Constrained Rigid Rod

A thin rigid rod is viewed as a rigid body in three dimensions; the end points of the rigid rod are constrained to move on a fixed, rigid sphere. An inertially fixed Euclidean frame is constructed so that its origin is located at the center of the fixed sphere. A body-fixed Euclidean frame is constructed so that its origin is located at the centroid of the rod and the third body-fixed axis is along the minor principal axis of the rod. The length of the rod along its minor principal axis is L. The radius of the fixed sphere is r and we assume that $r > \frac{L}{2}$. The two ends of the rigid rod are constrained to move in contact with the fixed sphere. This gives rise to two scalar constraints that the ends of the rigid rod maintain contact with the fixed sphere. A schematic of the constrained rigid rod is given in Figure 2.13.

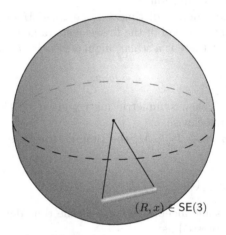

Fig. 2.13 A constrained rigid rod

Let $R \in \mathsf{SO}(3)$ denote the attitude of the rigid rod and let $x \in \mathbb{R}^3$ denote the position vector of the center point of the rod in the fixed Euclidean frame. The two constraint equations that encode the contact between the ends of the rod and the sphere are

$$\|x\|^2 - r^2 + \left(\frac{L}{2}\right)^2 = 0, \tag{2.26}$$

$$x^T R e_3 = 0. \tag{2.27}$$

The position vector to one end of the rod is $x_1 = x + \frac{L}{2}R^T e_3 \in \mathbb{R}^3$ while the position vector to the other end of the rod is $x_2 = x - \frac{L}{2}R^T e_3 \in \mathbb{R}^3$. Simple computations show that $\|x_1\|^2 = r^2$ and $\|x_2\|^2 = r^2$ so that each end of the rod is in contact with the sphere.

Thus, the configuration manifold is

$$M = \left\{ (R, x) \in \mathsf{SE}(3) : \|x\|^2 - r^2 + \left(\frac{L}{2}\right)^2 = 0,\ x^T R e_3 = 0 \right\}. \qquad (2.28)$$

It can be shown that the dimension of the configuration manifold is four so that the rigid rod, constrained to move on the sphere, has four degrees of freedom.

Consider a motion of the constrained rigid rod given by $t \to (R, x) \in M$. This motion implies that there is an angular velocity vector $\omega \in \mathbb{R}^3$ and a translational velocity vector $v \in \mathbb{R}^3$ such that

$$\dot{R} = RS(\omega), \qquad (2.29)$$
$$\dot{x} = Rv, \qquad (2.30)$$

using the 3×3 skew-symmetric matrix function given by (1.8). Here, $\omega \in \mathbb{R}^3$ is the angular velocity vector of the rigid rod in the body-fixed frame and $v \in \mathbb{R}^3$ is the translational velocity vector of the rigid rod in the body-fixed frame.

Differentiating the constraints (2.26) and (2.27) and using the kinematics results in

$$x^T R v = 0, \qquad (2.31)$$
$$e_3^T v + e_3^T R^T S(x) R \omega = 0. \qquad (2.32)$$

Thus, if $(\dot{R}, \dot{x})$ satisfy (2.29) and (2.30) and the velocities $(\omega, v) \in \mathbb{R}^6$ satisfy (2.31) and (2.32), then $(\dot{R}, \dot{x}) \in \mathsf{T}_{(R,x)}M$. The rotational and translational kinematics of the constrained rigid rod can be viewed through the evolution of $(R, x, \dot{R}, \dot{x}) \in \mathsf{T}M$ in the tangent bundle of M or through the evolution of $(R, x, \omega, v) \in M \times \mathbb{R}^6$ where (ω, v) are subject to the constraints (2.31) and (2.32). These differential equations define the constrained kinematics on the configuration manifold M.

Suppose that the angular velocity vector and the translational velocity vector $\omega \in \mathbb{R}^3$, $v \in \mathbb{R}^3$ are smooth functions of the configuration $(R, x) \in M$ that satisfy (2.31) and (2.32). It can be shown that if the initial conditions $R(t_0) = R_0 \in \mathsf{SO}(3)$, $x(t_0) = x_0 \in \mathbb{R}^3$ satisfy $(R_0, x_0) \in M$, then there exists a unique solution $(R(t), x(t)) \in M$ for all t.

2.8 Problems

2.1. A particle moves on a curve embedded in $\mathbb{R}^2$ described by the manifold $M = \left\{ x \in \mathbb{R}^2 : x_1 - (x_2)^2 = 0 \right\}$.

(a) Describe the geometry of the manifold.
(b) Describe the translational kinematics of the particle on the curve in terms of a single kinematics parameter.
(c) Interpret the geometric significance of the kinematics parameter you selected.

2.2. A particle moves on a plane embedded in $\mathbb{R}^3$ described by the manifold $M = \left\{ x \in \mathbb{R}^3 : x_1 + x_2 + x_3 - 1 = 0 \right\}$.

(a) Describe the geometry of the manifold.
(b) Describe the translational kinematics of the particle on the plane in terms of two kinematics parameters.
(c) Interpret the geometric significance of the two kinematics parameters you selected.

2.3. A particle moves on a plane embedded in $\mathbb{R}^3$ described by the manifold $M = \left\{ x \in \mathbb{R}^3 : x_1 - x_2 + x_3 - 1 = 0 \right\}$.

(a) Describe the geometry of the manifold.
(b) Describe the translational kinematics of the particle on the plane in terms of two kinematics parameters.
(c) Interpret the geometric significance of the two kinematics parameters you selected.

2.4. A particle moves on a surface embedded in $\mathbb{R}^3$ described by the manifold $M = \left\{ x \in \mathbb{R}^3 : (x_1)^2 - x_2 + x_3 = 0 \right\}$.

(a) Describe the geometry of the manifold.
(b) Describe the translational kinematics of the particle on the surface in terms of two kinematics parameters.
(c) Interpret the geometric significance of the two kinematics parameters you selected.

2.5. A particle is constrained to move on a surface embedded in $\mathbb{R}^3$ described by the manifold $M = \left\{ x \in \mathbb{R}^3 : (x_1)^2 + (x_2)^2 - x_3 = 0 \right\}$.

(a) Describe the geometry of the manifold.
(b) Describe the translational kinematics of the particle on the surface in terms of two kinematics parameters.
(c) Interpret the geometric significance of the two kinematics parameters you selected.

2.6. A particle is constrained to move on a line embedded in $\mathbb{R}^3$ described by the manifold $M = \left\{ x \in \mathbb{R}^3 : x_1 - x_2 = 0, \ x_1 + x_2 - x_3 = 0 \right\}$.

(a) Describe the geometry of the manifold.
(b) Describe the translational kinematics of the particle on the line in terms of one kinematics parameter.
(c) Interpret the geometric significance of the one kinematics parameter you selected.

2.7. A particle is constrained to move on a curve embedded in $\mathbb{R}^3$ described by the manifold $M = \left\{x \in \mathbb{R}^3 : x_1 - x_2 = 0,\ (x_1)^2 + (x_2)^2 - x_3 = 0\right\}$.

(a) Describe the geometry of the manifold.
(b) Describe the translational kinematics of the particle on the curve in terms of one kinematics parameter.
(c) Interpret the geometric significance of the one kinematics parameter you selected.

2.8. Let $R > 0$ and $L > 0$. A particle moves on a curve embedded in $\mathbb{R}^3$ given by the manifold $M = \left\{x \in \mathbb{R}^3 : x_1 = R\cos\left(\frac{2\pi x_3}{L}\right),\ x_2 = R\sin\left(\frac{2\pi x_3}{L}\right)\right\}$.

(a) Describe the geometry of the manifold. What is the geometric interpretation of the parameters R and L?
(b) Describe the translational kinematics of the particle on the curve in terms of one kinematics parameter.
(c) Interpret the geometric significance of the one kinematics parameter you selected.

2.9. Let $a > 0$, $b > 0$. A particle is constrained to move on an elliptical curve embedded in $\mathbb{R}^2$ that is given by $M = \left\{q \in \mathbb{R}^2 : \{\frac{q_1}{a}\}^2 + \{\frac{q_2}{b}\}^2 - 1 = 0\right\}$. Show that the kinematics of the particle can be expressed in terms of the kinematics on S^1 using the global diffeomorphism $\phi : \mathsf{S}^1 \to M$ given by $\phi(q) = (\frac{q_1}{a}, \frac{q_2}{b})$.

2.10. Let $a > 0$, $b > 0$, $c > 0$. A particle is constrained to move on an ellipsoidal surface given by $M = \left\{q \in \mathbb{R}^3 : \{\frac{q_1}{a}\}^2 + \{\frac{q_2}{b}\}^2 + \{\frac{q_3}{c}\}^2 - 1 = 0\right\}$. Show that the kinematics of the particle can be expressed in terms of the kinematics on S^2 using the global diffeomorphism $\phi : \mathsf{S}^2 \to M$ given by $\phi(q) = (\frac{q_1}{a}, \frac{q_2}{b}, \frac{q_3}{c})$.

2.11. Show the following results hold.

(a) The angular velocity $\omega \in \mathbb{R}^1$ of the planar pendulum satisfies $\omega = -q^T S \dot{q}$.
(b) The angular velocity $\omega \in \mathbb{R}^3$ of the spherical pendulum satisfies $\omega = S(q)\dot{q}$.
(c) The angular velocities $\omega_1 \in \mathbb{R}^1$ and $\omega_2 \in \mathbb{R}^3$ for the connection of a planar pendulum and a spherical pendulum satisfy $\omega_1 = -q_1^T S \dot{q}_1$ and $\omega_2 = S(q_2)\dot{q}_2$.
(d) The angular velocity $\omega \in \mathbb{R}^3$ of the rotating free rigid body satisfies $\omega = (R^T \dot{R})^\vee$, where $(\cdot)^\vee$ denotes the inverse of the hat map defined by (1.8).

2.12. Show that the following results are valid for the planar pendulum, the spherical pendulum, the double planar pendulum, and the double spherical pendulum.

(a) The velocity vector of any material point on the link of a planar pendulum is proportional to $\dot{q} \in T_q S^1$.
(b) The velocity vector of any material point on the link of a spherical pendulum is proportional to $\dot{q} \in T_q S^2$.
(c) The velocity vector of any material point on either link of a double planar pendulum is a linear combination of $\dot{q}_1 \in T_{q_1} S^1$ and $\dot{q}_2 \in T_{q_2} S^1$.
(d) The velocity vector of any material point on either link of a double spherical pendulum is a linear combination of $\dot{q}_1 \in T_{q_1} S^2$ and $\dot{q}_2 \in T_{q_2} S^2$.

2.13. Consider a rotating rigid body with a body-fixed point located at the origin of the body-fixed frame; this body-fixed point is also assumed to be fixed in the inertial frame at the origin of the inertial frame. Assuming the rotational motion of the rigid body is given by $t \to R(t) \in SO(3)$, define the vector $q(t) = R(t)a \in \mathbb{R}^3$, where $a \in \mathbb{R}^3$ denotes an identified point on the rigid body in the body-fixed frame.

(a) Confirm that $q \in \mathbb{R}^3$ is the position vector of the identified point on the rigid body in the inertial frame.
(b) Determine an expression for the angular velocity vector of this identified point in the inertial frame.
(c) Show that this identified point moves on the surface of a sphere in the inertial frame.

2.14. Use the rotational kinematics of a rigid body to show that the angular velocity vector $\omega \in \mathbb{R}^3$, in the body-fixed frame, is constant in time if and only if the angular velocity vector $R\omega \in \mathbb{R}^3$, in the inertial frame, is constant in time.

2.15. Verify that the rotational kinematics of a rotating rigid body, expressed in terms of the transpose of the rows of the attitude matrix $r_i = R^T e_i$, $i = 1, 2, 3$, satisfy $r_i \in S^2$, $i = 1, 2, 3$, and

$$\dot{r}_i = S(r_i)\omega, \quad i = 1, 2, 3.$$

2.16. In contrast with the prior development in this chapter, show that the rotational and translational kinematics of a rigid body constrained to undergo planar Euclidean motion can be described as follows. The configuration manifold is $SE(2) = SO(2) \times \mathbb{R}^2$; the configuration is $(R, x) \in SE(2)$ where $R \in SE(2)$ denotes the planar rotation matrix of the body and $x \in \mathbb{R}^2$ denotes the position vector of a selected point in the body expressed in a fixed two-dimensional Euclidean frame. Show that the rotational and translational kinematics are given by

$$\dot{R} = RS\omega,$$
$$\dot{x} = Rv,$$

where $\omega \in \mathbb{R}^1$ is the scalar angular velocity of the rigid body and $v \in \mathbb{R}^2$ is the velocity vector of the selected point on the body expressed in the body-fixed Euclidean frame.

2.17. Consider the three-dimensional translational kinematics of a rigid link with ends constrained to slide along a straight line and a sphere of radius r; assume the straight line passes through the center of the sphere and the length of the link is $L > r$.

(a) Show that the configuration manifold

$$M = \{(q, x) \in \mathsf{S}^2 \times \mathbb{R}^1 : \|rq - xe_1\|^2 - L^2 = 0\}$$

is a differentiable manifold under the given assumptions.
(b) What are the resulting kinematics of the rigid link on M expressed in terms of the angular velocity vector?

2.18. Consider the planar pendulum with rotational kinematics given by

$$\dot{q} = \omega S q,$$

where $q \in \mathsf{S}^1$, $\omega \in \mathbb{R}^1$ and

$$S = \begin{bmatrix} 0 & -1 \\ 1 & 0 \end{bmatrix}.$$

Suppose that the angular velocity is given in terms of the configuration by $\omega = \sin q_2$. This defines a *closed loop* kinematics system.

(a) What are the closed loop kinematics? Show that they define a continuous vector field on S^1.
(b) Show that there are two equilibrium solutions.
(c) Determine linearized equations at each equilibrium solution. What are the stability properties of each equilibrium solution?
(d) Describe the physical motions of the planar pendulum as governed by the closed loop kinematics.

2.19. Consider the spherical pendulum with rotational kinematics given by

$$\dot{q} = S(\omega) q,$$

where $q \in \mathsf{S}^2$, $\omega \in \mathbb{R}^3$, and

$$S(\omega) = \begin{bmatrix} 0 & -\omega_3 & \omega_2 \\ \omega_3 & 0 & -\omega_1 \\ -\omega_2 & \omega_1 & 0 \end{bmatrix}.$$

Suppose that the angular velocity vector is given in terms of the configuration by

$$\omega = \begin{bmatrix} 0 \\ \sin q_3 \\ -\sin q_2 \end{bmatrix}.$$

This defines a *closed loop* kinematics system.

(a) What are the closed loop kinematics? Show that they define a continuous vector field on S^2.
(b) Show that there are two equilibrium solutions.
(c) Determine linearized equations at each equilibrium solution. What are the stability properties of each equilibrium solution?
(d) Describe the physical motions of the spherical pendulum as governed by the closed loop kinematics.

2.20. Consider the rotational kinematics of a rigid body given by

$$\dot{R} = RS(\omega),$$

where $R \in \mathsf{SO}(3)$, $\omega \in \mathbb{R}^3$ and

$$S(\omega) = \begin{bmatrix} 0 & -\omega_3 & \omega_2 \\ \omega_3 & 0 & -\omega_1 \\ -\omega_2 & \omega_1 & 0 \end{bmatrix}.$$

Suppose that the angular velocity vector is given in terms of the configuration by

$$\omega = \sum_{i=1}^{3} e_i \times R^T e_i,$$

where e_1, e_2, e_3 denote the standard unit basis vectors in $\mathbb{R}^3$. This defines a *closed loop* kinematics system.

(a) What are the closed loop kinematics? Show that they define a continuous vector field on $\mathsf{SO}(3)$.
(b) Show that there are four equilibrium solutions.
(c) Determine linearized equations at each equilibrium solution. What are the stability properties of each equilibrium solution?
(d) Describe the rotational motions of the rigid body as governed by the closed loop kinematics.

2.21. Consider the translational kinematics of a rigid link with ends constrained to slide along a straight line and a circle in a fixed plane. The radius of the circle is r and the length of the rigid link is $L > r$.

(a) Assume the angular velocity of the end of the link constrained to the circle is identically zero. Describe the possible configurations of the rigid link.

(b) Assume the angular velocity of the end of the link constrained to the circle is a nonzero constant. Describe the motion of the link. What is the translational motion of the end of the rigid link that is constrained to move along a straight line?

(c) Assume the translational velocity of the end of the link constrained to the straight line is sinusoidal. Describe the motion of the link. What is the rotational motion of the end of the link that is constrained to move along the circle?

2.22. Consider the kinematics of a rigid link that is constrained to move within a fixed plane described by a two-dimensional Euclidean frame. Let A, B, and C denote three fixed points on the rigid link: point A of the link is constrained to translate along the x-axis of the Euclidean frame while point B of the link is constrained to translate along the y-axis of the Euclidean frame. The distance between points A and B is denoted by L, while the distance between point B and C is denoted by D. This mechanism is referred to as the *Trammel of Archimedes* [4] .

(a) Let $(x, y) \in \mathbb{R}^2$ denote the position vector of point C on the link in the Euclidean frame. What is the algebraic constraint that this position vector must satisfy? What is the configuration manifold embedded in $\mathbb{R}^2$? Describe the geometry of the configuration manifold.

(b) Describe the kinematics relationship of point C on the link, by expressing the time derivative of the configuration in terms of the scalar angular velocity $\omega \in \mathbb{R}^1$ of the rigid link and the configuration.

(c) Suppose the angular velocity of the link is constant; describe the resulting motion of point C on the link.

2.23. A knife-edge can slide on a horizontal plane without friction; the knife-edge is assumed to have a single point of contact with the plane. The motion of the point of contact of the knife-edge is constrained so that its velocity vector is always in the direction of the axis of the knife-edge. This constraint on the direction of the velocity vector is an example of a nonholonomic or non-integrable constraint . The motion of the knife-edge is controlled by the axial speed of the knife-edge and the rotation rate of the knife-edge about its point of contact. A two-dimensional Euclidean frame is introduced for the horizontal plane, so that $x \in \mathbb{R}^2$ denotes the position vector of the contact point of the knife-edge; let $q \in \mathsf{S}^1$ denote the direction vector of the knife-edge in the horizontal plane. Let $V \in \mathbb{R}^1$ be the scalar speed of the knife-edge and let $\omega \in \mathbb{R}^1$ be the scalar rotation rate of the knife-edge.

(a) Show that the nonholonomic constraint can be expressed as

$$\dot{x} = Vq,$$

and the rotational kinematics of the knife-edge are

$$\dot{q} = \omega Sq.$$

(b) Show that these equations of motion can be written in the standard nonlinear control form

$$\begin{bmatrix} \dot{x} \\ \dot{q} \end{bmatrix} = g_1(x, q)V + g_2(x, q)\omega.$$

This is an example of a drift-free nonlinear control system with control
vector fields $g_1(x, q)$ and $g_2(x, q)$ defined on the manifold $\{x, q) : x \in \mathbb{R}^2, q \in S^1\}$; what are the control vector fields?

(c) Suppose that the speed of the knife-edge is a positive constant and the
rotation rate of the knife-edge is zero. Describe the motion of the point
of contact of the knife-edge in the horizontal plane.

(d) Suppose that the speed of the knife-edge is a positive constant and the
rotation rate of the knife-edge is a positive constant. Describe the motion
of the point of contact of the knife-edge in the horizontal plane.

Chapter 3
Classical Lagrangian and Hamiltonian Dynamics

In this chapter, we introduce Lagrangian dynamics and Hamiltonian dynamics that evolve on the vector space $\mathbb{R}^n$ by using classical variational calculus. The Euler–Lagrange equations and Hamilton's equations are obtained. There are many excellent treatments that go far beyond the brief discussion in this chapter. References that emphasize geometric features in the development are [5, 10, 16, 65, 69, 70, 72] while traditional developments are given in [30, 32, 45, 99].

3.1 Configurations as Elements in $\mathbb{R}^n$

We consider systems that can be described by Lagrangian dynamics and by Hamiltonian dynamics with the classical assumption that the configuration is described by the so-called generalized coordinates that constitute a configuration vector $q = [q_1, ..., q_n]^T \in \mathbb{R}^n$. Since the configurations lie in a manifold that is the vector space $\mathbb{R}^n$, it follows that there are n degrees of freedom. Further, the time derivative $\dot{q} = [\dot{q}_1, \ldots, \dot{q}_n]^T \in \mathsf{T}_q\mathbb{R}^n$ lies in the tangent space which is diffeomorphic to $\mathbb{R}^n$ and $(q, \dot{q}) \in \mathsf{T}\mathbb{R}^n$ lies in the tangent bundle of $\mathbb{R}^n$ which is diffeomorphic to $\mathbb{R}^{2n}$. The simple linear geometry of these vector spaces is reflected in the subsequent variational analysis.

Such classical descriptions using configurations in $\mathbb{R}^n$ arise locally if the Lagrangian or Hamiltonian dynamics evolve on any manifold, since any finite-dimensional manifold can be described *locally* using configurations in $\mathbb{R}^n$ for some n. In fact, this is the common practice [5, 30, 32, 69, 70, 72, 99] in the study of Lagrangian and Hamiltonian dynamics on a manifold: first introduce local coordinates on the manifold; then formulate and analyze the dynamics in terms of these local coordinates defined in an open subset of $\mathbb{R}^n$. Although this is common practice, there are many deficiencies in this approach; the

© Springer International Publishing AG 2018   89
T. Lee et al., *Global Formulations of Lagrangian and Hamiltonian Dynamics on Manifolds*, Interaction of Mechanics and Mathematics,
DOI 10.1007/978-3-319-56953-6_3

subsequent chapters of this book introduce a fundamentally different approach that avoids the use of local coordinates on the configuration manifold.

3.2 Lagrangian Dynamics on $\mathbb{R}^n$

The action integral is defined in terms of a Lagrangian, and Hamilton's variational principle is used to derive the Euler–Lagrange equations on $\mathbb{R}^n$ using variational calculus.

3.2.1 Lagrangian Function

The Lagrangian function $L : T\mathbb{R}^n \rightarrow \mathbb{R}^1$ is a function of the configurations, or generalized coordinates, and their time derivatives, sometimes referred to as the generalized velocities. The Lagrangian function is the system kinetic energy, expressed in terms of the configurations and their time derivatives, minus the system potential energy, expressed in terms of the configurations. That is, the Lagrangian function is

$$L(q, \dot{q}) = T(q, \dot{q}) - U(q),$$

where $T(q, \dot{q})$ is the kinetic energy function and $U(q)$ is the potential energy function. The Lagrangian function is defined on the tangent bundle of the configuration manifold $T\mathbb{R}^n$ which is diffeomorphic to $\mathbb{R}^{2n}$.

The Lagrangian formulation provides an effective means to obtain the equations of motion since they require knowledge only of the Lagrangian function which can be obtained using energy concepts. This provides a user-friendly approach since the kinetic energy and the potential energy can often be obtained using elementary concepts of physics. Additionally, they provide a coordinate-independent description of mechanics, allowing one to obtain equations of motion even in terms of coordinates with respect to non-inertial reference frames. This is in contrast to approaches based on Newton's second law, which requires great care in obtaining correct expressions for forces and acceleration in a non-inertial frame.

3.2.2 Variations on $\mathbb{R}^n$

Suppose that a curve $q : [t_0, t_f] \rightarrow \mathbb{R}^n$ describes a motion. We introduce the variation of the curve $q(t)$, which is the ϵ-parametrized family of curves $q^\epsilon(t)$ taking values in $\mathbb{R}^n$, where $\epsilon \in (-c, c)$ for some $c > 0$, $q^0(t) = q(t)$, and the endpoints are fixed, that is, $q^\epsilon(t_0) = q(t_0)$, and $q^\epsilon(t_f) = q(t_f)$. Then, by expanding $q^\epsilon(t)$ as a power series in ϵ, we obtain

$$q^\epsilon(t) = q(t) + \epsilon\delta q(t) + \mathcal{O}(\epsilon^2).$$

By the equality of mixed partial derivatives, the variation of the time derivative of the motion is

$$\dot{q}^\epsilon(t) = \dot{q}(t) + \epsilon \delta \dot{q}(t) + \mathcal{O}(\epsilon^2).$$

The variation of a motion is given by $(q^\epsilon, \dot{q}^\epsilon) : [t_0, t_f] \to \mathbb{R}^{2n}$. The infinitesimal variations of the motion $(q, \dot{q})$ are

$$\frac{d}{d\epsilon} q^\epsilon \Big|_{\epsilon=0} = \delta q, \qquad (3.1)$$

$$\frac{d}{d\epsilon} \dot{q}^\epsilon \Big|_{\epsilon=0} = \delta \dot{q}, \qquad (3.2)$$

where the infinitesimal variations satisfy $(\delta q, \delta \dot{q}) : [t_0, t_f] \to T\mathbb{R}^n$. Due to the fixed endpoint conditions, $\delta q(t_0) = \delta q(t_f) = 0$.

This framework allows us to introduce the action integral and Hamilton's principle which, using methods of variational calculus, provide a coordinate-independent method for deriving the equations of motion.

3.2.3 Hamilton's Variational Principle

The action integral is the integral of the Lagrangian function along a motion of the system over a fixed time period, which is taken as the interval $[t_0, t_f]$:

$$\mathfrak{G} = \int_{t_0}^{t_f} L(q, \dot{q}) \, dt.$$

The action integral along a variation of a motion of the system is

$$\mathfrak{G}^\epsilon = \int_{t_0}^{t_f} L(q^\epsilon, \dot{q}^\epsilon) \, dt.$$

The varied value of the action integral corresponding to a variation of a motion can be expressed as a power series in ϵ as

$$\mathfrak{G}^\epsilon = \mathfrak{G} + \epsilon \delta \mathfrak{G} + \mathcal{O}(\epsilon^2),$$

where the infinitesimal variation of the action integral is

$$\delta \mathfrak{G} = \frac{d}{d\epsilon} \mathfrak{G}^\epsilon \Big|_{\epsilon=0}.$$

Hamilton's principle states that the action integral is stationary, that is the infinitesimal variation of the action integral along any motion of the system is zero:

$$\delta\mathfrak{G} = \frac{d}{d\epsilon}\mathfrak{G}^\epsilon\bigg|_{\epsilon=0} = 0, \tag{3.3}$$

for all possible variations $q^\epsilon(t)$ of $q(t)$ with fixed endpoints, or equivalently, all differentiable curves $(\delta q, \delta\dot{q}) : [t_0, t_f] \to \mathsf{T}\mathbb{R}^n$ satisfying $\delta q(t_0) = \delta q(t_f) = 0$. Motions, that is functions $q : [t_0, t_f] \to \mathbb{R}^n$ that satisfy (3.3), are sometimes referred to as extremals.

Hamilton's variational principle is the foundational cornerstone of Lagrangian and Hamiltonian dynamics. It is discussed in detail in many of the books on classical and geometric treatments of Lagrangian and Hamiltonian dynamics that have already been cited. It is also the foundation for the developments in subsequent chapters of this book.

3.2.4 Euler–Lagrange Equations

Because the configuration manifold is the vector space $\mathbb{R}^n$, the variation of the action integral is easily determined by differentiating under the integral sign to obtain

$$\delta\mathfrak{G} = \frac{d}{d\epsilon}\mathfrak{G}^\epsilon\bigg|_{\epsilon=0} = \int_{t_0}^{t_f}\left\{\frac{\partial L(q,\dot{q})}{\partial \dot{q}}\cdot\delta\dot{q} + \frac{\partial L(q,\dot{q})}{\partial q}\cdot\delta q\right\}dt = 0.$$

We integrate the first term in the integral by parts to obtain

$$\delta\mathfrak{G} = \frac{\partial L(q,\dot{q})}{\partial \dot{q}}\cdot\delta q\bigg|_{t_0}^{t_f} + \int_{t_0}^{t_f}\left\{-\frac{d}{dt}\left(\frac{\partial L(q,\dot{q})}{\partial \dot{q}}\right) + \frac{\partial L(q,\dot{q})}{\partial q}\right\}\cdot\delta q\, dt.$$

We now invoke Hamilton's principle that $\delta\mathfrak{G} = 0$ for all possible variations with fixed endpoints. Using the fact that the infinitesimal variations δq at t_0 and at t_f vanish, the fundamental lemma of the calculus of variations (given in the Appendix A) can be applied to obtain the Euler–Lagrange equations:

$$\frac{d}{dt}\left(\frac{\partial L(q,\dot{q})}{\partial \dot{q}}\right) - \frac{\partial L(q,\dot{q})}{\partial q} = 0. \tag{3.4}$$

With appropriate assumptions on the Lagrangian function, the following result can be stated for the associated initial-value problem. For each initial condition $(q(t_0), \dot{q}(t_0)) = (q_0, \dot{q}_0) \in \mathsf{T}\mathbb{R}^n$, there exists a unique solution of the Euler–Lagrange equation (3.4) denoted by $(q(t), \dot{q}(t)) \in \mathsf{T}\mathbb{R}^n$ that evolves on the tangent bundle of $\mathbb{R}^n$. The Lagrangian flow is the time evolution of $(q, \dot{q}) \in \mathsf{T}\mathbb{R}^n$ corresponding to the Lagrangian vector field on $\mathsf{T}\mathbb{R}^n$ determined by the above Euler–Lagrange equations.

The Lagrangian function for an interacting set of ideal particles and for many other mechanical and physical systems has a special form: the kinetic energy function is quadratic in the time derivatives of the configuration variables.

Consider the Lagrangian function

$$L(q, \dot{q}) = \frac{1}{2}\dot{q}^T M(q)\dot{q} - U(q), \tag{3.5}$$

where $q = (q_1, \ldots, q_n) \in \mathbb{R}^n$ and $\dot{q} = (\dot{q}_1, \ldots, \dot{q}_n) \in T_q\mathbb{R}^n$. The first term in the Lagrangian is the kinetic energy expressed as a positive-definite quadratic form and the second term is the potential energy which is only configuration dependent. In this formulation, $M(q) = [M_{ij}(q)]$ is a symmetric, positive-definite $n \times n$ matrix function of the configuration, with $M_{ij}(q) : \mathbb{R}^n \to \mathbb{R}^1$ denoting the scalar function in the ith row and j^{th} column of the matrix. This matrix function, that defines the kinetic energy, is sometimes referred to as the inertia matrix or the mass matrix.

The resulting Euler–Lagrange equations can be expressed in scalar form as

$$\sum_{j=1}^n M_{kj}(q)\ddot{q}_j + F_k(q, \dot{q}) + \frac{\partial U(q)}{\partial q_k} = 0, \quad k = 1, \ldots, n, \tag{3.6}$$

where

$$F_k(q, \dot{q}) = \sum_{i=1}^n \dot{M}_{ki}(q)\dot{q}_i - \frac{\partial}{\partial q_k} \sum_{i,j=1}^n \frac{1}{2}\dot{q}_i M_{ij}(q)\dot{q}_j, \quad k = 1, \ldots, n.$$

This function can be written in a conventional form as

$$\begin{aligned}
F_k(q, \dot{q}) &= \sum_{i,j=1}^n \dot{q}_i \frac{\partial M_{ki}(q)}{\partial q_j}\dot{q}_j - \sum_{i,j=1}^n \frac{1}{2}\dot{q}_i \frac{\partial M_{ij}(q)}{\partial q_k}\dot{q}_j \\
&= \sum_{i,j=1}^n \frac{1}{2}\dot{q}_i \left\{ \frac{\partial M_{ki}(q)}{\partial q_j} + \frac{\partial M_{ik}(q)}{\partial q_j} - \frac{\partial M_{ij}(q)}{\partial q_k} \right\}\dot{q}_j \\
&= \sum_{i,j=1}^n \frac{1}{2}\dot{q}_i \left\{ \frac{\partial M_{ik}(q)}{\partial q_j} + \frac{\partial M_{jk}(q)}{\partial q_i} - \frac{\partial M_{ij}(q)}{\partial q_k} \right\}\dot{q}_j,
\end{aligned}$$

where we have reindexed the second set of terms and used the symmetry of the inertia terms. This is commonly expressed as

$$F_k(q, \dot{q}) = \sum_{i,j=1}^n [ij, k]\dot{q}_i\dot{q}_j, \quad k = 1, \ldots, n, \tag{3.7}$$

where the Christoffel symbols are

$$[ij,k] = \frac{1}{2} \left\{ \frac{\partial M_{ik}(q)}{\partial q_j} + \frac{\partial M_{jk}(q)}{\partial q_i} - \frac{\partial M_{ij}(q)}{\partial q_k} \right\}, \quad i,j,k = 1,\ldots,n.$$

The Christoffel terms characterize the curvature associated with the kinetic energy metric defined by the inertia matrix $M(q) \in \mathbb{R}^{n \times n}$. Note that if the inertia matrix is independent of the configuration, then the Christoffel terms are all zero.

The Euler–Lagrange equations can be expressed in vector form as

$$M(q)\ddot{q} + F(q,\dot{q}) + \frac{\partial U(q)}{\partial q} = 0, \tag{3.8}$$

where $F(q,\dot{q}) = (F_1(q,\dot{q}),\ldots,F_n(q,\dot{q}))$. This form of the Euler–Lagrange equations describes the evolution of $(q,\dot{q}) \in \mathsf{T}\mathbb{R}^n$ on the tangent bundle of $\mathbb{R}^n$.

When the inertia matrix is constant, the Christoffel terms vanish, and these equations reduce to

$$M\ddot{q} + \frac{\partial U(q)}{\partial q} = 0,$$

or equivalently,

$$\frac{d}{dt}(M\dot{q}) = -\frac{\partial U(q)}{\partial q},$$

which can be interpreted as requiring that the time rate of change of the momentum equals the external force that arises from the potential, this latter interpretation being an expression of Newton's second law. Put another way, Hamilton's principle and the associated Euler–Lagrange equations provide a coordinate-independent characterization of the equations of motion. In particular, Hamilton's principle allows us to derive equations of motion in a non-inertial frame directly from the Lagrangian, given by the kinetic energy minus the potential energy in an inertial frame but expressed in the non-inertial coordinates. The associated Euler–Lagrange equations follow.

3.3 Hamiltonian Dynamics on $\mathbb{R}^n$

The Legendre transformation is introduced and used to determine the conjugate momentum. The Hamiltonian function is defined in terms of the phase variables, the configuration, and the momentum, which are viewed as evolving on the cotangent bundle $\mathsf{T}^*\mathbb{R}^n$. Hamilton's phase space variational principle is used to derive Hamilton's equations.

3.3.1 Legendre Transformation and the Hamiltonian

The Legendre transformation of the Lagrangian gives an equivalent Hamiltonian form of the equations of motion in terms of conjugate momentum covectors. For $q \in \mathbb{R}^n$, the velocity vector $\dot{q}$ lies in the tangent space $\mathsf{T}_q\mathbb{R}^n$ whereas the conjugate momentum covector p lies in the dual space $\mathsf{T}_q^*\mathbb{R}^n$. The tangent space $\mathsf{T}_q\mathbb{R}^n$ and its dual space $\mathsf{T}_q^*\mathbb{R}^n$ are identified by using the usual dot product in $\mathbb{R}^n$, which corresponds to identifying column vectors with row vectors. The Legendre transformation is based on the inner product relation

$$p \cdot \dot{q} = \frac{\partial L(q, \dot{q})}{\partial \dot{q}} \cdot \dot{q},$$

for all $\dot{q} \in \mathsf{T}_q\mathbb{R}^n$. Thus, the Legendre transformation $\mathbb{F}L : \mathsf{T}\mathbb{R}^n \to \mathsf{T}^*\mathbb{R}^n$ is defined as

$$\mathbb{F}L(q, \dot{q}) = (q, p),$$

where $p \in \mathsf{T}_q^*\mathbb{R}^n$ is the conjugate momentum given by

$$p = \frac{\partial L(q, \dot{q})}{\partial \dot{q}}. \tag{3.9}$$

It is assumed that the Lagrangian function has the property that expression (3.9) is invertible in the sense that $\dot{q}$ is uniquely expressible in terms of p and q. The property that the Legendre transformation is globally invertible is referred to as hyperregularity of the Lagrangian function, and if the Legendre transformation is non-invertible, we say that the Lagrangian function is degenerate. For the remainder of this book, we will focus exclusively on hyperregular Lagrangian functions, which in the case of Lagrangians of the form (3.5) correspond to requiring that the inertia matrix $M(q)$ is always full-rank.

The Hamiltonian dynamics can be described by introducing the Hamiltonian function defined on the cotangent bundle of $\mathbb{R}^n$ as $H : \mathsf{T}^*\mathbb{R}^n \to \mathbb{R}^1$:

$$H(q, p) = p \cdot \dot{q} - L(q, \dot{q}), \tag{3.10}$$

where $\dot{q}$ is viewed as a function of (q, p) by inverting the Legendre transformation (3.9).

3.3.2 Hamilton's Equations and Euler–Lagrange Equations

Based on the above definition of the Hamiltonian, we compute its derivative with respect to q and p to obtain,

$$\frac{\partial H(q,p)}{\partial q} = p \cdot \frac{\partial \dot{q}}{\partial q} - \frac{\partial L(q,\dot{q})}{\partial q} - \frac{\partial L(q,\dot{q})}{\partial \dot{q}} \frac{\partial \dot{q}}{\partial q}$$

$$= -\frac{\partial L(q,\dot{q})}{\partial q}$$

$$= -\frac{d}{dt}\left(\frac{\partial L(q,\dot{q})}{\partial \dot{q}}\right)$$

$$= -\dot{p},$$

$$\frac{\partial H(q,p)}{\partial p} = \dot{q} + p \cdot \frac{\partial \dot{q}}{\partial p} - \frac{\partial L(q,\dot{q})}{\partial \dot{q}} \frac{\partial \dot{q}}{\partial p}$$

$$= \dot{q},$$

where we used the Legendre transformation, the Euler–Lagrange equations, and the fact that the Legendre transformation, $p = \frac{\partial L(q,\dot{q})}{\partial \dot{q}}$ implicitly defines $\dot{q}$ in terms of (q,p). This yields Hamilton's equations,

$$\dot{q} = \frac{\partial H(q,p)}{\partial p},$$

$$\dot{p} = -\frac{\partial H(q,p)}{\partial q}.$$

Conversely, if we start with the Hamiltonian, the associated Legendre transformation $\mathbb{F}H : T^*\mathbb{R}^n \to T\mathbb{R}^n$ is defined as

$$\mathbb{F}H(q,p) = (q,\dot{q}),$$

where $\dot{q} \in T_q\mathbb{R}^n$ is given by,

$$\dot{q} = \frac{\partial H(q,p)}{\partial p}. \tag{3.11}$$

Then we can define the Lagrangian in terms of the Hamiltonian,

$$L(q,\dot{q}) = p \cdot \dot{q} - H(q,p),$$

where p is defined implicitly in terms of $(q,\dot{q})$ by inverting the Legendre transformation (3.11). We compute the derivative of the Lagrangian with respect to q and $\dot{q}$ to obtain,

$$\frac{\partial L(q,\dot{q})}{\partial q} = \frac{\partial p}{\partial q}\dot{q} - \frac{\partial H(q,p)}{\partial q} - \frac{\partial H(q,p)}{\partial p}\frac{\partial p}{\partial q}$$

$$= -\frac{\partial H(q,p)}{\partial p}$$

$$= \dot{p},$$

$$\frac{\partial L(q,\dot{q})}{\partial \dot{q}} = \frac{\partial p}{\partial \dot{q}}\dot{q} + p - \frac{\partial H(q,p)}{\partial p}\frac{\partial p}{\partial \dot{q}}$$

$$= p,$$

where we used the Legendre transformation, the Hamilton's equations, and the fact that the Legendre transformation, $\dot{q} = \frac{\partial H(q,p)}{\partial p}$ implicitly defines p in terms of $(q, \dot{q})$. Combining these two equations, we recover the Euler–Lagrange equations,

$$\frac{d}{dt}\left(\frac{\partial L(q,\dot{q})}{\partial \dot{q}}\right) - \frac{\partial L(q,\dot{q})}{\partial q} = 0.$$

3.3.3 Hamilton's Phase Space Variational Principle

Hamilton's equations can also be derived from *Hamilton's phase space variational principle*, based on the action integral expressed in the form

$$\mathfrak{G} = \int_{t_0}^{t_f} \{p \cdot \dot{q} - H(q,p)\}\, dt.$$

We consider variations of the motion $(q,p) \in \mathsf{T}^*\mathbb{R}^n$ given by

$$q^\epsilon = q + \epsilon \delta q + \mathcal{O}(\epsilon^2),$$
$$p^\epsilon = p + \epsilon \delta p + \mathcal{O}(\epsilon^2),$$

where the infinitesimal variations $(\delta q, \delta p) : [t_0, t_f] \to \mathsf{T}_{(q,p)}\mathsf{T}^*\mathbb{R}^n$ satisfy $\delta q(t_0) = \delta q(t_f) = 0$. The varied value of the action integral is given by

$$\mathfrak{G}^\epsilon = \int_{t_0}^{t_f} \{p^\epsilon \cdot \dot{q}^\epsilon - H(q^\epsilon, p^\epsilon)\}\, dt.$$

The varied value of the action integral can be expressed as a power series in ϵ as

$$\mathfrak{G}^\epsilon = \mathfrak{G} + \epsilon \delta \mathfrak{G} + \mathcal{O}(\epsilon^2),$$

where the infinitesimal variation of the action integral is

$$\delta \mathfrak{G} = \left. \frac{d}{d\epsilon} \mathfrak{G}^\epsilon \right|_{\epsilon=0}.$$

Hamilton's phase space variational principle is that the action integral is stationary, that is

$$\delta \mathfrak{G} = \left. \frac{d}{d\epsilon} \mathfrak{G}^\epsilon \right|_{\epsilon=0} = 0,$$

for all differentiable functions $(\delta q, \delta p) : [t_0, t_f] \to \mathsf{T}_{(q,p)}\mathsf{T}^*\mathbb{R}^n$ that satisfy $\delta q(t_0) = \delta q(t_f) = 0$. Note that only the infinitesimal variations in q are required to vanish at the endpoints.

3.3.4 Hamilton's Equations

As before, we compute the variation of the action integral,

$$\delta\mathfrak{G} = \frac{d}{d\epsilon}\mathfrak{G}^\epsilon\bigg|_{\epsilon=0} = \int_{t_0}^{t_f} \left\{ p \cdot \delta\dot{q} - \frac{\partial H(q,p)}{\partial q} \cdot \delta q + \left(\dot{q} - \frac{\partial H(q,p)}{\partial p} \right) \cdot \delta p \right\} dt.$$

We integrate the first term in the integral by parts to obtain

$$\delta\mathfrak{G} = p \cdot \delta q\bigg|_{t_0}^{t_f} + \int_{t_0}^{t_f} \left\{ \left(-\dot{p} - \frac{\partial H(q,p)}{\partial q} \right) \cdot \delta q + \left(\dot{q} - \frac{\partial H(q,p)}{\partial p} \right) \cdot \delta p \right\} dt.$$

We now invoke Hamilton's phase space variational principle that $\delta\mathfrak{G} = 0$ for all possible variations of q and p where the infinitesimal variations of q vanish at the endpoints. The boundary term vanishes since the infinitesimal variations δq at t_0 and at t_f are zero, and the fundamental lemma of the calculus of variations, as in Appendix A, yields Hamilton's equations,

$$\dot{q} = \frac{\partial H(q,p)}{\partial p}, \tag{3.12}$$

$$\dot{p} = -\frac{\partial H(q,p)}{\partial q}. \tag{3.13}$$

With appropriate assumptions on the Lagrangian function, the following result can be stated for the associated initial-value problem. For each initial condition $(q(t_0), p(t_0)) = (q_0, p_0) \in \mathsf{T}^*\mathbb{R}^n$ there exists a unique solution of Hamilton's equations (3.12) and (3.13) denoted by $(q(t), p(t)) \in \mathsf{T}^*\mathbb{R}^n$ that evolves on the cotangent bundle of $\mathbb{R}^n$. The Hamiltonian flow is the time evolution of $(q, p) \in \mathsf{T}^*\mathbb{R}^n$ corresponding to the Hamiltonian vector field on $\mathsf{T}^*\mathbb{R}^n$ determined by the above Hamilton's equations.

The following property follows directly from the above formulation of Hamilton's equations:

$$\begin{aligned} \frac{dH(q,p)}{dt} &= \frac{\partial H(q,p)}{\partial q} \cdot \dot{q} + \frac{\partial H(q,p)}{\partial p} \cdot \dot{p} \\ &= \frac{\partial H(q,p)}{\partial q} \cdot \frac{\partial H(q,p)}{\partial p} - \frac{\partial H(q,p)}{\partial p} \cdot \frac{\partial H(q,p)}{\partial q} \\ &= 0. \end{aligned}$$

This formulation exposes an important property of the Hamiltonian flow on the cotangent bundle: the Hamiltonian function is constant along each solution of Hamilton's equations. It should be emphasized that this property does not hold if the Hamiltonian function has a nontrivial explicit dependence on time.

We again consider the case that the Lagrangian function is quadratic in the time derivatives of the configuration:

$$L(q, \dot{q}) = \frac{1}{2} \dot{q}^T M(q) \dot{q} - U(q),$$

where $q = (q_1, \ldots, q_n) \in \mathbb{R}^n$ and $\dot{q} = (\dot{q}_1, \ldots, \dot{q}_n) \in T_q \mathbb{R}^n$.

In this case, the conjugate momentum is obtained from the Legendre transformation as

$$p = \frac{\partial L(q, \dot{q})}{\partial \dot{q}} = M(q)\dot{q},$$

where we view $p \in T_q^* \mathbb{R}^n$ in the cotangent space of $\mathbb{R}^n$.

The Hamiltonian function $H : \mathsf{T}^* \mathbb{R}^n \to \mathbb{R}^1$ can be shown to be

$$H(q, p) = \frac{1}{2} p^T M^I(q) p + U(q),$$

where, as before, $M^I(q) = M^{-1}(q)$ denotes the matrix inverse of $M(q)$ for each $q \in \mathbb{R}^n$. In this case, we see that the Hamiltonian function is the sum of the kinetic energy and the potential energy; that is, the Hamiltonian is interpreted as the total energy.

Thus, Hamilton's equations are expressed in the form

$$\dot{q} = M^I(q)p, \tag{3.14}$$

$$\dot{p} = \frac{\partial}{\partial q} \left\{ \frac{1}{2} p^T M^I(q) p \right\} - \frac{\partial U(q)}{\partial q}. \tag{3.15}$$

Hamilton's equations describe evolution of the Hamiltonian dynamics $(q, p) \in \mathsf{T}^* \mathbb{R}^n$ on the cotangent bundle of $\mathbb{R}^n$.

3.4 Flow Properties of Lagrangian and Hamiltonian Dynamics

The Lagrangian flow $(q, \dot{q}) \in \mathsf{T}\mathbb{R}^n$ defines a Lagrangian vector field on the tangent bundle of $\mathbb{R}^n$; it is equivalent to the Hamiltonian flow $(q, p) \in \mathsf{T}^* \mathbb{R}^n$ that defines a Hamiltonian vector field on the cotangent bundle of $\mathbb{R}^n$. Each formulation is useful and can provide insight into the dynamical flow.

It is natural to focus on solutions of an initial-value problem for the Euler–Lagrange equations or for Hamilton's equations, assuming that each initial-value problem has a unique solution and the solution is defined for all time. However, it is often useful to conceptualize a family of solutions of the Euler–Lagrange equations or Hamilton's equations defined for a corresponding family of initial conditions; this defines the Lagrangian flow or the Hamiltonian flow for the given Lagrangian or Hamiltonian function. The Euler–Lagrange equations and Hamilton's equations provide two different, but equivalent, perspectives on the dynamical flow.

The Lagrangian flow and the Hamiltonian flow exhibit special properties, a few of which are now summarized.

3.4.1 Energy Properties

Consider a Lagrangian function of the form (3.5), which is quadratic in the time derivatives of the configuration. In this case, the Hamiltonian (3.10) expressed in position and velocity form coincides with the total energy, which is the sum of the kinetic energy and the potential energy,

$$H(q, \dot{q}) = \frac{\partial L(q, \dot{q})}{\partial \dot{q}} \cdot \dot{q} - L(q, \dot{q}) = \frac{1}{2} \dot{q}^T M(q) \dot{q} + U(q).$$

This agrees with the Hamiltonian in position and momentum form,

$$H(q, p) = \frac{1}{2} p^T M^I(q) p + U(q),$$

as the Legendre transform gives $p = M(q)\dot{q}$. As shown previously, the Hamiltonian is constant along each solution of the dynamical flow.

More generally, this is true for any Lagrangian or Hamiltonian that does not explicitly depend on time. To show this directly, we compute the time derivative of the Hamiltonian, expressed in position and velocity form, along a solution of the Euler–Lagrange equations,

$$\frac{d}{dt} H(q, \dot{q}) = \frac{d}{dt} \left(\frac{\partial L(q, \dot{q})}{\partial \dot{q}} \cdot \dot{q} - L(q, \dot{q}) \right)$$

$$= \frac{d}{dt} \left(\frac{\partial L(q, \dot{q})}{\partial \dot{q}} \right) \dot{q} + \frac{\partial L(q, \dot{q})}{\partial \dot{q}} \ddot{q} - \frac{\partial L(q, \dot{q})}{\partial q} \dot{q} - \frac{\partial L(q, \dot{q})}{\partial \dot{q}} \ddot{q}$$

$$= \left\{ \frac{d}{dt} \left(\frac{\partial L(q, \dot{q})}{\partial \dot{q}} \right) - \frac{\partial L(q, \dot{q})}{\partial q} \right\} \dot{q}$$

$$= 0.$$

Similarly, computing the time derivative of the Hamiltonian, expressed in position and momentum form, along a solution of Hamilton's equations yields,

$$\frac{d}{dt} H(q,p) = \frac{\partial H(q,p)}{\partial q} \dot{q} + \frac{\partial H(q,p)}{\partial p} \dot{p}$$

$$= \frac{\partial H(q,p)}{\partial q} \frac{\partial H(q,p)}{\partial p} + \frac{\partial H(q,p)}{\partial p} \left(-\frac{\partial H(q,p)}{\partial q} \right)$$

$$= 0.$$

These results have important implications. For example, for any real constant c the set $\{(q,\dot{q}) \in \mathsf{T}\mathbb{R}^n : H(q,\dot{q}) = c\}$ is an invariant subset of the tangent bundle $\mathsf{T}\mathbb{R}^n$; similarly, the set $\{(q,p) \in \mathsf{T}^*\mathbb{R}^n : H(q,p) = c\}$ is an invariant subset of the cotangent bundle $\mathsf{T}^*\mathbb{R}^n$. That is, for any initial condition in such a set, the solution remains in the set.

These results can also be used to characterize a number of stability properties; see [76] for details of such developments.

3.4.2 Cyclic Coordinates, Conserved Quantities, and Classical Reduction

In classical terms, a scalar configuration variable q_k for a fixed index k is cyclic, sometimes referred to as ignorable, if the Lagrangian function is independent of this configuration variable. If this is true, then it follows from the Euler–Lagrange equations that

$$\frac{d}{dt} \left(\frac{\partial L(q,\dot{q})}{\partial \dot{q}_k} \right) = 0.$$

This implies that the conjugate momentum,

$$p_k = \frac{\partial L(q,\dot{q})}{\partial \dot{q}_k},$$

is constant along each solution of the Lagrangian flow. Similarly, p_k is constant along each solution of the corresponding Hamiltonian flow. In this case, we obtain additional conserved quantities or integrals of the motion which are constant along the dynamical flow.

The existence of a cyclic generalized coordinate allows reduction of the dynamics in the sense that the reduced dynamics of the noncyclic coordinates can be defined by fixing the value of the conserved quantity. In classical terms, this leads to a reduced Lagrangian model, using the method of Routh reduction. The advantage is that the reduced equations are expressed in terms of fewer configuration variables; on the other hand, it is often the case that the reduced equations are more analytically complicated.

We now briefly describe the procedure for classical Routh reduction, which is applicable for the case of one or more cyclic variables. Consider a Lagrangian of the form $L(q_1, \ldots q_s, \dot{q}_1, \ldots, \dot{q}_s, \dot{q}_{s+1}, \ldots, \dot{q}_n)$, where $s < n$, which is invariant under a shift in the $q_{s+1}, \ldots q_n$ variables. We then introduce the Routhian, which can be viewed as a partial Legendre transformation, with respect to the cyclic variables, of the Lagrangian,

$$R(q_1, \ldots q_n, \dot{q}_1, \ldots, \dot{q}_s, p_{s+1}, \ldots, p_n)$$

$$= \sum_{i=s+1}^{n} p_i \dot{q}_i - L(q_1, \ldots q_n, \dot{q}_1, \ldots, \dot{q}_s, \dot{q}_{s+1}, \ldots, \dot{q}_n),$$

where we recognize that the Lagrangian, and hence the Routhian, does not depend on the cyclic variables. By taking the variation of both sides of the equation, we obtain,

$$\sum_{i=1}^{n} \frac{\partial R}{\partial q_i} \delta q_i + \sum_{i=1}^{s} \frac{\partial R}{\partial \dot{q}_i} \delta \dot{q}_i + \sum_{i=s+1}^{n} \frac{\partial R}{\partial p_i} \delta p_i$$

$$= \sum_{i=s+1}^{n} \dot{q}_i \delta p_i + \sum_{i=s+1}^{n} p_i \delta \dot{q}_i - \sum_{i=1}^{n} \frac{\partial L}{\partial q_i} \delta q_i - \sum_{i=1}^{n} \frac{\partial L}{\partial \dot{q}_i} \delta \dot{q}_i$$

$$= \sum_{i=s+1}^{n} \dot{q}_i \delta p_i - \sum_{i=1}^{n} \frac{\partial L}{\partial q_i} \delta q_i - \sum_{i=1}^{s} \frac{\partial L}{\partial \dot{q}_i} \delta \dot{q}_i,$$

where we used the fact that $p_i = \frac{\partial L}{\partial \dot{q}_i}$. Equating the coefficients for δq_i, $\delta \dot{q}_i$, and δp_i on both sides, we obtain

$$\frac{\partial R}{\partial q_i} = -\frac{\partial L}{\partial q_i}, \qquad \frac{\partial R}{\partial \dot{q}_i} = -\frac{\partial L}{\partial \dot{q}_i}, \qquad i = 1, \ldots, s;$$

$$\frac{\partial R}{\partial q_i} = -\dot{p}_i, \qquad \frac{\partial R}{\partial p_i} = \dot{q}_i, \qquad i = s+1, \ldots, n,$$

where we used the fact that $\frac{\partial L}{\partial q_i} = \dot{p}_i$.

This leads to a restatement of the Euler–Lagrange equations. The equations for the cyclic variables have the form of Hamilton's equations where the Routhian plays the role of the Hamiltonian, and the equations for the noncyclic variables have the form of the Euler–Lagrange equations where the Routhian plays the role of the Lagrangian. This should perhaps come as no surprise due to the fact that the full Legendre transformation of the Lagrangian leads to the Hamiltonian and Hamilton's equations.

Now, we make use of the fact that the Lagrangian, and hence the Routhian, does not depend on the cyclic variables, which means that $\dot{p}_i = -\frac{\partial R}{\partial q_i} = 0$, for $i = s+1, \ldots, n$, which is to say that the conjugate momenta to the cyclic variables are constant. If we specify the values of these constant momenta, by letting $p_i = \mu_i$, for $i = s + 1, \ldots, n$, and define

$$R^\mu(q_1, \ldots q_s, \dot{q}_1, \ldots, \dot{q}_s) = \sum_{i=s+1}^{n} \mu_i \dot{q}_i - L(q_1, \ldots q_s, \dot{q}_1, \ldots, \dot{q}_s, \dot{q}_{s+1}, \ldots, \dot{q}_n),$$

it follows that the reduced dynamics in the noncyclic variables are given by the Euler–Lagrange equations in terms of R^μ,

$$\frac{d}{dt}\left(\frac{\partial R^\mu}{\partial \dot{q}_i}\right) - \frac{\partial R^\mu}{\partial q_i} = 0, \qquad i = 1, \ldots, s.$$

If one desires, after solving this equation for the time evolution of the non-cyclic variables, the full dynamics in all the variables can be recovered by using the fact that the conjugate momenta to the cyclic variables are constant.

The presence of one or more cyclic configuration variables is sufficient to demonstrate the existence of a conserved quantity, but it is not necessary. If there are symmetry actions defined in terms of a Lie group action that leave the Lagrangian function invariant, then there are corresponding conserved quantities of the flow known as momentum maps. The notion of reduction can also be generalized to deal with symmetries of the Lagrangian described in terms of Lie group actions.

In the case of cyclic configuration variables, the symmetry transformation only acts on each cyclic variable separately, and the components of the velocity are not affected. To understand what it means for a Lagrangian to be invariant under a more general symmetry action, we first describe how elements of the tangent bundle transform. The action of the symmetry group on the configuration manifold induces a lifted action on the tangent bundle to the manifold. In particular, tangent vectors are transformed by the linearization of the symmetry transformation on the configuration manifold. While it is possible to consider symmetry groups acting on tangent bundles that are not induced by a symmetry action on the underlying configuration manifold, we will not discuss such group actions in this book. For the remainder of our discussion, we restrict ourselves to symmetries that are point transformations, which act on the tangent or cotangent bundles by the lift of a symmetry action on the configuration manifold.

To understand how momentum maps generalize the notion of conjugate momentum associated with a cyclic variable, we first make the observation that the conjugate momentum is the component of the image of the Legendre transformation in the symmetry (or cyclic) direction. The generalization to Lie group symmetries involves considering the Lie group acting on a configuration point, which generates a group orbit along which the Lagrangian has the same value. Then, the tangent space to this group orbit gives a set of symmetry directions and the components of the momentum in these symmetry directions are again invariant. The various momentum components are combined to yield the momentum map. This relationship between Lie group symmetries of the Lagrangian and the invariance of the associated momentum maps is referred to as Noether's theorem.

It is possible to show, for example, that the spatial angular momentum arises as the momentum map associated with the left action of the rotation group on $\mathbb{R}^3$ (see page 390 of [70]). This means that by Noether's theorem, a Lagrangian which is invariant under the lifted left action of rotations will lead to a dynamical flow in which the spatial angular momentum is conserved. Similarly, linear momentum can be viewed as the momentum map associated with translations in $\mathbb{R}^3$, and the conservation of linear momentum is a consequence of Noether's theorem for Lagrangians that are invariant under the lifted action of the translation group. These momentum maps can be used to carry out a reduction process, by restricting the dynamics to a level set of the momentum map; this is often referred to as geometric reduction.

We do not go further into these issues here. In many of the examples treated in later chapters, we point out the existence of conserved quantities, but we do not develop reduced equations in any of those examples. For a discussion of reduction theory, the reader is referred to [1, 17, 70, 74].

3.4.3 Symplectic Property

The Hamiltonian flow on the cotangent bundle (the phase space) $\mathsf{T}^*\mathbb{R}^n$ is symplectic; this means that a symplectic form, which is a closed, nondegenerate differential two-form, is conserved along the Hamiltonian flow. In the case of Hamilton's equations, the flow preserves the canonical symplectic form $\Omega_{can} = \sum_{i=1}^n dq_i \wedge dp_i$, which measures the sum of the areas projected onto matched position-momentum planes in the phase space. The expression above for the canonical symplectic form is expressed in terms of the exterior calculus of differential forms, but the symplectic form can be described using a local coordinate expression. For an in-depth discussion of the exterior calculus of differential forms, the reader is referred to [2, 91].

The symplectic form Ω_{can} can be viewed as an alternating bilinear form that takes two vectors $v, w \in \mathsf{T}_{(q,p)}\mathsf{T}^*\mathbb{R}^n$ and returns the scalar

$$\Omega_{can}(v, w) = v^T \mathbb{J} w,$$

where $\mathbb{J} = \begin{bmatrix} 0_{n \times n} & I_{n \times n} \\ -I_{n \times n} & 0_{n \times n} \end{bmatrix}$ is the symplectic matrix.

By Darboux's theorem, given any symplectic form, there are local coordinates for which the symplectic form can be written locally as the canonical symplectic form. We say that a map $F : \mathsf{T}^*\mathbb{R}^n \to \mathsf{T}^*\mathbb{R}^n$ is symplectic with respect to the canonical symplectic structure, if for all $z \in \mathsf{T}^*\mathbb{R}^n$,

$$\left(\frac{\partial F(z)}{\partial z}\right)^T \mathbb{J}\left(\frac{\partial F(z)}{\partial z}\right) = \mathbb{J},$$

where $\left(\frac{\partial F(z)}{\partial z}\right)$ is the Jacobian of the map F.

One consequence of symplecticity is Liouville's theorem, which states that the Hamiltonian flow preserves the phase volume: for any open set $D \subset T^*\mathbb{R}^n$, the volume of $\{(q(t), p(t)) : (q(0), p(0)) \in D\}$ is invariant in time, where $(q(t), p(t)) \in T^*\mathbb{R}^n$ denotes a trajectory of Hamilton's equations. This is a consequence of the fact that the n-fold wedge product of the canonical symplectic form, $\Omega_{can}^n = \Omega_{can} \wedge \cdots \wedge \Omega_{can}$ is the usual volume form on the phase space $T^*\mathbb{R}^n$.

Lagrangian flows and Hamiltonian flows can be shown to be invariant under time reversals, that is the Lagrangian dynamics and the Hamiltonian dynamics do not change if the direction of time is reversed.

These properties have important implications in terms of analysis of the detailed flow characteristics.

3.5 Lagrangian and Hamiltonian Dynamics with Holonomic Constraints

Constraints restrict the set of admissible configurations and velocities in a dynamical system, and can be expressed in terms of an algebraic equation in the configuration and velocity variables. Holonomic constraints refer to a special class of constraints that can be expressed only in terms of the configuration variables in $\mathbb{R}^n$, or they can be integrated to obtain such a form. For example, the constraint

$$x_1 \dot{x}_1 + x_2 \dot{x}_2 = 0,$$

in $\mathbb{R}^2$ is an example of a holonomic constraint, since it can be integrated to yield

$$x_1^2 + x_2^2 = r^2,$$

which is an algebraic constraint involving only the configuration variables. Equivalently, we say that a constraint is holonomic if there is an equation, expressed in terms of configuration variables only, which when differentiated yields the original constraint equations.

Without loss of generality, we restrict ourselves to the case where the holonomic constraint is expressed in terms of the zero level set of a constraint function involving only the configuration variables. Several possible approaches can be followed to describe the Lagrangian and Hamiltonian dynamics for such holonomically constrained systems.

1. The algebraic constraints are solved, perhaps only locally, to obtain a reduced set of configuration variables, in a lower-dimensional vector space, without constraints. In the context of embedded manifolds, this corresponds to choosing local coordinates on the manifold. Standard variational methods are then used to obtain Euler–Lagrange differential equations on this reduced vector space of configurations. This approach has a number of practical deficiencies and it is only effective for simple constraint functions.
2. The algebraic constraint functions are appended to the Lagrangian using Lagrange multipliers; this does not change the value of the Lagrangian on the constraint manifold (assuming the constraint manifold is described by the zero level set of the constraint function), so that the methods of variational calculus can be applied in a rather direct way. This leads to Euler–Lagrange differential equations that include the Lagrange multipliers; these differential equations must be considered in conjunction with the algebraic constraint equations.
3. The set of configurations that satisfy the holonomic constraints are viewed as defining a configuration manifold, sometimes referred to as the constraint manifold, that is embedded in $\mathbb{R}^n$. The geometry of this configuration manifold may be significantly different from the geometry of $\mathbb{R}^n$. Nevertheless, variational methods may be developed so long as the variations are constrained to respect the geometry of the configuration manifold.

In this chapter, and throughout the rest of this book, we make use of each approach. The choice of approach should be selected to best fit the features of each case. The first two approaches indicated above are well known in the classical literature on variational methods [30, 32, 45, 99]. The second approach is summarized in the following paragraphs with proofs of the results given in Chapter 8. The third approach, where holonomic constraints are used to define the configuration manifold, is not so well known and its development and illustration constitute one of the main contributions of this book.

To briefly illustrate the second approach, assume that the configurations $q \in \mathbb{R}^n$ are required to satisfy the constraints $f_i(q) = 0$, $i = 1, \ldots, m$, where $f_i : \mathbb{R}^n \to \mathbb{R}^1$, $i = 1, \ldots, m$ are real-valued differentiable functions with linearly independent gradients.

The constraint manifold is

$$M = \{q \in \mathbb{R}^n : f_i(q) = 0, \, i = 1, \ldots, m\}.$$

The tangent space of M at $q \in M$ is given by

$$\mathsf{T}_q M = \left\{ \dot{q} \in \mathbb{R}^n : \left(\frac{\partial f_i(q)}{\partial q} \right)^T \dot{q} = 0, \, i = 1, \ldots, m \right\},$$

and

$$\mathsf{T}M = \left\{ (q, \dot{q}) \in \mathbb{R}^{2n} : q \in M, \, \dot{q} \in \mathsf{T}_q M \right\}$$

is the tangent bundle of the constraint manifold M.

Introduce Lagrange multipliers $\lambda = [\lambda_1, \ldots, \lambda_m]^T \in \mathbb{R}^m$; it can be shown that the resulting Euler–Lagrange equations are given by the differential equations

$$\frac{d}{dt} \left(\frac{\partial L(q, \dot{q})}{\partial \dot{q}} \right) - \frac{\partial L(q, \dot{q})}{\partial q} + \sum_{i=1}^{m} \lambda_i \frac{\partial f_i(q)}{\partial q} = 0, \qquad (3.16)$$

together with algebraic equations that define the constraints

$$f_i(q) = 0, \quad i = 1, \ldots, m. \qquad (3.17)$$

These differential-algebraic equations can be shown to have index two, so that they define a Lagrangian vector field on $\mathsf{T}M$, which is the tangent bundle of the constraint manifold M. A vector field on $\mathsf{T}M$ is a vertical map from $\mathsf{T}M$ to $\mathsf{TT}M$, which means that it takes $(q, \dot{q}) \in \mathsf{T}M$ to $(q, \dot{q}, \ddot{q}) \in \mathsf{TT}M$.

These Euler–Lagrange equations can be written in a simpler form by introducing the augmented Lagrangian function $L^a : \mathsf{T}^*\mathbb{R}^n \times \mathbb{R}^m \to \mathbb{R}^1$ that is given by

$$L^a(q, \dot{q}, \lambda) = L(q, \dot{q}) + \sum_{i=1}^{m} \lambda_i f_i(q).$$

The constrained Euler–Lagrange equations on $\mathbb{R}^n$ can be expressed as

$$\frac{d}{dt} \left(\frac{\partial L^a(q, \dot{q}, \lambda)}{\partial \dot{q}} \right) - \frac{\partial L^a(q, \dot{q}, \lambda)}{\partial q} = 0. \qquad (3.18)$$

The augmented Euler–Lagrange equations, together with the m algebraic constraint equations given in (3.17), can be shown to be index two differential-algebraic equations. They guarantee that the constrained Lagrangian dynamics described by $(q, \dot{q})$ evolve on $\mathsf{T}M$, the tangent bundle of the constraint manifold M.

Consider the Legendre transformation associated with the augmented Lagrangian

$$p = \frac{\partial L^a(q, \dot{q}, \lambda)}{\partial \dot{q}},$$

and assume that $\dot{q} \in \mathsf{T}_q\mathbb{R}^n$ can be expressed uniquely in terms of $q \in \mathbb{R}^n$, $p \in \mathsf{T}_q^*\mathbb{R}^n$ and $\lambda \in \mathbb{R}^m$. Define the augmented Hamiltonian function $H^a : \mathsf{T}^*\mathbb{R}^n \times \mathbb{R}^m \to \mathbb{R}^1$ as

$$H^a(q, p, \lambda) = p \cdot \dot{q} - L^a(q, \dot{q}, \lambda),$$

so that we obtain Hamilton's equations in the form

$$\dot{q} = \frac{\partial H^a(q, p, \lambda)}{\partial p}, \tag{3.19}$$

$$\dot{p} = -\frac{\partial H^a(q, p, \lambda)}{\partial q}. \tag{3.20}$$

These equations of motion are augmented by the m algebraic constraint equations (3.17), consistent with the introduction of the m real-valued Lagrange multipliers. The augmented Hamilton's equations, together with the constraint equations, are also index two differential-algebraic equations that guarantee that the constrained Hamiltonian dynamics described by (q, p) evolve on T^*M, which is the cotangent bundle of the constraint manifold M.

The following interpretation for each of the Lagrange multipliers is standard: the constraint force $\lambda_i \frac{\partial f_i(q)}{\partial q} \in \mathbb{R}^n$ appears in the Euler–Lagrange equations and in Hamilton's equations to guarantee satisfaction of the constraint function $f_i(q) = 0$. Alternatively, it can also be said that the Lagrange multipliers λ_i are chosen so that the Euler–Lagrange vector field and the Hamiltonian vector field take values in $\mathsf{TT}M$ and TT^*M, respectively. This is always possible, since the set of gradient vectors $\left\{ \frac{\partial f_i(q)}{\partial q} \in \mathbb{R}^n : i = 1, \ldots, m \right\}$ span a complementary subspace to $\mathsf{T}_q M$. If the constraints are satisfied, then the constraint force can be shown to satisfy the variational condition $\delta f_i(q) = \frac{\partial f_i(q)}{\partial q} \cdot \delta q = 0$, $i = 1, \ldots, m$; thus, the forces that arise from the holonomic constraints *do no work*.

3.6 Lagrange–d'Alembert Principle

We describe a modification of Hamilton's principle to incorporate the effects of external forces; these external forces may or may not be derivable from a potential. We assume that the dynamics evolves on a configuration manifold $\mathbb{R}^n$. This modification is usually referred to as the Lagrange–d'Alembert principle. It states that the infinitesimal variation of the action integral over a fixed time period equals the negative of the work done by the external forces, corresponding to an infinitesimal variation of the configuration, during this same time period. Obviously, this reduces to Hamilton's principle when there are no external forces. This version of the variational principle requires determining the virtual work that corresponds to an infinitesimal variation of the configuration.

In particular, the external forces are described by a time-dependent vertical mapping $F : [t_0, t_f] \times \mathbb{R}^n \to \mathsf{T}^*\mathbb{R}^n$ from the configuration space $\mathbb{R}^n$ to the cotangent bundle $\mathsf{T}^*\mathbb{R}^n$. In this context, the vertical assumption means that $F(t, q) \in \mathsf{T}_q^*\mathbb{R}^n$ and implies that the kinematics are unchanged, while the external forces affect only the dynamics. Because the map is vertical, we can identify the covector $F(t, q)$ with a vector based at the point q, and describe the external force by a map $F : [t_0, t_f] \times \mathbb{R}^n \to \mathbb{R}^n$. Thus, the virtual work along an infinitesimal variation of the configuration is given by

$$\int_{t_0}^{t_f} F(q)^T \delta q \, dt,$$

where we have suppressed the time-dependence of F for notational brevity. The Lagrange–d'Alembert principle states that

$$\delta \int_{t_0}^{t_f} L(q, \dot{q}) dt = - \int_{t_0}^{t_f} F(q)^T \delta q \, dt,$$

holds for all possible infinitesimal variations $\delta q : [t_0, t_f] \to \mathbb{R}^n$ that vanish at the endpoints, that is, $\delta q(t_0) = \delta q(t_f) = 0$.

Following the prior development, this leads to the forced Euler–Lagrange equations

$$\frac{d}{dt}\left(\frac{\partial L(q, \dot{q})}{\partial \dot{q}}\right) - \frac{\partial L(q, \dot{q})}{\partial q} = F(q), \tag{3.21}$$

that include the external forces.

With appropriate assumptions on the Lagrangian function, the following result can be stated for the associated initial-value problem. Suppose that the external forces are specified by the map $F : [t_0, t_f] \times \mathbb{R}^n \to \mathbb{R}^n$. For each initial condition $(q(t_0), \dot{q}(t_0)) = (q_0, \dot{q}_0) \in \mathsf{T}\mathbb{R}^n$ there exists a unique solution of the forced Euler–Lagrange equation (3.21) denoted by $(q(t), \dot{q}(t)) \in \mathsf{T}\mathbb{R}^n$.

It is easy to see that Hamilton's equations can also be modified to incorporate external forces,

$$\dot{q} = \frac{\partial H(q, p)}{\partial p}, \tag{3.22}$$

$$\dot{p} = -\frac{\partial H(q, p)}{\partial q} + F(q). \tag{3.23}$$

Many, but not all, of the results described for the autonomous case without inclusion of external forces hold for this case. But it is important to be careful. For instance, if there are external forces it is not necessarily true that the Hamiltonian is conserved.

In practice, this version of the Euler–Lagrange equations and Hamilton's equations that include external forces are important. This is the mechanism whereby physical effects such as friction and other forms of energy dissipation can be incorporated into the equations of motion. This is also one way in which time-varying effects can be incorporated into the equations of motion. Finally, external forces can be used to model control effects and external disturbances. Such modifications significantly broaden the physical application of the Lagrangian and Hamiltonian approaches to dynamics.

3.7 Classical Particle Dynamics

In this section, several Lagrangian and Hamiltonian systems defined in terms of ideal particles, with mass, are introduced. In each case, the physical description and assumptions are made clear and it is shown that the configurations can be identified with $\mathbb{R}^n$. The Euler–Lagrange equations and Hamilton's equations are obtained.

3.7.1 Dynamics of a Particle in Uniform, Constant Gravity

Consider a particle that moves in three dimensions under the influence of uniform, constant gravity. The mass of the particle is m and the constant gravitational acceleration is g. An inertial Euclidean frame is selected so that the third axis of the inertial frame is vertical.

Let $q \in \mathbb{R}^3$ denote the position vector of the particle in the inertial frame. We take the configuration manifold to be the vector space $\mathbb{R}^3$. Thus, there are three degrees of freedom.

3.7.1.1 Euler–Lagrange Equations

The Lagrangian function $L : \mathsf{T}\mathbb{R}^3 \to \mathbb{R}^1$ is given by

$$L(q, \dot{q}) = \frac{1}{2} m \dot{q}^T \dot{q} - m g e_3^T q.$$

The first term is the kinetic energy function of the particle while the second term is the negative of the gravitational potential energy.

The Euler–Lagrange equation (3.6) yields the equation of motion

$$m\ddot{q} + mge_3 = 0. \tag{3.24}$$

Thus, the acceleration of the particle is due to gravity and is given by the constant vector $-ge_3$. This vector differential equation defines the Lagrangian dynamics of the particle in terms of $(q, \dot{q}) \in \mathsf{T}\mathbb{R}^3$ on the tangent bundle of $\mathbb{R}^3$.

3.7.1.2 Hamilton's Equations

By defining the conjugate momentum of the particle as $p = \frac{\partial L(q,\dot{q})}{\partial \dot{q}} = m\dot{q} \in \mathsf{T}_q^*\mathbb{R}^3$ using the Legendre transformation, we obtain the Hamiltonian function

$$H(q,p) = \frac{1}{2m}p^T p + mge_3^T q.$$

Thus, Hamilton's equations, from (3.14) and (3.15), are

$$\dot{q} = \frac{p}{m},$$

$$\dot{p} = -mge_3.$$

These vector differential equations define the Hamiltonian dynamics of the particle in terms of $(q,p) \in T^*\mathbb{R}^3$ on the cotangent bundle of $\mathbb{R}^3$.

3.7.1.3 Conservation Properties

The Hamiltonian of the particle

$$H = \frac{1}{2}m\dot{q}^T \dot{q} + mge_3^T q,$$

which coincides with the total energy E in this case, is constant along each solution of the dynamical flow.

It is easy to show that the horizontal components of the linear momentum, given by $me_1^T \dot{q}$ and $me_2^T \dot{q}$, are constant along each solution of the dynamical flow. This can be viewed as a consequence of Noether's theorem and the fact that the Lagrangian function is invariant with respect to the lifted action of translations in the e_1 and e_2 directions.

3.7.1.4 Equilibrium Properties

There are no equilibrium solutions of this dynamical system.

3.7.1.5 Modification to Include the Force of Air Resistance

In the above analysis, gravitational effects are included through the gravitational potential energy, giving rise to the constant force of gravity. An important external force, often included in such analyses, is the aerodynamic drag force due to the resistance of air. This force is not derivable from a potential, so it must be included separately as an external force. A common expression for the aerodynamic drag is

$$F = -C_D \|\dot{q}\| \dot{q},$$

where C_D is an aerodynamic drag coefficient. This force acts opposite to the direction of the velocity vector and is proportional to the square of the magnitude of the velocity vector.

The forced Euler–Lagrange equations are

$$m\ddot{q} + C_D \left\| \dot{q} \right\| \dot{q} + mge_3 = 0, \tag{3.25}$$

and the Lagrangian flow evolves on the tangent bundle $\mathsf{T}\mathbb{R}^3$.

The forced Hamilton's equations are

$$\dot{q} = \frac{p}{m}, \tag{3.26}$$

$$\dot{p} = -\frac{C_D}{m^2} \left\| p \right\| p - mge_3, \tag{3.27}$$

and the Hamiltonian flow evolves on the cotangent bundle $\mathsf{T}^*\mathbb{R}^3$.

The inclusion of aerodynamics drag into the equations of motion for a particle is relatively easy, but the resulting dynamics are more complicated.

3.7.2 Dynamics of a Particle, Constrained to an Inclined Plane, in Uniform, Constant Gravity

A particle is constrained to move, without friction, on an inclined plane in $\mathbb{R}^3$ under the influence of gravity. The mass of the particle is m. The position of the particle is denoted by $x = [x_1, x_2, x_3]^T \in \mathbb{R}^3$ and expressed with respect to an inertial Euclidean frame whose third axis is in the vertical direction; g denotes the constant acceleration of gravity. The plane is described by the linear constraint manifold

$$M = \left\{ x \in \mathbb{R}^3 : x_1 + x_2 + x_3 - 1 = 0 \right\}.$$

A schematic of the particle on an inclined plane is shown in Figure 3.1.

Since the position vector of the particle $x \in M$, the linear manifold M is viewed as the configuration manifold. The dimension of the configuration manifold is two, and the translational motion of a particle in the plane is said to have two degrees of freedom.

As previously in Chapter 2, a constant basis for the tangent space $\mathsf{T}_x M$ is selected as $\left\{ [1, -1, 0]^T, [0, -1, 1]^T \right\}$. Then, $x \in M$ can be written as

$$x = q_1 \begin{bmatrix} 1 \\ -1 \\ 0 \end{bmatrix} + q_2 \begin{bmatrix} 0 \\ -1 \\ 1 \end{bmatrix} + \begin{bmatrix} 0 \\ 0 \\ 1 \end{bmatrix}, \tag{3.28}$$

for $q = [q_1, q_2]^T \in \mathbb{R}^2$. This also implies that the time derivative $\dot{x} \in \mathsf{T}_x M$ can be written as

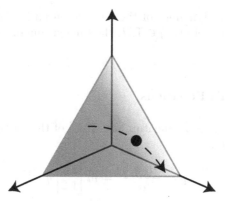

Fig. 3.1 Particle on an inclined plane

$$\dot{x} = \dot{q}_1 \begin{bmatrix} 1 \\ -1 \\ 0 \end{bmatrix} + \dot{q}_2 \begin{bmatrix} 0 \\ -1 \\ 1 \end{bmatrix}. \tag{3.29}$$

These expressions can be used to characterize the tangent plane and the tangent bundle of the configuration manifold M. It is clear that M is diffeomorphic to $\mathbb{R}^2$.

3.7.2.1 Euler–Lagrange Equation

The Lagrangian function $L : \mathsf{T}M \rightarrow \mathbb{R}^1$ is given by

$$L(x, \dot{x}) = \frac{1}{2} m \dot{x}^T \dot{x} - mg e_3^T x,$$

which can be rewritten, with abuse of notation, as the Lagrangian function $L : \mathsf{T}\mathbb{R}^2 \rightarrow \mathbb{R}^1$

$$L(q, \dot{q}) = \frac{1}{2} m \dot{q}^T \begin{bmatrix} 2 & 1 \\ 1 & 2 \end{bmatrix} \dot{q} - mg(q_2 + 1).$$

The first term is the kinetic energy function of the particle while the second term is the negative of the gravitational potential energy.

Viewing the configuration as $q \in \mathbb{R}^2$, the Euler–Lagrange equations, obtained from (3.6), yield the equations of motion

$$m \begin{bmatrix} 2 & 1 \\ 1 & 2 \end{bmatrix} \ddot{q} + mg \begin{bmatrix} 0 \\ 1 \end{bmatrix} = \begin{bmatrix} 0 \\ 0 \end{bmatrix}. \tag{3.30}$$

This vector differential equation defines the Lagrangian dynamics of the particle on an inclined plane in terms of $(q, \dot{q}) \in \mathsf{T}\mathbb{R}^2$ on the tangent bundle of

$\mathbb{R}^2$. The Lagrangian dynamics of the particle on an inclined plane can also be expressed in terms of $(x, \dot{x}) \in TM$, the tangent bundle of M, using (3.28) and (3.29).

3.7.2.2 Hamilton's Equations

We define the conjugate momentum $p \in T_q^*\mathbb{R}^2$ of the particle using the Legendre transformation

$$p = \frac{\partial L(q, \dot{q})}{\partial \dot{q}} = m \begin{bmatrix} 2 & 1 \\ 1 & 2 \end{bmatrix} \dot{q}.$$

From this, the Hamiltonian function is

$$H(q, p) = \frac{1}{6m} p^T \begin{bmatrix} 2 & -1 \\ -1 & 2 \end{bmatrix} p + mg(q_2 + 1).$$

Thus, Hamilton's equations, from (3.14) and (3.15), are

$$\dot{q} = \frac{1}{3m} \begin{bmatrix} 2 & -1 \\ -1 & 2 \end{bmatrix} p, \tag{3.31}$$

$$\dot{p} = -mg \begin{bmatrix} 0 \\ 1 \end{bmatrix}. \tag{3.32}$$

These vector differential equations define the Hamiltonian dynamics of the particle in terms of $(q, p) \in T^*\mathbb{R}^2$ on the cotangent bundle of $\mathbb{R}^2$. These equations can be used to determine the Hamiltonian dynamics expressed on the cotangent bundle T^*M.

3.7.2.3 Conservation Properties

The Hamiltonian of the particle

$$H = \frac{1}{2} m \dot{x}^T \dot{x} + mge_3^T x,$$

which coincides with the total energy E in this case, is constant along each solution of the dynamical flow. The Hamiltonian can, of course, also be expressed in terms of $(q, \dot{q}) \in T\mathbb{R}^2$ as

$$H = \frac{1}{2} m \dot{q}^T \begin{bmatrix} 2 & 1 \\ 1 & 2 \end{bmatrix} \dot{q} + mg(q_2 + 1),$$

or in terms of $(q, p) \in T^*\mathbb{R}^2$.

3.7.2.4 Equilibrium Properties

There are no equilibrium solutions of this dynamical system.

3.7.3 Dynamics of a Particle, Constrained to a Hyperbolic Paraboloid, in Uniform, Constant Gravity

A particle is constrained to move, without friction, on a smooth surface in $\mathbb{R}^3$ under the influence of gravity. The mass of the particle is m. The position of the particle is denoted by $x = [x_1, x_2, x_3]^T \in \mathbb{R}^3$ expressed with respect to an inertial Euclidean frame whose third axis is in the vertical direction; g denotes the constant acceleration of gravity. The surface is a hyperbolic paraboloid described by the embedded manifold

$$M = \left\{ x \in \mathbb{R}^3 : -x_1^2 + x_2^2 - x_3 = 0 \right\}.$$

The surface has a particularly simple mathematical description in that M can also be described by

$$M = \left\{ [x_1, x_2, -x_1^2 + x_2^2]^T \in \mathbb{R}^3 : [x_1, x_2]^T \in \mathbb{R}^2 \right\}.$$

Thus, M and $\mathbb{R}^2$ are diffeomorphic.

Consequently, the vector $[x_1, x_2]^T \in \mathbb{R}^2$ parameterizes any point on M and can be viewed as the configuration vector, and $\mathbb{R}^2$ can be viewed as the configuration manifold. Since the dimension of the configuration manifold is two, the translational motion of a particle on the surface is said to have two degrees of freedom.

Introducing the notation, $q = [q_1, q_2]^T = [x_1, x_2]^T \in \mathbb{R}^2$, a vector $x \in M$ can be expressed as

$$x = \begin{bmatrix} q_1 \\ q_2 \\ -q_1^2 + q_2^2 \end{bmatrix}, \tag{3.33}$$

and $\dot{x} \in \mathsf{T}_x M$ can be expressed as

$$\dot{x} = \dot{q}_1 \begin{bmatrix} 1 \\ 0 \\ -2q_1 \end{bmatrix} + \dot{q}_2 \begin{bmatrix} 0 \\ 1 \\ 2q_2 \end{bmatrix}. \tag{3.34}$$

3.7.3.1 Euler–Lagrange Equation

The Lagrangian function $L : \mathsf{T}M \to \mathbb{R}^1$ is given by

$$L(x, \dot{x}) = \frac{1}{2} m \dot{x}^T \dot{x} - mg e_3^T x,$$

which can be rewritten, with abuse of notation, as the Lagrangian function $L : \mathsf{T}\mathbb{R}^2 \to \mathbb{R}^1$

$$L(q, \dot{q}) = \frac{1}{2} m \dot{q}^T \begin{bmatrix} 1 + 4q_1^2 & -4q_1 q_2 \\ -4q_1 q_2 & 1 + 4q_2^2 \end{bmatrix} \dot{q} - mg(-q_1^2 + q_2^2).$$

The first term is the kinetic energy function of the particle while the second term is the negative of the gravitational potential energy.

The Euler–Lagrange equations on $\mathbb{R}^2$, obtained from (3.6), yield the equations of motion

$$m \begin{bmatrix} 1 + 4q_1^2 & -4q_1 q_2 \\ -4q_1 q_2 & 1 + 4q_2^2 \end{bmatrix} \ddot{q} + m \begin{bmatrix} 4q_1(\dot{q}_1^2 - \dot{q}_2^2) \\ 4q_2(\dot{q}_2^2 - \dot{q}_1^2) \end{bmatrix} + mg \begin{bmatrix} -2q_1 \\ 2q_2 \end{bmatrix} = \begin{bmatrix} 0 \\ 0 \end{bmatrix}. \qquad (3.35)$$

This vector differential equation defines the Lagrangian dynamics of the particle on the hyperbolic paraboloid surface in terms of $(q, \dot{q}) \in \mathsf{T}\mathbb{R}^2$ on the tangent bundle of $\mathbb{R}^2$. The Lagrangian dynamics of the particle on the surface can also be expressed in terms of $(x, \dot{x}) \in \mathsf{T}M$, the tangent bundle of M, using (3.33) and (3.34).

3.7.3.2 Hamilton's Equations

Using the Legendre transformation, we define the conjugate momentum $p \in \mathsf{T}_q^* \mathbb{R}^2$ of the particle to be

$$p = \frac{\partial L(q, \dot{q})}{\partial \dot{q}} = m \begin{bmatrix} 1 + 4q_1^2 & -4q_1 q_2 \\ -4q_1 q_2 & 1 + 4q_2^2 \end{bmatrix} \dot{q}.$$

The Hamiltonian function is

$$H(q, p) = \frac{1}{2m} \frac{1}{(1 + 4(q_1^2 + q_2^2))} p^T \begin{bmatrix} 1 + 4q_2^2 & 4q_1 q_2 \\ 4q_1 q_2 & 1 + 4q_1^2 \end{bmatrix} p + mg(-q_1^2 + q_2^2).$$

Thus, Hamilton's equations, obtained from (3.14) and (3.15), are

$$\dot{q} = \frac{1}{m(1 + 4(q_1^2 + q_2^2))} \begin{bmatrix} 1 + 4q_2^2 & 4q_1 q_2 \\ 4q_1 q_2 & 1 + 4q_1^2 \end{bmatrix} p, \qquad (3.36)$$

$$\dot{p} = \frac{4(p_1 q_1 - p_2 q_2)}{m(1 + 4(q_1^2 + q_2^2))^2} \begin{bmatrix} -4p_1 q_2^2 - 4p_2 q_1 q_2 - p_1 \\ 4p_2 q_1^2 + 4p_1 q_1 q_2 + p_2 \end{bmatrix} + mg \begin{bmatrix} 2q_1 \\ -2q_2 \end{bmatrix}. \qquad (3.37)$$

These vector differential equations define the Hamiltonian dynamics of the particle in terms of $(q, p) \in \mathsf{T}^*\mathbb{R}^2$ on the cotangent bundle of $\mathbb{R}^2$. The Hamiltonian dynamics can also be described in terms of the evolution on the cotangent bundle T^*M.

3.7.3.3 Conservation Properties

The Hamiltonian of the particle

$$H = \frac{1}{2}m\dot{x}^T\dot{x} + mge_3^T x,$$

which coincides with the total energy E in this case, is constant along each solution of the dynamical flow. The Hamiltonian can, of course, also be expressed in terms of $(q, \dot{q}) \in \mathsf{T}\mathbb{R}^2$ as

$$H = \frac{1}{2}m\dot{q}^T \begin{bmatrix} 1 + 4q_1^2 & -4q_1q_2 \\ -4q_1q_2 & 1 + 4q_2^2 \end{bmatrix} \dot{q} + mg(-q_1^2 + q_2^2),$$

or in terms of $(q, p) \in \mathsf{T}^*\mathbb{R}^2$.

3.7.3.4 Equilibrium Properties

There is a single equilibrium solution at $[0, 0]^T \in \mathsf{T}\mathbb{R}^2$. Viewing $[q_1, q_2]^T \in \mathbb{R}^2$ as local coordinates for $\mathbb{R}^2$, the linearized vector field using (3.35) can be described by

$$m\dot{\xi}_1 - 2mg\xi_1 = 0,$$
$$m\dot{\xi}_2 + 2mg\xi_2 = 0.$$

The eigenvalues are easily computed to be: $\sqrt{2g}$, $-\sqrt{2g}$, $j\sqrt{2g}$, $-j\sqrt{2g}$. Since one eigenvalue is real and positive, the equilibrium $[0, 0]^T \in \mathsf{T}\mathbb{R}^2$, and thus, the equilibrium $[0, 0]^T \in \mathsf{T}M$, is unstable.

3.7.4 Keplerian Dynamics of a Particle in Orbit

We consider the restricted two-body problem in orbital mechanics. A particle is in orbit about a large spherical body acted on by the gravitational force of the large spherical body. The mass of the orbiting body is m and the mass M of the spherical body is sufficiently large that it can be assumed to be fixed; this is the restricted two-body problem.

The origin of an inertial Euclidean frame is fixed at the center of mass of the large spherical body. The configuration vector of the orbiting body is $q \in \mathbb{R}^3$ and $\mathbb{R}^3$ is the configuration manifold. A schematic of this three degrees of freedom system is shown in Figure 3.2.

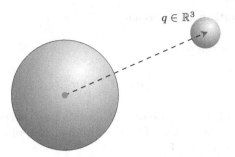

Fig. 3.2 Particle in Keplerian orbit

3.7.4.1 Euler–Lagrange Equations

The kinetic energy of the orbiting body, viewed as a particle, is given by

$$T(q, \dot{q}) = \frac{1}{2} m \, \|\dot{q}\|^2 .$$

The gravitational potential energy of the particle in orbit is obtained as the integral of the potential energy for each mass increment in the central body over the material points in the central body. As shown in [80], this potential energy is the same as if all mass of the central body is concentrated at its center of mass, that is

$$U(q) = -\frac{m\mu}{\|q\|},$$

where μ is the positive gravitational constant depending on the mass of the large spherical body.

Thus, the Lagrangian function $L : \mathsf{T}\mathbb{R}^3 \to \mathbb{R}^1$ is

$$L(q, \dot{q}) = \frac{1}{2} m \, \|\dot{q}\|^2 + \frac{m\mu}{\|q\|} .$$

The Euler–Lagrange equation of motion, obtained from (3.6), is given by

$$m\ddot{q} + m\mu \frac{q}{\|q\|^3} = 0. \tag{3.38}$$

The mass of the orbiting body is thus seen to be irrelevant to the orbital dynamics.

Note that there is a singularity in the gravitational term at the origin so that the equations are motion are valid only on $\mathbb{R}^3 - \{0\}$. This singularity arises from the gravitational model. Excluding the origin, the vector differential equation describes the Lagrangian flow of Kepler's dynamics in terms of $(q, \dot{q}) \in \mathsf{T}\mathbb{R}^3$ on the tangent bundle of $\mathbb{R}^3$.

3.7.4.2 Hamilton's Equations

Hamilton's equations of motion are obtained from the Lagrangian function by using the Legendre transformation to define the orbital momentum $p = \frac{\partial L(q,\dot{q})}{\partial \dot{q}} = m\dot{q} \in T_q^* \mathbb{R}^3$ and introducing the Hamiltonian function

$$H(q,p) = \frac{1}{2m} \|p\|^2 - \frac{m\mu}{\|q\|}.$$

From (3.14) and (3.15) we obtain Hamilton's equations of motion

$$\dot{q} = \frac{p}{m}, \tag{3.39}$$

$$\dot{p} = -m\mu \frac{q}{\|q\|^3}. \tag{3.40}$$

These differential equations define the Hamiltonian flow of Kepler's dynamics described in terms of $(q,p) \in T^*\mathbb{R}^3$ on the cotangent bundle of $\mathbb{R}^3$.

3.7.4.3 Conservation Properties

There are several conserved quantities. First, the Hamiltonian, which coincides with the total energy E in this case, is conserved:

$$H = \frac{1}{2}m \|\dot{q}\|^2 - \frac{m\mu}{\|q\|},$$

is constant along each solution of the Lagrangian flow.

Second, it is easily verified that the spatial angular momentum

$$h = q \times m\dot{q}$$

is constant along each solution of the Lagrangian flow. This guarantees that each orbital motion lies in a fixed orbital plane in $\mathbb{R}^3$. This arises as a consequence of Noether's theorem, due to the invariance of the Lagrangian under the lifted left action of the rotation group.

In addition, the Runge–Lenz vector

$$r = \dot{q} \times h - \frac{m\mu q}{\|q\|}$$

can be shown to be constant along each solution of the Lagrangian flow. This conservation law is more subtle and does not arise from a symmetry of the Lagrangian, at least if we only consider symmetry groups acting on the configuration manifold. Rather, the Runge–Lenz vector is related to the invariance of the Newtonian potential $-k/\|q\|$ under the lifted action of the Lorentz group on spacetime.

3.7.4.4 Equilibrium Properties

There are no equilibrium solutions. However, the conserved functions, which
are constant along each orbital motion, can be used to derive Kepler's three
laws for orbiting bodies. These can be used to show that the orbital paths
in $\mathbb{R}^3$ are conic sections; that is the Keplerian orbital paths are circular,
elliptical, parabolic, or hyperbolic.

3.7.5 Dynamics of a Particle Expressed in a Rotating Euclidean Frame

The dynamics of a particle with mass m moving under the influence of a potential in an inertial Euclidean frame constitute one of the simplest problems
in Lagrangian dynamics. The equations of motion are easily determined using
Newton's laws or the Euler–Lagrange equations for a configuration vector in
$\mathbb{R}^3$.

There are many situations, such as in imaging or target tracking, for which
the dynamics of a particle with mass m moving under the influence of a
potential are required to be expressed in a non-inertial, rotating Euclidean
frame. Such equations of motion can be obtained using the Euler–Lagrange
equations on $\mathbb{R}^3$. In this section, these basic results are presented; they are
of interest in their own right and they demonstrate the general approach in
obtaining the equations of motion expressed in a rotating Euclidean frame
using variational methods.

A rotating Euclidean frame and an inertial Euclidean frame are assumed
to have identical origins. The position vector of the particle is $q \in \mathbb{R}^3$ in
the rotating frame; the position vector of the particle in the inertial frame is
$x \in \mathbb{R}^3$. The vector $q \in \mathbb{R}^3$ is taken as the configuration vector, so that $\mathbb{R}^3$ is
the configuration manifold. The particle has three degrees of freedom.

3.7.5.1 Euler–Lagrange Equations

The non-inertial, rotating frame is described by the known map: $\mathsf{R} : [0, \infty) \to$
$\mathsf{SO}(3)$, which is the transformation between vectors in the rotating frame and
corresponding vectors in the inertial frame, as described in Chapter 2. As in
(2.16), there exists an angular velocity vector $\omega : [0, \infty) \to \mathbb{R}^3$ such that
$\dot{R} = RS(\omega)$.

Thus, the position vector of the particle in the inertial frame is related to
the position vector of the particle in the rotating frame by

$$x = Rq.$$

The velocity vector of the particle in the inertial frame is

$$\dot{x} = R(\dot{q} + S(\omega)q),$$

and the kinetic energy of the particle is given by

$$T(q, \dot{q}) = \frac{1}{2}m \|\dot{x}\|^2$$
$$= \frac{1}{2}m \|\dot{q} + S(\omega)q\|^2.$$

The potential energy of the particle is given by the scalar function $U(x) = U(Rq)$.

Thus, the Lagrangian function $L : \mathsf{T}\mathbb{R}^3 \to \mathbb{R}^1$ is

$$L(q, \dot{q}) = \frac{1}{2}m \|\dot{q} + S(\omega)q\|^2 - U(Rq),$$

which is the difference of the kinetic and potential energy in the inertial frame, expressed in non-inertial coordinates. The partial derivatives are

$$\frac{\partial L(q, \dot{q})}{\partial \dot{q}} = m(\dot{q} + S(\omega)q),$$
$$\frac{\partial L(q, \dot{q})}{\partial q} = -mS(\omega)(\dot{q} + S(\omega)q) - \frac{\partial U(Rq)}{\partial q},$$

and the Euler–Lagrange equation of motion, obtained from (3.6), is given by

$$m\left\{\ddot{q} + 2S(\omega)\dot{q} + S(\omega)S(\omega)q + S(\dot{\omega})q\right\} + \frac{\partial U(Rq)}{\partial q} = 0. \qquad (3.41)$$

This can also be expressed using the cross-product notation as

$$m\left\{\ddot{q} + 2\omega \times \dot{q} + \omega \times (\omega \times q) + \dot{\omega} \times q\right\} + \frac{\partial U(Rq)}{\partial q} = 0. \qquad (3.42)$$

The second and third terms in these equations are often referred to as the Coriolis and centripetal terms. Note that if $\omega \equiv 0$, then the rotating frame is in fact stationary and the above equations reduce to describe the simple dynamics of a particle moving in a potential, expressed in an inertial frame.

These vector differential equations define the Lagrangian dynamics of a particle in a potential in terms of $(q, \dot{q}) \in \mathsf{T}\mathbb{R}^3$ on the tangent bundle of $\mathbb{R}^3$. Here, the configuration vector is the position vector of the particle in an arbitrary rotating, non-inertial Euclidean frame.

These Euler–Lagrange equations for particle dynamics in a rotating frame appear in many textbooks and in many research publications. The publication [24] contains a detailed description of the response properties, in the case that the potential $U(q) = \frac{1}{2} k \|q\|^2$ with $k > 0$, so that the particle dynamics, ignoring rotation of the frame, corresponds to a spatial harmonic oscillator.

3.7.5.2 Hamilton's Equations

Using the Legendre transformation, define the conjugate momentum $p \in T_q^* \mathbb{R}^3$ of the particle to be

$$p = \frac{\partial L(q, \dot{q})}{\partial \dot{q}} = m(\dot{q} + S(\omega)q).$$

The Hamiltonian function is

$$H(q, p) = \frac{\|p\|^2}{2m} - p \cdot (S(\omega)q) + U(Rq).$$

Thus, Hamilton's equations, obtained from (3.14) and (3.15), are

$$\dot{q} = \frac{p}{m} - S(\omega)q, \tag{3.43}$$

$$\dot{p} = S(\omega)p - \frac{\partial U(Rq)}{\partial q}. \tag{3.44}$$

These vector differential equations define the Hamiltonian dynamics of the particle in terms of $(q, p) \in T^* \mathbb{R}^3$ on the cotangent bundle of $\mathbb{R}^3$.

3.7.5.3 Conservation Properties

The Hamiltonian of the particle in the potential

$$H = \frac{1}{2} m \|\dot{q} + S(\omega)q\|^2 - m(\dot{q} + S(\omega)q) \cdot (S(\omega)q) + U(Rq)$$

is time dependent, since it depends on R, which is a function of time. Consequently, the Hamiltonian is not constant along each solution of the dynamical flow. However, the total energy in the inertial frame expressed in non-inertial coordinates

$$E = \frac{1}{2} m \|\dot{q} + S(\omega)q\|^2 + U(Rq)$$

is constant along each solution of the dynamical flow.

There may be additional conserved properties if the potential function has certain symmetry properties.

3.8 Problems

3.1. Consider the dynamics of a particle, with mass m, in three dimensions. The particle acts under a gravitational field given, in an inertial frame, by $G : \mathbb{R}^3 \to \mathsf{T}\mathbb{R}^3$. That is, the gravitational force on a particle, located at $x \in \mathbb{R}^3$ is given by $mG(x)$.

(a) What are the Euler–Lagrange equations for the particle?
(b) What are Hamilton's equations for the particle?
(c) What are conditions that guarantee existence of an equilibrium solution of the particle in a gravitational field?
(d) Suppose that the gravitational field is constant. What are conditions for existence of an equilibrium solution of the particle?

3.2. Consider the dynamics of a charged particle, with mass m, in three dimensions. The particle acts under an electric field given, in an inertial frame, by $E : \mathbb{R}^3 \to \mathsf{T}\mathbb{R}^3$ and a magnetic field given, in the inertial frame, by $B : \mathbb{R}^3 \to \mathsf{T}\mathbb{R}^3$. The Lorentz force on a particle, located at $x \in \mathbb{R}^3$, is given by $Q(E(x) + \dot{x} \times B(x))$, where Q denotes the constant charge on the particle.

(a) What are the Euler–Lagrange equations for the particle?
(b) What are Hamilton's equations for the particle?
(c) What are conditions that guarantee existence of an equilibrium solution of the particle in an electric field and a magnetic field?
(d) Suppose that the electric field and the magnetic field are constant. What are the conditions for existence of an equilibrium solution of the particle?

3.3. Consider the dynamics of a particle, with mass m, in three dimensions. The particle is constrained to move on an inclined plane described by

$$M = \left\{ x \in \mathbb{R}^3 : x_1 + x_2 + x_3 - 1 = 0 \right\},$$

with uniform constant gravity acting along the negative direction of the third axis of the inertial frame.

(a) View the constant gravity force as an external force, instead of arising from a potential. Show that the forced Euler–Lagrange equations (3.21) and forced Hamilton's equations (3.22) and (3.23) can be used to recover the Euler–Lagrange equations (3.30) and the Hamilton's equations (3.31) and (3.32).
(b) Derive an expression for the contact force of the plane on the particle, expressed in terms of $(x, \dot{x}) \in \mathsf{T}M$.

3.4. Consider the dynamics of a particle, with mass m, in three dimensions. The particle is constrained to move, without friction, on a surface defined by a hyperbolic paraboloid described by

$$M = \left\{ x \in \mathbb{R}^3 : -x_1^2 + x_2^2 - x_3 = 0 \right\}.$$

Uniform constant gravity acts along the negative direction of the third axis of the inertial frame.

(a) View the constant gravity force as an external force, instead of arising from a potential. Show that the forced Euler–Lagrange equations (3.21) and forced Hamilton's equations (3.22) and (3.23) can be used to recover the Euler–Lagrange equations (3.35) and the Hamilton's equations (3.36) and (3.37).
(b) Derive an expression for the contact force of the surface on the particle, expressed in terms of $(x, \dot{x}) \in \mathsf{T}M$.

3.5. A particle of mass m is constrained to move, without friction, on a plane embedded in $\mathbb{R}^3$ that is given by

$$M = \left\{x \in \mathbb{R}^3 : x_1 - x_2 + x_3 - 1 = 0\right\},$$

under the influence of constant gravity that acts along the negative direction of the third axis of the inertial frame.

(a) Show that M and $\mathbb{R}^2$ are diffeomorphic.
(b) Show that the Lagrangian function can be expressed as a function on the tangent bundle $\mathsf{T}\mathbb{R}^2$.
(c) What are the resulting Euler–Lagrange equations?
(d) What are the resulting Hamilton's equations?
(e) Describe the flow on the tangent bundle $\mathsf{T}M$ in terms of the flow on the tangent bundle $\mathsf{T}\mathbb{R}^2$.
(f) What are conservation properties of the dynamical flow on $\mathsf{T}M$?
(g) What are the equilibrium solutions of the dynamical flow on $\mathsf{T}M$?

3.6. A particle of mass m is constrained to move, without friction, on a surface embedded in $\mathbb{R}^3$ that is given by

$$M = \left\{x \in \mathbb{R}^3 : (x_1)^2 + (x_2)^2 - x_3 = 0\right\},$$

under the influence of constant gravity that acts along the negative direction of the third axis of the inertial frame.

(a) Show that M and $\mathbb{R}^2$ are diffeomorphic.
(b) Show that the Lagrangian function can be expressed as a function on the tangent bundle $\mathsf{T}\mathbb{R}^2$.
(c) What are the resulting Euler–Lagrange equations?
(d) What are the resulting Hamilton's equations?
(e) Describe the flow on the tangent bundle $\mathsf{T}M$ in terms of the flow on the tangent bundle $\mathsf{T}\mathbb{R}^2$.
(f) What are conservation properties of the dynamical flow on $\mathsf{T}M$?
(g) What are the equilibrium solutions of the dynamical flow on $\mathsf{T}M$?

3.7. A particle of mass m is constrained to move, without friction, on a line embedded in $\mathbb{R}^3$ that is given by

$$M = \left\{x \in \mathbb{R}^3 : x_1 - x_2 = 0, \ x_1 + x_2 - x_3 = 0\right\},$$

under the influence of constant gravity that acts along the negative direction of the third axis of the inertial frame.

(a) Show that M and $\mathbb{R}^1$ are diffeomorphic.
(b) Show that the Lagrangian function can be expressed as a function on the tangent bundle $\mathsf{T}\mathbb{R}^1$.
(c) What are the resulting Euler–Lagrange equations?
(d) What are the resulting Hamilton's equations?
(e) Describe the flow on the tangent bundle $\mathsf{T}M$ in terms of the flow on the tangent bundle $\mathsf{T}\mathbb{R}^1$.
(f) What are conservation properties of the dynamical flow on $\mathsf{T}M$?
(g) What are the equilibrium solutions of the dynamical flow on $\mathsf{T}M$?

3.8. A particle of mass m is constrained to move, without friction, on a curve embedded in $\mathbb{R}^3$ that is given by

$$M = \left\{x \in \mathbb{R}^3 : x_1 - x_2 = 0, \ (x_1)^2 + (x_2)^2 - x_3 = 0\right\},$$

under the influence of constant gravity that acts along the negative direction of the third axis of the inertial frame.

(a) Show that M and $\mathbb{R}^1$ are diffeomorphic.
(b) Show that the Lagrangian function can be expressed as a function on the tangent bundle $\mathsf{T}\mathbb{R}^1$.
(c) What are the resulting Euler–Lagrange equations?
(d) What are the resulting Hamilton's equations?
(e) Describe the flow on the tangent bundle $\mathsf{T}M$ in terms of the flow on the tangent bundle $\mathsf{T}\mathbb{R}^1$.
(f) What are conservation properties of the dynamical flow on $\mathsf{T}M$?
(g) What are the equilibrium solutions of the dynamical flow on $\mathsf{T}M$?

3.9. A particle of mass m is constrained to move, without friction, on a plane that rotates at a constant angular rate $\Omega \in \mathbb{R}^1$ about an inertially fixed line that is orthogonal to the rotating plane. No potential forces or external forces act on the particle. Let $q \in \mathbb{R}^2$ denote the position vector of the particle with respect to a two-dimensional Euclidean frame fixed to the rotating plane; let $x \in \mathbb{R}^3$ denote the position vector of the particle with respect to a three-dimensional Euclidean frame that is inertial. The position vector, in the inertial frame, can be expressed in terms of the position vector $q \in \mathbb{R}^2$ in the rotating frame by

$$x = \begin{bmatrix} \cos \Omega t & -\sin \Omega t \\ \sin \Omega t & \cos \Omega t \\ 0 & 0 \end{bmatrix} q.$$

(a) What is the Lagrangian function on the tangent bundle $T\mathbb{R}^2$, where the configuration is the position vector in the rotating frame?
(b) What are the resulting Euler–Lagrange equations?
(c) What are the resulting Hamilton's equations?
(d) What are conserved quantities?
(e) What are the equilibrium solutions of the dynamical flow on $T\mathbb{R}^2$?

3.10. Two particles, each of mass m, are connected by a massless link of length L and act under uniform, constant gravity. Let $x_1 \in \mathbb{R}^3$ and $x_2 \in \mathbb{R}^3$ denote the position vectors of the two particles with respect to an inertial Euclidean frame. Gravity acts along the negative direction of the third axis of the inertial frame. The link length constraint can be written as $\|x_1 - x_2\|^2 - L^2 = 0$ and this can be used to define the constraint manifold and the augmented Lagrangian function on the tangent bundle $T\mathbb{R}^6$.

(a) Describe the constraint manifold M as an embedded manifold in $\mathbb{R}^6$.
(b) What is the augmented Lagrangian function expressed as a function on the tangent bundle $T\mathbb{R}^6$?
(c) What are the resulting Euler–Lagrange equations, expressed using Lagrange multipliers?
(d) What are the resulting Hamilton's equations, expressed using Lagrange multipliers?
(e) Show that the Euler–Lagrange differential-algebraic equations have index two on the tangent bundle of the constraint manifold; show that Hamilton's differential-algebraic equations have index two on the cotangent bundle of the constraint manifold.
(f) What are conservation properties of the dynamical flow on TM?
(g) What are equilibrium solutions of the dynamical flow on TM?

3.11. Consider the restricted three-body problem in orbital mechanics. Two particles are in orbit in three dimensions about a massive spherical body; the two particles are acted on by the Newtonian gravitational force of a massive spherical body and by their mutual Newtonian gravitational force. The masses of the two orbiting bodies are m_1 and m_2 and the mass M of the spherical body is sufficiently large that it can be assumed to be inertially fixed. The configuration of the restricted three-body problem is $(q_1, q_2) \in \mathbb{R}^3 \times \mathbb{R}^3$, denoting the position vectors of the two particles from the center of the large spherical body, defined with respect to a Euclidean frame whose origin is located at the center of the fixed spherical body. Thus, the configuration manifold is $T\mathbb{R}^6$.

(a) What is the Lagrangian function defined on the tangent bundle $T\mathbb{R}^6$?
(b) What are the Euler–Lagrange equations for the two particles?
(c) What are Hamilton's equations for the two particles?
(d) Define the location of the center of mass of the two particles. Describe the dynamics of the center of mass of the two particles.
(e) What are conservation properties for the dynamics of the two particles?

3.12. Consider the separable Lagrangian function $L : T\mathbb{R}^n \to \mathbb{R}^1$ given by

$$L(q, \dot{q}) = \frac{1}{2} \sum_{i=1}^{n} m_i(q_i)\dot{q}_i^2 - \sum_{i=1}^{n} U_i(q_i),$$

where $q = [q_1, \ldots, q_n]^T \in \mathbb{R}^n$ and $m_i(q_i) > 0$, $i = 1, \ldots, n$ and $U_i(q_i)$, $i = 1, \ldots, n$ are real scalar functions.

(a) What are the Euler–Lagrange equations for this separable Lagrangian?
(b) What are Hamilton's equations for this separable Lagrangian?
(c) Show that there are n conserved quantities, namely

$$E_i = \frac{1}{2} m_i(q_i)\dot{q}_i^2 + U_i(q_i), \quad i = 1, \ldots, n,$$

that are constant along the dynamical flow.

3.13. Consider n identical particles, each of mass m, that translate in $\mathbb{R}^3$. Each particle is constrained to translate, without friction, on a straight line embedded in $\mathbb{R}^3$. That is, the position vector x_i of the i-th particle in an inertial Euclidean frame is constrained to lie in the one-dimensional linear manifold

$$M_i = \left\{ x \in \mathbb{R}^3 : x \in \text{span}\{a_i\} + b_i \right\},$$

where $a_i \in \mathbb{R}^3$, $a_i \neq 0$, $b_i \in \mathbb{R}^3$, $i = 1, \ldots, n$. We assume that none of the manifolds $M_i, \ldots, M_n$ intersect. The configuration manifold is given by $M = M_1 \times \cdots \times M_n$ so that the configuration vector is $x = (x_1, \ldots, x_n) \in M$. The n particles act under the influence of a mutual potential function given by

$$U(x) = \sum_{i=1}^{n} \sum_{j \neq i} K \|x_i - x_j\|^2,$$

where $K > 0$ is constant.

(a) Show that the configuration manifold M is globally diffeomorphic to $\mathbb{R}^n$.
(b) Use this diffeomorphism to show that the Lagrangian function $L : TM \to \mathbb{R}^1$ can be expressed as a function defined on the tangent bundle $T\mathbb{R}^n$.
(c) What are the resulting Euler–Lagrange equations?
(d) What is the Legendre transformation?
(e) What is the Hamiltonian function?
(f) What are the resulting Hamilton's equations?
(g) What are conservation properties of the dynamical flow?
(h) What are equilibrium solutions of the dynamical flow?

3.14. This problem treats the dynamics of a particle moving in a potential, expressing the equations of motion in terms of a configuration that is the position vector of the particle in a translating and rotating, non-inertial Euclidean frame. The position vector of a particle, with mass m, is described with respect to a translating and rotating Euclidean frame by the vector $q \in \mathbb{R}^3$. The position vector of the particle in the inertial frame is given by $x = z + Rq \in \mathbb{R}^3$, where $z : [0, \infty) \to \mathbb{R}^3$ denotes the (possibly time-varying) position vector of the origin of the non-inertial frame and $R : [0, \infty) \to \mathsf{SO}(3)$ denotes the (possibly time-varying) transformation matrix from the inertial frame to the moving frame; the motion of the moving frame, characterized by $(z, R) \in \mathsf{SE}(3)$, is assumed to be given. We also assume $\dot{R} = RS(\omega)$, where $\omega : [0, \infty) \to \mathbb{R}^3$ is the given angular velocity vector of the non-inertial frame. The particle acts under the influence of a potential energy $U(x) = U(z + Rq)$ that depends only on the position vector of the particle in the non-inertial frame. The configuration vector can be selected as the position vector of the particle $q \in \mathbb{R}^3$ in the reference frame, so that $\mathbb{R}^3$ is the configuration manifold.

(a) Obtain an expression for the Lagrangian function $L : \mathsf{T}\mathbb{R}^3 \to \mathbb{R}^1$.
(b) What are the Euler–Lagrange equations for the particle expressed in terms of $(q, \dot{q}) \in \mathsf{T}\mathbb{R}^3$?
(c) Define the conjugate momentum $p \in \mathsf{T}_q^*\mathbb{R}^3$ according to the Legendre transformation. What is the Hamiltonian function $H : \mathsf{T}^*\mathbb{R}^3 \to \mathbb{R}^1$?
(d) What are Hamilton's equations for the particle expressed in terms of $(q, p) \in \mathsf{T}^*\mathbb{R}^3$?

3.15. Consider the problem of finding the curve(s) $[0, 1] \to \mathbb{R}^3$ of shortest length that connect two fixed points in $\mathbb{R}^3$. Such curves are referred to as geodesic curve(s) on $\mathbb{R}^3$.

(a) If the curve is parameterized by $t \to q(t) \in \mathbb{R}^3$, show that the incremental arc length of the curve is $ds = \sqrt{\|dq\|^2}$ so that the geodesic curve(s) minimize $\int_0^1 \sqrt{\|\dot{q}\|^2} dt$.
(b) Show that the geodesic curve(s) necessarily satisfy the variational property $\delta \int_0^1 \|\dot{q}\| \, dt = 0$ for all smooth curves $t \to q(t) \in \mathbb{R}^3$ that satisfy the boundary conditions $q(0) = q_0 \in \mathbb{R}^3$, $q(1) = q_1 \in \mathbb{R}^3$.
(c) What are the Euler–Lagrange equations and Hamilton's equations that geodesic curves in $\mathbb{R}^3$ must satisfy?
(d) Use the equations and boundary conditions for the geodesic curves to show that for each $q_0 \in \mathbb{R}^3$, $q_1 \in \mathbb{R}^3$, there is a unique geodesic curve in $\mathbb{R}^3$. Describe this geodesic curve. Show that the geodesic curve is actually a minimum of $\int_0^1 \|\dot{q}\| \, dt$.
(e) For each $q_0 \in \mathbb{R}^3$, $q_1 \in \mathbb{R}^3$, view the geodesic curve as a one-dimensional manifold embedded in $\mathbb{R}^3$. Describe the geometrical properties of the tangent bundle along the geodesic curve.

3.16. Generalize the results of the prior problem to geodesic curves in $\mathbb{R}^n$. Consider the problem of finding the curve(s) $[0, 1] \to \mathbb{R}^n$ of shortest length that connect two fixed points in $\mathbb{R}^n$. Such curves are referred to as geodesic curve(s) on $\mathbb{R}^n$.

(a) If the curve is parameterized by $t \to q(t) \in \mathbb{R}^n$, show that the incremental arc length of the curve is $ds = \sqrt{\|dq\|^2}$ so that the geodesic curve(s) minimize $\int_0^1 \sqrt{\|\dot{q}\|^2} dt$.

(b) Show that the geodesic curve(s) necessarily satisfy the variational property $\delta \int_0^1 \|\dot{q}\| \, dt = 0$ for all smooth curves $t \to q(t) \in \mathbb{R}^n$ that satisfy the boundary conditions $q(0) = q_0 \in \mathbb{R}^n$, $q(1) = q_1 \in \mathbb{R}^n$.

(c) What are the Euler–Lagrange equations and Hamilton's equations that geodesic curves in $\mathbb{R}^n$ must satisfy?

(d) Use the equations and boundary conditions for the geodesic curves to show that for each $q_0 \in \mathbb{R}^n$, $q_1 \in \mathbb{R}^n$, there is a unique geodesic curve in $\mathbb{R}^n$. Describe this geodesic curve. Show that the geodesic curve is actually a minimum of $\int_0^1 \|\dot{q}\| \, dt$.

(e) For each $q_0 \in \mathbb{R}^n$, $q_1 \in \mathbb{R}^n$, view the geodesic curve as a one-dimensional manifold embedded in $\mathbb{R}^n$. Describe the geometrical properties of the tangent bundle along the geodesic curve.

3.17. Suppose the configuration manifold is $\mathbb{R}^n$ and the kinetic energy has the form of a general quadratic function in the time derivatives of the configuration variables, so that the Lagrangian function $L : \mathsf{T}\mathbb{R}^n \to \mathbb{R}^1$ is given by

$$L(q, \dot{q}) = \frac{1}{2} \sum_{i=1}^n \sum_{j=1}^n m_{ij}(q)\dot{q}_i\dot{q}_j + \sum_{i=1}^n a_i(q)^T \dot{q}_i - U(q),$$

where $q = [q_1, \ldots, q_n]^T \in \mathbb{R}^n$ and $m_{ij}(q) = m_{ji}(q) > 0, i = 1, \ldots, n$, $a_i(q), i = 1, \ldots, n$ are vector functions and $U(q)$ is a real scalar function.

(a) What are the Euler–Lagrange equations for this Lagrangian?

(b) What is the Hamiltonian function?

(c) What are the Hamilton's equations for the Hamiltonian associated with this Lagrangian?

Chapter 4
Lagrangian and Hamiltonian Dynamics on $(\mathsf{S}^1)^n$

This chapter introduces Lagrangian and Hamiltonian dynamics defined on the configuration manifold $(\mathsf{S}^1)^n$, the product of n copies of the one-sphere embedded in $\mathbb{R}^2$. Euler–Lagrange equations and Hamilton's equations are developed for systems that evolve on $(\mathsf{S}^1)^n$. This development is fundamentally different from the common approach in most of the published literature that makes use of angle coordinates for $(\mathsf{S}^1)^n$; in particular, the development here makes use of the differential geometry of the configuration manifold and leads to results that are globally valid everywhere on the configuration manifold. This chapter also serves as an introduction to variational methods and methods of analysis on configuration manifolds that are further developed in subsequent chapters.

The results in this chapter are illustrated by several formulations of Lagrangian dynamics and Hamiltonian dynamics on the configuration manifold $(\mathsf{S}^1)^n$. The key ideas of the development were first presented in published form in [46, 54] for the configuration manifold $(\mathsf{S}^2)^n$; those ideas are developed for the configuration manifold $(\mathsf{S}^1)^n$ in this chapter.

4.1 Configurations as Elements in $(\mathsf{S}^1)^n$

We develop Euler–Lagrange equations for Lagrangian systems evolving on the configuration manifold $(\mathsf{S}^1)^n$ that is a product of n copies of unit spheres in $\mathbb{R}^2$. Since the dimension of the configuration manifold is n, there are n degrees of freedom. The basic differential geometry for such a configuration manifold is now introduced.

© Springer International Publishing AG 2018
T. Lee et al., *Global Formulations of Lagrangian and Hamiltonian Dynamics on Manifolds*, Interaction of Mechanics and Mathematics,
DOI 10.1007/978-3-319-56953-6_4

As background, recall that the one-sphere, as an embedded manifold in $\mathbb{R}^2$, is

$$S^1 = \{q \in \mathbb{R}^2 : \|q\|^2 = 1\}.$$

The product of n spheres in $\mathbb{R}^2$, denoted by $(S^1)^n = S^1 \times \cdots \times S^1$, consists of all ordered n-tuples of vectors $q = (q_1, \ldots, q_n)$, with $q_i \in S^1$, $i = 1, \ldots, n$. This manifold is also described as

$$(S^1)^n = \{q \in \mathbb{R}^{2n} : q_i \in S^1, \, i = 1, \ldots, n\}.$$

As previously discussed, for $q \in (S^1)^n$,

$$T_q(S^1)^n = \{\xi \in \mathbb{R}^{2n} : (q_i \cdot \xi_i) = 0, \, i = 1, \ldots, n\},$$

denotes the tangent space to $(S^1)^n$ at $q \in (S^1)^n$; any vector $\xi \in T_q(S^1)^n$ is referred to as a tangent vector to $(S^1)^n$ at $q \in (S^1)^n$. Also,

$$T(S^1)^n = \{(q, \xi) \in \mathbb{R}^{2n} \times \mathbb{R}^{2n} : q \in (S^1)^n, \, \xi \in T_q(S^1)^n\},$$

denotes the tangent bundle of $(S^1)^n$.

For each $q \in (S^1)^n$, the set of all linear functionals defined on the tangent space $T_q(S^1)^n$ is the dual of $T_q(S^1)^n$ denoted by

$$T_q^*(S^1)^n = (T_q(S^1)^n)^*,$$

and it is the cotangent space to $(S^1)^n$ at $q \in (S^1)^n$; any element $\zeta \in T_q^*(S^1)^n$ is referred to as a covector to $(S^1)^n$ at $q \in (S^1)^n$. Also

$$T^*(S^1)^n = \{(q, \zeta) \in \mathbb{R}^{2n} \times (\mathbb{R}^{2n})^* : q \in (S^1)^n, \, \zeta \in T_q^*(S^1)^n\},$$

is the cotangent bundle of $(S^1)^n$. These definitions are important in our subsequent development of variational calculus on $(S^1)^n$.

4.2 Kinematics on $(\mathbf{S}^1)^n$

Consider a time-parameterized curve $t \to q(t) \in S^1$; its time derivative satisfies $q \cdot \dot{q} = 0$ for all t. This fact implies that there exists a scalar-valued angular velocity function $t \to \omega(t) \in \mathbb{R}^1$ such that the time derivative of this curve can be written as

$$\dot{q} = \omega S q,$$

where the 2×2 skew-symmetric matrix

$$S = \begin{bmatrix} 0 & -1 \\ 1 & 0 \end{bmatrix},$$

given in (1.6) acts by matrix multiplication to rotate a vector by $\frac{\pi}{2}$ counterclockwise.

It also follows that a time-parameterized curve $t \to q = (q_1, \ldots, q_n) \in (S^1)^n$ and its time derivative $\dot{q} = (\dot{q}_1, \ldots, \dot{q}_n) \in T_q(S^1)^n$ satisfy $(q_i \cdot \dot{q}_i) = 0$ for $i = 1, \ldots, n$ and all t. This implies that there are scalar-valued angular velocity functions $t \to \omega_i \in \mathbb{R}^1$, $i = 1, \ldots, n$, such that the time derivatives can be written as

$$\dot{q}_i = \omega_i S q_i, \quad i = 1, \ldots, n. \tag{4.1}$$

These equations describe the rotational kinematics on the configuration manifold $(S^1)^n$.

Taking the inner product of each equation of (4.1) with $S q_i$, it follows that the scalar angular velocities can be expressed as

$$\omega_i = q_i^T S^T \dot{q}_i, \quad i = 1, \ldots, n.$$

We subsequently use the notation $\omega = (\omega_1, \ldots, \omega_n) \in \mathbb{R}^n$ for the angular velocity vector.

4.3 Lagrangian Dynamics on $(S^1)^n$

We introduce a Lagrangian function, and we derive the Euler–Lagrange equations that make the integral of the Lagrangian over time, called the action integral, stationary; that is, the variation of the action integral is zero. The Euler–Lagrange equations are expressed in terms of the configuration vector and the time derivative of the configuration vector. A second form of the Euler–Lagrange equations is obtained in terms of a modified Lagrangian expressed in terms of the configuration vector and the angular velocity vector. In each case, these Euler–Lagrange equations are simplified for the important case that the kinetic energy function is a quadratic function of the time derivative of the configuration vector or the angular velocity vector.

4.3.1 Hamilton's Variational Principle in Terms of $(q, \dot{q})$

The Lagrangian $L : T(S^1)^n \to \mathbb{R}^1$ is a real-valued function defined on the tangent bundle of the configuration manifold $(S^1)^n$; we assume that the Lagrangian function

$$L(q, \dot{q}) = T(q, \dot{q}) - U(q),$$

is given by the difference between a kinetic energy function $T(q, \dot{q})$, defined on the tangent bundle, and a configuration-dependent potential energy function $U(q)$.

We first describe variations of curves or functions with values in $(S^1)^n$. We can express the variation of a curve with values in S^1 using a 2×2 matrix exponential map. This observation allows us to develop expressions for the infinitesimal variations of curves on $(S^1)^n$.

Let $q = (q_1, \dots, q_n) : [t_0, t_f] \to (S^1)^n$ be a differentiable curve. The variation of q_i is an ϵ-parameterized differentiable curve $q_i^\epsilon(t)$ taking values in S^1, where $\epsilon \in (-c, c)$ for some $c > 0$, $q_i^0(t) = q_i(t)$ for any $t \in [t_0, t_f]$, and the endpoints are fixed, that is, $q_i^\epsilon(t_0) = q_i(t_0)$, $q_i^\epsilon(t_f) = q(t_f)$ for any $\epsilon \in (-c, c)$.

If $q = (q_1, \dots, q_n) : [t_0, t_f] \to (S^1)^n$ is a differentiable curve on $(S^1)^n$, then its variation is $q^\epsilon = (q_1^\epsilon, \dots, q_n^\epsilon) : [t_0, t_f] \to (S^1)^n$. Similarly, the time derivative is $\dot{q} = (\dot{q}_1, \dots, \dot{q}_n) \in T_q(S^1)^n$, and its variation is $\dot{q}^\epsilon = (\dot{q}_1^\epsilon, \dots, \dot{q}_n^\epsilon) : [t_0, t_f] \to T_q(S^1)^n$.

Since the variational derivation of the Euler–Lagrange equations depends on the infinitesimal variation, there is no loss of generality in expressing the variation in terms of the matrix exponential map as follows:

$$q_i^\epsilon(t) = e^{\epsilon S \gamma_i(t)} q_i(t), \quad i = 1, \dots, n,$$

for differentiable curves $\gamma_i : [t_0, t_f] \to \mathbb{R}^1$, satisfying $\gamma_i(t_0) = \gamma_i(t_f) = 0$, $i = 1, \dots, n$. Since the exponent $\epsilon S \gamma_i$ is a 2×2 skew-symmetric matrix, it is easy to show that $\gamma_i \in \mathbb{R}^1 \to e^{\epsilon S \gamma_i} \in SO(2)$ is a local diffeomorphism, and that $e^{\epsilon S \gamma_i(t)} q_i(t) \in S^1$.

The infinitesimal variations are computed as

$$\delta q_i(t) = \frac{d}{d\epsilon} q_i^\epsilon(t) \Big|_{\epsilon=0}$$
$$= \gamma_i(t) S q_i(t), \quad i = 1, \dots, n, \tag{4.2}$$

where $\gamma_i(t_0) = \gamma_i(t_f) = 0$, $i = 1, \dots, n$. The infinitesimal variations vanish at the end points of the time interval since $\gamma_i(t_0) = \gamma_i(t_f) = 0$, $i = 1, \dots, n$.

Since the variation and time differentiation commute, the infinitesimal variations of the time derivative are given by

$$\delta \dot{q}_i(t) = \frac{d}{d\epsilon} \dot{q}_i^\epsilon(t) \Big|_{\epsilon=0}$$
$$= \dot{\gamma}_i(t) S q_i(t) + \gamma_i(t) S \dot{q}_i(t), \quad i = 1, \dots, n. \tag{4.3}$$

These expressions characterize the infinitesimal variations for a vector-valued function of time $(q, \dot{q}) = (q_1, \dots, q_n, \dot{q}_1, \dots, \dot{q}_n) : [t_0, t_f] \to T(S^1)^n$. The infinitesimal variations are important ingredients to derive the Euler–Lagrange equations on $(S^1)^n$. We subsequently suppress the time argument, thereby simplifying the notation.

The action integral is the integral of the Lagrangian function along a motion of the system over a fixed time period. The variations are taken over all differentiable curves with values in $(S^1)^n$ for which the initial and final values are fixed. The action integral along a motion is

$$\mathfrak{G} = \int_{t_0}^{t_f} L(q_1, \ldots, q_n, \dot{q}_1, \ldots, \dot{q}_n) \, dt.$$

The action integral along a varied motion is

$$\mathfrak{G}^\epsilon = \int_{t_0}^{t_f} L(q_1^\epsilon, \ldots, q_n^\epsilon, \dot{q}_1^\epsilon, \ldots, \dot{q}_n^\epsilon) \, dt.$$

The value of the action integral along a variation of a motion can be expressed as a power series in ϵ as

$$\mathfrak{G}^\epsilon = \mathfrak{G} + \epsilon \delta \mathfrak{G} + \mathcal{O}(\epsilon^2),$$

where the infinitesimal variation of the action integral is

$$\delta \mathfrak{G} = \frac{d}{d\epsilon} \mathfrak{G}^\epsilon \bigg|_{\epsilon=0}.$$

Hamilton's principle states that the infinitesimal variation of the action integral along any motion is zero:

$$\delta \mathfrak{G} = \frac{d}{d\epsilon} \mathfrak{G}^\epsilon \bigg|_{\epsilon=0} = 0, \tag{4.4}$$

for all possible differentiable functions $\gamma_i : [t_0, t_f] \to \mathbb{R}^1$ satisfying $\gamma_i(t_0) = \gamma_i(t_f) = 0$, $i = 1, \ldots, n$.

The infinitesimal variation of the action integral can be expressed in terms of the infinitesimal variation of the configuration as

$$\delta \mathfrak{G} = \int_{t_0}^{t_f} \sum_{i=1}^{n} \left\{ \frac{\partial L(q, \dot{q})}{\partial \dot{q}_i} \cdot \delta \dot{q}_i + \frac{\partial L(q, \dot{q})}{\partial q_i} \cdot \delta q_i \right\} dt.$$

We now substitute the expressions for the infinitesimal variations of the motion (4.2) and (4.3) into the above expression for the infinitesimal variation of the action integral. We then simplify the result to obtain the Euler–Lagrange equations expressed in terms of $(q, \dot{q}) \in \mathsf{T}(S^1)^n$.

4.3.2 Euler–Lagrange Equations in Terms of $(q, \dot{q})$

Substituting (4.2) and (4.3) we obtain

$$\delta \mathfrak{G} = -\int_{t_0}^{t_f} \sum_{i=1}^{n} \left\{ q_i \cdot \mathsf{S} \frac{\partial L(q, \dot{q})}{\partial \dot{q}_i} \dot{\gamma}_i + \dot{q}_i \cdot \mathsf{S} \frac{\partial L(q, \dot{q})}{\partial \dot{q}_i} \gamma_i + q_i \cdot \mathsf{S} \frac{\partial L(q, \dot{q})}{\partial q_i} \gamma_i \right\} dt.$$

Integrating the first term in the integral by parts, the infinitesimal variation of the action integral is given by

$$\delta\mathfrak{G} = -\sum_{i=1}^{n} q_i \cdot \mathrm{S}\frac{\partial L(q,\dot{q})}{\partial \dot{q}_i}\gamma_i\bigg|_{t_0}^{t_f}$$

$$+\sum_{i=1}^{n}\int_{t_0}^{t_f} q_i \cdot \mathrm{S}\left\{\frac{d}{dt}\left(\frac{\partial L(q,\dot{q})}{\partial \dot{q}_i}\right) - \frac{\partial L(q,\dot{q})}{\partial q_i}\right\}\gamma_i\, dt.$$

According to Hamilton's principle, $\delta\mathfrak{G} = 0$ for all continuous infinitesimal variations $\gamma_i : [t_0, t_f] \to \mathbb{R}^1$ that vanish at t_0 and t_f, $i = 1, \ldots, n$. Since $\gamma_i(t_0) = \gamma_i(t_f) = 0$, the boundary terms vanish, and the fundamental lemma of the calculus of variations, as described in Appendix A, implies that

$$q_i \cdot \mathrm{S}\left\{\frac{d}{dt}\left(\frac{\partial L(q,\dot{q})}{\partial \dot{q}_i}\right) - \frac{\partial L(q,\dot{q})}{\partial q_i}\right\} = 0, \quad i = 1, \ldots, n. \tag{4.5}$$

This means that q_i and the expression inside the braces are collinear. Hence, there are differentiable curves $c_i : [t_0, t_f] \to \mathbb{R}^1$ for $i = 1, \ldots, n$, such that

$$\frac{d}{dt}\left(\frac{\partial L(q,\dot{q})}{\partial \dot{q}_i}\right) - \frac{\partial L(q,\dot{q})}{\partial q_i} = c_i(t)q_i, \quad i = 1, \ldots, n.$$

Taking the dot product of the equation above with q_i, we obtain

$$c_i = q_i \cdot \left\{\frac{d}{dt}\left(\frac{\partial L(q,\dot{q})}{\partial \dot{q}_i}\right) - \frac{\partial L(q,\dot{q})}{\partial q_i}\right\}, \quad i = 1, \ldots, n.$$

This leads to the following proposition.

Proposition 4.1 *The Euler–Lagrange equations for a Lagrangian function* $L : \mathsf{T}(\mathsf{S}^1)^n \to \mathbb{R}^1$ *are*

$$(I_{2\times 2} - q_i q_i^T)\left\{\frac{d}{dt}\left(\frac{\partial L(q,\dot{q})}{\partial \dot{q}_i}\right) - \frac{\partial L(q,\dot{q})}{\partial q_i}\right\} = 0, \quad i = 1, \ldots, n. \tag{4.6}$$

The matrix $(I_{2\times 2} - q_i q_i^T)$ is a projection of $\mathbb{R}^2$ onto $\mathsf{T}_{q_i}(\mathsf{S}^1)$ in the sense that for any $q_i \in \mathsf{S}^1$ and any $\dot{q}_i \in \mathsf{T}_{q_i}(\mathsf{S}^1)$:

$$(I_{2\times 2} - q_i q_i^T)q_i = 0,$$
$$(I_{2\times 2} - q_i q_i^T)\dot{q}_i = \dot{q}_i.$$

This form of the Euler–Lagrange equations is reminiscent of the classical Euler–Lagrange equations from Chapter 3. The presence of the projection matrix in equation (4.6) reflects the fact that the configuration vector q_i is not an independent vector in $\mathbb{R}^2$ but is rather constrained to satisfy $q_i \in \mathsf{S}^1$, $i = 1, \ldots, n$. In this way, the Euler–Lagrange equations reflect the geometry of the configuration manifold. Alternatively, equation (4.6) can be viewed as stating that only the projection of the Euler–Lagrange equations onto the tangent space to the constraint manifold is satisfied, and this point of view will be expanded upon in Chapter 8.

Quadratic Kinetic Energy We now consider the important case that the kinetic energy is a quadratic function of the time derivative of the configuration vector, that is the Lagrangian function $L : \mathsf{T}(\mathsf{S}^1)^n \to \mathbb{R}^1$ is

$$L(q, \dot{q}) = \frac{1}{2} \sum_{j=1}^{n} \sum_{k=1}^{n} \dot{q}_j^T m_{jk}(q) \dot{q}_k - U(q), \tag{4.7}$$

where the scalar inertia terms $m_{jk} : (\mathsf{S}^1)^n \to \mathbb{R}^1$ satisfy the symmetry condition $m_{jk}(q) = m_{kj}(q)$ and the quadratic form in the time derivative of the configuration vector is positive-definite on $(\mathsf{S}^1)^n$.

We first determine the derivatives of the Lagrangian function

$$\frac{\partial L(q, \dot{q})}{\partial \dot{q}_i} = \sum_{j=1}^{n} m_{ij}(q) \dot{q}_j,$$

$$\frac{\partial L(q, \dot{q})}{\partial q_i} = \frac{1}{2} \frac{\partial}{\partial q_i} \sum_{j=1}^{n} \sum_{k=1}^{n} \dot{q}_j^T m_{jk}(q) \dot{q}_k - \frac{\partial U(q)}{\partial q_i},$$

and thus

$$\frac{d}{dt} \left\{ \frac{\partial L(q, \dot{q})}{\partial \dot{q}_i} \right\} = \sum_{j=1}^{n} m_{ij}(q) \ddot{q}_j + \sum_{j=1}^{n} \dot{m}_{ij}(q) \dot{q}_j.$$

It follows from (4.6) that the Euler–Lagrange equations can be written in the form

$$(I_{2 \times 2} - q_i q_i^T) \left\{ \sum_{j=1}^{n} m_{ij}(q) \ddot{q}_j + \sum_{j=1}^{n} \dot{m}_{ij}(q) \dot{q}_j \right.$$

$$\left. - \frac{1}{2} \frac{\partial}{\partial q_i} \sum_{j=1}^{n} \sum_{k=1}^{n} \dot{q}_j^T m_{jk}(q) \dot{q}_k + \frac{\partial U(q)}{\partial q_i} \right\} = 0, \quad i = 1, \ldots, n. \tag{4.8}$$

Since $q_i^T \dot{q}_i = 0$, it follows that $\frac{d}{dt}(q_i^T \dot{q}_i) = (q_i^T \ddot{q}_i) + \|\dot{q}_i\|^2 = 0$; thus we obtain

$$(I_{2 \times 2} - q_i q_i^T) \ddot{q}_i = \ddot{q}_i - (q_i q_i^T) \ddot{q}_i$$

$$= \ddot{q}_i + \|\dot{q}_i\|^2 q_i, \quad i = 1, \ldots, n.$$

Thus, we can expand the time derivative expression from (4.8) to obtain the following Euler–Lagrange equations on $\mathsf{T}(\mathsf{S}^1)^n$:

$$m_{ii}(q)\ddot{q}_i + m_{ii}(q)\|\dot{q}_i\|^2 q_i + (I_{2\times2} - q_i q_i^T)\sum_{\substack{j=1 \\ j\neq i}}^{n} m_{ij}(q)\ddot{q}_j$$

$$+ (I_{2\times2} - q_i q_i^T)F_i(q,\dot{q}) + (I_{2\times2} - q_i q_i^T)\frac{\partial U(q)}{\partial q_i} = 0, \quad i = 1,\ldots,n, \quad (4.9)$$

where the vector-valued functions

$$F_i(q,\dot{q}) = \sum_{j=1}^{n} \dot{m}_{ij}(q)\dot{q}_j - \frac{1}{2}\frac{\partial}{\partial q_i}\sum_{j=1}^{n}\sum_{k=1}^{n} \dot{q}_j^T m_{jk}(q)\dot{q}_k, \quad i = 1,\ldots,n,$$

can be expressed as a quadratic function of the time derivative of the configuration vector by using Christoffel symbols (see equation (3.7)). Note that if the inertia terms are constants, independent of the configuration, then $F_i(q,\dot{q}) = 0$, $i = 1,\ldots,n$.

These Euler–Lagrange equations (4.9) describe the dynamical flow of $(q,\dot{q}) \in \mathsf{T}(\mathsf{S}^1)^n$ on the tangent bundle of the configuration manifold $(\mathsf{S}^1)^n$.

Assuming that the inertia terms and the potential terms in (4.9) are globally defined on $(\mathbb{R}^2)^n$, then the domain of definition of (4.9) on $\mathsf{T}(\mathsf{S}^1)^n$ can be extended to $\mathsf{T}(\mathbb{R}^2)^n$. This extension is natural and useful in that it defines a Lagrangian vector field on the tangent bundle $\mathsf{T}(\mathbb{R}^2)^n$. Alternatively, the manifold $\mathsf{T}(\mathsf{S}^1)^n$ can be viewed as an invariant manifold of this Lagrangian vector field on $\mathsf{T}(\mathbb{R}^2)^n$ and its restriction to this invariant manifold describes the Lagrangian flow of (4.9) on $\mathsf{T}(\mathsf{S}^1)^n$.

4.3.3 Hamilton's Variational Principle in Terms of (q, ω)

We now give an alternative expression for the Euler–Lagrange equations in terms of angular velocities as introduced in (4.1).

By making use of the kinematics, the Lagrangian function can also be expressed in terms of the angular velocities. We write

$$\tilde{L}(q,\omega) = L(q,\dot{q}),$$

where the time derivative of the configuration vector is given by the kinematics (4.1). This is referred to as the modified Lagrangian function and we view it as also being defined on the tangent bundle of $(\mathsf{S}^1)^n$.

The infinitesimal variation of the modified action integral can be written as

$$\delta\tilde{\mathfrak{G}} = \int_{t_0}^{t_f}\sum_{i=1}^{n}\left\{\frac{\partial\tilde{L}(q,\omega)}{\partial\omega_i}\delta\omega_i + \frac{\partial\tilde{L}(q,\omega)}{\partial q_i}\cdot\delta q_i\right\}dt.$$

Note that the first term in the integral on the right-hand side is the product of two scalars, so the dot product notation is not utilized; the second term in the integral on the right-hand side is the dot product of two vectors.

The infinitesimal variations of the motion are given by

$$\delta q_i = \gamma_i S q_i, \quad i = 1, \ldots, n, \tag{4.10}$$

$$\delta \omega_i = \dot{\gamma}_i, \quad i = 1, \ldots, n, \tag{4.11}$$

for curves $\gamma_i : [t_0, t_f] \to \mathbb{R}^1$ satisfying $\gamma_i(t_0) = \gamma_i(t_f) = 0$, $i = 1, \ldots, n$. The first expression was previously given, while the second expression can be easily derived from the kinematics and the condition that $q_i \cdot \dot{q}_i = 0$.

4.3.4 Euler–Lagrange Equations in Terms of (q, ω)

Substituting (4.10) and (4.11) into the expression for the infinitesimal variation of the modified action integral we obtain

$$\delta \tilde{\mathfrak{G}} = \int_{t_0}^{t_f} \sum_{i=1}^{n} \left\{ \frac{\partial \tilde{L}(q, \omega)}{\partial \omega_i} \dot{\gamma}_i + \frac{\partial \tilde{L}(q, \omega)}{\partial q_i} \cdot \gamma_i S q_i \right\} dt.$$

Integrating the first term in the integral by parts, the infinitesimal variation of the modified action integral is given by

$$\delta \tilde{\mathfrak{G}} = \sum_{i=1}^{n} \frac{\partial \tilde{L}(q, \omega)}{\partial \omega_i} \gamma_i \Big|_{t_0}^{t_f}$$

$$+ \sum_{i=1}^{n} \int_{t_0}^{t_f} \left\{ -\frac{d}{dt} \left(\frac{\partial \tilde{L}(q, \omega)}{\partial \omega_i} \right) - q_i^T S \frac{\partial \tilde{L}(q, \omega)}{\partial q_i} \right\} \gamma_i \, dt.$$

According to Hamilton's principle, $\delta \tilde{\mathfrak{G}} = 0$ for all differentiable functions $\gamma_i : [t_0, t_f] \to \mathbb{R}^1$, $i = 1, \ldots, n$ that vanish at t_0 and t_f, $i = 1, \ldots, n$. Since $\gamma_i(t_0) = \gamma_i(t_f) = 0$, the boundary terms vanish, and the fundamental lemma of the calculus of variations, as described in Appendix A, implies the following proposition.

Proposition 4.2 *The Euler–Lagrange equations for a modified Lagrangian function $\tilde{L} : \mathsf{T}(\mathsf{S}^1)^n \to \mathbb{R}^1$ are*

$$\frac{d}{dt} \left(\frac{\partial \tilde{L}(q, \omega)}{\partial \omega_i} \right) + q_i^T S \frac{\partial \tilde{L}(q, \omega)}{\partial q_i} = 0, \quad i = 1, \ldots, n. \tag{4.12}$$

Thus, the evolution on the tangent bundle $\mathsf{T}(\mathsf{S}^1)^n$ is described by the kinematics equations (4.1) and the Euler–Lagrange equations (4.12).

This form of the Euler–Lagrange equations on $(S^1)^n$, expressed in terms of angular velocities, can be obtained in a different way directly from the Euler–Lagrange equations given in (4.6). The kinematics (4.1) can be viewed as defining a change of variables $(q, \dot{q}) \in T(S^1)^n \to (q, \omega) \in (S^1)^n \times \mathbb{R}^n$. This approach can be used to show the equivalence of the Euler–Lagrange equations (4.6) and the Euler–Lagrange equations (4.12) and the kinematics (4.1).

Quadratic Kinetic Energy We now consider the important case that the Lagrangian is a quadratic function of the angular velocities as in (4.7); that is, the modified Lagrangian function is

$$\tilde{L}(q, \omega) = \frac{1}{2} \sum_{i=1}^{n} \sum_{j=1}^{n} \omega_i q_i^T m_{ij}(q) q_j \omega_j - U(q). \tag{4.13}$$

The modified Lagrangian function can also be written as

$$\tilde{L}(q, \omega) = \frac{1}{2} \sum_{i=1}^{n} m_{ii}(q) \omega_i^2 + \frac{1}{2} \sum_{i=1}^{n} \sum_{\substack{j=1 \\ j \neq i}}^{n} \omega_i q_i^T m_{ij}(q) q_j \omega_j - U(q).$$

Now we show that the Euler–Lagrange equations can be expressed in terms of the angular velocity vector $\omega = (\omega_1, \ldots, \omega_n) \in \mathbb{R}^n$.

We first determine the derivatives of the modified Lagrangian function

$$\frac{\partial \tilde{L}(q, \omega)}{\partial \omega_i} = \sum_{j=1}^{n} q_i^T m_{ij}(q) q_j \omega_j$$

$$= m_{ii}(q) \omega_i + \sum_{\substack{j=1 \\ j \neq i}}^{n} q_i^T m_{ij}(q) q_j \omega_j,$$

$$\frac{\partial \tilde{L}(q, \omega)}{\partial q_i} = \frac{1}{2} \frac{\partial}{\partial q_i} \sum_{j=1}^{n} \sum_{k=1}^{n} \omega_j q_j^T m_{jk}(q) q_k \omega_k - \frac{\partial U(q)}{\partial q_i},$$

and thus

$$\frac{d}{dt} \left\{ \frac{\partial \tilde{L}(q, \omega)}{\partial \omega_i} \right\} = m_{ii}(q) \dot{\omega}_i + \sum_{\substack{j=1 \\ j \neq i}}^{n} q_i^T m_{ij}(q) q_j \dot{\omega}_j$$

$$+ \sum_{j=1}^{n} q_i^T \dot{m}_{ij}(q) q_j \omega_j + \sum_{\substack{j=1 \\ j \neq i}}^{n} m_{ij}(q)(\dot{q}_i^T q_j + q_i^T \dot{q}_j) \omega_j.$$

The Euler–Lagrange equations (4.12) can be expressed as

$$m_{ii}(q)\dot\omega_i + \sum_{\substack{j=1\\j\neq i}}^{n} q_i^T m_{ij}(q)q_j\dot\omega_j + \sum_{j=1}^{n} q_i^T \dot m_{ij}(q)q_j\omega_j + \sum_{\substack{j=1\\j\neq i}}^{n} m_{ij}(q)(\dot q_i^T q_j + q_i^T \dot q_j)\omega_j$$

$$+ q_i^T S\left\{\frac{1}{2}\frac{\partial}{\partial q_i}\sum_{j=1}^{n}\sum_{k=1}^{n}\omega_j q_j^T m_{jk}(q)q_k\omega_k - \frac{\partial U(q)}{\partial q_i}\right\} = 0.$$

We use the kinematics (4.1) to write

$$\sum_{\substack{j=1\\j\neq i}}^{n} m_{ij}(q)(\dot q_i^T q_j + q_i^T \dot q_j)\omega_j = \sum_{\substack{j=1\\j\neq i}}^{n} m_{ij}(q)q_i^T S q_j(\omega_j - \omega_i)\omega_j.$$

Thus, it follows from (4.12) that

$$m_{ii}(q)\dot\omega_i + \sum_{\substack{j=1\\j\neq i}}^{n} q_i^T m_{ij}(q)q_j\dot\omega_j + \sum_{j=1}^{n} q_i^T \dot m_{ij}(q)q_j\omega_j$$

$$+ \sum_{\substack{j=1\\j\neq i}}^{n} m_{ij}(q)q_i^T S q_j(\omega_j - \omega_i)\omega_j$$

$$+ q_i^T S\left\{\frac{1}{2}\frac{\partial}{\partial q_i}\sum_{j=1}^{n}\sum_{k=1}^{n}\omega_j q_j^T m_{jk}(q)q_k\omega_k - \frac{\partial U(q)}{\partial q_i}\right\} = 0.$$

Using the observation that $q_i^T q_j = -q_i^T SSq_j$, we rewrite the third term to obtain a convenient form of the Euler–Lagrange equations on $T(S^1)^n$:

$$m_{ii}(q)\dot\omega_i + \sum_{\substack{j=1\\j\neq i}}^{n} q_i^T m_{ij}(q)q_j\dot\omega_j + q_i^T S\sum_{\substack{j=1\\j\neq i}}^{n} m_{ij}(q)\omega_j^2 q_j$$

$$- q_i^T S\left\{\sum_{\substack{j=1\\j\neq i}}^{n}\omega_i m_{ij}(q)\omega_j q_j + F_i(q,\omega) + \frac{\partial U(q)}{\partial q_i}\right\} = 0, \quad i = 1,\ldots,n, \quad (4.14)$$

where

$$F_i(q,\omega) = \sum_{j=1}^{n} \dot m_{ij}(q)S q_j\omega_j - \frac{1}{2}\frac{\partial}{\partial q_i}\sum_{j=1}^{n}\sum_{k=1}^{n}\omega_j q_j^T m_{jk}(q)q_k\omega_k, \quad i = 1,\ldots,n,$$

can be shown to be quadratic in the angular velocities. Note that if the inertial terms are constants, independent of the configuration, then $F_i(q,\omega) = 0$, $i = 1,\ldots,n$.

This version of the Euler–Lagrange differential equations (4.14) and the kinematics equations (4.1), expressed in terms of the angular velocities, describe the Lagrangian flow of $(q, \dot{q}) \in \mathsf{T}(\mathsf{S}^1)^n$ on the tangent bundle of the configuration manifold $(\mathsf{S}^1)^n$.

If the inertia terms and the potential terms in (4.14) are globally defined on $(\mathbb{R}^2)^n$, then the domain of definition of (4.14) on $\mathsf{T}(\mathsf{S}^1)^n$ can be extended to $\mathsf{T}(\mathbb{R}^2)^n$. This extension is natural in that it defines a Lagrangian vector field on the tangent bundle $\mathsf{T}(\mathbb{R}^2)^n$. Alternatively, the manifold $\mathsf{T}(\mathsf{S}^1)^n$ is an invariant manifold of this Lagrangian vector field on $\mathsf{T}(\mathbb{R}^2)^n$ and its restriction to this invariant manifold describes the Lagrangian flow of (4.14) on $\mathsf{T}(\mathsf{S}^1)^n$.

Equations (4.9) and the kinematics (4.1) can be shown to be equivalent to (4.14) by viewing the kinematics as defining a transformation from $(q, \dot{q})$ to (q, ω). This provides an alternate derivation of (4.14).

4.4 Hamiltonian Dynamics on $(\mathsf{S}^1)^n$

Here, the Legendre transformation is introduced to derive Hamilton's equations for dynamics that evolve on $(\mathsf{S}^1)^n$. The derivation is based on the phase space variational principle, a natural modification of Hamilton's principle for Lagrangian dynamics. Two forms of Hamilton's equations are obtained. One form is expressed in terms of momentum covectors $(\mu_1, \ldots, \mu_n) \in \mathsf{T}_q^*(\mathsf{S}^1)^n$ that are conjugate to the velocities $(\dot{q}_1, \ldots, \dot{q}_n) \in \mathsf{T}_q(\mathsf{S}^1)^n$, where $q \in (\mathsf{S}^1)^n$. The other form of Hamilton's equations is expressed in terms of the momentum $\pi = (\pi_1, \ldots, \pi_n)$ that is conjugate to the angular velocity vector $(\omega_1, \ldots, \omega_n)$.

4.4.1 Hamilton's Phase Space Variational Principle in Terms of (q, μ)

As in the prior section, we begin with a Lagrangian function $L : \mathsf{T}(\mathsf{S}^1)^n \to \mathbb{R}^1$, which is a real-valued function defined on the tangent bundle of the configuration manifold $(\mathsf{S}^1)^n$; we assume that the Lagrangian function

$$L(q, \dot{q}) = T(q, \dot{q}) - U(q),$$

is given by the difference between a kinetic energy function $T(q, \dot{q})$, defined on the tangent bundle, and a configuration-dependent potential energy function $U(q)$.

The Legendre transformation of the Lagrangian function $L(q, \dot{q})$ provides the basis for obtaining a Hamiltonian form of the equations of motion in terms of a conjugate momentum. The Legendre transformation $(\dot{q}_1, \ldots, \dot{q}_n) \in T_q(S^1)^n \to (\mu_1, \ldots, \mu_n) \in T_q^*(S^1)^n$ satisfies

$$\mu_i \cdot \dot{q}_i = \frac{\partial L(q, \dot{q})}{\partial \dot{q}_i} \cdot \dot{q}_i, \quad i = 1, \ldots, n,$$

for all $\dot{q}_i \in T_{q_i} S^1$, $i = 1, \ldots, n$. The momentum μ_i is an element of the dual of the tangent space $T_{q_i} S^1$, and its action on a tangent vector $\dot{q}_i$ via the inner product is independent of any component that is collinear with q_i. This results in an ambiguity in the representation of the momentum as a vector, and we address this by requiring that the vector representing μ_i is orthogonal to q_i; that is μ_i is equal to the orthogonal projection of $\frac{\partial L(q, \dot{q})}{\partial \dot{q}_i}$ onto the tangent space $T_{q_i} S^1$. Thus,

$$\mu_i = \frac{\partial L(q, \dot{q})}{\partial \dot{q}_i} - \left(q_i \cdot \frac{\partial L(q, \dot{q})}{\partial \dot{q}_i} \right) q_i, \quad i = 1, \ldots, n,$$

which can be written as

$$\mu_i = (I_{2 \times 2} - q_i q_i^T) \frac{\partial L(q, \dot{q})}{\partial \dot{q}_i}, \quad i = 1, \ldots, n, \tag{4.15}$$

where the operators $(I_{2 \times 2} - q_i q_i^T)$ project $\mathbb{R}^2$ onto $T_{q_i} S^1$ for $i = 1, \ldots, n$.

We assume that the Lagrangian function is hyperregular; that is the Legendre transformation, viewed as a map $T_q(S^1)^n \to T_q^*(S^1)^n$, is invertible.

The Hamiltonian function $H : T^*(S^1)^n \to \mathbb{R}^1$ is given by

$$H(q, \mu) = \sum_{i=1}^{n} \mu_i \cdot \dot{q}_i - L(q, \dot{q}), \tag{4.16}$$

where the right-hand side can be evaluated by expressing $(q, \dot{q})$ in terms of (q, μ) by inverting the Legendre transformation (4.15).

The Legendre transformation can be viewed as defining a transformation $(q, \dot{q}) \in T(S^1)^n \to (q, \mu) \in T^*(S^1)^n$, which implies that the Euler–Lagrange equations can be written in terms of the transformed variables; this is effectively Hamilton's equations. However, Hamilton's equations can also be obtained using Hamilton's phase space variational principle, and this approach is now introduced.

Consider the action integral in the form,

$$\mathfrak{G} = \int_{t_0}^{t_f} \left\{ \sum_{i=1}^{n} \mu_i \cdot \dot{q}_i - H(q, \mu) \right\} dt.$$

The infinitesimal variation of the action integral is given by

$$\delta \mathfrak{G} = \sum_{i=1}^{n} \int_{t_0}^{t_f} \left\{ \mu_i \cdot \delta \dot{q}_i - \frac{\partial H(q, \mu)}{\partial q_i} \cdot \delta q_i + \left(\dot{q}_i - \frac{\partial H(q, \mu)}{\partial \mu_i} \right) \cdot \delta \mu_i \right\} dt.$$

Hamilton's phase space variational principle states that the infinitesimal variation of the action integral is zero for all admissible variations of $(q, \mu) \in$ $\mathsf{T}^*(\mathsf{S}^1)^n$. By integrating the first term on the right-hand side by parts, we can express the variational principle as

$$\delta \mathfrak{G} = \sum_{i=1}^n \mu_i \cdot \delta q_i \bigg|_{t_0}^{t_f}$$

$$+ \sum_{i=1}^n \int_{t_0}^{t_f} \left\{ \left(-\dot{\mu}_i - \frac{\partial H(q, \mu)}{\partial q_i} \right) \cdot \delta q_i + \left(\dot{q}_i - \frac{\partial H(q, \mu)}{\partial \mu_i} \right) \cdot \delta \mu_i \right\} dt = 0,$$

for all possible differentiable curves $(\delta q_i, \delta \mu_i) : [t_0, t_f] \to \mathsf{T}_{(q_i, \mu_i)} \mathsf{T}^* \mathsf{S}^1$, satisfying $\delta q_i(t_0) = \delta q_i(t_f) = 0$ for all $i = 1, \ldots, n$.

From the definition of the conjugate momenta μ_i given by (4.15), we have $q_i \cdot \mu_i = 0$, which implies that $\delta q_i \cdot \mu_i + q_i \cdot \delta \mu_i = 0$. To impose this constraint on the variations, it is convenient to decompose $\delta \mu_i$ into a component collinear with q_i, namely $\delta \mu_i^C = q_i q_i^T \delta \mu_i$, and a component orthogonal to q_i, namely $\delta \mu_i^M = (I_{2\times 2} - q_i q_i^T) \delta \mu_i$. Thus, the constraint on the momentum induces the following constraint on the infinitesimal variations, $q_i^T \delta \mu_i^C = q_i^T \delta \mu_i = -\mu_i^T \delta q_i$, but $\delta \mu_i^M = (I_{2\times 2} - q_i q_i^T) \delta \mu_i$ is otherwise unconstrained.

Recall the prior expression for the infinitesimal variations on $(\mathsf{S}^1)^n$:

$$\delta q_i = \gamma_i S q_i, \quad i = 1, \ldots, n,$$

for differentiable curves $\gamma_i : [t_0, t_f] \to \mathbb{R}^1$, $i = 1, \ldots, n$ satisfying $\gamma_i(t_0) = \gamma_i(t_f) = 0$, $i = 1, \ldots, n$.

These results can be summarized as: the infinitesimal variation of the action integral is zero, for all possible differentiable curves $\gamma_i : [t_0, t_f] \to \mathbb{R}^1$ and $\delta \mu_i^M : [t_0, t_f] \to \mathbb{R}^2$, with $\delta \mu_i^M \cdot q_i = 0$ and $\gamma_i(t_0) = \gamma_i(t_f) = 0$, for $i = 1, \ldots, n$.

4.4.2 Hamilton's Equations in Terms of (q, μ)

Substituting the expressions for the infinitesimal variations of the configuration and using the fact that the infinitesimal variation of the configuration vanishes at the endpoints, the infinitesimal variation of the action integral can be rewritten as

$$\delta \mathfrak{G} = \sum_{i=1}^n \int_{t_0}^{t_f} \left\{ \left(-\dot{\mu}_i - \frac{\partial H(q, \mu)}{\partial q_i} \right) \cdot \delta q_i + \left(q_i q_i^T \left(\dot{q}_i - \frac{\partial H(q, \mu)}{\partial \mu_i} \right) \right) \cdot \delta \mu_i^C \right.$$

$$\left. + \left((I_{2\times 2} - q_i q_i^T) \left(\dot{q}_i - \frac{\partial H(q, \mu)}{\partial \mu_i} \right) \right) \cdot \delta \mu_i^M \right\} dt$$

$$= \sum_{i=1}^{n} \int_{t_0}^{t_f} \left\{ \left(-\dot{\mu}_i - \frac{\partial H(q,\mu)}{\partial q_i} \right) \cdot \delta q_i - \left(q_i^T \frac{\partial H(q,\mu)}{\partial \mu_i} \right) \cdot (q_i^T \delta \mu_i^C) \right.$$

$$\left. + \left(\dot{q}_i - (I_{2\times 2} - q_i q_i^T) \frac{\partial H(q,\mu)}{\partial \mu_i} \right) \cdot \delta \mu_i^M \right\} dt$$

$$= \sum_{i=1}^{n} \int_{t_0}^{t_f} \left\{ \left(-\dot{\mu}_i - \frac{\partial H(q,\mu)}{\partial q_i} \right) \cdot \delta q_i + \left(q_i^T \frac{\partial H(q,\mu)}{\partial \mu_i} \right) (\mu_i \cdot \delta q_i) \right.$$

$$\left. + \left(\dot{q}_i - (I_{2\times 2} - q_i q_i^T) \frac{\partial H(q,\mu)}{\partial \mu_i} \right) \cdot \delta \mu_i^M \right\} dt$$

$$= \sum_{i=1}^{n} \int_{t_0}^{t_f} \left\{ q_i^T \mathsf{S} \left(-\dot{\mu}_i - \frac{\partial H(q,\mu)}{\partial q_i} + \mu_i q_i^T \frac{\partial H(q,\mu)}{\partial \mu_i} \right) \gamma_i \right.$$

$$\left. + \left(\dot{q}_i - (I_{2\times 2} - q_i q_i^T) \frac{\partial H(q,\mu)}{\partial \mu_i} \right) \cdot \delta \mu_i^M \right\} dt = 0.$$

We now invoke Hamilton's phase space variational principle that $\delta \mathfrak{G} = 0$ for all possible differentiable curves $\gamma_i : [t_0, t_f] \to \mathbb{R}^1$ and $\delta \mu_i^M : [t_0, t_f] \to \mathbb{R}^2$ that are orthogonal to q_i, for $i = 1, \ldots, n$.

According to the fundamental lemma of the calculus of variations, as described in Appendix A, the first condition gives

$$q_i^T \mathsf{S} \left(\dot{\mu}_i + \frac{\partial H(q,\mu)}{\partial q_i} - \mu_i q_i^T \frac{\partial H(q,\mu)}{\partial \mu_i} \right) = 0, \quad i = 1, \ldots, n.$$

By the same arguments used to derive equation (4.6), we obtain

$$(I_{2\times 2} - q_i q_i^T) \left(\dot{\mu}_i + \frac{\partial H(q,\mu)}{\partial q_i} - \mu_i q_i^T \frac{\partial H(q,\mu)}{\partial \mu_i} \right) = 0, \quad i = 1, \ldots, n.$$

These are incomplete since the first equation only determines the component of $\dot{\mu}_i$ in the tangent space $\mathsf{T}_{q_i} \mathsf{S}^1$. The component of $\dot{\mu}_i$ that is orthogonal to this tangent space, that is collinear with q_i, is determined as follows. The time derivative of $q_i \cdot \mu_i = 0$ gives $q_i \cdot \dot{\mu}_i = -\dot{q}_i \cdot \mu_i$, which allows computation of the component $\dot{\mu}_i^C$ that is collinear with q_i.

Since both terms multiplying $\delta \mu_i^M$ in the above variational expression are necessarily orthogonal to q_i, and the second condition implies that the component orthogonal to q_i vanishes, it follows that

$$\dot{q}_i - (I_{2\times 2} - q_i q_i^T) \frac{\partial H(q,\mu)}{\partial \mu_i} = 0, \quad i = 1, \ldots, n.$$

By combining these, Hamilton's equations on the configuration manifold $(\mathsf{S}^1)^n$ are given as follows.

Proposition 4.3 *Hamilton's equations for a Hamiltonian function* H :
$T^*(S^1)^n \to \mathbb{R}^1$ *are*

$$\dot{q}_i = (I_{2\times 2} - q_i q_i^T) \frac{\partial H(q, \mu)}{\partial \mu_i}, \quad i = 1, \ldots, n, \tag{4.17}$$

$$\dot{\mu}_i = \left(\mu_i q_i^T - q_i \mu_i^T\right) \left(\frac{\partial H(q, \mu)}{\partial \mu_i}\right) - \left(I_{2\times 2} - q_i q_i^T\right) \frac{\partial H(q, \mu)}{\partial q_i}, \quad i = 1, \ldots, n. \tag{4.18}$$

The second term in the last equation comes from the fact that the component of $\dot{\mu}_i$ in the q_i direction is given by $-(\dot{q}_i \cdot \mu_i) q_i$. Thus, equations (4.17) and (4.18) describe the Hamiltonian flow in terms of $(q, \mu) \in T^*(S^1)^n$ on the cotangent bundle $T^*(S^1)^n$.

The following property follows directly from the above formulation of Hamilton's equations on $(S^1)^n$:

$$\begin{aligned}
\frac{dH(q, \mu)}{dt} &= \sum_{i=1}^{n} \frac{\partial H(q, \mu)}{\partial q_i} \cdot \dot{q}_i + \frac{\partial H(q, \mu)}{\partial \mu_i} \cdot \dot{\mu}_i \\
&= \sum_{i=1}^{n} \frac{\partial H(q, \mu)}{\partial \mu_i} \cdot \left\{ \mu_i q_i^T \frac{\partial H(q, \mu)}{\partial \mu_i} - \mu_i^T \frac{\partial H(q, \mu)}{\partial \mu_i} q_i \right\} \\
&= \sum_{i=1}^{n} \frac{\partial H(q, \mu)}{\partial \mu_i} \cdot \left\{ (\mu_i q_i^T - q_i \mu_i^T) \frac{\partial H(q, \mu)}{\partial \mu_i} \right\} \\
&= 0.
\end{aligned}$$

This formulation exposes an important property of the Hamiltonian flow on the cotangent bundle: the Hamiltonian function is constant along each solution of Hamilton's equation. As previously mentioned, this property does not hold if the Hamiltonian function has a nontrivial explicit time dependence.

Quadratic Kinetic Energy We now consider the important case where the Lagrangian is a quadratic function of the time derivative of the configuration vector that was introduced previously in equation (4.7) and given by

$$L(q, \dot{q}) = \frac{1}{2} \sum_{j=1}^{n} \sum_{k=1}^{n} \dot{q}_j^T m_{jk}(q) \dot{q}_k - U(q), \tag{4.19}$$

where the scalar inertia terms $m_{jk}(q) : (S^1)^n \to \mathbb{R}^1$ satisfy the symmetry condition $m_{jk}(q) = m_{kj}(q)$ and the quadratic form is positive-definite.

Thus, the conjugate momentum is defined by the Legendre transformation

$$\mu_i = (I_{2\times2} - q_i q_i^T) \frac{\partial L(q, \dot{q})}{\partial \dot{q}_i}$$

$$= (I_{2\times2} - q_i q_i^T) \sum_{j=1}^{n} m_{ij}(q) \dot{q}_j$$

$$= m_{ii}(q) \dot{q}_i + (I_{2\times2} - q_i q_i^T) \sum_{\substack{j=1 \\ j \neq i}}^{n} m_{ij}(q) \dot{q}_j, \quad i = 1, \ldots, n. \tag{4.20}$$

The algebraic equations (4.20), viewed as defining a linear mapping from $(\dot{q}_1, \ldots, \dot{q}_n) \in \mathsf{T}_q(S^1)^n$ to $(\mu_1, \ldots, \mu_n) \in \mathsf{T}_q^*(S^1)^n$, can be inverted and expressed in the form

$$\dot{q}_i = (I_{2\times2} - q_i q_i^T) \sum_{j=1}^{n} m_{ij}^I(q) \mu_j, \quad i = 1, \ldots, n,$$

where $m_{ij}^I : (S^1)^n \to \mathbb{R}^{2\times2}$, for $i = 1, \ldots, n$, $j = 1, \ldots, n$ are the entries in the inverse matrix from (4.20).

There is no loss of generality in including the indicated projection in the above expression since the inverse necessarily guarantees that if $(\mu_1, \ldots, \mu_n) \in \mathsf{T}_q^*(S^1)^n$ then $(\dot{q}_1, \ldots, \dot{q}_n) \in \mathsf{T}_q(S^1)^n$.

The Hamiltonian function can be written as

$$H(q, \mu) = \sum_{i=1}^{n} \dot{q}_i \cdot \mu_i - \frac{1}{2} \sum_{i=1}^{n} \sum_{j=1}^{n} \dot{q}_i^T m_{ij}(q) \dot{q}_j + U(q)$$

$$= \sum_{i=1}^{n} \dot{q}_i \cdot \mu_i - \frac{1}{2} \sum_{i=1}^{n} \dot{q}_i \cdot \sum_{j=1}^{n} m_{ij}(q) \dot{q}_j + U(q)$$

$$= \sum_{i=1}^{n} \dot{q}_i \cdot \mu_i - \frac{1}{2} \sum_{i=1}^{n} \dot{q}_i \cdot (I_{2\times2} - q_i q_i^T) \sum_{j=1}^{n} m_{ij}(q) \dot{q}_j + U(q)$$

$$= \frac{1}{2} \sum_{i=1}^{n} \dot{q}_i \cdot \mu_i + U(q).$$

The third equality uses the fact that the inner product $\dot{q}_i \cdot \sum_{j=1}^{n} m_{ij}(q) \dot{q}_j$ is not changed by projecting the vector defined by the sum onto the tangent space $\mathsf{T}_{q_i} S^1$.

Thus, the Hamiltonian function can be expressed as a quadratic function of the conjugate momenta

$$H(q, \mu) = \frac{1}{2} \sum_{i=1}^{n} \sum_{j=1}^{n} \mu_i^T m_{ij}^I(q) \mu_j + U(q).$$

Consequently, Hamilton's equations, expressed in terms of (q, μ), are:

$$\dot{q}_i = m_{ii}^I(q)\mu_i + (I_{2\times 2} - q_i q_i^T) \sum_{\substack{j=1 \\ j\neq i}}^{n} m_{ij}^I(q)\mu_j, \qquad i = 1, \ldots, n, \qquad (4.21)$$

$$\dot{\mu}_i = \sum_{j=1}^{n} (\mu_i q_i^T m_{ij}^I(q)\mu_j - \mu_i^T m_{ij}^I(q)\mu_j q_i)$$

$$- (I_{2\times 2} - q_i q_i^T) \frac{1}{2} \frac{\partial}{\partial q_i} \sum_{j=1}^{n} \sum_{k=1}^{n} \mu_j^T m_{jk}^I(q)\mu_k$$

$$- (I_{2\times 2} - q_i q_i^T) \frac{\partial U(q)}{\partial q_i}, \qquad\qquad i = 1, \ldots, n. \qquad (4.22)$$

The first two sets of summation terms on the right-hand side of (4.22) are necessarily quadratic in the conjugate momenta.

Hamilton's equations (4.21) and (4.22) describe the Hamiltonian flow in terms of $(q, \mu) \in \mathsf{T}^*(S^1)^n$ on the cotangent bundle of $(S^1)^n$.

Assuming the Legendre transformation is globally invertible and the potential terms in (4.21) and (4.22) are globally defined on $(\mathbb{R}^2)^n$, then the domain of definition of (4.21) and (4.22) on $\mathsf{T}^*(S^1)^n$ can be extended to $\mathsf{T}^*(\mathbb{R}^2)^n$. This extension is natural in that it defines a Hamiltonian vector field on the cotangent bundle $\mathsf{T}^*(\mathbb{R}^2)^n$. Alternatively, the manifold $\mathsf{T}^*(S^1)^n$ is an invariant manifold of this Hamiltonian vector field on $\mathsf{T}^*(\mathbb{R}^2)^n$ and its restriction to this invariant manifold describes the Hamiltonian flow of (4.21) and (4.22) on $\mathsf{T}^*(S^1)^n$.

4.4.3 Hamilton's Phase Space Variational Principle in Terms of (q, π)

We now present an alternate version of Hamilton's equations using the Legendre transformation defined with respect to the modified Lagrangian function $\tilde{L}(q, \omega)$. The modified Lagrangian function $\tilde{L}(q, \omega)$ is expressed in terms of the angular velocity vector, which satisfies the kinematics (4.1). The Legendre transformation $\omega = (\omega_1, \ldots, \omega_n) \to \pi = (\pi_1, \ldots, \pi_n)$ is defined by

$$\pi_i = \frac{\partial \tilde{L}(q, \omega)}{\partial \omega_i}, \quad i = 1, \ldots, n. \qquad (4.23)$$

The Lagrangian function is assumed to be hyperregular; that is the Legendre transformation, viewed as a map $\mathsf{T}_q(S^1)^n \to \mathsf{T}_q^*(S^1)^n$, is invertible.

The modified Hamiltonian function is given by

$$\tilde{H}(q, \pi) = \sum_{i=1}^{n} \pi_i \omega_i - \tilde{L}(q, \omega),$$

where the right-hand side is expressed in terms of (q, π) by inverting the Legendre transformation (4.23). Since $\pi_i \in \mathbb{R}^1$ and $\omega_i \in \mathbb{R}^1$ are scalars, there is no need to introduce a projection in the definition of the momenta.

Consider the modified action integral of the form,

$$\tilde{\mathfrak{G}} = \int_{t_0}^{t_f} \left\{ \sum_{i=1}^{n} \pi_i \omega_i - \tilde{H}(q, \pi) \right\} dt.$$

The infinitesimal variation of the modified action integral is given by

$$\delta\tilde{\mathfrak{G}} = \sum_{i=1}^{n} \int_{t_0}^{t_f} \left\{ \pi_i \delta\omega_i - \frac{\partial \tilde{H}(q, \pi)}{\partial q_i} \cdot \delta q_i + \left(\omega_i - \frac{\partial \tilde{H}(q, \pi)}{\partial \pi_i} \right) \delta\pi_i \right\} dt.$$

Recall that the infinitesimal variations δq_i and $\delta\omega_i$ can be written as

$$\delta q_i = \gamma_i \mathsf{S} q_i, \quad i = 1, \ldots, n,$$
$$\delta\omega_i = \dot{\gamma}_i, \quad i = 1, \ldots, n,$$

for differentiable curves $\gamma_i : [t_0, t_f] \rightarrow \mathbb{R}^1$, $i = 1, \ldots, n$ satisfying $\gamma_i(t_0) = \gamma_i(t_f) = 0$, $i = 1, \ldots, n$.

4.4.4 Hamilton's Equations in Terms of (q, π)

We now substitute the expressions for δq_i and $\delta\omega_i$ into the expression for the infinitesimal variation of the modified action integral, integrate by parts, and use the boundary conditions on the variations to obtain

$$\delta\tilde{\mathfrak{G}} = \sum_{i=1}^{n} \int_{t_0}^{t_f} \left\{ \left(\omega_i - \frac{\partial \tilde{H}(q, \pi)}{\partial \pi_i} \right) \delta\pi_i + \left(-\dot{\pi}_i - \frac{\partial \tilde{H}(q, \pi)}{\partial q_i} \cdot \mathsf{S} q_i \right) \gamma_i \right\} dt.$$

We invoke Hamilton's phase space variational principle that $\delta\tilde{\mathfrak{G}} = 0$ for all possible differentiable functions $\gamma_i : [t_0, t_f] \rightarrow \mathbb{R}^1$ and $\delta\pi_i : [t_0, t_f] \rightarrow \mathbb{R}^1$ satisfying $\gamma_i(t_0) = \gamma_i(t_f) = 0$, $i = 1, \ldots, n$. The fundamental lemma of the calculus of variations, as in Appendix A, implies that the expressions in each of the above parentheses should be zero. This leads to Hamilton's equations.

Proposition 4.4 *Hamilton's equations for a modified Hamiltonian function* $\tilde{H} : \mathsf{T}^*(\mathsf{S}^1)^n \rightarrow \mathbb{R}^1$ *are*

$$\dot{q}_i = \mathsf{S} q_i \frac{\partial \tilde{H}(q, \pi)}{\partial \pi_i}, \quad i = 1, \ldots, n, \tag{4.24}$$

$$\dot{\pi}_i = q_i^T \mathsf{S} \frac{\partial \tilde{H}(q, \pi)}{\partial q_i}, \quad i = 1, \ldots, n. \tag{4.25}$$

Thus, equations (4.24) and (4.25) describe the Hamiltonian flow in terms of $(q, \pi) \in (S^1 \times \mathbb{R}^1)^n$ as it evolves on the cotangent bundle of $(S^1)^n$.

The following property follows directly from the above formulation of Hamilton's equations on the configuration manifold $(S^1)^n$:

$$
\frac{d\tilde{H}(q,\pi)}{dt} = \sum_{i=1}^{n} \frac{\partial \tilde{H}(q,\pi)}{\partial q_i} \cdot \dot{q}_i + \frac{\partial \tilde{H}(q,\pi)}{\partial \pi_i} \dot{\pi}_i
$$

$$
= \sum_{i=1}^{n} \frac{\partial \tilde{H}(q,\pi)}{\partial \pi_i} \left\{ \left(\frac{\partial \tilde{H}(q,\pi)}{\partial q_i} \right)^T Sq_i + q_i^T S \frac{\partial \tilde{H}(q,\pi)}{\partial q_i} \right\}
$$

$$
= 0.
$$

This formulation exposes the fact that the modified Hamiltonian function is constant along the Hamiltonian flow on the cotangent bundle. We again note that this property does not hold if the modified Hamiltonian function has a nontrivial explicit time dependence.

Now consider the case that the modified Lagrangian function is a quadratic function of the angular velocity vector that was previously introduced in equation (4.13) and given by

$$
\tilde{L}(q,\omega) = \frac{1}{2} \sum_{i=1}^{n} \sum_{j=1}^{n} \omega_i q_i^T m_{ij}(q) q_j \omega_j - U(q). \tag{4.26}
$$

The conjugate momentum is defined by the Legendre transformation

$$
\pi_i = \frac{\partial \tilde{L}(q,\omega)}{\partial \omega_i}
$$

$$
= \sum_{j=1}^{n} q_i^T m_{ij}(q) q_j \omega_j
$$

$$
= m_{ii}(q)\omega_i + \sum_{\substack{j=1 \\ j \neq i}}^{n} m_{ij}(q) S(q_i)^T S(q_j)\omega_j, \quad i = 1, \ldots, n. \tag{4.27}
$$

These algebraic equations, viewed as a linear mapping from $\omega = (\omega_1, \ldots, \omega_n)$ to $\pi = (\pi_1, \ldots, \pi_n)$, can be inverted and expressed in the form

$$
\omega_i = \sum_{j=1}^{n} m_{ij}^I(q)\pi_j, \quad i = 1, \ldots, n,
$$

where $m_{ij}^I : (S^1)^n \to \mathbb{R}^1$, for $i = 1, \ldots, n$, $j = 1, \ldots, n$ are the entries in the matrix inverse obtained from (4.27).

Thus, the modified Hamiltonian function can be written as

$$\tilde{H}(q,\pi) = \frac{1}{2} \sum_{i=1}^{n} \sum_{j=1}^{n} \pi_i m_{ij}^I(q)\pi_j + U(q). \tag{4.28}$$

Hamilton's equations are:

$$\dot{q}_i = Sq_i \left\{ \sum_{j=1}^{n} m_{ij}^I(q)\pi_j \right\}, \qquad\qquad i=1,\ldots,n, \tag{4.29}$$

$$\dot{\pi}_i = q_i^T S \left\{ \frac{1}{2} \frac{\partial}{\partial q_i} \sum_{j=1}^{n} \sum_{k=1}^{n} \pi_j m_{jk}^I(q)\pi_k + \frac{\partial U(q)}{\partial q_i} \right\}, \quad i=1,\ldots,n. \tag{4.30}$$

Hamilton's equations (4.29) and (4.30) describe the Hamiltonian flow as it evolves on the cotangent bundle $T^*(S^1)^n$ of the configuration manifold.

If the Legendre transformation is globally invertible and the potential terms in (4.29) and (4.30) are globally defined on $(\mathbb{R}^2)^n$, then the domain of definition of (4.29) and (4.30) on $T^*(S^1)^n$ can be extended to $T^*(\mathbb{R}^2)^n$. This extension defines a Hamiltonian vector field on the cotangent bundle $T^*(\mathbb{R}^2)^n$. Alternatively, the manifold $T^*(S^1)^n$ is an invariant manifold of this Hamiltonian vector field on $T^*(\mathbb{R}^2)^n$ and its restriction to this invariant manifold describes the Hamiltonian flow of (4.29) and (4.30) on $T^*(S^1)^n$.

4.5 Linear Approximations of Dynamics on $(S^1)^n$

We have developed a geometric form of the Euler–Lagrange equations and Hamilton's equations on the configuration manifold $(S^1)^n$ that provides insight into the global dynamics of the associated Lagrangian vector field on $T(S^1)^n$ or the Hamiltonian vector field on $T^*(S^1)^n$. We emphasize the Lagrangian vector field in the sequel, although a similar development holds for the Hamiltonian vector field.

Let $(q_e, 0) \in T(S^1)^n$ be an equilibrium solution of the Lagrangian vector field. It is possible to construct a linear vector field that approximates the Lagrangian vector field on $T(S^1)^n$, at least locally in a neighborhood of $(q_e, 0) \in T(S^1)^n$. Such linear approximations are closely related to the use of local coordinates on $T(S^1)^n$.

Although our main emphasis throughout this book is on global methods, we do make use of local coordinates as a way of describing a linear vector field that approximates a vector field on a manifold, at least in the neighborhood of an equilibrium solution. This approach is used subsequently in the chapter to study the local flow properties near an equilibrium. As further background for the subsequent development, Appendix B summarizes a linearization procedure for a Lagrangian vector field defined on TS^1.

4.6 Dynamics of Systems on $(S^1)^n$

In this section, several specific examples of Lagrangian and Hamiltonian dynamics that evolve on the configuration manifold $(S^1)^n$ are introduced. In each case, the physical description and assumptions are made clear. The Euler–Lagrange equations are expressed in two different forms and Hamilton's equations are obtained in two different forms. These follow directly from the expression for the Lagrangian function in each example, and the general form of the equations of motion for dynamics on $(S^1)^n$ developed earlier in this chapter. Special features of these equations are described.

4.6.1 Dynamics of a Planar Pendulum

The planar pendulum is perhaps the most commonly studied example in dynamics. The planar pendulum, viewed here as an idealized massless link with mass concentrated at a fixed location on the link, is constrained to rotate in a fixed vertical plane about a fixed frictionless joint axis or pivot under the influence of uniform gravity.

The common treatment of a planar pendulum is based on defining the attitude configuration of the pendulum as a single angle that a body-fixed direction makes with a fixed direction vector in a two-dimensional inertial frame. Although this choice of configuration is natural for small angle motions of the planar pendulum where the configuration space can be viewed as an interval of the real line, this choice of configuration and the resulting pendulum model is problematic in characterizing large rotational motions of the pendulum, due to the need to identify angles that differ by 2π.

Here, we introduce a preferred notion of the attitude of a planar pendulum as the direction vector of the pendulum link, referred to as the attitude of the link. This direction vector is defined with respect to an inertially fixed two-dimensional Euclidean frame. For simplicity, we introduce a two-dimensional frame selected so that the two axes define the plane of rotation of the pendulum, with the first axis in a horizontal direction and the second axis in the vertical direction. The origin of the frame is located at the pivot. We then develop Lagrangian and Hamiltonian dynamics for the planar pendulum on this configuration manifold; this pendulum model is globally defined and allows analysis and computation of extreme pendulum dynamics. A schematic of a planar pendulum is shown in Figure 4.1.

The vector $q \in S^1$ is the attitude of the pendulum link with respect to the two-dimensional frame; it defines the configuration vector of the planar pendulum. Thus, the configuration manifold is S^1. The planar pendulum has one degree of freedom.

Uniform gravity acts on the concentrated mass of the pendulum and g denotes the constant acceleration of gravity. The distance from the pivot to

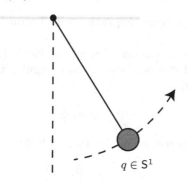

Fig. 4.1 Planar pendulum

the pendulum mass is L and m is the mass of the pendulum. No forces, other than gravity, act on the planar pendulum.

To illustrate the several possible formulations of the equations of motion, we present two equivalent versions of the Euler–Lagrange equations and two equivalent versions of Hamilton's equations.

4.6.1.1 Euler–Lagrange Equations in Terms of $(q, \dot{q})$

The expression for the kinetic energy of the planar pendulum is

$$T(q, \dot{q}) = \frac{1}{2}mL^2 \|\dot{q}\|^2.$$

The gravitational potential energy of the planar pendulum is

$$U(q) = mgLe_2^T q.$$

The vectors e_1, e_2 denote the standard basis vectors in $\mathbb{R}^2$. The Lagrangian $L : TS^1 \to \mathbb{R}^1$ is

$$L(q, \dot{q}) = \frac{1}{2}mL^2 \|\dot{q}\|^2 - mgLe_2^T q.$$

The Euler–Lagrange equations can be expressed in terms of the attitude configuration and its time derivative. Based on (4.9), the Euler–Lagrange equation is

$$mL^2\ddot{q} + mL^2 \|\dot{q}\|^2 q + mgL(I_{2\times 2} - qq^T)e_2 = 0. \tag{4.31}$$

Thus, the Lagrangian dynamics of the planar pendulum are described by the evolution of $(q, \dot{q}) \in TS^1$ on the tangent bundle of S^1.

4.6.1.2 Euler–Lagrange Equations in Terms of (q, ω)

A somewhat more convenient form of the Euler–Lagrange equations is obtained in terms of the pendulum scalar angular velocity that satisfies the rotational kinematics

$$\dot{q} = \omega S q. \tag{4.32}$$

As previously discussed, S denotes the 2×2 skew-symmetric matrix

$$S = \begin{bmatrix} 0 & -1 \\ 1 & 0 \end{bmatrix},$$

which rotates a vector by $\frac{\pi}{2}$ counterclockwise.

The modified Lagrangian function can be expressed in terms of the angular velocity of the planar pendulum as

$$\tilde{L}(q, \omega) = \frac{1}{2} m L^2 \omega^2 - m g L e_2^T q.$$

The Euler–Lagrange equation, expressed in terms of the angular velocity, is obtained from (4.14); the resulting Euler–Lagrange equation is

$$m L^2 \dot{\omega} + m g L e_1^T q = 0. \tag{4.33}$$

Thus, the Lagrangian dynamics of the planar pendulum are described by the kinematics equation (4.32) and the Euler–Lagrange equation (4.33) as they evolve on TS^1. This form of the pendulum dynamics can be identified with the dynamics of $(q, \dot{q}) \in \mathsf{TS}^1$ on the tangent bundle of S^1 via the kinematics equation (4.32).

4.6.1.3 Hamilton's Equations in Terms of (q, μ)

The conjugate momentum

$$\mu = (I_{2 \times 2} - q q^T) \frac{\partial L(q, \dot{q})}{\partial \dot{q}} = m L^2 \dot{q} \in \mathsf{T}_q^* \mathsf{S}^1,$$

is defined using the Legendre transformation. Thus, the Hamiltonian function is

$$H(q, \mu) = \frac{1}{2 m L^2} \|\mu\|^2 + m g L e_2^T q.$$

Hamilton's equations of motion for the planar pendulum are obtained from (4.21) and (4.22) as

$$\dot{q} = \frac{\mu}{mL^2}, \tag{4.34}$$

$$\dot{\mu} = -\frac{1}{mL^2}\|\mu\|^2 q - mgL(I_{2\times2} - qq^T)e_2. \tag{4.35}$$

Thus, the Hamiltonian flow of the planar pendulum is described by (4.34) and (4.35) in terms of the evolution of $(q, \mu) \in T^*S^1$ on the cotangent bundle of S^1.

4.6.1.4 Hamilton's Equations in Terms of (q, π)

We introduce the scalar momentum conjugate to the scalar angular velocity to obtain another form of Hamilton's equations on the cotangent bundle. Define the scalar momentum by

$$\pi = \frac{\partial \tilde{L}(q, \omega)}{\partial \omega} = mL^2\omega,$$

according to the Legendre transformation. Thus, the momentum is conjugate to the angular velocity. The modified Hamiltonian function in this case is

$$\tilde{H}(q, \pi) = \frac{1}{2mL^2}\pi^2 + mgLe_2^T q.$$

Hamilton's equations of motion for the planar pendulum are obtained from (4.24) and (4.25) as

$$\dot{q} = \frac{\pi}{mL^2}Sq, \tag{4.36}$$

$$\dot{\pi} = -mgLe_1^T q. \tag{4.37}$$

Thus, the Hamiltonian flow of the planar pendulum is described by (4.36) and (4.37). These dynamics can be identified with the dynamics of $(q, \mu) \in T^*S^1$ on the cotangent bundle TS^1 via the relationship $\mu = \pi Sq$.

4.6.1.5 Conservation Properties

The Euler–Lagrange equation (4.31) suggests that the dynamics of the planar pendulum do not depend on the mass value of the pendulum but do depend on the gravity to length ratio $\frac{g}{L}$.

It is easy to show that the Hamiltonian of the planar pendulum

$$H = \frac{1}{2}mL^2\|\dot{q}\|^2 + mgLe_2^T q,$$

which coincides with the total energy E in this case, is constant along each solution of the Lagrangian flow of the planar pendulum.

4.6.1.6 Equilibrium Properties

An important feature of the dynamics of the planar pendulum is its equilibrium configurations in S^1. The conditions for an equilibrium are that the time derivatives of the configuration, or equivalently the angular velocity vector or the conjugate momentum, is zero and the equilibrium configuration satisfies:

$$(I_{2\times 2} - qq^T)e_2 = 0,$$

which implies that the time derivatives of the angular velocity or momentum vanish as well. Hence, there are two equilibrium solutions given by $(-e_2, 0) \in \mathsf{TS}^1$ and by $(e_2, 0) \in \mathsf{TS}^1$. A configuration is an equilibrium if and only if it is collinear with the direction of gravity. The equilibrium $(-e_2, 0) \in \mathsf{TS}^1$ is referred to as the hanging equilibrium of the planar pendulum; the equilibrium $(e_2, 0) \in \mathsf{TS}^1$ is the inverted equilibrium of the planar pendulum.

The stability properties of each equilibrium are studied in turn using (4.31). We first linearize (4.31) about the inverted equilibrium $(e_2, 0) \in \mathsf{TS}^1$. The resulting linearized equations can be shown to be

$$mL^2 \ddot{\xi}_1 - mgL\xi_1 = 0,$$

when restricted to the two-dimensional tangent space of TS^1 at $(e_2, 0) \in \mathsf{TS}^1$. This linear dynamics can be used to approximate the local dynamics of the planar pendulum in a neighborhood of the inverted equilibrium $(e_2, 0) \in \mathsf{TS}^1$. The eigenvalues are easily determined to be $+\sqrt{\frac{g}{L}}, -\sqrt{\frac{g}{L}}$. Since there is a positive eigenvalue, the inverted equilibrium $(e_2, 0) \in \mathsf{TS}^1$ is unstable.

We now linearize (4.31) about the hanging equilibrium $(-e_2, 0) \in \mathsf{TS}^1$. The resulting linearized equations can be shown to be

$$mL^2 \ddot{\xi}_1 + mgL\xi_1 = 0,$$

when restricted to the two-dimensional tangent space of TS^1 at $(-e_2, 0) \in \mathsf{TS}^1$. These linear dynamics can be used to approximate the local dynamics of the planar pendulum in a neighborhood of the hanging equilibrium $(-e_2, 0) \in \mathsf{TS}^1$. The eigenvalues are easily determined to be $+j\sqrt{\frac{g}{L}}, -j\sqrt{\frac{g}{L}}$. Since the eigenvalues are purely imaginary no conclusion can be drawn about the stability of the hanging equilibrium.

We now describe a Lyapunov method to demonstrate the stability of the hanging equilibrium. The total energy of the planar pendulum can be shown to have a strict local minimum at the hanging equilibrium on TS^1, and the sublevel sets in TS^1 in a neighborhood of the hanging equilibrium are compact. Furthermore, the time derivative of the total energy along the flow given by (4.31) is zero, that is the total energy does not increase along the flow. According to standard Lyapunov stability results, this guarantees that the hanging equilibrium $(-e_2, 0) \in TS^1$ is stable.

These stability results provide an analytical confirmation of the physically intuitive conclusions about the local flow properties of these two equilibrium solutions.

4.6.2 Dynamics of a Particle Constrained to a Circular Hoop That Rotates with Constant Angular Velocity

Consider a particle, with mass m, that is constrained to move, without friction, on a rigid circular hoop of radius $L > 0$. The circular hoop lies in a plane that rotates about an inertially fixed diameter at a constant angular velocity vector with angular speed $\Omega \in \mathbb{R}^1$. This example can also be interpreted as a planar pendulum whose pivot, and hence the plane of rotation of the pendulum, rotates at a constant angular velocity. Usually, pendulum models include gravity effects, whereas the particle on a circular hoop does not include gravity or other external forces in the current formulation. Consequently, we interpret the subsequent development in terms of a particle constrained to a rotating circular hoop.

Since we intend to obtain global representations for the dynamics of the particle, we do not make use of angles to describe the configuration of the particle. Rather, the two-dimensional plane that contains the circular hoop defines the configuration manifold S^1 embedded in $\mathbb{R}^2$, where this two-dimensional plane rotates in three dimensions about a fixed axis that is a diameter of the hoop. Hence, the motion of the particle is in three dimensions.

The position vector of the particle, with respect to a two-dimensional hoop-fixed Euclidean frame, is denoted by $Lq \in \mathbb{R}^2$, where the direction vector of the position vector $q \in S^1$. We assume the origin of this frame is located at the center of the circular hoop and the second axis of this hoop-fixed frame is in the direction of the angular velocity vector of the hoop. A hoop-fixed three-dimensional Euclidean frame is obtained by a natural extension of the two-dimensional frame defined by the plane of the hoop. In the hoop-fixed frame, the position vector of the particle is given by LCq, where

$$C = \begin{bmatrix} 1 & 0 \\ 0 & 0 \\ 0 & 1 \end{bmatrix}.$$

The origin of the three-dimensional inertial frame is located at the fixed center of the circular hoop; the third inertial axis is in the fixed direction of the angular velocity vector of the hoop in three dimensions. The first two axes of the inertial frame are selected so that the inertial frame is a right-hand

Euclidean frame. We denote the position vector of the particle, with respect to the three-dimensional inertial frame, by $x \in \mathbb{R}^3$. Thus, the position vector $x \in \mathbb{R}^3$ is related to $LCq \in \mathbb{R}^3$ through the rotation matrix from the hoop-fixed frame to the inertial frame.

We develop Lagrangian and Hamiltonian dynamics for the particle on a rotating hoop; this one degree of freedom model is globally defined and allows analysis and computation of extreme dynamics of the particle. A schematic of a particle on a rotating circular hoop is shown in Figure 4.2.

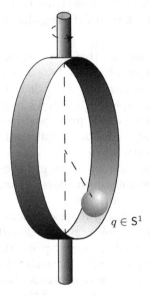

Fig. 4.2 Particle on a rotating circular hoop

To illustrate the several possible formulations of the equations of motion, we present two equivalent versions of the Euler–Lagrange equations and two equivalent versions of Hamilton's equations.

4.6.2.1 Euler–Lagrange Equations in Terms of $(q, \dot{q})$

Assume initially the two axes of the hoop-fixed frame are aligned with the first and third axes of the inertial frame. The position vector of the particle in the three-dimensional inertial frame can be expressed in terms of the configuration vector of the particle by

$$x = L \begin{bmatrix} \cos \Omega t & -\sin \Omega t & 0 \\ \sin \Omega t & \cos \Omega t & 0 \\ 0 & 0 & 1 \end{bmatrix} Cq,$$

where the indicated rotation matrix reflects the constant rotation rate about the third inertial axis.

The velocity vector of the particle in the inertial frame is given by

$$
\dot{x} = \begin{bmatrix} \cos \Omega t & -\sin \Omega t & 0 \\ \sin \Omega t & \cos \Omega t & 0 \\ 0 & 0 & 1 \end{bmatrix} (LC\dot{q} + LS(\Omega e_3)Cq).
$$

Since the relative velocity vector of the particle with respect to the hoop-fixed frame and the velocity vector of the point on the hoop where the particle is located are orthogonal as reflected in this equation, the kinetic energy of the particle on a rotating hoop can be expressed as

$$
T(q, \dot{q}) = \frac{1}{2}m \left\| \dot{x} \right\|^2 = \frac{1}{2}mL^2 \left\{ \left\| \dot{q} \right\|^2 + \Omega^2 (e_1^T q)^2 \right\},
$$

where the vectors e_1, e_2 denote the standard basis vectors in $\mathbb{R}^2$. There is no potential energy so that the Lagrangian $L : \mathsf{TS}^1 \to \mathbb{R}^1$ is

$$
L(q, \dot{q}) = \frac{1}{2}mL^2 \left\{ \left\| \dot{q} \right\|^2 + \Omega^2 (e_1^T q)^2 \right\}.
$$

Note that, although there is no potential, the second term in the Lagrangian, arising from the angular velocity of the circular hoop, depends only on the configuration and is effectively equivalent to a potential.

The Euler–Lagrange equations can be expressed in terms of the configuration vector and its time derivative. Based on (4.9), the Euler–Lagrange equation is

$$
mL^2\ddot{q} + mL^2 \left\| \dot{q} \right\|^2 q - mL^2\Omega^2(I_{2\times 2} - qq^T)e_1 e_1^T q = 0. \tag{4.38}
$$

Thus, the Lagrangian dynamics of the particle on a rotating circular hoop are described by the evolution of $(q, \dot{q}) \in \mathsf{TS}^1$ on the tangent bundle of S^1. The dynamics can also be used to describe the motion of the particle with respect to the inertial frame using the expression for x in terms of q given previously.

4.6.2.2 Euler–Lagrange Equations in Terms of (q, ω)

A convenient form of the Euler–Lagrange equation is obtained in terms of the scalar angular velocity that satisfies the rotational kinematics

$$
\dot{q} = \omega Sq. \tag{4.39}
$$

As previously discussed, S denotes the 2×2 skew-symmetric matrix

$$S = \begin{bmatrix} 0 & -1 \\ 1 & 0 \end{bmatrix},$$

which rotates a vector by $\frac{\pi}{2}$ counterclockwise in a plane.

The modified Lagrangian function can be expressed in terms of the angular velocity of the particle on the hoop as

$$\tilde{L}(q,\omega) = \frac{1}{2} m L^2 \left\{ \omega^2 + \Omega^2 (e_1^T q)^2 \right\}.$$

The Euler–Lagrange equation, expressed in terms of the angular velocity, is obtained from (4.14); the resulting Euler–Lagrange equation is

$$m L^2 \dot{\omega} + m L^2 \Omega^2 q^T e_2 e_1^T q = 0. \tag{4.40}$$

Thus, the Lagrangian dynamics of the particle on a rotating hoop are described by the kinematics equation (4.39) and the Euler–Lagrange equation (4.40) in terms of the evolution of $(q,\omega) \in TS^1$. This can be identified with the dynamics of $(q,\dot{q}) \in TS^1$ via the kinematics equation (4.39). The dynamics can also be used to describe the motion of the particle with respect to the inertial frame.

4.6.2.3 Hamilton's Equations in Terms of (q,μ)

The conjugate momentum

$$\mu = (I_{2\times 2} - qq^T)\frac{\partial L(q,\dot{q})}{\partial \dot{q}} = m L^2 \dot{q},$$

is defined using the Legendre transformation. Thus, $\mu \in T_q^* S^1$ is conjugate to $\dot{q} \in T_q S^1$. The Hamiltonian function is

$$H(q,\mu) = \frac{1}{2m L^2} \|\mu\|^2 - \frac{1}{2} m L^2 \Omega^2 (e_1^T q)^2.$$

Hamilton's equations of motion for the particle on a rotating hoop are obtained from (4.21) and (4.22) as

$$\dot{q} = \frac{\mu}{m L^2}, \tag{4.41}$$

$$\dot{\mu} = -\frac{1}{m L^2} \|\mu\|^2 q + m L^2 \Omega^2 (I_{2\times 2} - qq^T) e_1 e_1^T q. \tag{4.42}$$

Thus, the Hamiltonian flow of the particle on a rotating hoop is described by (4.41) and (4.42) in terms of the evolution of $(q,\mu) \in T^*S^1$ on the cotangent bundle of S^1.

4.6.2.4 Hamilton's Equations in Terms of (q, π)

We introduce the scalar momentum conjugate to the scalar angular velocity to obtain Hamilton's equations on the cotangent bundle. Define the momentum by

$$\pi = \frac{\partial \tilde{L}(q, \omega)}{\partial \omega} = mL^2 \omega,$$

according to the Legendre transformation. Thus, the momentum is conjugate to the angular velocity. The modified Hamiltonian function in this case is

$$\tilde{H}(q, \pi) = \frac{1}{2mL^2} \pi^2 - \frac{1}{2} mL^2 \Omega^2 (e_1^T q)^2.$$

Hamilton's equations of motion for the particle on a rotating hoop are obtained from (4.24) and (4.25), as

$$\dot{q} = \frac{\pi}{mL^2} Sq, \tag{4.43}$$

$$\dot{\pi} = -mL^2 \Omega^2 q^T e_2 e_1^T q. \tag{4.44}$$

Thus, the Hamiltonian flow of the particle on a rotating hoop is described by (4.43) and (4.44) in terms of the evolution of $(q, \pi) \in \mathsf{T}^* \mathsf{S}^1$. This can be identified with the dynamics of $(q, \mu) \in \mathsf{T}^* \mathsf{S}^1$ on the cotangent bundle of S^1 via the relationship, $\mu = \pi Sq$.

4.6.2.5 Conservation Properties

The Euler–Lagrange equation (4.38) suggests that the dynamics of the particle on a rotating hoop do not depend on the mass value or radius of the hoop but do depend on the constant angular velocity Ω of the rotating hoop.

It is easy to show that the Hamiltonian of the particle on a rotating hoop, expressed in terms of $(q, \dot{q}) \in \mathsf{T} \mathsf{S}^1$, is

$$H = \frac{1}{2} mL^2 \left\{ \|\dot{q}\|^2 - \Omega^2 (e_1^T q)^2 \right\},$$

which coincides with the total energy E in this case, and it is constant along each solution of the dynamical flow of the particle on a rotating hoop.

4.6.2.6 Equilibrium Properties

The equilibria of the particle on a rotating hoop are important features of the dynamics. The equilibria correspond to solutions with constant configurations of the particle in the rotating Euclidean frame. The conditions for such a equilibrium are that the time derivative of the configuration, or equivalently the angular velocity or the momentum, is zero and the equilibrium configuration satisfies:

$$-mL^2\Omega^2(I_{2\times2} - qq^T)e_1e_1^Tq = 0,$$

which implies that the time derivatives of the angular velocity or momentum vanish as well. This can be simplified to

$$\begin{bmatrix} -q_2^2q_1 \\ q_1^2q_2 \end{bmatrix} = \begin{bmatrix} 0 \\ 0 \end{bmatrix}.$$

Hence, there are four equilibrium solutions given by $(e_1, 0) \in \mathsf{TS}^1$, $(-e_1, 0) \in \mathsf{TS}^1$, $(e_2, 0) \in \mathsf{TS}^1$, and $(-e_2, 0) \in \mathsf{TS}^1$. A configuration vector is an equilibrium if and only if it is collinear or orthogonal to the angular velocity vector of the circular hoop. In physical terms, the first equilibrium solution corresponds to circular motion $x = (L\cos\Omega t, L\sin\Omega t, 0)^T$ of the particle in the inertial frame, the second equilibrium solution corresponds to the circular motion $x = (-L\cos\Omega t, L\sin\Omega t, 0)^T$ of the particle in the inertial frame, the third equilibrium solution corresponds to the constant motion $x = (0, 0, L)^T$ of the particle in the inertial frame, and the fourth equilibrium solution corresponds to the constant motion $x = (0, 0, -L)^T$ of the particle in the inertial frame; the first two motions are periodic in the inertial frame, but fixed in the rotating frame, so they are sometime referred to as relative equilibrium solutions.

The stability properties of each equilibrium solution are studied in turn using the Euler–Lagrange equation (4.38). We first linearize (4.38) about the equilibrium $(e_1, 0) \in \mathsf{TS}^1$. The resulting linearized equations are

$$\ddot{\xi}_2 + \Omega^2\xi_2 = 0,$$

when restricted to the tangent space of TS^1 at $(e_1, 0) \in \mathsf{TS}^1$. These linear dynamics approximate the local dynamics of the particle on a circular hoop in a neighborhood of the relative equilibrium $(e_1, 0) \in \mathsf{TS}^1$. The eigenvalues of the local dynamics are $j\Omega$, $-j\Omega$. Since the eigenvalues are purely imaginary, no conclusion can be drawn about stability from the linear stability analysis.

It can be shown that the local dynamics of the particle on a circular hoop in a neighborhood of the equilibrium $(-e_1, 0) \in \mathsf{TS}^1$ are also given by

$$\ddot{\xi}_2 + \Omega^2\xi_2 = 0.$$

As before, this corresponds to purely imaginary eigenvalues, and we cannot draw a conclusion about the stability of this equilibrium.

We now linearize (4.38) about the equilibrium $(e_2, 0) \in \mathsf{TS}^1$. This gives the resulting linearized equations

$$\ddot{\xi}_1 - \Omega^2 \xi_1 = 0,$$

when restricted to the tangent space of TS^1 at $(e_2, 0) \in \mathsf{TS}^1$. These linear dynamics approximate the local dynamics of the particle on a rotating hoop in a neighborhood of the equilibrium $(e_2, 0) \in \mathsf{TS}^1$. The eigenvalues are easily determined to be $+\Omega$, $-\Omega$. Since there is a positive eigenvalue, this equilibrium is unstable.

It can be shown that the local dynamics of the particle on a circular hoop in a neighborhood of the equilibrium $(-e_2, 0) \in \mathsf{TS}^1$ are also given by

$$\ddot{\xi}_1 - \Omega^2 \xi_1 = 0,$$

with eigenvalues $+\Omega$, $-\Omega$. Since there is a positive eigenvalue, the equilibrium $(-e_2, 0) \in \mathsf{TS}^1$ is also unstable.

We now describe a Lyapunov method to demonstrate the stability of the equilibrium $(e_1, 0) \in \mathsf{TS}^1$. The total energy of the system can be shown to have a strict local minimum at the equilibrium on TS^1, and the sublevel sets in TS^1 in a neighborhood of the equilibrium $(e_1, 0) \in \mathsf{TS}^1$ are compact. Further the time derivative of the total energy along the flow given by (4.38) is zero, that is the total energy does not increase along the flow. According to standard Lyapunov stability results, this guarantees that the equilibrium $(e_1, 0) \in \mathsf{TS}^1$ is stable. The same argument shows that the equilibrium $(-e_1, 0) \in \mathsf{TS}^1$ is stable.

4.6.3 Dynamics of Two Elastically Connected Planar Pendulums

Two identical pendulums are attached to a common inertially fixed supporting point by a frictionless pivot. Each pendulum is constrained to rotate in a common two-dimensional plane. Gravitational effects are ignored, but a rotational spring exerts an elastic restoring moment on each pendulum with an elastic potential energy that is proportional to $(1 - \cos \theta)$ where θ is the angle between the two pendulums. Each pendulum is assumed to be a thin rigid rod with concentrated mass m located a distance L from the pivot. A schematic of the two elastically connected planar pendulums is shown in Figure 4.3.

Let $q_i \in \mathsf{S}^1$, $i = 1, 2$, denote the attitude of the i-th pendulum in the inertial frame. Consequently, the configuration of the two pendulums is given by $q = (q_1, q_2) \in (\mathsf{S}^1)^2$. Thus, the configuration manifold is $(\mathsf{S}^1)^2$ and the dynamics of the elastically connected planar pendulums have two degrees of freedom.

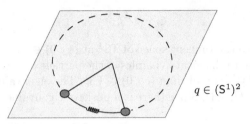

Fig. 4.3 Two elastically connected planar pendulums

4.6.3.1 Euler–Lagrange Equations in Terms of $(q, \dot{q})$

The total kinetic energy is the sum of the kinetic energies of the two planar pendulums

$$T(q, \dot{q}) = \frac{1}{2} mL^2 \|\dot{q}_1\|^2 + \frac{1}{2} mL^2 \|\dot{q}_2\|^2 .$$

The elastic potential energy is assumed to be of the form

$$U(q) = \kappa(1 - \cos\theta),$$

where κ is a positive elastic constant and θ is the angle between the two pendulums. Since $\cos\theta = q_1^T q_2$, the elastic potential energy can be expressed in terms of the configuration as

$$U(q) = \kappa(1 - q_1^T q_2).$$

Note that the elastic potential energy has zero gradient and hence exerts no force when the two pendulum are collinear, that is the angle between the two pendulums is either 0 radians or π radians.

The Lagrangian function $L : \mathsf{T}(\mathsf{S}^1)^2 \to \mathbb{R}^1$ for the two elastically connected planar pendulums is thus given by

$$L(q, \dot{q}) = \frac{1}{2} mL^2 \|\dot{q}_1\|^2 + \frac{1}{2} mL^2 \|\dot{q}_2\|^2 - \kappa(1 - q_1^T q_2).$$

The inertia matrix is constant so that the Euler–Lagrange equations, according to (4.9), are

$$mL^2 \ddot{q}_1 + mL^2 \|\dot{q}_1\|^2 q_1 - \kappa(I_{2\times2} - q_1 q_1^T)q_2 = 0, \qquad (4.45)$$

$$mL^2 \ddot{q}_2 + mL^2 \|\dot{q}_2\|^2 q_2 - \kappa(I_{2\times2} - q_2 q_2^T)q_1 = 0. \qquad (4.46)$$

This version of the Euler–Lagrange equations describes the Lagrangian dynamics of the two elastically connected planar pendulums in terms of $(q, \dot{q}) \in \mathsf{T}(\mathsf{S}^1)^2$ on the tangent bundle of the configuration manifold.

4.6.3.2 Euler–Lagrange Equations in Terms of (q, ω)

An alternative version of the Euler–Lagrange equations of motion for the double spherical pendulum is expressed in terms of the scalar angular velocities of the two pendulum links. The rotational kinematics are given by

$$\dot{q}_1 = \omega_1 S q_1, \tag{4.47}$$

$$\dot{q}_2 = \omega_2 S q_2, \tag{4.48}$$

where $\omega = (\omega_1, \omega_2) \in \mathbb{R}^2$. As previously discussed, S denotes the 2×2 skew-symmetric matrix

$$S = \begin{bmatrix} 0 & -1 \\ 1 & 0 \end{bmatrix},$$

which rotates a vector by $\frac{\pi}{2}$ counterclockwise. Thus, the modified Lagrangian can be expressed in terms of the angular velocities as

$$\tilde{L}(q, \omega) = \frac{1}{2} mL^2 \omega_1{}^2 + \frac{1}{2} mL^2 \omega_2{}^2 - \kappa(1 - q_1^T q_2).$$

Following the prior development in this chapter, the Euler–Lagrange equations for the elastically connected spherical pendulums, expressed in terms of the angular velocities, are obtained from (4.14) as

$$mL^2 \dot{\omega}_1 + \kappa q_1^T S q_2 = 0, \tag{4.49}$$

$$mL^2 \dot{\omega}_2 + \kappa q_2^T S q_1 = 0. \tag{4.50}$$

Equations (4.47), (4.48), (4.49), and (4.50) also describe the Lagrangian dynamics of the two elastically connected planar pendulums in terms of $(q, \omega) \in \mathsf{T}(S^1)^2$. This can be identified with the dynamics of $(q, \dot{q}) \in \mathsf{T}(S^1)^2$ on the tangent bundle of $(S^1)^2$ via the kinematics equations (4.47) and (4.48).

4.6.3.3 Hamilton's Equations in Terms of (q, μ)

Hamilton's equations on the cotangent bundle $\mathsf{T}^*(S^1)^2$ are obtained by defining the conjugate momentum $\mu = (\mu_1, \mu_2) \in \mathsf{T}_q^*(S^1)^2$ according to the Legendre transformation

$$\mu_1 = (I_{2\times 2} - q_1 q_1^T) \frac{\partial L(q, \dot{q})}{\partial \dot{q}_1} = mL^2 \dot{q}_1,$$

$$\mu_2 = (I_{2\times 2} - q_2 q_2^T) \frac{\partial L(q, \dot{q})}{\partial \dot{q}_2} = mL^2 \dot{q}_2.$$

The Hamiltonian function $H : \mathsf{T}^*(S^1)^2 \to \mathbb{R}^1$ is

$$H(q,\mu) = \frac{1}{2}\frac{\|\mu_1\|^2}{mL^2} + \frac{1}{2}\frac{\|\mu_2\|^2}{mL^2} + \kappa(1 - q_1^T q_2).$$

Hamilton's equations of motion, obtained from (4.21) and (4.22), are given by

$$\dot{q}_1 = \frac{\mu_1}{mL^2}, \tag{4.51}$$

$$\dot{q}_2 = \frac{\mu_2}{mL^2}, \tag{4.52}$$

and

$$\dot{\mu}_1 = -\frac{1}{mL^2}\|\mu_1\|^2 q_1 + \kappa(I_{2\times2} - q_1 q_1^T)q_2, \tag{4.53}$$

$$\dot{\mu}_2 = -\frac{1}{mL^2}\|\mu_2\|^2 q_2 + \kappa(I_{2\times2} - q_2 q_2^T)q_1. \tag{4.54}$$

Hamilton's equations (4.51), (4.52), (4.53), and (4.54) describe the Hamiltonian dynamics of the elastically connected planar pendulums in terms of $(q,\mu) \in T^*(S^1)^2$ on the cotangent bundle of $(S^1)^2$.

4.6.3.4 Hamilton's Equations in Terms of (q,π)

A different form of Hamilton's equations on the cotangent bundle $T^*(S^1)^2$ can be obtained in terms of the momentum $\pi = (\pi_1, \pi_2) \in \mathbb{R}^2$ which is conjugate to the angular velocity vector $\omega = (\omega_1, \omega_2) \in \mathbb{R}^2$. The Legendre transformation gives

$$\pi_1 = \frac{\partial \tilde{L}(q,\omega)}{\partial \omega_1} = mL^2\omega_1,$$

$$\pi_2 = \frac{\partial \tilde{L}(q,\omega)}{\partial \omega_2} = mL^2\omega_2.$$

The modified Hamiltonian function is

$$\tilde{H}(q,\pi) = \frac{1}{2}\frac{\pi_1^2}{mL^2} + \frac{1}{2}\frac{\pi_2^2}{mL^2} + \kappa(1 - q_1^T q_2).$$

Thus, Hamilton's equations of motion for the elastically connected planar pendulums, according to (4.24) and (4.25), are

$$\dot{q}_1 = \frac{\pi_1}{mL^2}Sq_1, \tag{4.55}$$

$$\dot{q}_2 = \frac{\pi_2}{mL^2}Sq_2, \tag{4.56}$$

and

$$\dot{\pi}_1 = -\kappa q_1^T S q_2, \tag{4.57}$$

$$\dot{\pi}_2 = -\kappa q_2^T S q_1. \tag{4.58}$$

The Hamiltonian flow of the elastically connected planar pendulums is described by equations (4.55), (4.56), (4.57), and (4.58) in terms of $(q, \pi) \in T(S^1)^2$ on the cotangent bundle. This can be identified with the dynamics of $(q, \mu) \in T^*(S^1)^2$ via the relationship, $\mu_i = \pi_i S q_i$, $i = 1, 2$.

4.6.3.5 Conservation Properties

It is easy to show that the Hamiltonian of the two elastically connected planar pendulums is

$$H = \frac{1}{2} m L^2 \|\dot{q}_1\|^2 + \frac{1}{2} m L^2 \|\dot{q}_2\|^2 + \kappa(1 - q_1^T q_2),$$

which coincides with the total energy E in this case, and that it is constant along each solution of the dynamical flow.

Further, the total scalar angular momentum of the two elastically connected planar pendulums

$$\pi_1 + \pi_2 = m L^2 (\omega_1 + \omega_2)$$

can be shown to be constant along each solution of the dynamical flow. This arises as a consequence of Noether's theorem, due to the invariance of the Lagrangian with respect to the lifted action of rotations about the pivot.

4.6.3.6 Equilibrium Properties

The equilibrium solutions of the elastically connected planar pendulums occur when the time derivative of the configuration vector or equivalently the angular velocity vector or momentum is zero and the equilibrium configuration satisfies:

$$(I_{2 \times 2} - q_1 q_1^T) q_2 = 0,$$
$$(I_{2 \times 2} - q_2 q_2^T) q_1 = 0,$$

which implies that the time derivatives of the angular velocity or momentum vanish as well. This is equivalent to

$$q_1^T S q_2 = 0.$$

Consequently, equilibrium solutions occur when the pendulum links are stationary in an arbitrary attitude but with the angle between them either 0

radians or π radians. There are two disjoint manifolds of equilibrium configurations given by

$$\{(q_1, q_2) \in (S^1)^2 : q_1 = q_2\},$$
$$\{(q_1, q_2) \in (S^1)^2 : q_1 = -q_2\}.$$

In the former case, the pendulum links are said to be *in phase*, while in the
latter case they are said to be *out of phase*.

The stability properties of two typical equilibrium solutions are studied
using (4.45) and (4.46). Without loss of generality, we first develop a linearization of (4.45) and (4.46) at the equilibrium $(e_2, -e_2, 0, 0) \in T(S^1)^2$;
this equilibrium corresponds to the pendulum links being out of phase. We
use the notation $q_i = (q_{i1}, q_{i2}) \in S^1$, $i = 1, 2$ so that the first index of each
double subscript refers to the pendulum index.

The resulting linearized equations about an out of phase equilibrium can
be shown to be

$$mL^2\ddot{\xi}_{11} - \kappa(\xi_{21} + \xi_{11}) = 0,$$
$$mL^2\ddot{\xi}_{21} - \kappa(\xi_{11} + \xi_{21}) = 0,$$

when restricted to the four-dimensional tangent space of $T(S^1)^2$ at the equilibrium $(e_2, -e_2, 0, 0) \in T(S^1)^2$. These linear dynamics approximate the local
dynamics of the elastically connected planar pendulums in a neighborhood of
the out of phase equilibrium $(e_2, -e_2, 0, 0) \in T(S^1)^2$. The eigenvalues for this
out of phase equilibrium are given by $+\sqrt{\frac{2\kappa}{mL^2}}$, $-\sqrt{\frac{2\kappa}{mL^2}}$, 0, 0. Since there is
a positive eigenvalue, this equilibrium with out of phase pendulum links is
unstable. Similarly, it can be shown that all out of phase equilibrium solutions
are unstable.

Without loss of generality, we now develop a linearization of (4.45) and
(4.46) at the equilibrium $(e_2, e_2, 0, 0) \in T(S^1)^2$; this equilibrium corresponds
to the pendulum links being in phase.

The linearized equations can be shown to be

$$mL^2\ddot{\xi}_{11} - \kappa(\xi_{21} - \xi_{11}) = 0,$$
$$mL^2\ddot{\xi}_{21} - \kappa(\xi_{11} - \xi_{21}) = 0,$$

when restricted to the four-dimensional tangent space of $T(S^1)^2$ at the equilibrium $(e_2, e_2, 0, 0) \in T(S^1)^2$. These linear dynamics approximate the local
dynamics of the elastically connected planar pendulums in a neighborhood
of the in phase equilibrium $(e_2, e_2, 0, 0) \in T(S^1)^2$. The eigenvalues for this
in phase equilibrium are $+j\sqrt{\frac{2\kappa}{mL^2}}$, $-j\sqrt{\frac{2\kappa}{mL^2}}$, 0, 0. Since the eigenvalues are
purely imaginary or zero, no conclusion can be made about the stability of
this equilibrium on the basis of this analysis.

We mention that a Lyapunov approach, using the total energy of the elastically connected planar pendulums, does not provide a positive result in this case since the in phase equilibrium is not a strict local minimum of the total energy function on the tangent bundle $T(S^1)^2$. This is because there is a one-parameter family of in phase equilibrium solutions which have the same minimum energy.

Relative equilibrium solutions occur when the angular velocities of the two pendulums are identically equal and constant and the angle between them is either 0 radians or π radians. This corresponds to a constant planar rotation of the pendulum links about their common axis, as a *rigid* system. An analysis of the stability of these relative equilibrium solutions could follow the developments in [70].

4.6.4 Dynamics of a Double Planar Pendulum

The double planar pendulum is an interconnection of two rigid bodies, with each body constrained to rotate in a common fixed vertical plane under uniform gravity. Here, each rigid body is idealized as a massless link of fixed length with a rigidly attached concentrated mass at the end. The first link rotates about a fixed frictionless pivot or joint while the second link is connected by a frictionless joint to the end of the first link.

The configuration vector of a double planar pendulum consists of a pair of direction vectors for the attitudes of the pendulum links, each in the unit circle S^1. Thus, the configuration vector of the double planar pendulum is $q = (q_1, q_2) \in (S^1)^2$. These two attitude vectors are defined with respect to an inertially fixed two-dimensional Euclidean frame, where the two axes define the plane of rotation of the two pendulums, with the second axis in the vertical direction. The origin of the frame is located at the pivot. A schematic of the double planar pendulum is shown in Figure 4.4.

We develop Lagrangian and Hamiltonian dynamics for the double planar pendulum on the configuration manifold $(S^1)^2$; this double planar pendulum model is globally defined and allows analysis and computation of the global dynamics of a double pendulum. The double planar pendulum has two degrees of freedom.

Uniform gravity acts on the mass elements of the two pendulum links; g is the constant acceleration of gravity. The distance from the fixed pivot to the pendulum mass element of the first link is L_1 and m_1 denotes its mass. The distance from the pivot connecting the two links to the mass element of the second link is L_2 and m_2 denotes its mass. No forces, other than gravity, act on the double planar pendulum.

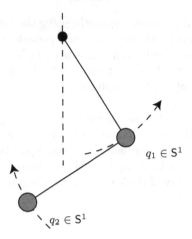

Fig. 4.4 Double planar pendulum

4.6.4.1 Euler–Lagrange Equations in Terms of $(q, \dot{q})$

The expression for the kinetic energy of the double planar pendulum is

$$T(q, \dot{q}) = \frac{1}{2} m_1 \left\| \dot{x}_1 \right\|^2 + \frac{1}{2} m_2 \left\| \dot{x}_2 \right\|^2,$$

where $x_1 = L_1 q_1 \in \mathbb{R}^2$ is the position vector of the mass element of the first link and $x_2 = L_1 q_1 + L_2 q_2 \in \mathbb{R}^2$ is the position vector of the mass element of the second link with respect to the two-dimensional inertial Euclidean frame.

Thus, the expression for the kinetic energy of the double planar pendulum is

$$T(q, \dot{q}) = \frac{1}{2} \begin{bmatrix} \dot{q}_1 \\ \dot{q}_2 \end{bmatrix}^T \begin{bmatrix} (m_1 + m_2) L_1^2 I_{2 \times 2} & m_2 L_1 L_2 I_{2 \times 2} \\ m_2 L_1 L_2 I_{2 \times 2} & m_2 L_2^2 I_{2 \times 2} \end{bmatrix} \begin{bmatrix} \dot{q}_1 \\ \dot{q}_2 \end{bmatrix}.$$

The gravitational potential energy of the double planar pendulum is

$$\begin{aligned} U(q) &= m_1 g e_2^T x_1 + m_2 g (e_2^T x_1 + e_2^T x_2) \\ &= (m_1 + m_2) g L_1 e_2^T q_1 + m_2 g L_2 e_2^T q_2, \end{aligned}$$

so that the Lagrangian function $L : T(S^1)^2 \to \mathbb{R}^1$ is

$$\begin{aligned} L(q, \dot{q}) = \frac{1}{2} \begin{bmatrix} \dot{q}_1 \\ \dot{q}_2 \end{bmatrix}^T &\begin{bmatrix} (m_1 + m_2) L_1^2 I_{2 \times 2} & m_2 L_1 L_2 I_{2 \times 2} \\ m_2 L_1 L_2 I_{2 \times 2} & m_2 L_2^2 I_{2 \times 2} \end{bmatrix} \begin{bmatrix} \dot{q}_1 \\ \dot{q}_2 \end{bmatrix} \\ &- (m_1 + m_2) g L_1 e_2^T q_1 - m_2 g L_2 e_2^T q_2. \end{aligned}$$

The Euler–Lagrange equations, according to (4.9), can be expressed in terms of the attitude configuration and its time derivative as

$$
\begin{bmatrix} (m_1 + m_2)L_1^2 I_{2\times2} & m_2 L_1 L_2 (I_{2\times2} - q_1 q_1^T) \\ m_2 L_1 L_2 (I_{2\times2} - q_2 q_2^T) & m_2 L_2^2 I_{2\times2} \end{bmatrix} \begin{bmatrix} \ddot{q}_1 \\ \ddot{q}_2 \end{bmatrix} +
$$
$$
\begin{bmatrix} (m_1 + m_2)L_1^2 \left\| \dot{q}_1 \right\|^2 q_1 \\ m_2 L_2^2 \left\| \dot{q}_2 \right\|^2 q_2 \end{bmatrix} + \begin{bmatrix} (m_1 + m_2)gL_1(I_{2\times2} - q_1 q_1^T)e_2 \\ m_2 g L_2 (I_{2\times2} - q_2 q_2^T)e_2 \end{bmatrix} = \begin{bmatrix} 0 \\ 0 \end{bmatrix}. \quad (4.59)
$$

These equations of motion (4.59) describe the dynamics of the double planar pendulum in terms of $(q, \dot{q}) \in \mathsf{T}(\mathsf{S}^1)^2$ on the tangent bundle of $(\mathsf{S}^1)^2$.

4.6.4.2 Euler–Lagrange Equations in Terms of (q, ω)

An alternative form of the Euler–Lagrange equations is expressed in terms of the angular velocities $\omega = (\omega_1, \omega_2) \in \mathbb{R}^2$ of the two pendulum links. The rotational kinematics are

$$
\dot{q}_1 = \omega_1 S q_1, \quad (4.60)
$$
$$
\dot{q}_2 = \omega_2 S q_2. \quad (4.61)
$$

As previously discussed, S denotes the 2×2 skew-symmetric matrix

$$
S = \begin{bmatrix} 0 & -1 \\ 1 & 0 \end{bmatrix},
$$

which rotates a vector by $\frac{\pi}{2}$ counterclockwise. Thus, the modified Lagrangian can be expressed in terms of the angular velocities as

$$
\tilde{L}(q, \omega) = \frac{1}{2} \begin{bmatrix} \omega_1 \\ \omega_2 \end{bmatrix}^T \begin{bmatrix} (m_1 + m_2)L_1^2 & m_2 L_1 L_2 q_1^T q_2 \\ m_2 L_1 L_2 q_2^T q_1 & m_2 L_2^2 \end{bmatrix} \begin{bmatrix} \omega_1 \\ \omega_2 \end{bmatrix}
$$
$$
- (m_1 + m_2)gL_1 e_2^T q_1 - m_2 g L_2 e_2^T q_2.
$$

Following the prior results in (4.14), the Euler–Lagrange equations can be expressed in terms of the angular velocities of the two pendulum links as

$$
\begin{bmatrix} (m_1 + m_2)L_1^2 & m_2 L_1 L_2 q_1^T q_2 \\ m_2 L_1 L_2 q_1^T q_2 & m_2 L_2^2 \end{bmatrix} \begin{bmatrix} \dot{\omega}_1 \\ \dot{\omega}_2 \end{bmatrix} + \begin{bmatrix} m_2 L_1 L_2 \omega_2 (\omega_2 - \omega_1) q_1^T S q_2 \\ m_2 L_1 L_2 \omega_1 (\omega_1 - \omega_2) q_2^T S q_1 \end{bmatrix}
$$
$$
+ \begin{bmatrix} (m_1 + m_2)gL_1 e_1^T q_1 \\ m_2 g L_2 e_1^T q_2 \end{bmatrix} = \begin{bmatrix} 0 \\ 0 \end{bmatrix}. \quad (4.62)
$$

Thus, the Lagrangian dynamics of the double planar pendulum are described by the kinematics equations (4.60) and (4.61) and the Euler–Lagrange equations (4.62) in terms of $(q, \omega) \in \mathsf{T}(\mathsf{S}^1)^2$ on the tangent bundle. This can be identified with the dynamics of $(q, \dot{q}) \in \mathsf{T}(S^1)^2$ via the kinematics equations (4.60) and (4.61).

4.6.4.3 Hamilton's Equations in Terms of (q, μ)

Hamilton's equations on the cotangent bundle $\mathsf{T}^*(S^1)^2$ are obtained by defining the conjugate momentum $\mu = (\mu_1, \mu_2) \in \mathsf{T}_q^*(S^1)^2$ according to the Legendre transformation,

$$\begin{bmatrix} \mu_1 \\ \mu_2 \end{bmatrix} = \begin{bmatrix} (I_{2\times2} - q_1 q_1^T) \frac{\partial L(q,\dot{q})}{\partial \dot{q}_1} \\ (I_{2\times2} - q_2 q_2^T) \frac{\partial L(q,\dot{q})}{\partial \dot{q}_2} \end{bmatrix}$$

$$= \begin{bmatrix} (m_1 + m_2)L_1^2 I_{2\times2} & m_2 L_1 L_2 (I_{2\times2} - q_1 q_1^T) \\ m_2 L_1 L_2 (I_{2\times2} - q_2 q_2^T) & m_2 L_2^2 I_{2\times2} \end{bmatrix} \begin{bmatrix} \dot{q}_1 \\ \dot{q}_2 \end{bmatrix}.$$

Define the 4×4 matrix inverse

$$\begin{bmatrix} m_{11}^I & m_{12}^I \\ m_{21}^I & m_{22}^I \end{bmatrix} = \begin{bmatrix} (m_1 + m_2)L_1^2 I_{2\times2} & m_2 L_1 L_2 (I_{2\times2} - q_1 q_1^T) \\ m_2 L_1 L_2 (I_{2\times2} - q_2 q_2^T) & m_2 L_2^2 I_{2\times2} \end{bmatrix}^{-1},$$

so that the Hamiltonian function $H : \mathsf{T}^*(S^1)^2 \to \mathbb{R}^1$ can be expressed as

$$H(q, \mu) = \frac{1}{2} \begin{bmatrix} \mu_1 \\ \mu_2 \end{bmatrix}^T \begin{bmatrix} m_{11}^I & m_{12}^I \\ m_{21}^I & m_{22}^I \end{bmatrix} \begin{bmatrix} \mu_1 \\ \mu_2 \end{bmatrix} + (m_1 + m_2)gL_1 e_2^T q_1 + m_2 g L_2 e_2^T q_2.$$

Thus, the prior results in (4.21) and (4.22) give Hamilton's equations of motion for the double planar pendulum as

$$\begin{bmatrix} \dot{q}_1 \\ \dot{q}_2 \end{bmatrix} = \begin{bmatrix} m_{11}^I & m_{12}^I (I_{2\times2} - q_1 q_1^T) \\ m_{21}^I (I_{2\times2} - q_2 q_2^T) & m_{22}^I \end{bmatrix} \begin{bmatrix} \mu_1 \\ \mu_2 \end{bmatrix}, \qquad (4.63)$$

and

$$\begin{bmatrix} \dot{\mu}_1 \\ \dot{\mu}_2 \end{bmatrix} = \begin{bmatrix} \mu_1 q_1^T m_{11}^I \mu_1 - \mu_1^T m_{11}^I \mu_1 q_1 + \mu_1 q_1^T m_{12}^I \mu_2 - \mu_1^T m_{12}^I \mu_2 q_1 \\ \mu_2 q_2^T m_{21}^I \mu_1 - \mu_2^T m_{21}^I \mu_1 q_2 + \mu_2 q_2^T m_{22}^I \mu_2 - \mu_2^T m_{22}^I \mu_2 q_2 \end{bmatrix}$$
$$- \begin{bmatrix} (m_1 + m_2)gL_1(I_{2\times2} - q_1 q_1^T)e_2 \\ m_2 g L_2 (I_{2\times2} - q_2 q_2^T)e_2 \end{bmatrix},$$

which can be simplified to

$$\begin{bmatrix} \dot{\mu}_1 \\ \dot{\mu}_2 \end{bmatrix} = \begin{bmatrix} \mu_1^T \left((m_{11}^I \mu_1 + m_{12}^I \mu_2)^T - (m_{11}^I \mu_1 + m_{12}^I \mu_2) \right) q_1 \\ \mu_2^T \left((m_{21}^I \mu_1 + m_{22}^I \mu_2)^T - (m_{21}^I \mu_1 + m_{22}^I \mu_2) \right) q_2 \end{bmatrix}$$
$$- \begin{bmatrix} (m_1 + m_2)gL_1(I_{2\times2} - q_1 q_1^T)e_2 \\ m_2 g L_2 (I_{2\times2} - q_2 q_2^T)e_2 \end{bmatrix}. \qquad (4.64)$$

The Hamiltonian flow of the double planar pendulum is described by equations (4.63) and (4.64) in terms of the evolution of $(q, \mu) \in T^*(S^1)^2$ on the cotangent bundle of $(S^1)^2$.

4.6.4.4 Hamilton's Equations in Terms of (q, π)

A different form of Hamilton's equations on the cotangent bundle $T^*(S^1)^2$ can be obtained in terms of the momentum $\pi = (\pi_1, \pi_2) \in \mathbb{R}^2$ that is conjugate to the angular velocity vector $\omega = (\omega_1, \omega_2) \in \mathbb{R}^2$. The Legendre transformation yields

$$
\begin{bmatrix} \pi_1 \\ \pi_2 \end{bmatrix} = \begin{bmatrix} \frac{\partial \tilde{L}(q,\omega)}{\partial \omega_1} \\ \frac{\partial \tilde{L}(q,\omega)}{\partial \omega_2} \end{bmatrix} = \begin{bmatrix} (m_1 + m_2)L_1^2 & m_2 L_1 L_2 q_1^T q_2 \\ m_2 L_1 L_2 q_2^T q_1 & m_2 L_2^2 \end{bmatrix} \begin{bmatrix} \omega_1 \\ \omega_2 \end{bmatrix}.
$$

The 2×2 inverse matrix is

$$
\begin{bmatrix} m_{11}^I & m_{12}^I \\ m_{21}^I & m_{22}^I \end{bmatrix} = \begin{bmatrix} (m_1 + m_2)L_1^2 & m_2 L_1 L_2 q_1^T q_2 \\ m_2 L_1 L_2 q_2^T q_1 & m_2 L_2^2 \end{bmatrix}^{-1},
$$

and the modified Hamiltonian function can be written as

$$
\tilde{H}(q, \pi) = \frac{1}{2} \begin{bmatrix} \pi_1 \\ \pi_2 \end{bmatrix}^T \begin{bmatrix} m_{11}^I & m_{12}^I \\ m_{21}^I & m_{22}^I \end{bmatrix} \begin{bmatrix} \pi_1 \\ \pi_2 \end{bmatrix} + (m_1 + m_2)g L_1 e_2^T q_1 + m_2 g L_2 e_2^T q_2.
$$

Thus, Hamilton's equations of motion for the double planar pendulum, obtained from (4.24) and (4.25), are

$$
\begin{bmatrix} \dot{q}_1 \\ \dot{q}_2 \end{bmatrix} = \begin{bmatrix} (m_{11}^I(q)\pi_1 + m_{12}^I(q)\pi_2)Sq_1 \\ (m_{21}^I(q)\pi_1 + m_{22}^I(q)\pi_2)Sq_2 \end{bmatrix}, \tag{4.65}
$$

and

$$
\begin{bmatrix} \dot{\pi}_1 \\ \dot{\pi}_2 \end{bmatrix} = - \begin{bmatrix} (m_1 + m_2)g L_1 e_1^T q_1 \\ m_2 g L_2 e_1^T q_2 \end{bmatrix}. \tag{4.66}
$$

The Hamiltonian flow for the double planar pendulum is described by equations (4.65) and (4.66) in terms of the evolution of $(q, \pi) \in T^*(S^1)^2$ on the cotangent bundle. This can be identified with the dynamics of $(q, \mu) \in T^*(S^1)^2$ on the cotangent bundle.

4.6.4.5 Conservation Properties

It is easy to show that the Hamiltonian of the double planar pendulum

$$
H = \frac{1}{2} \begin{bmatrix} \dot{q}_1 \\ \dot{q}_2 \end{bmatrix}^T \begin{bmatrix} (m_1 + m_2)L_1^2 I_{2\times2} & m_2 L_1 L_2 I_{2\times2} \\ m_2 L_1 L_2 I_{2\times2} & m_2 L_2^2 I_{2\times2} \end{bmatrix} \begin{bmatrix} \dot{q}_1 \\ \dot{q}_2 \end{bmatrix}
$$
$$
+ (m_1 + m_2)gL_1 e_2^T q_1 + m_2 g L_2 e_2^T q_2,
$$

which coincides with the total energy E in this case, is constant along each solution of the dynamical flow.

4.6.4.6 Equilibrium Properties

The equilibrium solutions of the double planar pendulum are easily determined. They occur when the time derivative of the configuration vector, or equivalently the angular velocity vector, is zero, and the angle that each link makes with the basis vector e_2 is either 0 or π radians, which implies that the time derivative of the angular velocity vanishes as well. Thus, there are four distinct equilibrium solutions: $(-e_2, -e_2, 0, 0)$, $(e_2, -e_2, 0, 0)$, $(-e_2, e_2, 0, 0)$, and $(e_2, e_2, 0, 0)$ in $\mathsf{T}(\mathsf{S}^1)^2$.

The total energy can be shown to have a strict local minimum at the first equilibrium $(-e_2, -e_2, 0, 0) \in \mathsf{T}(\mathsf{S}^1)^2$ on the tangent bundle $\mathsf{T}(\mathsf{S}^1)^2$. Since the time derivative of the total energy is zero along the flow of (4.59), it follows that this equilibrium is stable.

The other three equilibrium solutions are unstable. We demonstrate this fact only for the equilibrium solution $(e_2, e_2, 0, 0) \in \mathsf{T}(\mathsf{S}^1)^2$ by analysis of the properties of the linearized equations about this equilibrium. These linearized equations are restricted to the four-dimensional tangent space of $\mathsf{T}(\mathsf{S}^1)^2$ at $(e_2, e_2, 0, 0) \in \mathsf{T}(\mathsf{S}^1)^2$ to obtain

$$
\begin{bmatrix} (m_1 + m_2)L_1^2 & m_2 L_1 L_2 \\ m_2 L_1 L_2 & m_2 L_2^2 \end{bmatrix} \begin{bmatrix} \ddot{\xi}_1 \\ \ddot{\xi}_2 \end{bmatrix} - \begin{bmatrix} (m_1 + m_2)gL_1 & 0 \\ 0 & m_2 g L_2 \end{bmatrix} \begin{bmatrix} \xi_1 \\ \xi_2 \end{bmatrix} = \begin{bmatrix} 0 \\ 0 \end{bmatrix}.
$$

These linearized equations can be used to characterize the local dynamics of the double planar pendulum in a neighborhood of the equilibrium $(e_2, e_2, 0, 0) \in \mathsf{T}(\mathsf{S}^1)^2$. The characteristic equation can be shown to be

$$
\det \begin{bmatrix} (m_1 + m_2)L_1^2 \lambda^2 - (m_1 + m_2)gL_1 & m_2 L_1 L_2 \lambda^2 \\ m_2 L_1 L_2 \lambda^2 & m_2 L_2^2 \lambda^2 - m_2 g L_2 \end{bmatrix} = 0.
$$

This characteristic equation is quadratic in λ^2 with sign changes in its coefficients. This guarantees that there are one or more eigenvalues with positive real parts. To see this, observe that for characteristic equations that are polynomials in λ^2, the eigenvalues come in $\pm\lambda$ pairs. So, the only way for all the eigenvalues to not have a positive real part is for the eigenvalues to be purely imaginary, which corresponds to λ^2 being either a negative real number or zero. But, this would contradict the fact that coefficients of the characteristic equation change sign. Consequently, this equilibrium solution $(e_2, e_2, 0, 0) \in T(S^1)^2$ is unstable. A similar analysis shows that the other equilibrium solutions, namely $(e_2, -e_2, 0, 0) \in T(S^1)^2$ and $(-e_2, e_2, 0, 0) \in T(S^1)^2$, are also unstable.

4.6.5 Dynamics of a Particle on a Torus

An ideal particle, of mass m, is constrained to move on the surface of a torus in $\mathbb{R}^3$ without friction and under uniform gravity, where g is the acceleration due to gravity. The torus is the surface of revolution generated by revolving a circle about an axis, coplanar with the circle, that does not touch the circle. Without loss of generality, the torus has major radius $R > 0$ which is the distance from the axis of the torus to the center of the circle and minor axis $0 < r < R$ which is the radius of the revolved circle.

An inertially fixed Euclidean frame is constructed so that the center of the circle is located at $(R, 0, 0) \in \mathbb{R}^3$, the circle lies in the plane defined by the first and third axes and the axis of the torus is the third axis of the Euclidean frame, which is assumed to be vertical.

The position vector of the particle on the torus, in the inertial frame, is denoted by $x = (x_1, x_2, x_3) \in \mathbb{R}^3$. The configuration of the particle on the torus can be selected as $q = (q_1, q_2) \in (S^1)^2$, which uniquely determines the position vector of the particle on the torus as is shown subsequently. Thus, the configuration manifold is $(S^1)^2$ and the particle has two degrees of freedom. We describe the Lagrangian dynamics and the Hamiltonian dynamics of the particle on the torus by expressing these dynamics in terms of the evolution on the tangent bundle of the configuration manifold and on the cotangent bundle of the configuration manifold.

A schematic of the particle on a torus is shown in Figure 4.5.

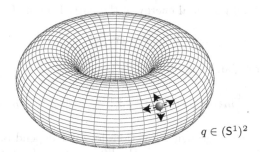

$q \in (S^1)^2$

Fig. 4.5 Particle constrained to a torus

As in Chapter 2, we express the position of the particle $x \in \mathbb{R}^3$ on the torus in terms of the configuration $q = (q_1, q_2) \in (S^1)^2$. The geometry of the torus implies that an arbitrary vector $x \in \mathbb{R}^3$ on the torus can be uniquely decomposed, in the Euclidean frame, into the sum of a vector from the origin to the center of the embedded circle on which x lies and a vector from the center of this embedded circle to x. This decomposition can be expressed as

$$x = \begin{bmatrix} (R + r(e_1^T q_2))(e_1^T q_1) \\ (R + r(e_1^T q_2))(e_2^T q_1) \\ r(e_2^T q_2) \end{bmatrix},$$

where e_1, e_2 denote the standard basis vectors in $\mathbb{R}^2$. This decomposition demonstrates that the position vector of the particle on the torus depends on the configuration of the particle and on the values of the major radius and minor radius of the torus.

4.6.5.1 Euler–Lagrange Equations in Terms of $(q, \dot{q})$

The expression for the kinetic energy of the particle on a torus is

$$T(q, \dot{q}) = \frac{1}{2} m \, \|\dot{x}\|^2 .$$

The velocity vector of the particle on the torus is

$$\dot{x} = \begin{bmatrix} (R + r(e_1^T q_2))e_1^T \\ (R + r(e_1^T q_2))e_2^T \\ 0_{1 \times 2} \end{bmatrix} \dot{q}_1 + \begin{bmatrix} r(e_1^T q_1)e_1^T \\ r(e_2^T q_1)e_1^T \\ r e_2^T \end{bmatrix} \dot{q}_2.$$

It can be shown that the kinetic energy of the particle on a torus can be expressed as

$$T(q, \dot{q}) = \frac{1}{2} m \{ (R + r(e_1^T q_2))^2 \, \|\dot{q}_1\|^2 + r^2 \, \|\dot{q}_2\|^2 \}.$$

The gravitational potential energy of the particle on a torus is

$$U(q) = mgr e_2^T q_2,$$

so that the Lagrangian function $L : \mathsf{T}(S^1)^2 \to \mathbb{R}^1$ is

$$L(q, \dot{q}) = \frac{1}{2} m \{ (R + r(e_1^T q_2))^2 \, \|\dot{q}_1\|^2 + r^2 \, \|\dot{q}_2\|^2 \} - mgr e_2^T q_2.$$

This shows that the Lagrangian function does not depend on $q_1 \in S^1$, that is the Lagrangian is invariant under the lifted action of rotations about the third axis of the Euclidean frame. This invariance implies the existence of an associated conserved quantity, corresponding to the component of angular momentum about the gravity direction.

Following the results in (4.9), the Euler–Lagrange equations can be expressed in terms of the configuration and its time derivative as

$$m(R + r(e_1^T q_2))^2 \ddot{q}_1 + m(R + r(e_1^T q_2))^2 \|\dot{q}_1\|^2 q_1$$
$$+ 2mr(R + r(e_1^T q_2))(e_1^T \dot{q}_2)\dot{q}_1 = 0, \tag{4.67}$$
$$mr^2 \ddot{q}_2 + mr^2 \|\dot{q}_2\|^2 q_2 - mr(R + r(e_1^T q_2))(I_{2\times 2} - q_2 q_2^T) \|\dot{q}_1\|^2 e_1$$
$$+ mgr(I_{2\times 2} - q_2 q_2^T)e_2 = 0. \tag{4.68}$$

The equations of motion (4.67) and (4.68) describe the dynamics of the particle on a torus in terms of $(q, \dot{q}) \in T(S^1)^2$ on the tangent bundle of $(S^1)^2$.

4.6.5.2 Euler–Lagrange Equations in Terms of (q, ω)

An alternative form of the Euler–Lagrange equations expresses the equations in terms of the configuration and the angular velocities $\omega = (\omega_1, \omega_2) \in \mathbb{R}^2$ of the particle on a torus. The rotational kinematics on the configuration manifold $(S^1)^2$ are

$$\dot{q}_1 = \omega_1 S q_1, \tag{4.69}$$
$$\dot{q}_2 = \omega_2 S q_2. \tag{4.70}$$

As previously discussed, S denotes the 2×2 skew-symmetric matrix

$$S = \begin{bmatrix} 0 & -1 \\ 1 & 0 \end{bmatrix},$$

which rotates a vector by $\frac{\pi}{2}$ counterclockwise. Thus, the modified Lagrangian can be expressed in terms of the angular velocities as

$$\tilde{L}(q, \omega) = \frac{1}{2}m\{(R + r(e_1^T q_2))^2 \omega_1^2 + r^2 \omega_2^2\} - mgre_2^T q_2.$$

The modified Lagrangian function is also invariant for rotations about the third axis of the Euclidean frame.

Following the prior results in (4.14), the Euler–Lagrange equations, expressed in terms of the angular velocities of the particle on a torus, are

$$m(R + r(e_1^T q_2))^2 \dot{\omega}_1 - 2mr(R + r(e_1^T q_2))e_2^T q_2 \omega_1 \omega_2 = 0, \tag{4.71}$$
$$mr^2 \dot{\omega}_2 + mr(R + r(e_1^T q_2))e_2^T q_2 \omega_1^2 + mgre_1^T q_2 = 0. \tag{4.72}$$

Thus, the Lagrangian dynamics of the particle on a torus are described by the kinematics equations (4.69) and (4.70) and the Euler–Lagrange equations (4.71) and (4.72) in terms of $(q, \omega) \in T(S^1)^2$. This can be identified with the dynamics of $(q, \dot{q}) \in T(S^1)^2$ via the kinematics equations (4.69) and (4.70).

4.6.5.3 Hamilton's Equations in Terms of (q, μ)

Hamilton's equations on the cotangent bundle $T^*(S^1)^2$ are obtained by defining the conjugate momentum $\mu = (\mu_1, \mu_2) \in T_q^*(S^1)^2$ according to the Legendre transformation,

$$\mu_1 = (I_{2\times2} - q_1 q_1^T) \frac{\partial L(q, \dot{q})}{\partial \dot{q}_1} = m(R + r(e_1^T q_2))^2 \dot{q}_1,$$

$$\mu_2 = (I_{2\times2} - q_2 q_2^T) \frac{\partial L(q, \dot{q})}{\partial \dot{q}_2} = mr^2 \dot{q}_2.$$

The Hamiltonian function $H : T^*(S^1)^2 \to \mathbb{R}^1$ can be expressed as

$$H(q, \mu) = \frac{1}{2} \left\{ \frac{\|\mu_1\|^2}{(m(R + r(e_1^T q_2))^2)} + \frac{\|\mu_2\|^2}{mr^2} \right\} + mgre_2^T q_2.$$

Thus, the prior results in (4.21) and (4.22) give Hamilton's equations of motion for the particle on a torus as

$$\dot{q}_1 = \frac{\mu_1}{m(R + r(e_1^T q_2))^2}, \tag{4.73}$$

$$\dot{q}_2 = \frac{\mu_2}{mr^2}, \tag{4.74}$$

and

$$\dot{\mu}_1 = -\frac{\|\mu_1\|^2}{m(R + r(e_1^T q_2))^2} q_1, \tag{4.75}$$

$$\dot{\mu}_2 = \frac{r \|\mu_1\|^2}{m(R + r(e_1^T q_2))^3} (I_{2\times2} - q_2 q_2^T) e_1 - \frac{\|\mu_2\|^2}{mr^2} q_2$$
$$- mgr(I_{2\times2} - q_2 q_2^T) e_2. \tag{4.76}$$

The Hamiltonian flow of the particle on a torus is described by equations (4.73), (4.74), (4.75), and (4.76) in terms of the evolution of $(q, \mu) \in T^*(S^1)^2$ on the cotangent bundle of $(S^1)^2$.

4.6.5.4 Hamilton's Equations in Terms of (q, π)

A different form of Hamilton's equations on the cotangent bundle $T^*(S^1)^2$ can be obtained by considering the momentum $\pi = (\pi_1, \pi_2) \in \mathbb{R}^2$ that is conjugate to the angular velocity vector $\omega = (\omega_1, \omega_2) \in \mathbb{R}^2$. The Legendre transformation gives

$$\pi_1 = \frac{\partial \tilde{L}(q, \omega)}{\partial \omega_1} = m(R + r(e_1^T q_2))^2 \omega_1,$$

$$\pi_2 = \frac{\partial \tilde{L}(q, \omega)}{\partial \omega_2} = mr^2 \omega_2.$$

The modified Hamiltonian function can be written as

$$\tilde{H}(q, \pi) = \frac{1}{2} \left\{ \frac{\pi_1^2}{m(R + r(e_1^T q_2))^2} + \frac{\pi_2^2}{mr^2} \right\} + mgre_2^T q_2.$$

Following the results in (4.24) and (4.25), Hamilton's equations of motion for a particle on a torus consist of the kinematics equations

$$\dot{q}_1 = \frac{\pi_1 S q_1}{m(R + r(e_1^T q_2))^2}, \tag{4.77}$$

$$\dot{q}_2 = \frac{\pi_2 S q_2}{mr^2}, \tag{4.78}$$

and the dynamics equations

$$\dot{\pi}_1 = 0, \tag{4.79}$$

$$\dot{\pi}_2 = -\frac{r\pi_1^2}{m(R + r(e_1^T q_2))^3} e_2^T q_2 - mgre_1^T q_2. \tag{4.80}$$

The Hamiltonian flow for the particle on a torus is described by equations (4.77), (4.78), (4.79), and (4.80) in terms of the evolution of $(q, \pi) \in T^*(S^1)^2$ on the cotangent bundle. This can be identified with the dynamics of $(q, \mu) \in T^*(S^1)^2$ via the relationship, $\mu_i = \pi_i S q_i$, $i = 1, 2$.

4.6.5.5 Conservation Properties

It is easy to show that the Hamiltonian of the particle on a torus

$$H = \frac{1}{2} m\{(R + r(e_1^T q_2))^2 \|\dot{q}_1\|^2 + r^2 \|\dot{q}_2\|^2\} + mgre_2^T q_2$$

which coincides with the total energy E in this case is constant along each solution of the dynamical flow.

Further, it follows that the scalar conjugate momentum

$$\pi_1 = m(R + r(e_1^T q_2))^2 \omega_1$$

is constant along each solution of the dynamical flow, which is a consequence of Noether's theorem and the rotational symmetry about the direction of gravity.

4.6.5.6 Equilibrium Properties

The equilibrium solutions of the dynamics of the particle on a torus are easily determined. They occur when the time derivatives of both the configuration vector and the angular velocity vector vanish. The equilibrium configurations $(q_1, q_2) \in (\mathsf{S}^1)^2$ are those for which $q_2 \in \mathsf{S}^1$ is either $e_2 \in \mathbb{R}^2$ or $-e_2 \in \mathsf{S}^1$, that is the attitude vector $q_2 \in \mathsf{S}^1$, characterizing the particle position vector on the torus, is either opposite to the direction of gravity or in the direction of gravity. Since the configuration component $q_1 \in \mathsf{S}^1$ is arbitrary, there are two distinct equilibrium manifolds

$$\left\{ (q_1, q_2) \in (\mathsf{S}^1)^2 : q_2 = e_2 \right\},$$
$$\left\{ (q_1, q_2) \in (\mathsf{S}^1)^2 : q_2 = -e_2 \right\}.$$

An equilibrium in the first manifold is referred to as a *top* equilibrium; an equilibrium in the second manifold is referred to as a *bottom* equilibrium.

We use the notation $q_i = (q_{i1}, q_{i2}) \in \mathsf{S}^1$, $i = 1, 2$ so that the first index of each double subscript refers to the configuration vector index.

Linearization of the differential equations (4.67) and (4.68) at the top equilibrium $(e_2, e_2, 0, 0) \in \mathsf{T}(\mathsf{S}^1)^2$, expressed in local coordinates, can be shown to be

$$mR^2 \ddot{\xi}_{11} = 0,$$
$$mr^2 \ddot{\xi}_{21} - mgr\xi_{21} = 0,$$

which are defined on the four-dimensional tangent space of $\mathsf{T}(\mathsf{S}^1)^2$ at the equilibrium point $(e_2, e_2, 0, 0) \in \mathsf{T}(\mathsf{S}^1)^2$. These linearized differential equations can be used to characterize the local dynamics of the particle on a torus in a neighborhood of the top equilibrium $(e_2, e_2, 0, 0) \in \mathsf{T}(\mathsf{S}^1)^2$. The eigenvalues are $+\sqrt{\frac{g}{r}}, -\sqrt{\frac{g}{r}}, 0, 0$. Since there is a positive eigenvalue, this equilibrium is unstable. Similarly, any equilibrium on the top of the torus is unstable.

Without loss of generality, we now develop a linearization of (4.67) and (4.68) at the bottom equilibrium $(e_2, -e_2, 0, 0) \in \mathsf{T}(\mathsf{S}^1)^2$. The linearized differential equations can be shown to be

$$mR^2 \ddot{\xi}_{11} = 0,$$
$$mr^2 \ddot{\xi}_{21} + mgr\xi_{21} = 0,$$

which are defined on the four-dimensional tangent space of $\mathsf{T}(\mathsf{S}^1)^2$ at the equilibrium point $(e_2, -e_2, 0, 0) \in \mathsf{T}(\mathsf{S}^1)^2$. These linearized differential equations can be used to characterize the local dynamics of the particle on a torus in a neighborhood of the bottom equilibrium $(e_2, -e_2, 0, 0) \in \mathsf{T}(\mathsf{S}^1)^2$. The eigenvalues are $+j\sqrt{\frac{g}{r}}, -j\sqrt{\frac{g}{r}}, 0, 0$. Since the eigenvalues are purely imaginary or zero, no conclusion can be made about the stability of this equilibrium on the basis of this analysis.

We mention that a Lyapunov approach using the total energy does not provide a positive result in this case since the equilibrium is not a strict local minimum of the total energy function on the tangent bundle $T(S^1)^2$.

4.6.6 Dynamics of a Furuta Pendulum

A Furuta pendulum [43] is a serial connection of two rigid links, where the first link is constrained by an inertially fixed pivot to rotate in a horizontal plane and the second link is constrained to rotate in a vertical plane orthogonal to the first link, under the influence of gravity.

Assume the first link is a thin, rigid body idealized as a massless link; one end is pinned to an inertial frame so that the first link is constrained to rotate in a horizontal plane; the length of the massless link is L_1 and a mass m_1 is concentrated at the end of the first link. The second link is also a thin, rigid body idealized as a massless link; the length of the massless link is L_2 and a mass m_2 is concentrated at the end of the second link. The second link is constrained to rotate in a vertical plane that is always orthogonal to the first link.

Introduce an inertial Euclidean frame for $\mathbb{R}^3$ where the origin is located at the inertially fixed pivot of the first link; the first two axes are assumed to be horizontal while the third axis is assumed to be vertical.

The following matrix notation is used subsequently:

$$C = \begin{bmatrix} 1 & 0 \\ 0 & 1 \\ 0 & 0 \end{bmatrix}, \quad D = \begin{bmatrix} 0 & 0 \\ 0 & 0 \\ 0 & 1 \end{bmatrix}, \quad S = \begin{bmatrix} 0 & -1 \\ 1 & 0 \end{bmatrix}.$$

A schematic of a Furuta pendulum is shown in Figure 4.6.

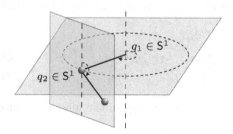

Fig. 4.6 Furuta pendulum

Since the mass at the end of the first link moves in a horizontal plane, the position vector $x_1 \in \mathbb{R}^3$ of the first mass element in the inertial frame is

$$x_1 = L_1 C q_1,$$

where $q_1 = (q_{11}, q_{12}) \in \mathsf{S}^1$ is the attitude vector of the first link in the horizontal plane within which the first mass element moves. This expression enforces the constraint that the first link rotates in a fixed horizontal plane.

The position vector $x_2 \in \mathbb{R}^3$ of the second mass element in the inertial frame is most easily described in terms of the basis for $\mathbb{R}^3$:

$$Cq_1, \quad CSq_1, \quad e_3.$$

This can be shown to be an orthonormal basis set for $\mathbb{R}^3$. This basis has the property that the direction of the first link is always along the first basis vector, while the direction of the second link lies in the span of the second and third basis vectors.

The position vector $x_2 \in \mathbb{R}^3$ of the second mass element can be expressed as

$$x_2 = x_1 + L_2(q_{21} C q_1 + q_{22} e_3),$$

where $q_2 = (q_{21}, q_{22}) \in \mathsf{S}^1$ is the attitude vector of the second link in its instantaneous plane of rotation. Thus,

$$
\begin{aligned}
x_2 &= L_1 C q_1 + L_2(e_1^T q_2 C S q_1 + D q_2) \\
&= L_1 C q_1 + L_2(C S q_1 e_1^T + D) q_2.
\end{aligned}
$$

Since the position vectors for the two mass elements that define the Furuta pendulum can be expressed in terms of $q = (q_1, q_2) \in (\mathsf{S}^1)^2$, the configuration manifold for the Furuta pendulum is $(\mathsf{S}^1)^2$; the dimension of the configuration manifold is two so there are two degrees of freedom.

4.6.6.1 Euler–Lagrange Equations in Terms of $(q, \dot{q})$

The velocity vectors of the two mass elements in the inertial frame are

$$
\begin{aligned}
\dot{x}_1 &= L_1 C \dot{q}_1, \\
\dot{x}_2 &= L_1 C \dot{q}_1 + L_2(e_1^T \dot{q}_2 C S q_1 + e_1^T q_2 C S \dot{q}_1 + D \dot{q}_2) \\
&= (L_1 C + L_2 e_1^T q_2 C S) \dot{q}_1 + L_2(C S q_1 e_1^T + D) \dot{q}_2.
\end{aligned}
$$

The expression for the kinetic energy of the mass element of the first link is

$$T_1(q, \dot{q}) = \frac{1}{2} m_1 \|\dot{x}_1\|^2 = \frac{1}{2} m_1 L_1^2 \|\dot{q}_1\|^2.$$

The kinetic energy of the mass element of the second link is

$$T_2(q, \dot{q}) = \frac{1}{2} m_2 \|\dot{x}_2\|^2$$

$$= \frac{1}{2} m_2 \left\| (L_1 C + L_2 e_1^T q_2 CS)\dot{q}_1 + L_2(CSq_1 e_1^T + D)\dot{q}_2 \right\|^2.$$

Thus, the expression for the kinetic energy of the Furuta pendulum is

$$T(q, \dot{q}) = \frac{1}{2} m_1 L_1^2 \|\dot{q}_1\|^2$$

$$+ \frac{1}{2} m_2 \left\| (L_1 C + L_2 e_1^T q_2 CS)\dot{q}_1 + L_2(CSq_1 e_1^T + D)\dot{q}_2 \right\|^2$$

$$= \frac{1}{2} \dot{q}_1^T m_{11} \dot{q}_1 + \dot{q}_1^T m_{12} \dot{q}_2 + \frac{1}{2} \dot{q}_2^T m_{22} \dot{q}_2,$$

where the inertia terms are

$$m_{11} = (m_1 + m_2)L_1^2 + m_2 L_2^2 (e_1^T q_2)^2,$$

$$m_{12} = m_2 L_2 \left\{ L_1 S + L_2 e_1^T q_2 I_{2\times 2} \right\} q_1 e_1^T,$$

$$m_{22} = m_2 L_2^2.$$

The gravitational potential energy of the Furuta pendulum is

$$U(q) = m_2 g L_2 e_3^T x_2 = m_2 g L_2 e_3^T D q_2,$$

where we have used the facts that $e_3^T x_1 = 0$ and $e_3^T x_2 = L_2 e_3^T D q_2$.

The Lagrangian function $L : \mathsf{T}(\mathsf{S}^1)^2 \to \mathbb{R}^1$ for the Furuta pendulum is given by

$$L(q, \dot{q}) = \frac{1}{2} \dot{q}_1^T m_{11} \dot{q}_1 + \dot{q}_1^T m_{12} \dot{q}_2 + \frac{1}{2} \dot{q}_2^T m_{22} \dot{q}_2 - m_2 g L_2 e_3^T D q_2.$$

Consequently, the Euler–Lagrange equations are given by

$$m_{11} \ddot{q}_1 + (I_{2\times 2} - q_1 q_1^T) m_{12} \ddot{q}_2 + m_{11} \|\dot{q}_1\|^2 q_1$$
$$+ (I_{2\times 2} - q_1 q_1^T) F_1(q, \dot{q}) = 0, \tag{4.81}$$

$$(I_{2\times 2} - q_2 q_2^T) m_{12}^T \ddot{q}_1 + m_{22} \ddot{q}_2 + m_{22} \|\dot{q}_2\|^2 q_2 + (I_{2\times 2} - q_2 q_2^T) F_2(q, \dot{q})$$
$$+ m_2 g L_2 (I_{2\times 2} - q_2 q_2^T) e_2 = 0. \tag{4.82}$$

Here, the terms F_1, F_2 are quadratic in the time derivative of the configuration vector. These vector functions are

$$F_1(q, \dot{q}) = \dot{m}_{11} \dot{q}_1 + \dot{m}_{12} \dot{q}_2 - \frac{\partial T(q, \dot{q})}{\partial q_1},$$

$$F_2(q, \dot{q}) = \dot{m}_{12}^T \dot{q}_1 - \frac{\partial T(q, \dot{q})}{\partial q_2}.$$

These Euler–Lagrange equations (4.81) and (4.82) describe the global evolution of the dynamics of the Furuta pendulum on the tangent bundle $\mathsf{T}(\mathsf{S}^1)^2$.

4.6.6.2 Euler–Lagrange Equations in Terms of (q, ω)

An alternative form of the Euler–Lagrange equations is obtained in terms of the configuration and the angular velocities $\omega = (\omega_1, \omega_2) \in \mathbb{R}^2$ of the two links. The rotational kinematics on the configuration manifold $(\mathsf{S}^1)^2$ are

$$\dot{q}_1 = \omega_1 S q_1, \tag{4.83}$$

$$\dot{q}_2 = \omega_2 S q_2. \tag{4.84}$$

Thus, the modified Lagrangian can be expressed in terms of the angular velocity vector as

$$\tilde{L}(q, \omega) = \frac{1}{2} m_{11} \omega_1^2 + q_1^T S^T m_{12} q_2 S q_2 \omega_1 \omega_2$$
$$+ \frac{1}{2} m_{22} \omega_2^2 - m_2 g L_2 e_3^T D q_2.$$

The resulting Euler–Lagrange equations consist of the kinematics equations (4.83) and (4.84) and

$$m_{11} \dot{\omega}_1 + q_1^T m_{12} q_2 \dot{\omega}_2 + q_1^T S m_{12} q_2 \|\omega_2\|^2 - q_1^T S \left\{ m_{12} q_2 \omega_1 \omega_2 + F_1(q, \dot{q}) \right\} = 0, \tag{4.85}$$

$$q_2^T m_{12}^T q_1 \dot{\omega}_1 + m_{22} \dot{\omega}_2 - q_2^T S \left\{ m_{12}^T q_1 \omega_1 \omega_2 + F_2(q, \dot{q}) + m_2 g L_2 e_2 \right\} = 0. \tag{4.86}$$

These Euler–Lagrange equations describe the global evolution of the dynamics of the Furuta pendulum on the tangent bundle $\mathsf{T}(\mathsf{S}^1)^2$.

4.6.6.3 Hamilton's Equations in Terms of (q, μ)

Hamilton's equations on the cotangent bundle $\mathsf{T}^*(\mathsf{S}^1)^2$ are obtained by defining the conjugate momentum $\mu = (\mu_1, \mu_2) \in \mathsf{T}_q^*(\mathsf{S}^1)^2$ according to the Legendre transformation,

$$\mu_1 = (I_{2 \times 2} - q_1 q_1^T) \frac{\partial L(q, \dot{q})}{\partial \dot{q}_1} = m_{11} \dot{q}_1 + (I_{2 \times 2} - q_1 q_1^T) m_{12} \dot{q}_2,$$

$$\mu_2 = (I_{2 \times 2} - q_2 q_2^T) \frac{\partial L(q, \dot{q})}{\partial \dot{q}_2} = (I_{2 \times 2} - q_2 q_2^T) m_{12}^T \dot{q}_1 + m_{22} \dot{q}_2.$$

The Hamiltonian function $H : \mathsf{T}^*(\mathsf{S}^1)^2 \to \mathbb{R}^1$ can be expressed as

$$H(q, \mu) = \frac{1}{2} \mu_1^T m_{11}^I \mu_1 + \mu_1^T m_{12}^I{}^T \mu_2 + \frac{1}{2} \mu_2^T m_{22}^I \mu_2 + m_2 g L_2 e_3^T D q_2,$$

where

$$\begin{bmatrix} m_{11}^I & m_{12}^I \\ m_{12}^{I\ T} & m_{22}^I \end{bmatrix} = \begin{bmatrix} m_{11}I_{2\times2} & m_{12} \\ m_{12}^T & m_{22}I_{2\times2} \end{bmatrix}^{-1}.$$

The resulting Hamilton's equations are given by

$$\dot{q}_1 = m_{11}^I\mu_1 + m_{12}^I\mu_2, \tag{4.87}$$

$$\dot{q}_2 = m_{12}^{I\ T}\mu_1 + m_{22}^I\mu_2, \tag{4.88}$$

$$\dot{\mu}_1 = \mu_1 q_1^T m_{12}^I\mu_2 - \mu_1^T m_{12}^I\mu_2 q_1 - (I_{2\times2} - q_1q_1^T)\frac{\partial\mu_1^T m_{12}^I\mu_2}{\partial q_1}, \tag{4.89}$$

$$\dot{\mu}_2 = \mu_2 q_2^T m_{12}^{I\ T}\mu_1 - \mu_2^T m_{12}^{I\ T}\mu_1 q_2$$
$$- (I_{2\times2} - q_2q_2^T)\left\{ \frac{\partial}{\partial q_2}\left(\frac{1}{2}\mu_1^T m_{11}^I\mu_1 + \mu_1^T m_{12}^I\mu_2\right) - m_2gL_2e_2 \right\}. \tag{4.90}$$

The Hamiltonian flow of the Furuta pendulum is described by equations (4.87), (4.88), (4.89), and (4.90) in terms of the evolution of $(q,\mu) \in \mathsf{T}^*(\mathsf{S}^1)^2$ on the cotangent bundle of $(\mathsf{S}^1)^2$.

4.6.6.4 Hamilton's Equations in Terms of (q, π)

A different form of Hamilton's equations on the cotangent bundle $\mathsf{T}^*(\mathsf{S}^1)^2$ can be obtained by defining the momentum $\pi = (\pi_1, \pi_2) \in \mathbb{R}^2$ that is conjugate to the angular velocity vector $\omega = (\omega_1, \omega_2) \in \mathbb{R}^2$. The Legendre transformation gives

$$\pi_1 = \frac{\partial\tilde{L}(q,\omega)}{\partial\omega_1} = m_{11}\omega_1 + q_1^T Sm_{12}q_2\omega_2,$$

$$\pi_2 = \frac{\partial\tilde{L}(q,\omega)}{\partial\omega_2} = q_2^T m_{12}^T q_1\omega_1 + m_{22}\omega_2.$$

The modified Hamiltonian function can be written as

$$H(q,\pi) = \frac{1}{2}m_{11}^I\pi_1^2 + q_1^T m_{12}^{I\ T}q_2\pi_1\pi_2 + \frac{1}{2}m_{22}^I\pi_2^2 + m_2gL_2e_3^T Dq_2,$$

where

$$\begin{bmatrix} m_{11}^I & m_{12}^I \\ m_{12}^{I\ T} & m_{22}^I \end{bmatrix} = \begin{bmatrix} m_{11}I_{2\times2} & m_{12} \\ m_{12}^T & m_{22}I_{2\times2} \end{bmatrix}^{-1}.$$

The resulting Hamilton's equations are given by

$$\dot{q}_1 = Sq_1\left(m^I_{11}\pi_1 + m^I_{12}\pi_2\right), \tag{4.91}$$

$$\dot{q}_2 = Sq_2\left(m^I_{12}{}^T\pi_1 + m^I_{22}\pi_2\right), \tag{4.92}$$

$$\dot{\pi}_1 = q^T_1 S\frac{\partial}{\partial q_1}\left(\frac{1}{2}m^I_{11}\pi^2_1 + q^T_1 m^I_{12}{}^T q_2\pi_1\pi_2 + \frac{1}{2}m^I_{22}\pi^2_2\right), \tag{4.93}$$

$$\dot{\pi}_2 = q^T_2 S\frac{\partial}{\partial q_2}\left(\frac{1}{2}m^I_{11}\pi^2_1 + q^T_1 m^I_{12}{}^T q_2\pi_1\pi_2 + \frac{1}{2}m^I_{22}\pi^2_2\right)$$
$$+ q^T_2 S m_2 g L_2 e_2. \tag{4.94}$$

The Hamiltonian flow of the Furuta pendulum is described by equations (4.91), (4.92), (4.93), and (4.94) in terms of the evolution of $(q, \pi) \in \mathsf{T}^*(\mathsf{S}^1)^2$.

4.6.6.5 Conservation Properties

It is easy to show that the Hamiltonian of the Furuta pendulum

$$H = \frac{1}{2}\dot{q}^T_1 m_{11}\dot{q}_1 + \dot{q}^T_1 m_{12}\dot{q}_2 + \frac{1}{2}\dot{q}^T_2 m_{22}\dot{q}_2 - mgL_2 e^T_3 D q_2$$

which coincides with the total energy E in this case is constant along each solution of the dynamical flow.

In addition, the vertical component of angular momentum

$$e^T_3\left(m_1 x_1 \times \dot{x}_1 + m_2 x_2 \times \dot{x}_2\right)$$

is preserved as a consequence of Noether's theorem, due to the invariance of the Lagrangian with respect to the lifted action of rotations about the gravity direction.

4.6.6.6 Equilibrium Properties

The equilibrium solutions of the Furuta pendulum are easily determined. They occur when the time derivative of the configuration vector is zero and the attitude of the second link satisfies $(I_{2\times 2} - q_2 q^T_2)e_2 = 0$, which implies that the time derivative of the angular momentum vanishes as well; the attitude of the first link is arbitrary. Consequently, there are two distinct equilibrium manifolds in $(\mathsf{S}^1)^2$ given by

$$\left\{(q_1, q_2) \in (\mathsf{S}^1)^2 : q_2 = e_2\right\},$$

referred to as the inverted equilibrium manifold, and

$$\left\{(q_1, q_2) \in (\mathsf{S}^1)^2 : q_2 = -e_2\right\},$$

referred to as the hanging equilibrium manifold. An equilibrium in the first manifold is referred to as an *inverted* equilibrium; an equilibrium in the second manifold is referred to as a *hanging* equilibrium.

Linearization of the differential equations (4.81) and (4.82) at an inverted equilibrium $(e_2, e_2, 0, 0) \in \mathsf{T}(S^1)^2$, expressed in local coordinates, can be shown to be

$$(m_1 + m_2)L_1^2 \ddot{\xi}_{11} = 0,$$

$$m_2 L_2^2 \ddot{\xi}_{21} - m_2 g L_2 \xi_{21} = 0,$$

which are defined on the four-dimensional tangent space of $\mathsf{T}(S^1)^2$ at the equilibrium $(e_2, e_2, 0, 0) \in \mathsf{T}(S^1)^2$. These linearized differential equations can be used to characterize the local dynamics of the Furuta pendulum in a neighborhood of the inverted equilibrium $(e_2, e_2, 0, 0) \in \mathsf{T}(S^1)^2$. The eigenvalues are $+\sqrt{\frac{g}{L_2}}$, $-\sqrt{\frac{g}{L_2}}$, 0, 0. Since there is a positive eigenvalue, this equilibrium is unstable. Similarly, any inverted equilibrium is unstable.

We now develop a linearization of (4.81) and (4.82) at the hanging equilibrium $(e_2, -e_2, 0, 0) \in \mathsf{T}(S^1)^2$. The linearized differential equations can be shown to be

$$(m_1 + m_2)L_1^2 \ddot{\xi}_{11} = 0,$$

$$m_2 L_2^2 \ddot{\xi}_{21} + m_2 g L_2 \xi_{21} = 0,$$

which are defined on the four-dimensional tangent space of $\mathsf{T}(S^1)^2$ at the equilibrium $(e_2, -e_2, 0, 0) \in \mathsf{T}(S^1)^2$. These linearized differential equations can be used to characterize the local dynamics of the Furuta pendulum in a neighborhood of the hanging equilibrium $(e_2, -e_2, 0, 0) \in \mathsf{T}(S^1)^2$. The eigenvalues are $+j\sqrt{\frac{g}{r}}$, $-j\sqrt{\frac{g}{r}}$, 0, 0. Since the eigenvalues are purely imaginary or zero, no conclusion can be made about the stability of this equilibrium on the basis of this analysis.

The Lyapunov approach, using the total energy again, does not provide a positive result in this case since the hanging equilibrium is not a strict local minimum of the total energy function on the tangent bundle $\mathsf{T}(S^1)^2$.

It is interesting to compare the results of the local dynamics of the Furuta pendulum and the local dynamics of the particle on a torus.

4.6.7 Dynamics of a Three-Dimensional Revolute Joint Robot

A three-dimensional robot consists of a cylindrical base connected to two rigid links by revolute joints. The cylindrical base can rotate about a fixed vertical axis. A first link rotates about a one degree of freedom joint fixed to the cylindrical base with a horizontal axis of rotation; a second link rotates

about a one degree of freedom joint fixed to the end of the first link with
its joint axis parallel to the axis of rotation of the first link. In other words,
the two links form a two link planar robot but the plane of the two links is
not fixed but rotates with the cylindrical base. Consequently, this is a three-
dimensional robot. This is a generalization of the nonplanar double pendulum
considered previously. The dynamics of this three-dimensional revolute joint
robot are presented within a geometric framework. A schematic of a three-
dimensional revolute joint robot is shown in Figure 4.7.

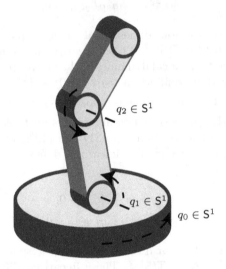

Fig. 4.7 Three-dimensional revolute joint robot

Large rotational motions of the cylindrical base and the two links are
allowed. It is assumed that there are no collisions between the links or with
the stationary base. Gravity forces act on the three-dimensional robot; no
other forces are included.

The challenging feature of this example is that the end of the third link
moves in three dimensions, rather than in a fixed plane. A geometric form of
the Euler–Lagrange equations is developed that avoids all representational
singularities and does not require the use of complicated trigonometry.

An inertially fixed Euclidean frame has its origin at the joint connecting
the first link and the axis of the rotating base. The first two axes of the fixed
frame lie in the horizontal plane and the third axis is vertical.

Let $q_0 \in \mathsf{S}^1$ be the attitude vector that describes the rotation of the cylindrical base about its vertical axis of rotation. Also, $q_1 \in \mathsf{S}^1$ is the attitude
vector that describes the attitude of the first link within its plane of rotation, $q_2 \in \mathsf{S}^1$ is the attitude vector that describes the attitude of the second

link within its plane of rotation. The configuration of the three-dimensional revolute joint robot is $q = (q_0, q_1, q_2) \in (\mathsf{S}^1)^3$. Since each of the rotations of the base and the joints is one-dimensional, the robot configuration manifold is $(\mathsf{S}^1)^3$, and the robot has three degrees of freedom.

The physical parameters of the robot are: L_1 is the length of the first link, L_2 is the length of the second link. Also, J is the scalar moment of inertia of the cylindrical base about its vertical axis of rotation, m_1 is the mass of the first link and m_2 is the mass of the second link; for simplicity each mass is assumed to be concentrated at the end of its link. As usual, g denotes the constant acceleration of gravity.

4.6.7.1 Euler–Lagrange Equations in Terms of $(q, \dot{q})$

Careful attention to the three-dimensional geometry of the robot leads to the following expression for the kinetic energy of the robot:

$$
\begin{aligned}
T(q, \dot{q}) &= \frac{1}{2} J \|\dot{q}_0\|^2 + \frac{1}{2} m_1 (\|L_1 \dot{q}_1\|^2 + \|e_1^T L_1 q_1 \dot{q}_0\|^2) \\
&\quad + \frac{1}{2} m_2 \left(\|L_1 \dot{q}_1 + L_2 \dot{q}_2\|^2 + \|(e_1^T (L_1 q_1 + L_2 q_2)) \dot{q}_0\|^2 \right) \\
&= \frac{1}{2} \begin{bmatrix} \dot{q}_0 \\ \dot{q}_1 \\ \dot{q}_2 \end{bmatrix}^T \begin{bmatrix} M_{00}(q) I_{2\times 2} & 0 & 0 \\ 0 & (m_1 + m_2) L_1^2 I_{2\times 2} & m_2 L_1 L_2 I_{2\times 2} \\ 0 & m_2 L_1 L_2 I_{2\times 2} & m_2 L_2^2 I_{2\times 2} \end{bmatrix} \begin{bmatrix} \dot{q}_0 \\ \dot{q}_1 \\ \dot{q}_2 \end{bmatrix},
\end{aligned}
$$

where $M_{00}(q) = J + m_1 (e_1^T L_1 q_1)^2 + m_2 (e_1^T (L_1 q_1 + L_2 q_2))^2$.

The gravitational potential energy of the robot is given by

$$
U(q) = m_1 g L_1 e_2^T q_1 + m_2 g e_2^T (L_1 q_1 + L_2 q_2).
$$

Thus, the Lagrangian function $L : \mathsf{T}(\mathsf{S}^1)^3 \to \mathbb{R}^1$ is

$$
\begin{aligned}
L(q, \dot{q}) &= \frac{1}{2} \begin{bmatrix} \dot{q}_0 \\ \dot{q}_1 \\ \dot{q}_2 \end{bmatrix}^T \begin{bmatrix} M_{00}(q) I_{2\times 2} & 0 & 0 \\ 0 & (m_1 + m_2) L_1^2 I_{2\times 2} & m_2 L_1 L_2 I_{2\times 2} \\ 0 & m_2 L_1 L_2 I_{2\times 2} & m_2 L_2^2 I_{2\times 2} \end{bmatrix} \begin{bmatrix} \dot{q}_0 \\ \dot{q}_1 \\ \dot{q}_2 \end{bmatrix} \\
&\quad - m_1 g L_1 e_2^T q_1 - m_2 g e_2^T (L_1 q_1 + L_2 q_2).
\end{aligned}
$$

Following the prior results in (4.9), the Euler–Lagrange equations for the robot dynamics are

$$
\begin{bmatrix} M_{00}(q) I_{2\times 2} & 0 & 0 \\ 0 & (m_1 + m_2) L_1^2 I_{2\times 2} & m_2 L_1 L_2 (I_{2\times 2} - q_1 q_1^T) \\ 0 & m_2 L_1 L_2 (I_{2\times 2} - q_2 q_2^T) & m_2 L_2^2 I_{2\times 2} \end{bmatrix} \begin{bmatrix} \ddot{q}_0 \\ \ddot{q}_1 \\ \ddot{q}_2 \end{bmatrix}
$$

$$+ \begin{bmatrix} M_{00}(q)\|\dot{q}_0\|^2 q_0 \\ (m_1+m_2)L_1^2\|\dot{q}_1\|^2 q_1 \\ m_2 L_2^2 \|\dot{q}_2\|^2 q_2 \end{bmatrix} + \begin{bmatrix} \dot{M}_{00}(q)(I_{2\times 2}-q_0 q_0^T)\dot{q}_0 \\ -(I_{2\times 2}-q_1 q_1^T)\frac{\partial}{\partial q_1}\dot{q}_0^T M_{00}(q)\dot{q}_0 \\ -(I_{2\times 2}-q_2 q_2^T)\frac{\partial}{\partial q_2}\dot{q}_0^T M_{00}(q)\dot{q}_0 \end{bmatrix}$$

$$+ \begin{bmatrix} 0 \\ (m_1+m_2)gL_1(I_{2\times 2}-q_1 q_1^T)e_2 \\ m_2 g L_2(I_{2\times 2}-q_2 q_2^T)e_2 \end{bmatrix} = \begin{bmatrix} 0 \\ 0 \\ 0 \end{bmatrix}. \tag{4.95}$$

These differential equations (4.95) describe the Lagrangian flow of the robot dynamics in terms of the evolution of $(q,\dot{q}) \in \mathsf{T}(S^1)^3$ on the tangent bundle of $(S^1)^3$.

4.6.7.2 Euler–Lagrange Equations in Terms of (q,ω)

The Euler–Lagrange equations can also be expressed in terms of the angular velocity vector. Here, $\omega_0 \in \mathbb{R}^1$ is the angular velocity of the cylindrical base, $\omega_1 \in \mathbb{R}^1$ is the angular velocity of the first link within its plane of rotation, and $\omega_2 \in \mathbb{R}^1$ is the angular velocity of the second link within its plane of rotation. Note that the relative angular velocity of the second link with respect to the first link, that is the rotation rate of the connecting joint, is $\omega_2 - \omega_1$.

The rotational kinematics equations are

$$\dot{q}_0 = \omega_0 S q_0, \tag{4.96}$$

$$\dot{q}_1 = \omega_1 S q_1, \tag{4.97}$$

$$\dot{q}_2 = \omega_2 S q_2. \tag{4.98}$$

As previously discussed, S denotes the 2×2 skew-symmetric matrix

$$S = \begin{bmatrix} 0 & -1 \\ 1 & 0 \end{bmatrix},$$

which rotates a vector by $\frac{\pi}{2}$ counterclockwise. Thus, the modified Lagrangian function can be expressed in terms of the angular velocity vector $\omega = (\omega_0, \omega_1, \omega_2)$ as

$$\tilde{L}(q,\omega) = \frac{1}{2} \begin{bmatrix} \omega_0 \\ \omega_1 \\ \omega_2 \end{bmatrix}^T \begin{bmatrix} M_{00}(q) & 0 & 0 \\ 0 & (m_1+m_2)L_1^2 & m_2 L_1 L_2 q_1^T q_2 \\ 0 & m_2 L_1 L_2 q_2^T q_1 & m_2 L_2^2 \end{bmatrix} \begin{bmatrix} \omega_0 \\ \omega_1 \\ \omega_2 \end{bmatrix}$$
$$- m_1 g L_1 e_2^T q_1 - m_2 g e_2^T (L_1 q_1 + L_2 q_2).$$

Using the prior results in (4.14), the resulting Euler–Lagrange equations, expressed in terms of the angular velocity vector, are:

$$\begin{bmatrix} M_{00}(q) & 0 & 0 \\ 0 & (m_1+m_2)L_1^2 & m_2L_1L_2q_1^Tq_2 \\ 0 & m_2L_1L_2q_2^Tq_1 & m_2L_2^2 \end{bmatrix} \begin{bmatrix} \dot\omega_0 \\ \dot\omega_1 \\ \dot\omega_2 \end{bmatrix}$$

$$+ \begin{bmatrix} 0 \\ m_2L_1L_2q_1^T S q_2\omega_2^2 \\ m_2L_1L_2q_2^T S q_1\omega_1^2 \end{bmatrix} + \begin{bmatrix} -\dot M_{00}(q)q_1^T S q_0\omega_0 \\ q_1^T S \frac{\partial}{\partial q_1} q_0^T M_{00}(q)q_0\omega_0^2 \\ q_2^T S \frac{\partial}{\partial q_2} q_0^T M_{00}(q)q_0\omega_0^2 \end{bmatrix}$$

$$+ \begin{bmatrix} 0 \\ (m_1+m_2)gL_1e_1^Tq_1 \\ m_2gL_2e_1^Tq_2 \end{bmatrix} = \begin{bmatrix} 0 \\ 0 \\ 0 \end{bmatrix}. \tag{4.99}$$

The differential equations (4.96), (4.97), (4.98), and (4.99) describe the dynamical flow of the robot dynamics on the tangent bundle $T(S^1)^3$. These dynamics can be identified with the dynamics of $(q,\dot q) \in T(S^1)^3$ via the kinematics equations (4.96), (4.97), and (4.98).

4.6.7.3 Hamilton's Equations in Terms of (q,μ)

The momentum $\mu = (\mu_0,\mu_1,\mu_2) \in T_q^*(S^1)^3$ is defined to be conjugate to $\dot q \in T_q(S^1)^3$ according to the Legendre transformation

$$\begin{bmatrix} \mu_0 \\ \mu_1 \\ \mu_2 \end{bmatrix} = \begin{bmatrix} (I_{2\times 2} - q_1q_1^T)\frac{\partial L(q,\dot q)}{\partial \dot q_1} \\ (I_{2\times 2} - q_2q_2^T)\frac{\partial L(q,\dot q)}{\partial \dot q_2} \\ (I_{2\times 2} - q_3q_3^T)\frac{\partial L(q,\dot q)}{\partial \dot q_3} \end{bmatrix},$$

$$= \begin{bmatrix} M_{00}(q)I_{2\times 2} & 0 & 0 \\ 0 & (m_1+m_2)L_1^2 I_{2\times 2} & m_2L_1L_2(I_{2\times 2} - q_1q_1^T) \\ 0 & m_2L_1L_2(I_{2\times 2} - q_2q_2^T) & m_2L_2^2 I_{2\times 2} \end{bmatrix} \begin{bmatrix} \dot q_0 \\ \dot q_1 \\ \dot q_2 \end{bmatrix}.$$

We assume that the 6×6 inverse matrix is

$$\begin{bmatrix} m_{00}^I(q) & 0 & 0 \\ 0 & m_{11}^I(q) & m_{12}^I(q) \\ 0 & m_{21}^I(q) & m_{22}^I(q) \end{bmatrix}$$

$$= \begin{bmatrix} M_{00}(q)I_{2\times 2} & 0 & 0 \\ 0 & (m_1+m_2)L_1^2 I_{2\times 2} & m_2L_1L_2(I_{2\times 2} - q_1q_1^T) \\ 0 & m_2L_1L_2(I_{2\times 2} - q_2q_2^T) & m_2L_2^2 I_{2\times 2} \end{bmatrix}^{-1}.$$

The Hamiltonian function $H : T^*(S^1)^3 \to \mathbb{R}^1$ is

$$H(q,\mu) = \frac{1}{2} \begin{bmatrix} \mu_0 \\ \mu_1 \\ \mu_2 \end{bmatrix}^T \begin{bmatrix} m_{00}^I(q) & 0 & 0 \\ 0 & m_{11}^I(q) & m_{12}^I(q) \\ 0 & m_{21}^I(q) & m_{22}^I(q) \end{bmatrix} \begin{bmatrix} \mu_0 \\ \mu_1 \\ \mu_2 \end{bmatrix}$$
$$+ m_1gL_1e_2^Tq_1 + m_2ge_2^T(L_1q_1 + L_2q_2).$$

Following the results in (4.21) and (4.22), Hamilton's equations can be shown to be

$$
\begin{bmatrix} \dot{q}_0 \\ \dot{q}_1 \\ \dot{q}_2 \end{bmatrix} = \begin{bmatrix} m_{00}^I(q) & 0 & 0 \\ 0 & m_{11}^I(q) & m_{12}^I(q) \\ 0 & m_{21}^I(q) & m_{22}^I(q) \end{bmatrix} \begin{bmatrix} \mu_0 \\ \mu_1 \\ \mu_2 \end{bmatrix}, \tag{4.100}
$$

and

$$
\begin{bmatrix} \dot{\mu}_0 \\ \dot{\mu}_1 \\ \dot{\mu}_2 \end{bmatrix} = \begin{bmatrix} -M_{00}(q) \left\| \dot{q}_0 \right\|^2 q_0 \\ -(m_1 + m_2)L_1^2 \left\| \dot{q}_1 \right\|^2 q_1 \\ -m_2 L_2^2 \left\| \dot{q}_2 \right\|^2 q_2 \end{bmatrix}
$$

$$
+ \begin{bmatrix} 0 \\ -m_2 L_1 L_2 \dot{q}_1 \dot{q}_2^T q_1 + (I_{2\times2} - q_1 q_1^T)\frac{\partial}{\partial q_1} \dot{q}_0^T M_{00}(q)\dot{q}_0 \\ -m_2 L_1 L_2 \dot{q}_2 \dot{q}_1^T q_2 + (I_{2\times2} - q_2 q_2^T)\frac{\partial}{\partial q_2} \dot{q}_0^T M_{00}(q)\dot{q}_0 \end{bmatrix}
$$

$$
+ \begin{bmatrix} 0 \\ -(m_1 + m_2)gL_1(I_{2\times2} - q_1 q_1^T)e_2 \\ -m_2 g L_2(I_{2\times2} - q_2 q_2^T)e_2 \end{bmatrix}, \tag{4.101}
$$

where the right-hand side of these equations can be expressed in terms of the momenta using the Legendre transformation. These differential equations define the Hamiltonian flow of the robot dynamics in terms of the evolution of (q, μ) on the cotangent bundle $T^*(S^1)^3$.

4.6.7.4 Hamilton's Equations in Terms of (q, π)

A different form of Hamilton's equations on the cotangent bundle $T^*(S^1)^3$ can be obtained by defining the momentum according to the Legendre transformation

$$
\begin{bmatrix} \pi_0 \\ \pi_1 \\ \pi_2 \end{bmatrix} = \begin{bmatrix} \frac{\partial \tilde{L}(q,\omega)}{\partial \omega_0} \\ \frac{\partial \tilde{L}(q,\omega)}{\partial \omega_1} \\ \frac{\partial \tilde{L}(q,\omega)}{\partial \omega_2} \end{bmatrix} = \begin{bmatrix} M_{00}(q) & 0 & 0 \\ 0 & (m_1 + m_2)L_1^2 & m_2 L_1 L_2 q_1^T q_2 \\ 0 & m_2 L_1 L_2 q_2^T q_1 & m_2 L_2^2 \end{bmatrix} \begin{bmatrix} \omega_0 \\ \omega_1 \\ \omega_2 \end{bmatrix},
$$

where the momentum $\pi = (\pi_0, \pi_1, \pi_2) \in \mathbb{R}^3$ is conjugate to the angular velocity vector $\omega = (\omega_0, \omega_1, \omega_2) \in \mathbb{R}^3$. Introduce the 3×3 matrix inverse

$$
\begin{bmatrix} m_{00}^I(q) & 0 & 0 \\ 0 & m_{11}^I(q) & m_{12}^I(q) \\ 0 & m_{21}^I(q) & m_{22}^I(q) \end{bmatrix} = \begin{bmatrix} M_{00}(q) & 0 & 0 \\ 0 & (m_1 + m_2)L_1^2 & m_2 L_1 L_2 q_1^T q_2 \\ 0 & m_2 L_1 L_2 q_2^T q_1 & m_2 L_2^2 \end{bmatrix}^{-1}.
$$

The modified Hamiltonian function $\tilde{H} : T^*(S^1)^3 \rightarrow \mathbb{R}^1$ can be written as

$$\tilde{H}(q, \pi) = \frac{1}{2} \begin{bmatrix} \pi_0 \\ \pi_1 \\ \pi_2 \end{bmatrix}^T \begin{bmatrix} m_{00}^I(q) & 0 & 0 \\ 0 & m_{11}^I(q) & m_{12}^I(q) \\ 0 & m_{21}^I(q) & m_{22}^I(q) \end{bmatrix} \begin{bmatrix} \pi_0 \\ \pi_1 \\ \pi_2 \end{bmatrix}$$
$$+ m_1 g L_1 e_2^T q_1 + m_2 g e_2^T (L_1 q_1 + L_2 q_2).$$

Following the results in (4.24) and (4.25), Hamilton's equations of motion for the revolute robot are

$$\begin{bmatrix} \dot{q}_0 \\ \dot{q}_1 \\ \dot{q}_2 \end{bmatrix} = \begin{bmatrix} m_{00}^I(q) S q_0 \pi_0 \\ (m_{11}^I(q)\pi_1 + m_{12}^I(q)\pi_2) S q_1 \\ (m_{21}^I(q)\pi_1 + m_{22}^I(q)\pi_2) S q_2 \end{bmatrix}, \tag{4.102}$$

and

$$\begin{bmatrix} \dot{\pi}_0 \\ \dot{\pi}_1 \\ \dot{\pi}_2 \end{bmatrix} = \begin{bmatrix} 0 \\ q_1^T S \frac{\partial \tilde{T}(q_1, q_2, \pi)}{\partial q_1} + (m_1 + m_2) g L_1 e_1^T q_1 \\ q_2^T S \frac{\partial \tilde{T}(q_1, q_2, \pi)}{\partial q_2} + m_2 g L_2 e_1^T q_2 \end{bmatrix}. \tag{4.103}$$

These equations, together with the rotational kinematics and the Legendre transformation, describe the Hamiltonian flow of the three-dimensional revolute robot on the cotangent bundle $T^*(S^1)^3$ of the configuration manifold.

4.6.7.5 Robot Kinematics

The rotational dynamics of the cylindrical base and of the first and second links are coupled in a complicated way, presumably giving rise to highly coupled nonlinear dynamics at least for large robot motions.

The three-dimensional position vector of the end of the second link in the inertial frame is of interest in robot applications; this position vector is given by

$$\begin{bmatrix} x_1 \\ x_2 \\ x_3 \end{bmatrix} = \begin{bmatrix} e_1^T (L_1 q_1 + L_2 q_2) e_1^T q_0 \\ e_1^T (L_1 q_1 + L_2 q_2) e_2^T q_0 \\ e_2^T (L_1 q_1 + L_2 q_2) \end{bmatrix}.$$

Consequently, this form of the *robot kinematics* for the three-dimensional translational velocity vector of the end of the second link can be expressed as a linear function of the time derivative of the configuration vector as

$$\begin{bmatrix} \dot{x}_1 \\ \dot{x}_2 \\ \dot{x}_3 \end{bmatrix} = \begin{bmatrix} e_1^T (L_1 q_1 + L_2 q_2) e_1^T & L_1 e_1^T q_0 e_1^T & L_2 e_1^T q_0 e_1^T \\ e_1^T (L_1 q_1 + L_2 q_2) e_2^T & L_1 e_2^T q_0 e_1^T & L_2 e_2^T q_0 e_1^T \\ 0 & L_1 e_2^T & L_2 e_2^T \end{bmatrix} \begin{bmatrix} \dot{q}_0 \\ \dot{q}_1 \\ \dot{q}_2 \end{bmatrix}.$$

These robot kinematics are globally defined and they demonstrate that the translational velocity vector of the end of the second link is a linear function of the link velocities. Configurations for which the above 3×3 matrix is rank deficient are known as the kinematic singularities. Since the equations are globally defined on $(S^1)^3$, these kinematic singularities are intrinsic and cannot be eliminated by a different choice of configuration variables.

The three-dimensional dynamics of the end of the second link of the robot depends on the complex, coupled dynamics of all three degrees of freedom of the robot. Our several descriptions of the robot dynamics on the tangent bundle $T(S^1)^3$ and on the cotangent bundle $T^*(S^1)^3$ are not traditional since we have avoided all use of angle coordinates which are, necessarily, only locally defined.

4.6.7.6 Conservation Properties

The Hamiltonian of the robot defined on the tangent bundle,

$$H = \frac{1}{2} \begin{bmatrix} \dot{q}_0 \\ \dot{q}_1 \\ \dot{q}_2 \end{bmatrix}^T \begin{bmatrix} M_{00}(q)I_{2\times2} & 0 & 0 \\ 0 & (m_1+m_2)L_1^2 I_{2\times2} & m_2 L_1 L_2 I_{2\times2} \\ 0 & m_2 L_1 L_2 I_{2\times2} & m_2 L_2^2 I_{2\times2} \end{bmatrix} \begin{bmatrix} \dot{q}_0 \\ \dot{q}_1 \\ \dot{q}_2 \end{bmatrix}$$
$$+ m_1 g L_1 e_2^T q_1 + m_2 g e_2^T (L_1 q_1 + L_2 q_2),$$

which coincides with the total energy E in this case, is constant along each solution of the robot dynamics.

It is observed that the Lagrangian and Hamiltonian functions do not depend on the attitude of the cylindrical base $q_0 \in S^1$; it is a cyclic configuration variable. According to Noether's theorem, it follows that the conjugate momentum

$$\pi_0 = M_{00}(q)\omega_0$$

is constant along each solution of the robot dynamics.

4.6.7.7 Equilibrium Properties

The equilibrium solutions of the robot dynamics are easily determined. They occur when the time derivative of the configuration vector, or equivalently the angular velocity vector or momentum, is zero and the two links are vertical, which implies that the time derivatives of angular velocity and momentum both vanish; the attitude of the cylindrical base is arbitrary. Consequently, there are four equilibrium manifolds given by

$$\left\{ (q_0, q_1, q_2) \in (S^1)^3 : q_1 = -e_2,\ q_2 = -e_2 \right\},$$
$$\left\{ (q_0, q_1, q_2) \in (S^1)^3 : q_1 = e_2,\ q_2 = -e_2 \right\},$$
$$\left\{ (q_0, q_1, q_2) \in (S^1)^3 : q_1 = -e_2,\ q_2 = e_2 \right\},$$
$$\left\{ (q_0, q_1, q_2) \in (S^1)^3 : q_1 = e_2,\ q_2 = e_2 \right\}.$$

It is possible to develop linearized equations of motion for any equilibrium solution and to assess stability properties of each equilibrium solutions using either the linearized equations or a Lyapunov approach. These details are not developed here.

4.6.7.8 Revolute Joint Robot with Actuated Joints

The above formulation and analysis of the dynamics for a revolute joint robot is based on the assumption of no external forces or moments other than those due to gravity. The common situation for industrial robots is that the three degrees of freedom are actuated in the sense that external moments are applied at each joint axis by motors. The incorporation of external moments into the equations of motion is easily followed from the prior development by replacing Hamilton's principle with the Lagrange–d'Alembert principle.

4.7 Problems

4.1. Starting with the Euler–Lagrange equations (4.6), view the kinematics equations (4.1) as defining a change of variables $(q, \dot{q}) \in T(S^1)^n \to (q, \omega) \in (S^1)^n \times \mathbb{R}^n$. Show that this change of variables can be used to derive the Euler–Lagrange equations (4.12) according to the following:

(a) Show that the transformation associated with this change of variables is a diffeomorphism. Is it a global diffeomorphism?

(b) Show that

$$\frac{\partial L(q, \dot{q})}{\partial \dot{q}_i} = S q_i \frac{\partial \tilde{L}(q, \omega)}{\partial \omega_i}, \qquad\qquad i = 1, \ldots, n,$$

$$\frac{\partial L(q, \dot{q})}{\partial q_i} = \frac{\partial \tilde{L}(q, \omega)}{\partial q_i} + \omega_i q_i \frac{\partial \tilde{L}(q, \omega)}{\partial \omega_i}, \quad i = 1, \ldots, n.$$

(c) Substitute these expressions into (4.6) and simplify.

(d) Take the inner product of each of the resulting vector equations with $S q_i$, $i = 1, \ldots, n$, and simplify to obtain the scalar Euler–Lagrange equations (4.12).

4.2. Show that the Euler–Lagrange equation (4.6) can be expressed as

$$\frac{d}{dt}\left(\frac{\partial L(q, \dot{q})}{\partial \dot{q}_i}\right) - q_i \frac{d}{dt}\left(q_i^T \frac{\partial L(q, \dot{q})}{\partial \dot{q}_i}\right)$$

$$- q_i \dot{q}_i^T \frac{\partial L(q, \dot{q})}{\partial \dot{q}_i} - \left(I_{2\times2} - q_i q_i^T\right) \frac{\partial L(q, \dot{q})}{\partial q_i} = 0, \quad i = 1, \ldots, n.$$

4.3. Consider the dynamics of a planar pendulum that is constrained to rotate in a fixed horizontal plane so that there are no gravitational forces. The pendulum mass is m and the pendulum length is L. The configuration manifold is S^1

(a) Determine the Lagrangian function $L : TS^1 \to \mathbb{R}^1$ defined on the tangent bundle of the configuration manifold.
(b) What are the Euler–Lagrange equations for the horizontal planar pendulum?
(c) Determine the Hamiltonian function $H : T^*S^1 \to \mathbb{R}^1$ defined on the cotangent bundle of the configuration manifold.
(d) What are Hamilton's equations for the horizontal planar pendulum?
(e) What are conserved quantities for the dynamical flow on TS^1?
(f) What are the equilibrium solutions of the horizontal planar pendulum?
(g) What are linear dynamics that approximate the dynamics of a horizontal planar pendulum in a neighborhood of an equilibrium solution?

4.4. Consider two identical particles, each of mass m, that translate in $\mathbb{R}^3$. Each particle is constrained to translate, without friction, on a planar circle embedded in $\mathbb{R}^3$. The two circles have common radius $R > 0$; the planes containing the circles are parallel and are separated by a distance $L > 0$. The position vector of the first particle in an inertial Euclidean frame is $x_1 = (Rq_1, 0) \in \mathbb{R}^3$ and the position vector of the second particle in the inertial Euclidean frame is $x_2 = (Rq_2, L) \in \mathbb{R}^3$. The configuration manifold is $(S^1)^2$. The two particles act under the influence of a mutual potential function given by

$$U(x) = K \left\| x_1 - x_2 \right\|^2 ,$$

where $K > 0$ is constant.

(a) Determine the Lagrangian function $L : T(S^1)^2 \to \mathbb{R}^1$ on the tangent bundle of the configuration manifold.
(b) What are the resulting Euler–Lagrange equations?
(c) Determine the Hamiltonian function $H : T^*(S^1)^2 \to \mathbb{R}^1$ defined on the cotangent bundle of the configuration manifold.
(d) What are the resulting Hamilton's equations?
(e) What are conservation properties of the dynamical flow on $T(S^1)^2$?
(f) What are the equilibrium solutions of the dynamical flow on $T(S^1)^2$?

4.5. Consider the dynamics of a particle, with mass m, constrained to move, without friction, on a circular hoop that rotates about a diameter with constant angular velocity. Introduce an inertial Euclidean frame. Assume uniform, constant gravity, acts in the direction $-e_3$ of the inertial frame and assume the constant angular velocity of the circular hoop is $\Omega e_3 \in \mathbb{R}^3$. With respect to a two-dimensional Euclidean frame defined by the plane of the hoop, the position vector of the particle is $q \in S^1$; the configuration manifold is S^1.

(a) Determine an expression for the position vector of the particle in the inertial frame, expressed in terms of the configuration vector.
(b) Determine the Lagrangian function $L : \mathsf{TS}^1 \to \mathbb{R}^1$ defined on the tangent bundle of the configuration manifold.
(c) What are the Euler–Lagrange equations for the particle constrained to a circular hoop that rotates with constant angular velocity under the action of uniform, constant gravity?
(d) Determine the Hamiltonian function $H : \mathsf{T}^*\mathsf{S}^1 \to \mathbb{R}^1$ defined on the cotangent bundle of the configuration manifold.
(e) What are Hamilton's equations for the particle constrained to a circular hoop that rotates with constant angular velocity under the action of uniform, constant gravity?
(f) What are conserved quantities for the dynamical flow on TS^1?
(g) What are the equilibrium solutions of the dynamical flow on TS^1?
(h) What are linear dynamics that approximate the dynamics of the particle constrained to a circular hoop that rotates with constant angular velocity in a neighborhood of an equilibrium solution?

4.6. Consider the dynamics of a particle, with mass m, constrained to move, without friction, on a circular hoop that rotates about a vertical vector, tangent to the hoop, through a fixed point on the hoop with constant angular velocity. Introduce an inertial Euclidean frame. Assume uniform, constant gravity, acts in the direction $-e_3$ of the inertial frame and assume the constant angular velocity of the circular hoop is $\Omega e_3 \in \mathbb{R}^3$. With respect to a two-dimensional Euclidean frame defined by the plane of the hoop, the position vector of the particle is $q \in \mathsf{S}^1$; the configuration manifold is S^1.

(a) Determine an expression for the position vector of the particle in the inertial frame, expressed in terms of the configuration vector.
(b) Determine the Lagrangian function $L : \mathsf{TS}^1 \to \mathbb{R}^1$ defined on the tangent bundle of the configuration manifold.
(c) What are the Euler–Lagrange equations for the particle constrained to a circular hoop that rotates with constant angular velocity under the action of uniform, constant gravity?
(d) Determine the Hamiltonian function $H : \mathsf{T}^*\mathsf{S}^1 \to \mathbb{R}^1$ defined on the cotangent bundle of the configuration manifold.
(e) What are Hamilton's equations for the particle constrained to a circular hoop that rotates with constant angular velocity under the action of uniform, constant gravity?
(f) What are conserved quantities for the dynamical flow on TS^1?
(g) What are the equilibrium solutions of the dynamical flow on TS^1?

4.7. Consider the dynamics of a planar double pendulum that is constrained to rotate in a fixed horizontal plane so that there are no gravitational forces. The pendulum masses are m_1 and m_2 and the pendulum lengths are L_1 and L_2. The configuration manifold is $(\mathsf{S}^1)^2$.

(a) Determine the Lagrangian function $L : \mathsf{T}(S^1)^2 \to \mathbb{R}^1$ on the tangent bundle of the configuration manifold.
(b) What are the Euler–Lagrange equations for the double pendulum in a horizontal plane?
(c) Determine the Hamiltonian function $H : \mathsf{T}^*(S^1)^2 \to \mathbb{R}^1$ defined on the cotangent bundle of the configuration manifold.
(d) What are Hamilton's equations for the double pendulum in a horizontal plane?
(e) What are conserved quantities for the dynamical flow on $\mathsf{T}(S^1)^2$?
(f) What are the equilibrium solutions of the dynamical flow on $\mathsf{T}(S^1)^2$?
(g) What are linear dynamics that approximate the dynamics of a double pendulum in a horizontal plane in a neighborhood of an equilibrium solution?

4.8. Consider the dynamics of a planar double pendulum that is constrained to rotate in a vertical plane with constant rotation rate. The vertical plane rotates about an inertially fixed axis with constant angular velocity $\Omega \in \mathbb{R}^1$; the first joint of the double planar pendulum lies on the axis of rotation of the vertical plane, at the origin of the inertial frame. There are no gravitational forces. The pendulum masses are m_1 and m_2 and the pendulum lengths are L_1 and L_2. The configuration manifold is $(S^1)^2$.

(a) Determine the Lagrangian function $L : \mathsf{T}(S^1)^2 \to \mathbb{R}^1$ on the tangent bundle of the configuration manifold.
(b) What are the Euler–Lagrange equations for the planar double pendulum in a vertical plane with constant rotation rate?
(c) Determine the Hamiltonian function $H : \mathsf{T}^*(S^1)^2 \to \mathbb{R}^1$ defined on the cotangent bundle of the configuration manifold.
(d) What are Hamilton's equations for the planar double pendulum in a vertical plane with constant rotation rate?
(e) What are conserved quantities for the dynamical flow on $\mathsf{T}(S^1)^2$?
(f) What are linear dynamics that approximate the dynamics of a planar double pendulum in a vertical plane with constant rotation rate in a neighborhood of an equilibrium solution?

4.9. Consider the dynamics of a nonplanar double pendulum and assume there are no gravitational forces. The pendulum masses are m_1 and m_2 and the pendulum lengths are L_1 and L_2. The configuration manifold is $(S^1)^2$.

(a) Determine the Lagrangian function $L : \mathsf{T}(S^1)^2 \to \mathbb{R}^1$ on the tangent bundle of the configuration manifold.
(b) What are the Euler–Lagrange equations for the nonplanar double pendulum?
(c) Determine the Hamiltonian function $H : \mathsf{T}^*S^1 \to \mathbb{R}^1$ defined on the cotangent bundle of the configuration manifold.
(d) What are Hamilton's equations for the nonplanar double pendulum?
(e) What are conserved quantities for the dynamical flow on $\mathsf{T}(S^1)^2$?

(f) What are the equilibrium solutions of the dynamical flow on $\mathsf{T}(\mathsf{S}^1)^2$?

(g) What are linear dynamics that approximate the dynamics of a nonplanar double pendulum in a neighborhood of an equilibrium solution?

4.10. Consider the dynamics of a modified Furuta pendulum; the first link is constrained to rotate in a fixed horizontal plane while the second link is constrained to rotate in a vertical plane that contains the first link. Include the gravitational forces. The pendulum masses are m_1 and m_2 and the pendulum lengths are L_1 and L_2. The configuration manifold is $(\mathsf{S}^1)^2$.

(a) Determine the Lagrangian function $L : \mathsf{T}(\mathsf{S}^1)^2 \to \mathbb{R}^1$ on the tangent bundle of the configuration manifold.

(b) What are the Euler–Lagrange equations for the modified Furuta pendulum?

(c) Determine the Hamiltonian function $H : \mathsf{T}^*(\mathsf{S}^1)^2 \to \mathbb{R}^1$ defined on the cotangent bundle of the configuration manifold.

(d) What are Hamilton's equations for the modified Furuta pendulum?

(e) What are conserved quantities for the dynamical flow on $\mathsf{T}(\mathsf{S}^1)^2$?

(f) What are the equilibrium solutions of the dynamical flow on $\mathsf{T}(\mathsf{S}^1)^2$?

(g) What are linear dynamics that approximate the dynamics of the modified Furuta pendulum in a neighborhood of an equilibrium solution?

4.11. Consider the dynamics of a particle on a torus and assume there are no gravitational forces. The mass of the particle is m and the torus has major axis R and minor axis $0 < r < R$. The configuration manifold is $(\mathsf{S}^1)^2$.

(a) Determine the Lagrangian function $L : \mathsf{T}(\mathsf{S}^1)^2 \to \mathbb{R}^1$ on the tangent bundle of the configuration manifold.

(b) What are the Euler–Lagrange equations for the particle on a torus, without gravity?

(c) Determine the Hamiltonian function $H : \mathsf{T}^*(\mathsf{S}^1)^2 \to \mathbb{R}^1$ defined on the cotangent bundle of the configuration manifold.

(d) What are Hamilton's equations for the particle on a torus, without gravity?

(e) What are conserved quantities for the dynamical flow on $\mathsf{T}(\mathsf{S}^1)^2$?

(f) What are the equilibrium solutions of the dynamical flow on $\mathsf{T}(\mathsf{S}^1)^2$?

(g) What are linear dynamics that approximate the dynamics of a particle on a torus in a neighborhood of an equilibrium solution?

4.12. Consider the dynamics of two elastically connected planar pendulums, with colocated inertial pivots, that are constrained to rotate in a fixed vertical plane under uniform, constant gravity. The pendulum masses are m_1 and m_2 and the pendulum lengths are L_1 and L_2. The potential for the elastic connection is given by $\kappa(1 - q_1^T q_2)$, where κ is an elastic stiffness constant. The configuration manifold is $(\mathsf{S}^1)^2$.

(a) Give a physical interpretation of this elastic potential energy. What are the configurations for which the potential energy is a maximum or a minimum?

(b) Determine the Lagrangian function $L : T(\mathsf{S}^1)^2 \to \mathbb{R}^1$ defined on the tangent bundle of the configuration manifold.

(c) What are the Euler–Lagrange equations for the two elastically connected planar pendulums in a vertical plane?

(d) Determine the Hamiltonian function $H : T^*(\mathsf{S}^1)^2 \to \mathbb{R}^1$ defined on the cotangent bundle of the configuration manifold.

(e) What are Hamilton's equations for the two elastically connected planar pendulums in a vertical plane?

(f) What are conserved quantities for the dynamical flow on $T(\mathsf{S}^1)^2$?

(g) What are the equilibrium solutions of the dynamical flow on $T(\mathsf{S}^1)^2$?

(h) What are linear dynamics that approximate the dynamics of the two elastically connected planar pendulums in a neighborhood of an equilibrium solution?

4.13. Consider the dynamics of two elastically connected planar pendulums, with colocated inertial pivots, that are constrained to rotate in a fixed horizontal plane so there are no gravitational forces. The pendulum masses are m_1 and m_2 and the pendulum lengths are L_1 and L_2. Assume the potential for the elastic connection is given by $\kappa(1 - q_1^T S^T q_2)$, where κ is an elastic stiffness constant. The configuration manifold is $(\mathsf{S}^1)^2$.

(a) Give a physical interpretation of this elastic potential energy. What are the configurations for which the potential energy is a maximum or a minimum?

(b) Determine the Lagrangian function $L : T(\mathsf{S}^1)^2 \to \mathbb{R}^1$ defined on the tangent bundle of the configuration manifold.

(c) What are the Euler–Lagrange equations for the two elastically connected planar pendulums in a horizontal plane?

(d) Determine the Hamiltonian function $H : T^*(\mathsf{S}^1)^2 \to \mathbb{R}^1$ defined on the cotangent bundle of the configuration manifold.

(e) What are Hamilton's equations for the two elastically connected planar pendulums in a horizontal plane?

(f) What are conserved quantities for the dynamical flow on $T(\mathsf{S}^1)^2$?

(g) What are the equilibrium solutions of the dynamical flow on $T(\mathsf{S}^1)^2$?

(h) What are linear dynamics that approximate the dynamics of the two elastically connected planar pendulums in a neighborhood of an equilibrium solution?

4.14. Consider the dynamics of a planar pendulum, with an elastic connection to an inertially fixed point. The planar pendulum is constrained to rotate in a fixed vertical plane under uniform, constant gravity. The pendulum mass is m and the pendulum length is L. The elastic connection consists of a linear elastic spring; one end of the spring is connected to an inertially fixed point that is a distance H directly above the pendulum pivot and the other end of the spring is connected to the end of the planar pendulum. Let κ denote the elastic stiffness constant of the spring. Assume the elastic spring always remains straight; the elastic spring has zero force when its end points are separated by the distance D. Assume $H > L + D$ so that the spring is always in tension. The configuration manifold is S^1.

(a) Describe the force exerted by the elastic spring expressed in terms of the configuration. Describe the moment on the planar pendulum exerted by the elastic spring expressed in terms of the configuration.
(b) Use the Lagrange–d'Alembert principle to determine the Euler–Lagrange equations for the planar pendulums, with an inertial elastic connection, in a vertical plane.
(c) Use the Lagrange–d'Alembert principle to determine Hamilton's equations for the planar pendulums, with an inertial elastic connection, in a vertical plane.
(d) What are conserved quantities for the dynamical flow on TS^1?
(e) What are algebraic equations that describe the equilibrium solutions on TS^1?

4.15. Consider the dynamics of two elastically connected planar pendulums, with different pivot locations; the pendulums are constrained to rotate in a fixed vertical plane under uniform, constant gravity. The two pivots of the planar pendulums lie in a horizontal plane separated by a distance H. The pendulum masses are m_1 and m_2 and the pendulum lengths are L_1 and L_2. The elastic connection consists of a linear elastic spring with ends connected to the ends of the two planar pendulums, where κ is the elastic stiffness constant. Assume the elastic spring always remains straight; the elastic spring has zero force when its end points are separated by the distance D. Assume $H > L_1 + L_2 + D$ so that the spring is always in tension. The configuration manifold is $(S^1)^2$.

(a) Describe the force exerted by the elastic spring expressed in terms of the configuration variables. Describe the moments on each pendulum exerted by the elastic spring expressed in terms of the configuration variables.
(b) Use the Lagrange–d'Alembert principle to determine the Euler–Lagrange equations for the two elastically connected planar pendulums in a vertical plane.
(c) Use the Lagrange–d'Alembert principle to determine Hamilton's equations for the two elastically connected planar pendulums in a vertical plane.
(d) What are conserved quantities for the dynamical flow on $T(S^1)^2$?
(e) What are algebraic equations that describe the equilibrium solutions on $T(S^1)^2$?

4.16. Consider the dynamics of a three-dimensional revolute joint robot and assume there are no gravitational forces. The moment of inertia of the cylindrical base is J; the two robot links have mass m_1 and m_2 and link lengths L_1 and L_2. The configuration manifold is $(S^1)^3$.

(a) Determine the Lagrangian function $L : \mathsf{T}(S^1)^3 \to \mathbb{R}^1$ defined on the tangent bundle of the configuration manifold.
(b) What are the Euler–Lagrange equations for the three-dimensional revolute joint robot, without gravity?
(c) Determine the Hamiltonian function $H : \mathsf{T}^*(S^1)^3 \to \mathbb{R}^1$ defined on the cotangent bundle of the configuration manifold.
(d) What are Hamilton's equations for the three-dimensional revolute joint robot, without gravity?
(e) What are conserved quantities for the dynamical flow on $(S^1)^3$?
(f) What are the equilibrium solutions of the dynamical flow on $(S^1)^3$?
(g) What are linear dynamics that approximate the dynamics of a three-dimensional revolute joint robot in a neighborhood of an equilibrium solution?

4.17. Consider a particle that moves, without friction, on a torus under uniform, constant gravity. The mass of the particle is m and the torus has major axis $R > 0$ and minor axis $0 < r < R$. In contrast to the prior assumption in this chapter, assume the major axis of the torus lies in a horizontal plane, say along the first axis of the inertial frame. The configuration manifold is $(S^1)^2$.

(a) Determine the Lagrangian function $L : \mathsf{T}(S^1)^2 \to \mathbb{R}^1$ for the particle on the torus defined on the tangent bundle of the configuration manifold.
(b) What are the Euler–Lagrange equations for the particle on the torus?
(c) Determine the Hamiltonian function $H : \mathsf{T}^*(S^1)^2 \to \mathbb{R}^1$ defined on the cotangent bundle of the configuration manifold.
(d) What are Hamilton's equations for the particle on the torus?
(e) What are conserved quantities for the dynamical flow on $\mathsf{T}(S^1)^2$?
(f) What are the equilibrium solutions of the dynamical flow on $\mathsf{T}(S^1)^2$?
(g) What are linear dynamics that approximate the dynamics of the particle on a torus in a neighborhood of an equilibrium solution?

4.18. A rigid circular wire can rotate, without friction, in three dimensions about a fixed vertical axis that is a diameter of the wire; the radius of the circular wire is r and the scalar moment of inertia of the wire, about its vertical axis, is J. An ideal particle is constrained to slide, without friction, on the wire; the mass of the particle is m. Assume constant uniform gravity. The configuration manifold is $(S^1)^2$.

(a) Determine the Lagrangian function $L : \mathsf{T}(S^1)^2 \to \mathbb{R}^1$ of the particle on a wire defined on the tangent bundle of the configuration manifold.
(b) What are the Euler–Lagrange equations for the wire and particle?
(c) Determine the Hamiltonian function $H : \mathsf{T}^*(S^1)^2 \to \mathbb{R}^1$ defined on the cotangent bundle of the configuration manifold.
(d) What are Hamilton's equations for the wire and particle?
(e) What are conserved quantities for the dynamical flow on $\mathsf{T}(S^1)^2$?
(f) What are the equilibrium solutions of the dynamical flow on $\mathsf{T}(S^1)^2$?

(g) What are linear dynamics that approximate the dynamics of the wire and particle in a neighborhood of an equilibrium solution?

4.19. Consider a flat plate that is constrained to rotate in three dimensions about a horizontal axis that is fixed in the plate. An ideal particle is constrained to translate on a circle fixed to the surface of the flat plate. Assume constant uniform gravity. The radius of the circle is r. The moment of inertia of the flat plate about its axis is J and m is the mass of the particle. The configuration manifold is $(\mathsf{S}^1)^2$.

(a) Determine the Lagrangian function $L : \mathsf{T}(\mathsf{S}^1)^2 \to \mathbb{R}^1$ of the flat plate and particle defined on the tangent bundle of the configuration manifold.
(b) What are the Euler–Lagrange equations for the flat plate and the particle?
(c) Determine the Hamiltonian function $H : \mathsf{T}^*(\mathsf{S}^1)^2 \to \mathbb{R}^1$ defined on the cotangent bundle of the configuration manifold.
(d) What are Hamilton's equations for the flat plate and the particle?
(e) What are conserved quantities for the dynamical flow on $\mathsf{T}(\mathsf{S}^1)^2$?
(f) What are the equilibrium solutions of the dynamical flow on $\mathsf{T}(\mathsf{S}^1)^2$?
(g) What are linear dynamics that approximate the dynamics of the flat plate and particle in a neighborhood of an equilibrium solution?

4.20. Consider the dynamics of a particle that is constrained to move on a circle in an inertially fixed plane under the influence of a gravitational field. Consider a two-dimensional inertial Euclidean frame for this plane, whose origin is located at the center of the circle. There is a gravitational field $G : \mathbb{R}^2 \to \mathsf{T}\mathbb{R}^2$. That is, the gravitational force on a particle, located at $x \in \mathbb{R}^2$ in the inertial frame, is given by $mG(x)$, where m denotes the constant mass of the particle. The configuration manifold is S^1.

(a) What are the Euler–Lagrange equations for the particle?
(b) What are Hamilton's equations for the particle?
(c) What are the conditions for an equilibrium solution of a particle in a gravitational field?
(d) Suppose that the gravitational field is constant. What are the conditions for an equilibrium solution of a particle?

4.21. Consider the dynamics of a charged particle, with mass m, that is constrained to move on a circle in an inertially fixed plane under the influence of an electric and a magnetic field. Consider a two-dimensional Euclidean frame for this plane, whose origin is located at the center of the circle. The charged particle moves in the presence of an electric field $E : \mathbb{R}^2 \to \mathsf{T}\mathbb{R}^2$ and a magnetic field given, in the inertial frame, by $B : \mathbb{R}^2 \to \mathbb{R}^1$ (This can be thought of as the magnitude of a magnetic field in three dimensions that is orthogonal to the plane of the constraining circle). The Lorentz force on a particle located at $x \in \mathbb{R}^2$ in the inertial frame is given by $Q(E(x) + S\dot{x}B(x))$, where Q denotes the constant charge on the particle. The configuration manifold is S^1.

(a) What are the Euler–Lagrange equations for the particle?
(b) What are Hamilton's equations for the particle?
(c) What are the conditions for an equilibrium solution of a particle in an electric field and a magnetic field?
(d) Suppose that the electric field and the magnetic field are constant. What are the conditions for an equilibrium solution of a particle?

4.22. Consider the problem of finding the curve(s) $[0,1] \to S^1$ of shortest length that connect two fixed points in S^1. Such curves are referred to as geodesic curve(s) on S^1.

(a) If the curve is parameterized by $t \to q(t) \in S^1$, show that the incremental arc length of the curve is $ds = \|dq\|$ so that the geodesic curve(s) minimize $\int_0^1 \sqrt{\|\dot{q}\|^2} dt$ among all curves on S^1 that connect the two fixed points.
(b) Show that the geodesic curve(s) necessarily satisfy the variational property $\delta \int_0^1 \|\dot{q}\| \, dt = 0$ for all smooth curves $t \to q(t) \in S^1$ that satisfy the boundary conditions $q(0) = q_0 \in S^1$, $q(1) = q_1 \in S^1$.
(c) What are the Euler–Lagrange equations and Hamilton's equations that geodesic curves in S^1 must satisfy?
(d) Suppose that $q_0 \in S^1$, $q_1 \in S^1$ do not lie on a common diameter of the sphere. Show that there is a unique geodesic curve in S^1. Describe the geodesic curve. Show that the geodesic curve is actually a minimum of $\int_0^1 \|\dot{q}\| \, dt$.
(e) Suppose that $q_0 \in S^1$, $q_1 \in S^1$ lie on a common diameter of the sphere. Show that there are two geodesic curves in S^1. Describe each geodesic curve. Show that each geodesic curve is actually a minimum of $\int_0^1 \|\dot{q}\| \, dt$.

4.23. Consider the problem of finding the geodesic curve(s) of shortest length that lies on the torus, with major radius $R > 0$ and minor radius $0 < r < R$, embedded in $\mathbb{R}^3$ and connects two fixed points on the torus. The position vector $x \in \mathbb{R}^3$ of a point on the torus is described with respect to a Euclidean frame using the configuration manifold $(S^1)^2$ as described previously in this chapter, where the torus is viewed as a manifold embedded in $\mathbb{R}^3$.

(a) If the curve is parameterized by $t \to q(t) \in (S^1)^2$, show that the incremental arc length of the curve is $ds = \sqrt{\|dx\|^2}$ so that the geodesic curve(s) minimize $\int_0^1 \sqrt{\|\dot{x}\|^2} dt$.
(b) Show that the geodesic curve(s) necessarily satisfy the variational property $\delta \int_0^1 \|\dot{x}\| \, dt = 0$ for all smooth curves $t \to q(t) \in (S^1)^2$ that satisfy the boundary conditions $q(0) = q_0 \in (S^1)^2$, $q(1) = q_1 \in (S^1)^2$.
(c) What are the Euler–Lagrange equations and Hamilton's equations that geodesic curves in S^1 must satisfy?
(d) What conditions on $q_0 \in (S^1)^2$, $q_1 \in (S^1)^2$ guarantee that there is a unique geodesic? Describe the geodesic curve.

(e) What conditions on $q_0 \in (S^1)^2$, $q_1 \in (S^1)^2$ guarantee that there are multiple geodesics? Describe the geodesic curves.

4.24. A planar four-bar linkage consists of three rigid links with one end of each of two of the links constrained to rotate in a fixed horizontal plane about a fixed pivot. The other link can rotate subject to the fact that its ends are pinned to the opposite ends of the first two links. The pivots and joint connections are assumed to be frictionless and to allow constrained rotation of the three links within a common horizontal plane. The two links with an inertially fixed end are assumed to have length L with mass m concentrated at the mid points of the links. The connecting link is assumed to have length $2.5\,L$ with mass $\frac{m}{2}$ concentrated at its mid point. The distance between the two fixed pivots is $2.5\,L$. We assume there are no collisions between links or with the inertial base. Choose an inertially fixed two-dimensional frame with origin located at the first fixed pivot point and with its first axis in the direction of the second fixed pivot point. With respect to the inertially fixed frame, let $q_i \in S^1$ denote the attitude as a direction vector of the i-th link with respect to the inertial frame for $i = 1, 2, 3$.

(a) Show that Lagrangian function, ignoring the holonomic constraint defined by the locations of the fixed pivots, can be written as

$$L(q, \dot{q}) = \frac{1}{2}m\left\|\frac{L}{2}\dot{q}_1\right\|^2 + \frac{1}{2}\left(\frac{m}{2}\right)\left\|\frac{5L}{2}\dot{q}_2\right\|^2 + \frac{1}{2}m\left\|\frac{L}{2}\dot{q}_3\right\|^2.$$

(b) Show that the holonomic constraint that arises from the fixed locations of the pivots can be written as

$$Lq_1 + 2.5\,Lq_2 + Lq_3 - 2.5\,Le_1 = 0.$$

(c) Describe the constraint manifold M embedded in $(S^1)^3$ and the augmented Lagrangian $L^a : TM \to \mathbb{R}^1$ on the tangent bundle of the constraint manifold.
(d) Introduce the angular velocity vector associated with the three attitude vectors and express the augmented Lagrangian function in terms of the angular velocity vector.
(e) What are the Euler–Lagrange equations, expressed in terms of the angular velocity vector?
(f) What are Hamilton's equations, expressed in terms of momenta conjugate to the angular velocity vector?
(g) What are conserved quantities for the dynamical flow on TM?
(h) What are the equilibrium solutions of the dynamical flow on TM?

4.25. Consider the dynamics of an ideal particle that evolves on a manifold M that is a smooth deformation of the embedded manifold S^1 in $\mathbb{R}^2$ in the sense that it is the image of a diffeomorphism $\phi : S^1 \to M$. Suppose that the

Lagrangian function is $L : TM \to \mathbb{R}^1$. Use the form for the Lagrangian dynamics on S^1 and this diffeomorphism to show that the Lagrangian dynamics on M satisfy

$$(I_{2\times2} - qq^T)\left(\frac{\partial\phi(q)}{\partial q}\right)\left\{\frac{d}{dt}\left(\frac{\partial L(x,\dot x)}{\partial \dot x}\right) - \frac{\partial L(x,\dot x)}{\partial x}\right\} = 0,$$

where $q = \phi^{-1}(x)$; here $\phi^{-1} : M \to S^1$ is the inverse of the diffeomorphism.

4.26. Consider an ideal particle, of mass m, that is constrained to move on an elliptical curve given by $M = \{q \in \mathbb{R}^2 : (\frac{q_1}{a})^2 + (\frac{q_2}{b})^2 - 1 = 0\}$ embedded in $\mathbb{R}^2$. Assume one axis of the ellipse is vertical and one axis of the ellipse is horizontal. Constant, uniform gravity acts on the particle.

(a) Determine the Lagrangian function $L : TM \to \mathbb{R}^1$ defined on the tangent bundle of the configuration manifold.
(b) Use the results of the previous problem; what are the Euler–Lagrange equations for the particle on the ellipse?
(c) Determine the Hamiltonian function $H : T^*M \to \mathbb{R}^1$ defined on the cotangent bundle of the configuration manifold.
(d) What are Hamilton's equations for the particle on the ellipse?
(e) What are conserved quantities for the dynamical flow on TM?
(f) What are the equilibrium solutions of the dynamical flow on TM?

4.27. Suppose the configuration manifold is $(S^1)^n$ and the kinetic energy has the form of a general quadratic function in the time derivative of the configuration vector so that the Lagrangian function $L : T(S^1)^n \to \mathbb{R}^1$ is given by

$$L(q,\dot q) = \frac{1}{2}\sum_{i=1}^{n}\sum_{j=1}^{n}m_{ij}(q)\dot q_i^T\dot q_j + \sum_{i=1}^{n}a_i(q)\dot q_i - U(q),$$

where $q = [q_1,\ldots,q_n]^T \in \mathbb{R}^n$ and $m_{ij}(q) = m_{ji}(q) > 0$, $i = 1,\ldots,n$, $j = 1,\ldots,n$, $a_i(q)$, $i = 1,\ldots,n$ are vector functions and $U(q)$ is a real scalar function.

(a) What are the Euler–Lagrange equations for this Lagrangian function?
(b) What are Hamilton's equations for this Lagrangian?
(c) Determine the modified Lagrangian function expressed in terms of angular velocity vector.
(d) What are the Euler–Lagrange equations for this modified Lagrangian?
(e) What are Hamilton's equations for the modified Hamiltonian associated with this modified Lagrangian?

Chapter 5
Lagrangian and Hamiltonian Dynamics on $(\mathsf{S}^2)^n$

This chapter introduces an important modification to the results presented in the preceding chapter. In particular, the configuration of a Lagrangian or Hamiltonian system is assumed to lie in the product of an arbitrary number of two-spheres in $\mathbb{R}^3$. The product of spheres in $\mathbb{R}^3$, denoted by $(\mathsf{S}^2)^n$, is a manifold with a conceptually simple geometry. Euler–Lagrange equations and Hamilton's equations are developed. The development in this chapter is somewhat similar to the development in Chapter 4, where the dynamics evolve on $(\mathsf{S}^1)^n$. In spite of the apparent similarities in the development, since the dynamics evolve on $(\mathsf{S}^2)^n$ here, there are important differences that make the development in this chapter more complicated. The results are illustrated by several examples of Lagrangian and Hamiltonian dynamics that evolve on $(\mathsf{S}^2)^n$.

The key ideas of the development in this chapter were first presented in published form in [19, 46, 54, 58]; those results are expanded and simplified in the presentation to follow. An earlier treatment of dynamics on spheres was given in [14].

5.1 Configurations as Elements in $(\mathsf{S}^2)^n$

We develop Euler–Lagrange equations for systems evolving on the configuration manifold $(\mathsf{S}^2)^n$. Since the dimension of the configuration manifold is $2n$, there are $2n$ degrees of freedom. A review of basic differential geometric concepts for $(\mathsf{S}^2)^n$ is now given.

The two-sphere, as a manifold in $\mathbb{R}^3$, is

$$\mathsf{S}^2 = \{q \in \mathbb{R}^3 : \|q\| = 1\}.$$

© Springer International Publishing AG 2018
T. Lee et al., *Global Formulations of Lagrangian and Hamiltonian Dynamics on Manifolds*, Interaction of Mechanics and Mathematics, DOI 10.1007/978-3-319-56953-6_5

The product of n spheres in $\mathbb{R}^3$, denoted by $(S^2)^n = S^2 \times \cdots \times S^2$, consists of all ordered n-tuples of vectors $q = (q_1, \ldots, q_n)$, with $q_i \in S^2$, $i = 1, \ldots, n$. This manifold is also described as

$$(S^2)^n = \{ q \in \mathbb{R}^{3n} : q_i \in S^2, \, i = 1, \ldots, n \}.$$

As discussed in Chapter 1, the tangent space of $(S^2)^n$ at $q \in (S^2)^n$ is

$$T_q(S^2)^n = \{ (\xi_1, \ldots, \xi_n) \in \mathbb{R}^{3n} : (q_i \cdot \xi_i) = 0, \, i = 1, \ldots, n \},$$

and any $\xi \in T_q(S^2)^n$ is referred to as a tangent vector to $(S^2)^n$ at $q \in (S^2)^n$. Also

$$T(S^2)^n = \{ (q, \xi) \in \mathbb{R}^{3n} : q \in (S^2)^n, \, \xi \in T_q(S^2)^n \},$$

denotes the tangent bundle of $(S^2)^n$. In addition,

$$T_q^*(S^2)^n = \{ \zeta \in (\mathbb{R}^{3n})^* : (\zeta \cdot \xi) \to \mathbb{R}^1, \, \xi \in T_q(S^2)^n \},$$

is the cotangent space to $(S^2)^n$ at $q \in (S^2)^n$; any $\zeta \in T_q^*(S^2)^n$ is referred to as a cotangent vector to $(S^2)^n$ at $q \in (S^2)^n$. Also

$$T^*(S^2)^n = \{ (q, \zeta) \in \mathbb{R}^{3n} \times (\mathbb{R}^{3n})^* : q \in (S^2)^n, \, \zeta \in T_q^*(S^2)^n \},$$

is the cotangent bundle of $(S^2)^n$. This geometry is important in our subsequent development of variational calculus on $(S^2)^n$.

5.2 Kinematics on $(S^2)^n$

Since the unit sphere in $\mathbb{R}^3$ is the set of points that are unit distance from the origin of $\mathbb{R}^3$, the tangent space $T_q S^2$ for $q \in S^2$ is the plane tangent to the sphere at the point $q \in S^2$. Thus, a time-parameterized curve $t \to q \in S^2$ and its time derivative $t \to \dot{q} \in T_q S^2$ satisfies $q \cdot \dot{q} = 0$ for all t. This fact implies that there exists an angular velocity vector function $t \to \omega \in \mathbb{R}^3$ such that the time derivative of this curve can be written as

$$\dot{q} = \omega \times q,$$

or equivalently

$$\dot{q} = S(\omega)q,$$

where $S(\omega)$ is the 3×3 skew symmetric matrix function defined in (1.8):

$$S(\omega) = \begin{bmatrix} 0 & -\omega_3 & \omega_2 \\ \omega_3 & 0 & -\omega_1 \\ -\omega_2 & \omega_1 & 0 \end{bmatrix}.$$

Without loss of generality, the angular velocity $t \to \omega \in \mathbb{R}^3$ can be constrained to be orthogonal to q, i.e., $q \cdot \omega = 0$. Therefore, it follows that q, $\dot{q}$ and ω are mutually orthogonal. This convention is followed throughout.

Consider a time-parameterized curve $t \to q = (q_1, \ldots, q_n) \in (S^2)^n$ and its time derivative $t \to \dot{q} = (\dot{q}_1, \ldots, \dot{q}_n) \in T_q(S^2)^n$; thus $q \cdot \dot{q} = \sum_{i=1}^{n}(q_i \cdot \dot{q}_i) = 0$ for all t. This is an important fact which implies that there is a vector function $t \to \omega = (\omega_1, \ldots, \omega_n) \in (\mathbb{R}^3)^n$ such that the time derivatives can be written as

$$\dot{q}_i = S(\omega_i)q_i, \quad i = 1, \ldots, n, \tag{5.1}$$

where the angular velocity vectors $\omega_i \in \mathbb{R}^3$, $i = 1, \ldots, n$, satisfy $q_i \cdot \omega_i = 0$, $i = 1, \ldots, n$. These equations describe the rotational kinematics on the configuration manifold $(S^2)^n$.

Premultiplying (5.1) by $S(q_i)$ and using the matrix identity $S(q_i)^2 = q_i q_i^T - I_{3 \times 3}$, it follows that the angular velocity vectors can be expressed as

$$\omega_i = S(q_i)\dot{q}_i, \quad i = 1, \ldots, n. \tag{5.2}$$

These relationships can be summarized as follows. For each $(q, \dot{q}) \in T(S^2)^n$, there exists an $\omega \in \mathbb{R}^{3n}$ such that $(q, S(\omega_1)q_1, \ldots, S(\omega_n)q_n) = (q, \dot{q})$. In this sense, we use the angular velocity vector as an alternative to the time derivative of the configuration vector and we can view $(q, \omega) \in T(S^2)^n$.

5.3 Lagrangian Dynamics on $(S^2)^n$

A Lagrangian function is introduced. Euler–Lagrange equations are derived using Hamilton's principle; that is, the variation of the action integral is zero. The Euler–Lagrange equations are first obtained in terms of a Lagrangian expressed in terms of the configuration vector and the time derivative of the configuration vector. A second form of the Euler–Lagrange equations is obtained in terms of a modified Lagrangian expressed in terms of the configuration vector and the angular velocity vector. In each case, these Euler–Lagrange equations are simplified for the important case that the kinetic energy function is a quadratic function of the time derivative of the configuration vector.

5.3.1 Hamilton's Variational Principle in Terms of $(q, \dot{q})$

The Lagrangian function $L : T(S^2)^n \to \mathbb{R}^1$ is defined on the tangent bundle of the configuration manifold. We assume that the Lagrangian function

$$L(q, \dot{q}) = T(q, \dot{q}) - U(q),$$

is given by the difference between a kinetic energy function $T(q, \dot{q})$, defined on the tangent bundle, and a configuration-dependent potential energy function $U(q)$.

The subsequent development describes variations of curves or functions with values in $(S^2)^n$. The unit sphere in $\mathbb{R}^3$ is not a Lie group, but the special orthogonal group $SO(3) = \{R \in GL(3) : R^T R = I, \det[R] = 1\}$ acts on the unit sphere transitively, i.e., for any $q_1, q_2 \in S^2$, there exists $R \in SO(3)$ such that $q_2 = Rq_1$. This is an example of a homogeneous manifold, which will be treated in greater generality in Chapter 8. Therefore, we can express the variation of a curve with values in S^2 in terms of a curve on $\mathfrak{so}(3)$ or a curve on $\mathbb{R}^3$ using the matrix exponential map. This observation allows us to develop expressions for variations of curves on $(S^2)^n$.

Let $q = (q_1, \ldots, q_n) : [t_0, t_f] \to (S^2)^n$ be a differentiable curve. The variation of q_i is defined by $q_i^\epsilon : (-c, c) \times [t_0, t_f] \to S^2$ for $c > 0$, such that $q_i^0(t) = q_i(t)$ for any $t \in [t_0, t_f]$ and $q_i^\epsilon(t_0) = q_i(t_0)$, $q_i^\epsilon(t_f) = q(t_f)$ for any $\epsilon \in (-c, c)$.

If $q = (q_1, \ldots, q_n) : [t_0, t_f] \to (S^2)^n$ is a differentiable curve on $(S^2)^n$, then its variation is $q^\epsilon = (q_1^\epsilon, \ldots, q_n^\epsilon) : [t_0, t_f] \to (S^2)^n$. Similarly, the time derivative is $\dot{q} = (\dot{q}_1, \ldots, \dot{q}_n) \in T_q(S^2)^n$, and its variation is $\dot{q}^\epsilon = (\dot{q}_1^\epsilon, \ldots, \dot{q}_n^\epsilon) : [t_0, t_f] \to T_q(S^2)^n$.

The variation can be expressed using the matrix exponential map as follows:

$$q_i^\epsilon(t) = e^{\epsilon S(\gamma_i(t))} q_i(t), \quad i = 1, \ldots, n,$$

for differentiable curves $\gamma_i : [t_0, t_f] \to \mathbb{R}^3$, satisfying $\gamma_i(t_0) = \gamma_i(t_f) = 0$, $i = 1, \ldots, n$. Since the exponent $\epsilon S(\gamma_i) \in \mathfrak{so}(3)$ for any $\gamma_i \in \mathbb{R}^3$, the exponential matrix is in $SO(3)$ and thus $e^{\epsilon S(\gamma_i)} : S^2 \to S^2$ is a local diffeomorphism. There is no loss of generality in requiring that $\gamma_i(t) \cdot q_i(t) = 0$ for all $t_0 \le t \le t_f$; that is, γ_i and q_i are orthogonal, $i = 1, \ldots, n$. The variations vanish at the end points of the time interval since $\gamma_i(t_0) = \gamma_i(t_f) = 0$, $i = 1, \ldots, n$.

The infinitesimal variations are computed as

$$\delta q_i(t) = \frac{d}{d\epsilon} q^\epsilon(t) \Big|_{\epsilon=0}$$
$$= S(\gamma_i(t)) q_i(t), \quad i = 1, \ldots, n, \tag{5.3}$$

where $\gamma_i(t_0) = \gamma_i(t_f) = 0$, $i = 1, \ldots, n$. The infinitesimal variations vanish at the end points of the time interval since $\gamma_i(t_0) = \gamma_i(t_f) = 0$, $i = 1, \ldots, n$.

Since the variation and differentiation commute, the expression for the infinitesimal variation of the time derivative is given by

$$\delta \dot{q}_i(t) = \frac{d}{d\epsilon} \dot{q}^\epsilon(t) \Big|_{\epsilon=0}$$
$$= S(\dot{\gamma}_i(t)) q_i(t) + S(\gamma_i(t)) \dot{q}_i(t). \tag{5.4}$$

These expressions define the infinitesimal variations for a vector function $(q,\dot{q}) = (q_1,\ldots,q_n,\dot{q}_1,\ldots,\dot{q}_n) : [t_0,t_f] \to \mathsf{T}(\mathsf{S}^2)^n$. The infinitesimal variations are important ingredients to derive the Euler–Lagrange equations on $(\mathsf{S}^2)^n$. To simplify the notation, we will subsequently suppress the time argument.

The action integral is the integral of the Lagrangian function along a motion of the system over a fixed time period. The variations are taken over all differentiable curves with values in $(\mathsf{S}^2)^n$ for which the initial and final values are fixed. The action integral along a motion is

$$\mathfrak{G} = \int_{t_0}^{t_f} L(q_1,\ldots,q_n,\dot{q}_1,\ldots,\dot{q}_n)\, dt.$$

The action integral along a variation of a motion is

$$\mathfrak{G}^\epsilon = \int_{t_0}^{t_f} L(q_1^\epsilon,\ldots,q_n^\epsilon,\dot{q}_1^\epsilon,\ldots,\dot{q}_n^\epsilon)\, dt.$$

The value of the action integral can be expressed as a power series in ϵ as

$$\mathfrak{G}^\epsilon = \mathfrak{G} + \epsilon\delta\mathfrak{G} + \mathcal{O}(\epsilon^2),$$

where the infinitesimal variation of the action integral is

$$\delta\mathfrak{G} = \frac{d}{d\epsilon}\mathfrak{G}^\epsilon\bigg|_{\epsilon=0}.$$

Hamilton's principle states that the infinitesimal variation of the action integral along any motion of the system is zero:

$$\delta\mathfrak{G} = \frac{d}{d\epsilon}\mathfrak{G}^\epsilon\bigg|_{\epsilon=0} = 0, \tag{5.5}$$

for all possible differentiable functions $\gamma_i : [t_0,t_f] \to \mathbb{R}^3$ satisfying $\gamma_i \cdot q_i = 0$ and $\gamma_i(t_0) = \gamma_i(t_f) = 0$, $i = 1,\ldots,n$.

The infinitesimal variation of the action integral can be expressed in terms of the infinitesimal variations of the motion as

$$\delta\mathfrak{G} = \int_{t_0}^{t_f}\left\{\sum_{i=1}^n \frac{\partial L(q,\dot{q})}{\partial\dot{q}_i}\cdot\delta\dot{q}_i + \sum_{i=1}^n \frac{\partial L(q,\dot{q})}{\partial q_i}\cdot\delta q_i\right\} dt.$$

We now substitute the expressions for the infinitesimal variations of the motion (5.3) and (5.4) into the above expression for the infinitesimal variation of the action integral. The result is simplified to obtain the Euler–Lagrange equations expressed in terms of $(q,\dot{q})$.

5.3.2 Euler–Lagrange Equations Expressed in Terms of $(q, \dot{q})$

Substitute (5.3) and (5.4) to obtain

$$\delta \mathfrak{G} = \int_{t_0}^{t_f} \sum_{i=1}^{n} \left\{ (S(\dot{\gamma}_i)q_i + S(\gamma_i)\dot{q}_i) \cdot \frac{\partial L(q, \dot{q})}{\partial \dot{q}_i} + (S(\gamma_i)q_i) \cdot \frac{\partial L(q, \dot{q})}{\partial q_i} \right\} dt.$$

This can be written as

$$\delta \mathfrak{G} = \int_{t_0}^{t_f} \sum_{i=1}^{n} \left\{ \dot{\gamma}_i \cdot \left(S(q_i) \frac{\partial L(q, \dot{q})}{\partial \dot{q}_i} \right) \right.$$
$$\left. + \gamma_i \cdot \left(S(\dot{q}_i) \frac{\partial L(q, \dot{q})}{\partial \dot{q}_i} + S(q_i) \frac{\partial L(q, \dot{q})}{\partial q_i} \right) \right\} dt.$$

Integrating the first term on the right by parts, the infinitesimal variation of the action integral is given by

$$\delta \mathfrak{G} = \sum_{i=1}^{n} \gamma_i \cdot \left(S(q_i) \frac{\partial L(q, \dot{q})}{\partial \dot{q}_i} \right) \Bigg|_{t_0}^{t_f}$$
$$- \sum_{i=1}^{n} \int_{t_0}^{t_f} \gamma_i \cdot \left(S(q_i) \left\{ \frac{d}{dt} \left(\frac{\partial L(q, \dot{q})}{\partial \dot{q}_i} \right) - \frac{\partial L(q, \dot{q})}{\partial q_i} \right\} \right) dt.$$

According to Hamilton's principle, $\delta \mathfrak{G} = 0$ for all continuous infinitesimal variations $\gamma_i : [t_0, t_f] \to \mathbb{R}^3$, that satisfy $(\gamma_i \cdot q_i) = 0$ and vanish at t_0 and t_f for $i = 1, \ldots, n$. The fundamental lemma of the calculus of variations, as described in Appendix A, implies that

$$S(q_i) \left\{ \frac{d}{dt} \left(\frac{\partial L(q, \dot{q})}{\partial \dot{q}_i} \right) - \frac{\partial L(q, \dot{q})}{\partial q_i} \right\} = 0, \quad i = 1, \ldots, n. \qquad (5.6)$$

Taking the cross product with q_i yields a preliminary form for the Euler–Lagrange equations

$$S(q_i)^2 \left\{ \frac{d}{dt} \left(\frac{\partial L(q, \dot{q})}{\partial \dot{q}_i} \right) - \frac{\partial L(q, \dot{q})}{\partial q_i} \right\} = 0, \quad i = 1, \ldots, n.$$

Using the cross-product matrix identity $S(q_i)^2 = q_i q_i^T - I_{3\times3}$, the Euler–Lagrange equations for a motion that evolves on $(S^2)^n$ can be written in the following form.

Proposition 5.1 *The Euler–Lagrange equations for a Lagrangian function* $L : \mathsf{T}(S^2)^n \to \mathbb{R}^1$ *are*

$$\left(I_{3\times3} - q_i q_i^T \right) \left\{ \frac{d}{dt} \left(\frac{\partial L(q, \dot{q})}{\partial \dot{q}_i} \right) - \frac{\partial L(q, \dot{q})}{\partial q_i} \right\} = 0, \quad i = 1, \ldots, n. \qquad (5.7)$$

The matrix $(I_{3\times3} - q_i q_i^T)$ is a projection of $\mathbb{R}^3$ onto $\mathsf{T}_{q_i}(\mathsf{S}^2)$ in the sense that for any $q_i \in \mathsf{S}^2$ and any $\dot{q}_i \in \mathsf{T}_{q_i}(\mathsf{S}^2)$:

$$(I_{3\times3} - q_i q_i^T)q_i = 0,$$
$$(I_{3\times3} - q_i q_i^T)\dot{q}_i = \dot{q}_i.$$

Equations (5.7) recall the classical Euler–Lagrange equations in Chapter 3, modified to reflect the fact that the configuration manifold is $(\mathsf{S}^2)^n$, and they also recall the analogous version of the Euler–Lagrange equations (4.6) defined on the configuration manifold $(\mathsf{S}^1)^n$ in Chapter 4.

We now consider the important case that the kinetic energy is a quadratic function of the derivative of the configuration vector, that is the Lagrangian function $L : \mathsf{T}(\mathsf{S}^2)^n \to \mathbb{R}^1$ is

$$L(q, \dot{q}) = \frac{1}{2} \sum_{j=1}^{n} \sum_{k=1}^{n} \dot{q}_j^T m_{jk}(q) \dot{q}_k - U(q), \tag{5.8}$$

where the scalar inertial terms $m_{jk} : (\mathsf{S}^2)^n \to \mathbb{R}^1$ satisfy the symmetry condition $m_{jk}(q) = m_{kj}(q)$ and the quadratic form in the time derivative of the configuration vector is positive-definite on $(\mathsf{S}^2)^n$.

We first determine the derivatives of the Lagrangian function

$$\frac{\partial L(q, \dot{q})}{\partial \dot{q}_i} = \sum_{j=1}^{n} m_{ij}(q)\dot{q}_j,$$

$$\frac{\partial L(q, \dot{q})}{\partial q_i} = \frac{1}{2} \frac{\partial}{\partial q_i} \sum_{j=1}^{n} \sum_{k=1}^{n} \dot{q}_j^T m_{jk}(q) \dot{q}_k - \frac{\partial U(q)}{\partial q_i},$$

and thus

$$\frac{d}{dt} \left\{ \frac{\partial L(q, \dot{q})}{\partial \dot{q}_i} \right\} = \sum_{j=1}^{n} m_{ij}(q)\ddot{q}_j + \sum_{j=1}^{n} \dot{m}_{ij}(q)\dot{q}_j.$$

It follows that the Euler–Lagrange equations can be written in a form that follows from (5.7):

$$(I_{3\times3} - q_i q_i^T)\left\{ \sum_{j=1}^{n} m_{ij}(q)\ddot{q}_j + \sum_{j=1}^{n} \dot{m}_{ij}(q)\dot{q}_j \right.$$
$$\left. - \frac{1}{2} \frac{\partial}{\partial q_i} \sum_{j=1}^{n} \sum_{k=1}^{n} \dot{q}_j^T m_{jk}(q) \dot{q}_k + \frac{\partial U(q)}{\partial q_i} \right\} = 0, \quad i = 1, \ldots, n. \tag{5.9}$$

Since $q_i^T \dot{q}_i = 0$, it follows that $\frac{d}{dt}(q_i^T \dot{q}_i) = (q_i^T \ddot{q}_i) + \|\dot{q}_i\|^2 = 0$; thus we obtain

$$(I_{3\times 3} - q_i q_i^T)\ddot{q}_i = \ddot{q}_i - (q_i q_i^T)\ddot{q}_i$$
$$= \ddot{q}_i + \|\dot{q}_i\|^2 q_i, \quad i = 1, \ldots, n.$$

Expand the time derivative expression from (5.9) to obtain the Euler–Lagrange equations on $T(S^2)^n$:

$$m_{ii}(q)\ddot{q}_i + (I_{3\times 3} - q_i q_i^T)\sum_{\substack{j=1 \\ j\neq i}}^{n} m_{ij}(q)\ddot{q}_j + m_{ii}(q)\|\dot{q}_i\|^2 q_i$$

$$+ (I_{3\times 3} - q_i q_i^T)F_i(q,\dot{q}) + (I_{3\times 3} - q_i q_i^T)\frac{\partial U(q)}{\partial q_i} = 0, \quad i = 1, \ldots, n, \quad (5.10)$$

where the vector-valued functions

$$F_i(q,\dot{q}) = \sum_{j=1}^{n} \dot{m}_{ij}(q)\dot{q}_j - \frac{1}{2}\frac{\partial}{\partial q_i}\sum_{j=1}^{n}\sum_{k=1}^{n} \dot{q}_j^T m_{jk}(q)\dot{q}_k, \quad i = 1, \ldots, n,$$

are quadratic in the time derivative of the configuration vector. As in Chapter 3, these functions can be expressed in terms of Christoffel terms. Note that if the inertia terms are constants independent of the configuration vector, then $F_i(q,\dot{q}) = 0$, $i = 1, \ldots, n$.

The Euler–Lagrange equations given by (5.10) describe the evolution of the dynamical flow $(q,\dot{q}) \in T(S^2)^n$ on the tangent bundle of the configuration manifold $(S^2)^n$.

If the inertia terms and the potential terms in (5.10) are globally defined on $(\mathbb{R}^3)^n$, then the domain of definition of (5.10) on $T(S^2)^n$ can be extended to $T(\mathbb{R}^3)^n$. This extension is natural and useful in that it defines a Lagrangian vector field on the tangent bundle $T(\mathbb{R}^3)^n$. Alternatively, the manifold $T(S^2)^n$ is an invariant manifold of this Lagrangian vector field on $T(\mathbb{R}^3)^n$ and its restriction to this invariant manifold describes the Lagrangian flow of (5.10) on $T(S^2)^n$.

5.3.3 Hamilton's Variational Principle in Terms of (q, ω)

An alternate expression for the Euler–Lagrange equations is now obtained in terms of the angular velocity vector introduced in (5.1).

We express the action integral in terms of the modified Lagrangian function

$$\tilde{L}(q,\omega) = L(q,\dot{q}),$$

where the kinematics equations are given by (5.1). We use the notation $\dot{q} = (\dot{q}_1, \ldots, \dot{q}_n) \in T_q(S^2)^n$ and, as described previously, we view $\omega =$

$(\omega_1, \ldots, \omega_n) \in \mathsf{T}_q(\mathsf{S}^2)^n$. Hence, the modified Lagrangian $\tilde{L}(q, \omega)$ is defined on the tangent bundle $\mathsf{T}(\mathsf{S}^2)^n$ using the kinematics (5.1).

The infinitesimal variation of the modified action integral can be written as

$$\delta\tilde{\mathfrak{G}} = \int_{t_0}^{t_f} \sum_{i=1}^{n} \left\{ \frac{\partial \tilde{L}(q, \omega)}{\partial \omega_i} \cdot \delta\omega_i + \frac{\partial \tilde{L}(q, \omega)}{\partial q_i} \cdot \delta q_i \right\} dt.$$

The infinitesimal variations of the motion are given by

$$\delta q_i = S(\gamma_i)q_i, \qquad\qquad i = 1, \ldots, n, \qquad\qquad (5.11)$$

$$\delta \dot{q}_i = S(\dot{\gamma}_i)q_i + S(\gamma_i)\dot{q}_i, \quad i = 1, \ldots, n \qquad\qquad (5.12)$$

for curves $\gamma_i : [t_0, t_f] \to \mathbb{R}^3$, $i = 1, \ldots, n$ that satisfy $(\gamma_i \cdot q_i) = 0$ and $\gamma_i(t_0) = \gamma_i(t_f) = 0$, $i = 1, \ldots, n$.

Next, we derive expressions for the infinitesimal variation of the angular velocity vector. Using the fact that $\omega_i = S(q_i)\dot{q}_i$, the infinitesimal variation of the angular velocity vectors is given by

$$\delta\omega_i = S(\delta q_i)\dot{q}_i + S(q_i)\delta\dot{q}_i, \quad i = 1, \ldots, n.$$

Substituting (5.3) and (5.4) and rearranging, we obtain

$$\delta\omega_i = (S(\gamma_i)q_i) \times \dot{q}_i + S(q_i)(S(\dot{\gamma}_i)q_i + S(\gamma_i)\dot{q}_i).$$

By expanding each term using the triple cross product vector identity, $S(x)(S(y)z) = (x \cdot z)y - (x \cdot y)z$ for any $x, y, z \in \mathbb{R}^3$, we obtain

$$\delta\omega_i = -(\dot{q}_i \cdot q_i)\gamma_i + (\dot{q}_i \cdot \gamma_i)q_i + (q_i \cdot q_i)\dot{\gamma}_i - (q_i \cdot \dot{\gamma}_i)q_i$$
$$+ (q_i \cdot \dot{q}_i)\gamma_i - (q_i \cdot \gamma_i)\dot{q}_i, \quad i = 1, \ldots, n.$$

Since $q_i \cdot q_i = 1$ and $q_i \cdot \dot{q}_i = q_i \cdot \gamma_i = 0$, this reduces to

$$\delta\omega_i = (\dot{q}_i \cdot \gamma_i)q_i + \dot{\gamma}_i - (q_i \cdot \dot{\gamma}_i)q_i, \quad i = 1, \ldots, n.$$

Substitute (5.1) and use the vector identity, $x \cdot (S(y)z) = y \cdot (S(z)x) = z \cdot (S(x)y)$ for any $x, y, z \in \mathbb{R}^3$ to obtain

$$\delta\omega_i = (\gamma_i \cdot (S(\omega_i)q_i))q_i + (I_{3\times3} - q_i q_i^T)\dot{\gamma}_i$$
$$= (q_i \cdot (S(\gamma_i)\omega_i))q_i + (I_{3\times3} - q_i q_i^T)\dot{\gamma}_i$$
$$= q_i q_i^T (S(\gamma_i)\omega_i) + (I_{3\times3} - q_i q_i^T)\dot{\gamma}_i, \quad i = 1, \ldots, n.$$

The matrix $q_i q_i^T$ corresponds to the orthogonal projection along q_i. Since both γ_i and ω_i are orthogonal to q_i, the infinitesimal variation of the angular velocity vector is

$$\delta\omega_i = -S(\omega_i)\gamma_i + (I_{3\times3} - q_iq_i^T)\dot\gamma_i, \quad i = 1,\ldots,n. \qquad (5.13)$$

The infinitesimal variation of ω_i is composed of two parts: the first term $\gamma_i \times \omega_i$ is collinear with q_i, and it represents the variations due to the change of q_i; the second term corresponds to the orthogonal projection of $\dot\gamma_i$ onto the space orthogonal to q_i, and it is due to the time rate change of the variation of q_i.

5.3.4 Euler–Lagrange Equations in Terms of (q, ω)

Substitute (5.11) and (5.13) to obtain

$$\delta\tilde{\mathfrak{G}} = \int_{t_0}^{t_f} \sum_{i=1}^{n} \left\{ (S(\gamma_i)\omega_i + (I_{3\times3} - q_iq_i^T)\dot\gamma_i) \cdot \frac{\partial\tilde{L}(q,\omega)}{\partial\omega_i} \right.$$
$$\left. + S(\gamma_i)q_i \cdot \frac{\partial\tilde{L}(q,\omega)}{\partial q_i} \right\} dt$$
$$= \int_{t_0}^{t_f} \left\{ \sum_{i=1}^{n} \left\{ S(q_i)\frac{\partial\tilde{L}(q,\omega)}{\partial q_i} + S(\omega_i)\frac{\partial\tilde{L}(q,\omega)}{\partial\omega_i} \right\} \cdot \gamma_i \right.$$
$$\left. + \left\{ (I_{3\times3} - q_iq_i^T)\frac{\partial\tilde{L}(q,\omega)}{\partial\omega_i} \right\} \cdot \dot\gamma_i \right\} dt.$$

Integrating the terms in the integral that multiply $\dot\gamma_i$ by parts, the infinitesimal variation of the modified action integral is given by

$$\delta\tilde{\mathfrak{G}} = \sum_{i=1}^{n} \gamma_i \cdot (I_{3\times3} - q_iq_i^T)\frac{\partial\tilde{L}(q,\omega)}{\partial\omega_i}\bigg|_{t_0}^{t_f}$$
$$+ \sum_{i=1}^{n} \int_{t_0}^{t_f} \gamma_i \cdot \left\{ -\frac{d}{dt}\left((I_{3\times3} - q_iq_i^T)\frac{\partial\tilde{L}(q,\omega)}{\partial\omega_i} \right) \right.$$
$$\left. + S(q_i)\frac{\partial\tilde{L}(q,\omega)}{\partial q_i} + S(\omega_i)\frac{\partial\tilde{L}(q,\omega)}{\partial\omega_i} \right\} dt.$$

According to Hamilton's principle, $\delta\tilde{\mathfrak{G}} = 0$ for all differentiable functions $\gamma_i : [t_0, t_f] \to \mathbb{R}^3$, $i = 1,\ldots,n$ that satisfy $(\gamma_i \cdot q_i) = 0$ and vanish at t_0 and t_f, $i = 1,\ldots,n$. The fundamental lemma of the calculus of variations, as described in Appendix A, implies that each expression in the braces should be orthogonal to the tangent space $\mathsf{T}_{q_i}\mathsf{S}^2$, or equivalently collinear with q_i. Thus

$$S(q_i)\left\{\frac{d}{dt}\left(S^2(q_i)\frac{\partial \tilde{L}(q,\omega)}{\partial \omega_i}\right) + S(\omega_i)\frac{\partial \tilde{L}(q,\omega)}{\partial \omega_i} + S(q_i)\frac{\partial \tilde{L}(q,\omega)}{\partial q_i}\right\} = 0,$$

for $i = 1,\ldots,n$. Multiply both sides of these equations by $S(q_i)$,

$$S(q_i)^4\frac{d}{dt}\left(\frac{\partial \tilde{L}(q,\omega)}{\partial \omega_i}\right) + S(q_i)^2 S(\omega_i)\frac{\partial \tilde{L}(q,\omega)}{\partial \omega_i}$$

$$+ S(q_i)^2\left\{S(\dot{q}_i)S(q_i) + S(q_i)S(\dot{q}_i)\right\}\frac{\partial \tilde{L}(q,\omega)}{\partial \omega_i} + S(q_i)^3\frac{\partial \tilde{L}(q,\omega)}{\partial q_i} = 0,$$

for $i = 1,\ldots,n$. Using $S(q_i)^3 = -S(q_i)$, this can be written as

$$-S(q_i)^2\frac{d}{dt}\left(\frac{\partial \tilde{L}(q,\omega)}{\partial \omega_i}\right) + S(q_i)^2 S(\omega_i)\frac{\partial \tilde{L}(q,\omega)}{\partial \omega_i}$$

$$+ S(q_i)^2\left\{S(\dot{q}_i)S(q_i) + S(q_i)S(\dot{q}_i)\right\}\frac{\partial \tilde{L}(q,\omega)}{\partial \omega_i} - S(q_i)\frac{\partial \tilde{L}(q,\omega)}{\partial q_i} = 0,$$

for $i = 1,\ldots,n$. To further simplify these expressions we use the matrix identities in Chapter 1 and the fact that $(\omega_i \cdot q_i) = 0$ to obtain

$$S(q_i)S(\omega_i)S(q_i) = S(q_i)\{-\omega_i^T q I_{3\times 3} + q_i\omega_i^T\} = 0.$$

From this it follows that

$$\begin{aligned}
S(q_i)S(\dot{q}_i) &= S(q_i)\{S(\omega_i)S(q_i) - S(q_i)S(\omega_i)\} \\
&= S(q_i)S(\omega_i)S(q_i) - S(q_i)^2 S(\omega_i) \\
&= -S(q_i)^2 S(\omega_i),
\end{aligned}$$

and

$$\begin{aligned}
S(\dot{q}_i)S(q_i) &= \{S(\omega_i)S(q_i) - S(q_i)S(\omega_i)\}S(q_i) \\
&= S(\omega_i)S(q_i)^2 - S(q_i)S(\omega_i)S(q_i) \\
&= S(\omega_i)S(q_i)^2.
\end{aligned}$$

Consequently, these results can be used to obtain

$$S(q_i)^2\{S(\dot{q}_i)S(q_i) + S(q_i)S(\dot{q}_i)\} = -S(q_i)^4 S(\omega_i) = S(q_i)^2 S(\omega_i).$$

Substituting this into the above expressions to simply the Euler–Lagrange equations we obtain

$$-S(q_i)^2 \frac{d}{dt}\left(\frac{\partial \tilde{L}(q,\omega)}{\partial \omega_i}\right) + 2S(q_i)^2 S(\omega_i)\frac{\partial \tilde{L}(q,\omega)}{\partial \omega_i}$$

$$- S(q_i)\frac{\partial \tilde{L}(q,\omega)}{\partial q_i} = 0, \quad i = 1,\dots,n.$$

Hence, the Euler–Lagrange equations take the following form.

Proposition 5.2 *The Euler–Lagrange equations for a modified Lagrangian function* $\tilde{L} : \mathsf{T}(\mathsf{S}^2)^n \to \mathbb{R}^1$ *are given by*

$$(I_{3\times 3} - q_i q_i^T)\left\{\frac{d}{dt}\left(\frac{\partial \tilde{L}(q,\omega)}{\partial \omega_i}\right) - 2S(\omega_i)\frac{\partial \tilde{L}(q,\omega)}{\partial \omega_i}\right\}$$

$$- S(q_i)\frac{\partial \tilde{L}(q,\omega)}{\partial q_i} = 0, \quad i = 1,\dots,n. \tag{5.14}$$

Thus, the Lagrangian flow on the tangent bundle $\mathsf{T}(\mathsf{S}^1)^n$ is obtained from the kinematics equations (5.1) and the Euler–Lagrange equations (5.14).

If $\frac{\partial \tilde{L}(q,\omega)}{\partial \omega_i}$ is orthogonal to q_i, the second term of the above equation vanishes to obtain

$$(I_{3\times 3} - q_i q_i^T)\frac{d}{dt}\left(\frac{\partial \tilde{L}(q,\omega)}{\partial \omega_i}\right) - S(q_i)\frac{\partial \tilde{L}(q,\omega)}{\partial q_i} = 0, \quad i = 1,\dots,n. \tag{5.15}$$

The Euler–Lagrange equations (5.14) on $(\mathsf{S}^2)^n$, expressed in terms of the angular velocity vector, can be obtained in a different way directly from the Euler–Lagrange equations given in (5.7) by viewing the kinematics (5.1) as defining a change of variables from the time derivative of the configuration vector to the angular velocity vector. This shows the equivalence of the Euler–Lagrange equations (5.14) and the Euler–Lagrange equations (5.7) and (5.1). The details of this derivation are not given.

Now consider the important case that the kinetic energy is a quadratic function of the time derivatives of the configuration where the Lagrangian function has the form given in (5.8). The expression for the modified Lagrangian function is

$$\tilde{L}(q,\omega) = \frac{1}{2}\sum_{i=1}^{n}\sum_{j=1}^{n}\omega_i^T S(q_i)^T m_{ij}(q)S(q_j)\omega_j - U(q). \tag{5.16}$$

Since $q_i^T \omega_i = 0$, it follows that

$$\omega_i^T S(q_i)^T S(q_i)\omega_i = \omega_i^T\left(I_{3\times 3} - q_i q_i^T\right)\omega_i = \omega_i^T \omega_i.$$

Thus, the modified Lagrangian can be expressed as

$$\tilde{L}(q,\omega) = \frac{1}{2}\sum_{i=1}^{n}\omega_i^T m_{ii}(q)\omega_i + \frac{1}{2}\sum_{i=1}^{n}\sum_{\substack{j=1\\j\neq i}}^{n}\omega_i^T S(q_i)^T m_{ij}(q)S(q_j)\omega_j - U(q).$$

We first determine the derivatives of the modified Lagrangian function

$$\frac{\partial \tilde{L}(q,\omega)}{\partial \omega_i} = \sum_{j=1}^{n} S(q_i)^T m_{ij}(q)S(q_j)\omega_j$$

$$= m_{ii}(q)\omega_i + \sum_{\substack{j=1\\j\neq i}}^{n} S(q_i)^T m_{ij}(q)S(q_j)\omega_j,$$

$$\frac{\partial \tilde{L}(q,\omega)}{\partial q_i} = \frac{1}{2}\frac{\partial}{\partial q_i}\sum_{j=1}^{n}\sum_{k=1}^{n}\omega_j^T S(q_j)^T m_{jk}(q)S(q_k)\omega_k - \frac{\partial U(q)}{\partial q_i}$$

$$= \sum_{j=1}^{n} S(\omega_i)^T m_{ij}(q)S(q_j)\omega_j$$

$$+ \frac{1}{2}\sum_{j=1}^{n}\sum_{k=1}^{n}(\omega_j^T S(q_j)^T S(q_k)\omega_k)\frac{\partial m_{jk}(q)}{\partial q_i} - \frac{\partial U(q)}{\partial q_i},$$

and thus

$$\frac{d}{dt}\left\{\frac{\partial \tilde{L}(q,\omega)}{\partial \omega_i}\right\} = m_{ii}(q)\dot{\omega}_i + \sum_{\substack{j=1\\j\neq i}}^{n} S(q_i)^T m_{ij}(q)S(q_j)\dot{\omega}_j$$

$$+ \sum_{j=1}^{n} S(q_i)^T \dot{m}_{ij}(q)S(q_j)\omega_j - \sum_{\substack{j=1\\j\neq i}}^{n} m_{ij}(q)\left\{S(\dot{q}_i)S(q_j) + S(q_i)S(\dot{q}_j)\right\}\omega_j.$$

We now use the relationship

$$(I_{3\times3} - q_i q_i^T)\dot{\omega}_i = \dot{\omega}_i,$$

and we use several identities to simplify:

$$(I_{3\times3} - q_i q_i^T)\left\{S(\dot{q}_i)S(q_j) + S(q_i)S(\dot{q}_j)\right\}\omega_j$$

$$= -S(q_i)^2\left\{S(\dot{q}_i)S(q_j) + S(q_i)S(\dot{q}_j)\right\}\omega_j$$

$$= -S(q_i)^2\left\{S(S(\omega_i)q_i)S(q_j) + S(q_i)S(S(\omega_j)q_j)\right\}\omega_j$$

$$= -S(q_i)^2\{S(\omega_i)S(q_i) - S(q_i)S(\omega_i)\}S(q_j)$$

$$+ S(q_i)\{S(\omega_j)S(q_j) - S(q_j)S(\omega_j)\}\omega_j.$$

We use the identity $S(\omega_i)S(q_i) = -\omega_i^T q_i I_{3\times 3} + q_i\omega_i^T$ to obtain

$$S(q_i)S(\omega_i)S(q_i) = S(q_i)(-\omega_i^T q_i I_{3\times 3} + q_i\omega_i^T) = 0,$$

so that

$$
\begin{aligned}
(I_{3\times 3} - q_i q_i^T) &\{S(\dot{q}_i)S(q_j) + S(q_i)S(\dot{q}_j)\} \omega_j \\
&= S(q_i)^3 S(\omega_i)S(q_j)\omega_j - S(q_i)^3 \{S(\omega_j)S(q_j) - S(q_j)S(\omega_j)\}\omega_j \\
&= -S(q_i)S(\omega_i)S(q_j)\omega_j + S(q_i)S(\omega_j)S(q_j)\omega_j \\
&= S(q_i)\{-S(\omega_i)S(q_j)\omega_j + S(\omega_j)S(q_j)\omega_j\} \\
&= S(q_i)\{-S(\omega_i)S(q_j)\omega_j + (-\omega_j^T q_j I_{3\times 3} + q_j\omega_j^T)\omega_j\} \\
&= S(q_i)\{-S(\omega_i)S(q_j)\omega_j + \|\omega_j\|^2 q_j\}.
\end{aligned}
$$

Using these relationships, we obtain

$$
\begin{aligned}
(I_{3\times 3} - q_i q_i^T)\frac{d}{dt}\left\{\frac{\partial \tilde{L}(q,\omega)}{\partial \omega_i}\right\} &= m_{ii}(q)\dot{\omega}_i \\
&+ \sum_{\substack{j=1 \\ j\neq i}}^{n} S(q_i)^T m_{ij}(q)S(q_j)\dot{\omega}_j + \sum_{j=1}^{n} S(q_i)^T \dot{m}_{ij}(q)S(q_j)\omega_j \\
&+ \sum_{\substack{j=1 \\ j\neq i}}^{n} m_{ij}(q)S(q_i)\left\{S(\omega_i)S(q_j)\omega_j - \|\omega_j\|^2 q_j\right\}.
\end{aligned}
$$

Also, we have

$$
(I_{3\times 3} - q_i q_i^T)S(\omega_i)\frac{\partial \tilde{L}(q,\omega)}{\partial \omega_i} = \sum_{\substack{j=1 \\ j\neq i}}^{n} m_{ij}(q)S(q_i)^2 S(\omega_i)S(q_i)S(q_j)\omega_j = 0,
$$

which follows since $S(q_i)S(\omega_i)S(q_i) = 0$. Substituting these expressions into (5.14) yields

$$
\begin{aligned}
m_{ii}(q)\dot{\omega}_i &+ \sum_{\substack{j=1 \\ j\neq i}}^{n} S(q_i)^T m_{ij}(q)S(q_j)\dot{\omega}_j + \sum_{j=1}^{n} S(q_i)^T \dot{m}_{ij}(q)S(q_j)\omega_j \\
&+ \sum_{\substack{j=1 \\ j\neq i}}^{n} m_{ij}(q)S(q_i)\left\{S(\omega_i)S(q_j)\omega_j - \|\omega_j\|^2 q_j\right\} \\
&- S(q_i)\sum_{j=1}^{n} S(\omega_i)^T m_{ij}(q)S(q_j)\omega_j \\
&- \frac{1}{2}S(q_i)\sum_{j=1}^{n}\sum_{k=1}^{n}(\omega_j^T S(q_j)^T S(q_k)\omega_k)\frac{\partial m_{jk}(q)}{\partial q_i} + S(q_i)\frac{\partial U(q)}{\partial q_i} = 0.
\end{aligned}
$$

This can be rearranged as

$$
m_{ii}(q)\dot{\omega}_i + \sum_{\substack{j=1 \\ j\neq i}}^{n} m_{ij}(q)S(q_i)^T S(q_j)\dot{\omega}_j
$$

$$
+ \sum_{\substack{j=1 \\ j\neq i}}^{n} m_{ij}(q)S(q_i)\left\{ S(\omega_i)S(q_j)\omega_j + S(\omega_i)S(\omega_j)q_j - \|\omega_j\|^2 q_j \right\}
$$

$$
+ m_{ii}(q)S(q_i)S(\omega_i)^2 q_i + S(q_i)\left\{ -\sum_{j=1}^{n} \dot{m}_{ij}(q)S(q_j)\omega_j \right.
$$

$$
\left. - \frac{1}{2}\sum_{j=1}^{n}\sum_{k=1}^{n}(\omega_j^T S(q_j)^T S(q_k)\omega_k)\frac{\partial m_{jk}(q)}{\partial q_i} + \frac{\partial U(q)}{\partial q_i} \right\} = 0.
$$

The result can be simplified using $S(q_i)S(\omega_i)^2 q_i = S(q_i)(-\|\omega_i\|^2 q_i) = 0$ and $S(q_j)\omega_j + S(\omega_j)q_j = 0$. This leads to a convenient form of the Euler–Lagrange equations on $\mathsf{T}(\mathsf{S}^2)^n$, which consists of the kinematics equations (5.1) and

$$
m_{ii}(q)\dot{\omega}_i + \sum_{\substack{j=1 \\ j\neq i}}^{n} S(q_i)^T m_{ij}(q)S(q_j)\dot{\omega}_j - \sum_{\substack{j=1 \\ j\neq i}}^{n} m_{ij}(q)S(q_i)\|\omega_j\|^2 q_j
$$

$$
+ S(q_i)\left\{ F_i(q,\omega) + \frac{\partial U(q)}{\partial q_i} \right\} = 0, \quad i=1,\ldots,n, \quad (5.17)
$$

where

$$
F_i(q,\omega) = \sum_{j=1}^{n} \dot{m}_{ij}(q)S(\omega_j)q_j
$$

$$
- \frac{1}{2}\sum_{j=1}^{n}\sum_{k=1}^{n}(q_j^T S(\omega_j)^T S(\omega_k)q_k)\frac{\partial m_{jk}(q)}{\partial q_i}, \quad i=1,\ldots,n,
$$

are quadratic in the angular velocity vector. If the inertial terms are constants and independent of the configuration, then $F_i(q,\omega) = 0$, $i=1,\ldots,n$.

This version of the Euler–Lagrange differential equations describe the dynamical flow $(q,\omega) \in \mathsf{T}(\mathsf{S}^2)^n$ on the tangent bundle $\mathsf{T}(\mathsf{S}^2)^n$.

Assuming that the inertia terms and the potential terms in (5.17) are globally defined on $(\mathbb{R}^3)^n$, then the domain of definition of (5.17) on $\mathsf{T}(\mathsf{S}^2)^n$ can be extended to $\mathsf{T}(\mathbb{R}^3)^n$. This extension is natural and useful in that it defines a Lagrangian vector field on the tangent bundle $\mathsf{T}(\mathbb{R}^3)^n$. Conversely, the manifold $\mathsf{T}(\mathsf{S}^2)^n$ is an invariant manifold of this Lagrangian vector field on $\mathsf{T}(\mathbb{R}^3)^n$ and its restriction to this invariant manifold describes the Lagrangian flow of (5.17) on $\mathsf{T}(\mathsf{S}^2)^n$.

Equations (5.17) and the kinematics (5.1) can be shown to be equivalent to (5.10) by viewing the kinematics as defining a transformation from $(q, \dot{q})$ to (q, ω). This can provide an alternate derivation of (5.17).

5.4 Hamiltonian Dynamics on $(S^2)^n$

The Legendre transformation is introduced to derive Hamilton's equations for the dynamics that evolve on the cotangent bundle $\mathsf{T}^*(S^2)^n$. The derivation is based on the phase space variational principle, a natural modification of Hamilton's principle for Lagrangian dynamics. Two forms of Hamilton's equations are obtained. One form is expressed in terms of a momentum $\mu = (\mu_1, \ldots, \mu_n) \in \mathsf{T}_q^*(S^2)^n$ that is conjugate to the velocity vector $\dot{q} = (\dot{q}_1, \ldots, \dot{q}_n) \in \mathsf{T}_q(S^2)^n$, where $q \in (S^2)^n$. The other form of Hamilton's equations are expressed in terms of a momentum $\pi = (\pi_1, \ldots, \pi_n) \in \mathsf{T}_q^*(S^2)^n$ that is conjugate to the angular velocity vector $\omega = (\omega_1, \ldots, \omega_n) \in \mathsf{T}_q(S^2)^n$.

5.4.1 Hamilton's Phase Space Variational Principle in Terms of (q, μ)

As in the prior section, we begin with a Lagrangian function $L : \mathsf{T}(S^2)^n \rightarrow \mathbb{R}^1$, which is a real-valued function defined on the tangent bundle of the configuration manifold $(S^2)^n$; we assume that the Lagrangian function

$$L(q, \dot{q}) = T(q, \dot{q}) - U(q),$$

is given by the difference between a kinetic energy function $T(q, \dot{q})$, defined on the tangent bundle, and a configuration-dependent potential energy function $U(q)$.

The Legendre transformation of the Lagrangian function $L(q, \dot{q})$ leads to the Hamiltonian form of the equations of motion in terms of a conjugate momentum. For $q \in (S^2)^n$, the corresponding conjugate momentum $\mu \in \mathsf{T}_q^*(S^2)^n$ lies in the dual space $\mathsf{T}^*(S^2)^n$. We identify the tangent space $\mathsf{T}_q(S^2)^n$ and its dual space $\mathsf{T}_q^*(S^2)^n$ by using the usual dot product in $\mathbb{R}^{3n}$.

The Legendre transformation maps $\dot{q} = (\dot{q}_1, \ldots, \dot{q}_n) \in \mathsf{T}_q(S^2)^n \rightarrow \mu = (\mu_1, \ldots, \mu_n) \in \mathsf{T}_q^*(S^2)^n$ and satisfies

$$\mu_i \cdot \dot{q}_i = \frac{\partial L(q, \dot{q})}{\partial \dot{q}_i} \cdot \dot{q}_i, \quad i = 1, \ldots, n.$$

Since the component of μ_i collinear with q_i has no effect on the inner product above, the vector representing μ_i is selected to be orthogonal to q_i; that is μ_i is equal to the projection of $\frac{\partial L(q,\dot{q})}{\partial \dot{q}_i}$ onto the cotangent space $\mathsf{T}_{q_i}^* S^2$. Thus

$$\mu_i = \frac{\partial L(q, \dot{q})}{\partial \dot{q}_i} - (q_i \cdot \frac{\partial L(q, \dot{q})}{\partial \dot{q}_i})q_i, \quad i = 1, \ldots, n,$$

which can be written as

$$\mu_i = (I_{3\times 3} - q_i q_i^T)\frac{\partial L(q, \dot{q})}{\partial \dot{q}_i}, \quad i = 1, \ldots, n, \tag{5.18}$$

using the projections $(I_{3\times 3} - q_i q_i^T)$ from $(\mathbb{R}^3)^*$ to $T_{q_i}^* S^2$ for $i = 1, \ldots, n$.

The Lagrangian function is assumed to be hyperregular; that is, the Legendre transformation, viewed as a map $T_q(S^2)^n \to T_q^*(S^2)^n$, is invertible.

The Hamiltonian function $H : T^*(S^2)^n \to \mathbb{R}^1$ is given by

$$H(q, \mu) = \sum_{i=1}^{n} \mu_i \cdot \dot{q}_i - L(q, \dot{q}), \tag{5.19}$$

where the right-hand side is expressed in terms of (q, μ) using the Legendre transformation (5.18).

The Legendre transformation can be viewed as defining a transformation $(q, \dot{q}) \in T(S^2)^n \to (q, \mu) \in T^*(S^2)^n$, which implies that the Euler–Lagrange equations can be written in terms of the transformed variables; this is effectively Hamilton's equations. However, Hamilton's equations can be obtained using Hamilton's phase space variational principle, and this approach is now introduced.

Consider the action integral in the form,

$$\mathfrak{G} = \int_{t_0}^{t_f} \left\{ \sum_{i=1}^{n} \mu_i \cdot \dot{q}_i - H(q, \mu) \right\} dt.$$

The infinitesimal variation of the action integral is given by

$$\delta\mathfrak{G} = \sum_{i=1}^{n} \int_{t_0}^{t_f} \left\{ \mu_i \cdot \delta\dot{q}_i - \frac{\partial H(q, \mu)}{\partial q_i} \cdot \delta q_i + \left(\dot{q}_i - \frac{\partial H(\mu, p)}{\partial \mu_i} \right) \cdot \delta\mu_i \right\} dt.$$

We can integrate the first term on the right-hand side by parts, so that Hamilton's phase space variational principle is:

$$\delta\mathfrak{G} = \sum_{i=1}^{n} \mu_i \cdot \delta q_i \Big|_{t_0}^{t_f}$$

$$+ \sum_{i=1}^{n} \int_{t_0}^{t_f} \left\{ \left(-\dot{\mu}_i - \frac{\partial H(q, \mu)}{\partial q_i} \right) \cdot \delta q_i + \left(\dot{q}_i - \frac{\partial H(\mu, p)}{\partial \mu_i} \right) \cdot \delta\mu_i \right\} dt = 0,$$

for all allowable differentiable variations satisfying $\delta q_i(t_0) = \delta q_i(t_f) = 0$ for all $i = 1, \ldots, n$.

According to the definition of the conjugate momenta μ_i given by (5.18), we have $q_i \cdot \mu_i = 0$, which implies that $\delta q_i \cdot \mu_i + q_i \cdot \delta \mu_i = 0$. To impose this constraint on the infinitesimal variations, we decompose $\delta \mu_i$ into the sum of two orthogonal components: one component collinear with q_i, namely $\delta \mu_i^C = q_i q_i^T \delta \mu_i$, and the other component orthogonal to q_i, namely $\delta \mu_i^M = (I_{3\times 3} - q_i q_i^T) \delta \mu_i$. Satisfaction of the above constraint implies that $q_i^T \delta \mu_i^C = q_i^T \delta \mu_i = -\mu_i^T \delta q_i$, and $\delta \mu_i^M = (I_{3\times 3} - q_i q_i^T) \delta \mu_i$ is otherwise unconstrained.

Recall the prior expression for the infinitesimal variation of a motion on $(\mathsf{S}^2)^n$:

$$\delta q_i = S(\gamma_i) q_i, \quad i = 1, \ldots, n,$$

for differentiable curves $\gamma_i : [t_0, t_f] \to \mathbb{R}^3$ satisfying $\gamma_i \cdot q_i = 0$ and $\gamma_i(t_0) = \gamma_i(t_f) = 0$ for $i = 1, \ldots, n$.

These results can be summarized as: the infinitesimal variation of the action integral is zero for all possible differentiable curves $\gamma_i : [t_0, t_f] \to \mathbb{R}^3$ and $\delta \mu_i^M : [t_0, t_f] \to \mathbb{R}^3$ satisfying $\gamma_i \cdot q_i = 0$, $\delta \mu_i^M \cdot q_i = 0$ and $\gamma_i(t_0) = \gamma_i(t_f) = 0$ for all $i = 1, \ldots, n$.

5.4.2 Hamilton's Equations in Terms of (q, μ)

Substituting the expressions for the infinitesimal variations of the motion and using the fact that the infinitesimal variations of the motion vanish at the endpoints, the infinitesimal variation of the action integral can be rewritten as

$$\delta \mathfrak{G} = \sum_{i=1}^n \int_{t_0}^{t_f} \left\{ \left(-\dot{\mu}_i - \frac{\partial H(q, \mu)}{\partial q_i} \right) \cdot \delta q_i + \left(q_i q_i^T \left(\dot{q}_i - \frac{\partial H(q, \mu)}{\partial \mu_i} \right) \right) \cdot \delta \mu_i^C \right.$$
$$\left. + \left((I_{3\times 3} - q_i q_i^T) \left(\dot{q}_i - \frac{\partial H(q, \mu)}{\partial \mu_i} \right) \right) \cdot \delta \mu_i^M \right\} dt$$
$$= \sum_{i=1}^n \int_{t_0}^{t_f} \left\{ \left(-\dot{\mu}_i - \frac{\partial H(q, \mu)}{\partial q_i} \right) \cdot \delta q_i - \left(q_i^T \frac{\partial H(q, \mu)}{\partial \mu_i} \right) \cdot (q_i^T \delta \mu_i^C) \right.$$
$$\left. + \left(\dot{q}_i - (I_{3\times 3} - q_i q_i^T) \frac{\partial H(q, \mu)}{\partial \mu_i} \right) \cdot \delta \mu_i^M \right\} dt.$$

Since $q_i^T \delta \mu_i^C = -\mu_i^T \delta q_i$ and $\delta q_i = S(\gamma_i) q_i = -S(q_i) \gamma_i$,

$$\delta \mathfrak{G} = \sum_{i=1}^n \int_{t_0}^{t_f} \left\{ \left(-\dot{\mu}_i - \frac{\partial H(q, \mu)}{\partial q_i} \right) \cdot \delta q_i + \left(q_i^T \frac{\partial H(q, \mu)}{\partial \mu_i} \right) \cdot (\mu_i^T \delta q_i) \right.$$

$$+ \left(\dot{q}_i - (I_{3\times 3} - q_i q_i^T) \frac{\partial H(q, \mu)}{\partial \mu_i} \right) \cdot \delta \mu_i^M \bigg\} \, dt$$

$$= \sum_{i=1}^{n} \int_{t_0}^{t_f} \left\{ S(q_i) \left(-\dot{\mu}_i - \frac{\partial H(q, \mu)}{\partial q_i} + \mu_i q_i^T \frac{\partial H(q, \mu)}{\partial \mu_i} \right) \cdot \gamma_i \right.$$

$$\left. + \left(\dot{q}_i - (I_{3\times 3} - q_i q_i^T) \frac{\partial H(q, \mu)}{\partial \mu_i} \right) \cdot \delta \mu_i^M \right\} \, dt.$$

We now invoke Hamilton's phase space variational principle that $\delta \mathfrak{G} = 0$ for all possible functions $\gamma_i : [t_0, t_f] \to \mathbb{R}^3$ satisfying $\gamma_i \cdot q_i = 0$ and $\delta \mu_i^M : [t_0, t_f] \to \mathbb{R}^3$ that are always orthogonal to q_i for $i = 1, \ldots, n$.

According to the fundamental lemma of the calculus of variations, in Appendix A, the first condition gives

$$S(q_i) \left(\dot{\mu}_i + \frac{\partial H(q, \mu)}{\partial q_i} - \mu_i q_i^T \frac{\partial H(q, \mu)}{\partial \mu_i} \right) = 0, \quad i = 1, \ldots, n.$$

We multiply this by $S(q_i)$ and use a matrix identity for $S(q_i)^2$ to obtain

$$(I_{3\times 3} - q_i q_i^T) \left(\dot{\mu}_i + \frac{\partial H(q, \mu)}{\partial q_i} - \mu_i q_i^T \frac{\partial H(q, \mu)}{\partial \mu_i} \right) = 0, \quad i = 1, \ldots, n.$$

Since both terms multiplying $\delta \mu_i^M$ in the above variational expression are necessarily orthogonal to q_i, it follows that

$$\dot{q}_i - (I_{3\times 3} - q_i q_i^T) \frac{\partial H(q, \mu)}{\partial \mu_i} = 0, \quad i = 1, \ldots, n.$$

We now determine an expression for $\dot{\mu}_i$. The above equation only determines the component of $\dot{\mu}_i$ that is in the dual of the tangent space $\mathsf{T}_{q_i}^* \mathsf{S}^1$. The component of $\dot{\mu}_i$ that is orthogonal to this tangent space, that is collinear with q_i, is determined as follows. The time derivative of $q_i \cdot \mu_i = 0$ gives $q_i \cdot \dot{\mu}_i = -\dot{q}_i \cdot \mu_i$ which allows computation of the component of $\dot{\mu}_i$ that is collinear with q_i. Thus, $\dot{\mu}_i$ is the sum of two components:

$$\dot{\mu}_i = (I_{3\times 3} - q_i q_i^T) \left(-\frac{\partial H(q, \mu)}{\partial q_i} + \mu_i q_i^T \frac{\partial H(q, \mu)}{\partial \mu_i} \right) - (\mu_i^T \dot{q}_i) q_i$$

$$= -(I_{3\times 3} - q_i q_i^T) \frac{\partial H(q, \mu)}{\partial q_i} + \mu_i q_i^T \frac{\partial H(q, \mu)}{\partial \mu_i} - \left(\mu_i^T \frac{\partial H(q, \mu)}{\partial \mu_i} \right) q_i$$

$$= -(I_{3\times 3} - q_i q_i^T) \frac{\partial H(q, \mu)}{\partial q_i} + \frac{\partial H(q, \mu)}{\partial \mu_i} \times (\mu_i \times q_i),$$

where the last step makes use of the triple cross product identity. In summary, Hamilton's equations are given as follows.

Proposition 5.3 *Hamilton's equations for a Hamiltonian function H :*
$T^*(S^2)^n \to \mathbb{R}^1$ *are*

$$\dot{q}_i = (I_{3\times3} - q_i q_i^T)\frac{\partial H(q,\mu)}{\partial \mu_i}, \quad i = 1, \ldots, n, \tag{5.20}$$

$$\dot{\mu}_i = -(I_{3\times3} - q_i q_i^T)\frac{\partial H(q,\mu)}{\partial q_i} + \frac{\partial H(q,\mu)}{\partial \mu_i} \times (\mu_i \times q_i), \quad i = 1, \ldots, n. \tag{5.21}$$

Thus, equations (5.20) and (5.21) describe the Hamiltonian flow in terms of
$(q,\mu) \in T^*(S^2)^n$ on the cotangent bundle of $(S^2)^n$.

The following property follows directly from the above formulation of
Hamilton's equations on the configuration manifold $(S^2)^n$:

$$\frac{dH(q,\mu)}{dt} = \sum_{i=1}^n \frac{\partial H(q,\mu)}{\partial q_i} \cdot \dot{q}_i + \frac{\partial H(q,\mu)}{\partial \mu_i} \cdot \dot{\mu}_i$$

$$= \sum_{i=1}^n \frac{\partial H(q,\mu)}{\partial \mu_i} \cdot \left\{ \frac{\partial H(q,\mu)}{\partial \mu_i} \times (\mu_i \times q_i) \right\}$$

$$= 0.$$

Thus, the Hamiltonian function is constant along each solution of Hamilton's
equation. As before, this property does not hold if the Hamiltonian function
has a nontrivial explicit dependence on time.

Now consider the important case that the kinetic energy function is a
quadratic function of the time derivative of the configuration vector so that
the Lagrangian function is given by

$$L(q,\dot{q}) = \frac{1}{2}\sum_{j=1}^n \sum_{k=1}^n \dot{q}_j^T m_{jk}(q)\dot{q}_k - U(q), \tag{5.22}$$

where the scalar inertial terms $m_{jk}(q) : (S^2)^n \to \mathbb{R}^1$ satisfy the symmetry
condition $m_{jk}(q) = m_{kj}(q)$ and the quadratic form is positive-definite on
$(S^2)^n$. Thus, the conjugate momentum is defined by the Legendre transformation

$$\mu_i = (I_{3\times3} - q_i q_i^T)\frac{\partial L(q,\dot{q})}{\partial \dot{q}_i}$$

$$= (I_{3\times3} - q_i q_i^T)\sum_{j=1}^n m_{ij}(q)\dot{q}_j$$

$$= m_{ii}(q)\dot{q}_i + (I_{3\times3} - q_i q_i^T)\sum_{\substack{j=1 \\ j \neq i}}^n m_{ij}(q)\dot{q}_j, \quad i = 1, \ldots, n. \tag{5.23}$$

We assume that the algebraic equations (5.23), viewed as a mapping from $\dot{q} = (\dot{q}_1, \ldots, \dot{q}_n) \in \mathsf{T}_q(S^2)^n$ to $\mu = (\mu_1, \ldots, \mu_n) \in \mathsf{T}_q^*(S^2)^n$, can be inverted and expressed in the form

$$\dot{q}_i = (I_{3\times 3} - q_i q_i^T) \sum_{j=1}^{n} m_{ij}^I(q)\mu_j, \quad i = 1, \ldots, n, \tag{5.24}$$

where $m_{ij}^I : (S^2)^n \to \mathbb{R}^{3\times 3}$, for $i = 1, \ldots, n$, $j = 1, \ldots, n$ are the entries in the matrix inverse obtained from (5.23). There is no loss of generality in including the indicated projection in (5.24) since the projection guarantees that if $(\mu_1, \ldots, \mu_n) \in \mathsf{T}_q^*(S^2)^n$ then $(\dot{q}_1, \ldots, \dot{q}_n) \in \mathsf{T}_q(S^2)^n$.

The Hamiltonian function can be written as

$$
\begin{aligned}
H(q, \mu) &= \sum_{i=1}^{n} \dot{q}_i \cdot \mu_i - \frac{1}{2}\sum_{i=1}^{n}\sum_{j=1}^{n} \dot{q}_i^T m_{ij}(q)\dot{q}_j + U(q) \\
&= \sum_{i=1}^{n} \dot{q}_i \cdot \mu_i - \frac{1}{2}\sum_{i=1}^{n} \dot{q}_i \cdot \sum_{j=1}^{n} m_{ij}(q)\dot{q}_j + U(q) \\
&= \sum_{i=1}^{n} \dot{q}_i \cdot \mu_i - \frac{1}{2}\sum_{i=1}^{n} \dot{q}_i \cdot (I_{3\times 3} - q_i q_i^T)\sum_{j=1}^{n} m_{ij}(q)\dot{q}_j + U(q) \\
&= \frac{1}{2}\sum_{i=1}^{n} \dot{q}_i \cdot \mu_i + U(q).
\end{aligned}
$$

The third step uses the fact that the inner product $\mu_i \cdot \sum_{j=1}^{n} m_{ij}(q)\dot{q}_j$ is not changed by projecting the vector defined by the sum onto the tangent space $\mathsf{T}_{q_i}S^2$. The Hamiltonian function can be expressed as

$$H(q, \mu) = \frac{1}{2}\sum_{j=1}^{n}\sum_{k=1}^{n} \mu_j^T m_{jk}^I(q)\mu_k + U(q).$$

From (5.20) and (5.21), Hamilton's equations, expressed in terms of the evolution of (q, μ) on the cotangent bundle $\mathsf{T}^*(S^2)^n$, are given by

$$\dot{q}_i = (I_{3\times 3} - q_i q_i^T)\sum_{j=1}^{n} m_{ij}^I(q)\mu_j, \quad i = 1, \ldots, n, \tag{5.25}$$

$$\dot{\mu}_i = \sum_{j=1}^{n} \left(m_{ij}^I(q)\mu_j\right) \times (\mu_i \times q_i) - (I_{3\times 3} - q_i q_i^T)\frac{1}{2}\frac{\partial}{\partial q_i}\sum_{j=1}^{n}\sum_{k=1}^{n} \mu_j^T m_{jk}^I(q)\mu_k$$

$$- (I_{3\times 3} - q_i q_i^T)\frac{\partial U(q)}{\partial q_i}, \quad i = 1, \ldots, n. \tag{5.26}$$

The summation terms on the right-hand side of (5.26) are quadratic in the conjugate momenta.

Hamilton's equations (5.25) and (5.26) describe the Hamiltonian flow in terms of $(q, \mu) \in T^*(S^2)^n$ on the cotangent bundle $T^*(S^2)^n$.

If the Legendre transformation is globally invertible and the potential terms in (5.25) and (5.26) are globally defined on $(\mathbb{R}^3)^n$, then the domain of definition of (5.25) and (5.26) on $T^*(S^2)^n$ can be extended to $T^*(\mathbb{R}^3)^n$. This extension is natural in that it defines a Hamiltonian vector field on the cotangent bundle $T^*(\mathbb{R}^3)^n$. Alternatively, the manifold $T^*(S^2)^n$ is an invariant manifold of this Hamiltonian vector field on $T^*(\mathbb{R}^3)^n$ and its restriction to this invariant manifold describes the Hamiltonian flow of (5.25) and (5.26) on $T^*(S^2)^n$.

5.4.3 Hamilton's Phase Space Variational Principle in Terms of (q, π)

We now present an alternate version of Hamilton's equations using the Legendre transformation of the modified Lagrangian function $\tilde{L}(q, \omega)$ to define the conjugate momentum. The Legendre transformation $\omega = (\omega_1, \ldots, \omega_n) \in T_q(S^2)^n \rightarrow \pi = (\pi_1, \ldots, \pi_n) \in T_q^*(S^2)^n$ is defined by

$$\pi_i = (I_{3\times3} - q_i q_i^T) \frac{\partial \tilde{L}(q, \omega)}{\partial \omega_i}, \quad i = 1, \ldots, n. \tag{5.27}$$

Assume the Lagrangian function is hyperregular; that is the Legendre transformation, viewed as a map $T_q(S^2)^n \rightarrow T_q^*(S^2)^n$, is invertible.

The modified Hamiltonian function given by

$$\tilde{H}(q, \pi) = \sum_{j=1}^{n} \pi_j \cdot \omega_j - \tilde{L}(q, \omega),$$

where the right-hand side is expressed in terms of (q, π) using the Legendre transformation (5.27).

Consider the modified action integral

$$\tilde{\mathfrak{G}} = \int_{t_0}^{t_f} \left\{ \sum_{j=1}^{n} \pi_j \cdot \omega_j - \tilde{H}(q, \omega) \right\} dt.$$

The infinitesimal variation of the action integral is given by

$$\delta\tilde{\mathfrak{G}} = \sum_{i=1}^{n} \int_{t_0}^{t_f} \left\{ \pi_i \cdot \delta\omega_i - \frac{\partial \tilde{H}(q, \pi)}{\partial q_i} \cdot \delta q_i + \left(\omega_i - \frac{\partial \tilde{H}(q, \pi)}{\partial \pi_i} \right) \cdot \delta\pi_i \right\} dt.$$

Recall from (5.3) and (5.13) that the infinitesimal variations can be written as

$$\delta q_i = S(\gamma_i) q_i, \qquad\qquad\qquad i = 1, \ldots, n,$$
$$\delta \omega_i = -S(\omega_i)\gamma_i + (I_{3\times 3} - q_i q_i^T)\dot\gamma_i, \quad i = 1, \ldots, n,$$

for differentiable curves $\gamma_i : [t_0, t_f] \to \mathbb{R}^3$, $i = 1, \ldots, n$ satisfying $\gamma_i \cdot q_i = 0$ and $\gamma_i(t_0) = \gamma_i(t_f) = 0$, $i = 1, \ldots, n$.

5.4.4 Hamilton's Equations in Terms of (q, π)

Substitute the expressions for δq_i and $\delta \omega_i$ into the infinitesimal variation of the modified action integral and integrate by parts to obtain

$$\delta \tilde{\mathfrak{G}} = \sum_{j=1}^{n} \int_{t_0}^{t_f} \left(\omega_i - \frac{\partial \tilde{H}(q, \pi)}{\partial \pi_i} \right) \cdot \delta\pi_i$$
$$+ \left(-\dot\pi_i + S(\omega)\pi_i - S(q_i)\frac{\partial \tilde{H}(q, \pi)}{\partial q_i} \right) \cdot \gamma_i \, dt,$$

using the fact that $(I_{3\times 3} - q_i q_i^T)\pi_i = \pi_i$ since π_i is orthogonal to q_i by the definition (5.27). The orthogonality condition $\pi_i \cdot q_i = 0$ also implies that $\delta q_i \cdot \pi_i + q_i \cdot \delta\pi_i = 0$. To impose this constraint on the infinitesimal variations, we decompose $\delta\pi_i$ into a component orthogonal to the tangent space $\mathsf{T}_{q_i}\mathsf{S}^2$, that is collinear with q_i, namely $\delta\pi_i^C = q_i q_i^T \delta\pi_i$, and a component in the tangent space $\mathsf{T}_{q_i}\mathsf{S}^2$, that is orthogonal to q_i, namely $\delta\pi_i^M = (I_{3\times 3} - q_i q_i^T)\delta\pi_i$. From the above constraint, we have $q_i^T \delta\pi_i = -\pi_i^T \delta q_i = -\pi_i^T S(\gamma_i)q_i = \pi_i^T S(q_i)\gamma_i$. Therefore $\delta\pi_i^C = q_i q_i^T \delta\pi_i = q_i \pi_i^T S(q_i)\gamma_i$.

Using these facts, the variation of the action integral can be rewritten as

$$\delta \tilde{\mathfrak{G}} = \sum_{i=1}^{n} \int_{t_0}^{t_f} \left\{ -\dot\pi_i + S(\omega_i)\pi_i - S(q_i)\frac{\partial \tilde{H}(q, \pi)}{\partial q_i} \right\} \cdot \gamma_i$$
$$+ \left\{ q_i q_i^T \left(\omega_i - \frac{\partial \tilde{H}(q, \pi)}{\partial \pi_i} \right) \right\} \cdot \delta\pi_i^C$$
$$+ \left\{ (I_{3\times 3} - q_i q_i^T) \left(\omega_i - \frac{\partial \tilde{H}(q, \pi)}{\partial \pi_i} \right) \right\} \cdot \delta\pi_i^M \, dt.$$

Since $q_i \cdot \omega_i = 0$ and $(I_{3\times 3} - q_i q_i^T)\omega_i = \omega_i$,

$$\delta\tilde{\mathfrak{G}} = \sum_{i=1}^{n}\int_{t_0}^{t_f}\left\{-\dot{\pi}_i + S(\omega_i)\pi_i - S(q_i)\frac{\partial\tilde{H}(q,\pi)}{\partial q_i} + S(q_i)\pi_i q_i^T\frac{\partial\tilde{H}(q,\pi)}{\partial\pi_i}\right\}\cdot\gamma_i$$

$$+\left\{\omega_i - (I_{3\times3} - q_i q_i^T)\frac{\partial\tilde{H}(q,\pi)}{\partial\pi_i}\right\}\cdot\delta\pi_i^M\,dt.$$

Hamilton's phase space variational principle gives: $\delta\tilde{\mathfrak{G}} = 0$ for all possible functions $\gamma_i : [t_0, t_f] \to \mathbb{R}^3$ and $\delta\pi_i^M : [t_0, t_f] \to \mathbb{R}^3$ that are in the cotangent space $\mathsf{T}_{q_i}^*\mathsf{S}^2$ and satisfy $\gamma_i(t_0) = \gamma_i(t_f) = 0$, $i = 1,\ldots,n$. This implies that the expression in each of the parentheses of the above equation should be collinear with q_i, or equivalently,

$$(I_{3\times3} - q_i q_i^T)\left(-\dot{\pi}_i + S(\omega_i)\pi_i - S(q_i)\frac{\partial\tilde{H}(q,\pi)}{\partial q_i} + S(q_i)\pi_i q_i^T\frac{\partial\tilde{H}(q,\pi)}{\partial\pi_i}\right) = 0,$$

$$\omega_i = (I_{3\times3} - q_i q_i^T)\frac{\partial\tilde{H}(q,\pi)}{\partial\pi_i} = -S(q_i)^2\frac{\partial\tilde{H}(q,\pi)}{\partial\pi_i}.$$

Using the fact that $(I_{3\times3} - q_i q_i^T)S(q_i) = -S(q_i)^3 = S(q_i)$ and $(I_{3\times3} - q_i q_i^T)S(\omega_i)\pi_i = -S(q_i)\{q_i \times (\omega_i \times \pi_i)\} = -S(q_i)\{(q_i\cdot\pi_i)\omega_i - (q_i\cdot\omega_i)\pi_i\} = 0$, the first equation of the above reduces to

$$-(I_{3\times3} - q_i q_i^T)\dot{\pi}_i - S(q_i)\frac{\partial\tilde{H}(q,\pi)}{\partial q_i} + S(q_i)\pi_i q_i^T\frac{\partial\tilde{H}(q,\pi)}{\partial\pi_i} = 0.$$

However, this is incomplete since it only determines the component of $\dot{\pi}_i$ that is orthogonal to q_i. The component of $\dot{\pi}_i$ that is collinear with q_i is determined by taking the time derivative of $q_i\cdot\pi_i = 0$ to obtain $q_i\cdot\dot{\pi}_i = -\dot{q}_i\cdot\pi_i$. Therefore, $q_i q_i^T\dot{\pi}_i = -q_i\pi_i^T\dot{q}_i$. By combining these, Hamilton's equations on the configuration manifold $(\mathsf{S}^2)^n$ can be written in terms of (q,π) as

$$\dot{q}_i = -S(q_i)\omega_i$$

$$= S(q_i)^3\frac{\partial\tilde{H}(q,\pi)}{\partial\pi_i}$$

$$= -S(q_i)\frac{\partial\tilde{H}(q,\pi)}{\partial\pi_i},$$

$$\dot{\pi}_i = -S(q_i)\frac{\partial\tilde{H}(q,\pi)}{\partial q_i} + S(q_i)\pi_i q_i^T\frac{\partial\tilde{H}(q,\pi)}{\partial\pi_i} + q_i\pi_i^T S(q_i)\frac{\partial\tilde{H}(q,\pi)}{\partial\pi_i}$$

$$= -S(q_i)\frac{\partial\tilde{H}(q,\pi)}{\partial q_i} + \frac{\partial\tilde{H}(q,\pi)}{\partial\pi_i}\times((S(q_i)\pi_i)\times q_i)$$

$$= -S(q_i)\frac{\partial\tilde{H}(q,\pi)}{\partial q_i} + \frac{\partial\tilde{H}(q,\pi)}{\partial\pi_i}\times(-S(q_i)^2\pi_i).$$

But $-S(q_i)^2\pi_i = \pi_i$ since π_i is orthogonal to q_i. In summary, Hamilton's equations are given as follows.

Proposition 5.4 *Hamilton's equations for a modified Hamiltonian function $\tilde{H} : \mathsf{T}^*(\mathsf{S}^2)^n \to \mathbb{R}^1$ are*

$$\dot{q}_i = -S(q_i)\frac{\partial \tilde{H}(q,\pi)}{\partial \pi_i}, \qquad\qquad i = 1,\ldots,n, \qquad (5.28)$$

$$\dot{\pi}_i = -S(q_i)\frac{\partial \tilde{H}(q,\pi)}{\partial q_i} + \frac{\partial \tilde{H}(q,\pi)}{\partial \pi_i} \times \pi_i, \quad i = 1,\ldots,n. \qquad (5.29)$$

Thus, equations (5.28) and (5.29) describe the Hamiltonian flow in terms of $(q,\pi) \in \mathsf{T}^*(\mathsf{S}^2)^n$ on the cotangent bundle $\mathsf{T}^*(\mathsf{S}^2)^n$.

The following property follows directly from the above formulation of Hamilton's equations on $(\mathsf{S}^2)^n$:

$$\begin{aligned}
\frac{d\tilde{H}(q,\pi)}{dt} &= \sum_{i=1}^{n} \frac{\partial \tilde{H}(q,\pi)}{\partial q_i} \cdot \dot{q}_i + \frac{\partial \tilde{H}(q,\pi)}{\partial \pi_i} \cdot \dot{\pi}_i \\
&= \sum_{i=1}^{n} \frac{\partial \tilde{H}(q,\pi)}{\partial \pi_i} \cdot \left\{ \frac{\partial \tilde{H}(q,\pi)}{\partial \pi_i} \times \pi \right\} \\
&= 0.
\end{aligned}$$

The modified Hamiltonian function is constant along each solution of Hamilton's equation. This property does not hold if the modified Hamiltonian function has a nontrivial explicit dependence on time.

Let the kinetic energy be a quadratic function of the angular velocity vector in the form that arises from the Lagrangian given by (5.22). Thus, the modified Lagrangian function, expressed in terms of the angular velocity vector, is

$$\tilde{L}(q,\omega) = \frac{1}{2}\sum_{j=1}^{n}\sum_{k=1}^{n} \omega_j^T S(q_j)^T m_{jk}(q) S(q_k)\omega_k - U(q). \qquad (5.30)$$

The conjugate momentum is defined by the Legendre transformation

$$\begin{aligned}
\pi_i &= (I_{3\times3} - q_i q_i^T)\frac{\partial \tilde{L}(q,\omega)}{\partial \omega_i} \\
&= (I_{3\times3} - q_i q_i^T)\sum_{j=1}^{n} S(q_i)^T m_{ij}(q) S(q_j)\omega_j \\
&= \sum_{j=1}^{n} S(q_i)^T m_{ij}(q) S(q_j)\omega_j \\
&= m_{ii}(q)\omega_i + \sum_{\substack{j=1 \\ j\neq i}}^{n} S(q_i)^T m_{ij}(q) S(q_j)\omega_j, \quad i = 1,\ldots,n. \qquad (5.31)
\end{aligned}$$

The algebraic equations (5.31), viewed as a linear mapping from $\omega = (\omega_1, \ldots, \omega_n) \in T_q(S^2)^n$ to $\pi = (\pi_1, \ldots, \pi_n) \in T_q^*(S^2)^n$, can be inverted and expressed in the form

$$\omega_i = \sum_{j=1}^{n} m_{ij}^I(q)\pi_j, \quad i = 1, \ldots, n,$$

where $m_{ij}^I : (S^2)^n \to \mathbb{R}^{3 \times 3}$, for $i = 1, \ldots, n$, $j = 1, \ldots, n$, are the entries of the matrix inverse obtained from (5.31).

The modified Hamiltonian function can be expressed as

$$\tilde{H}(q, \pi) = \frac{1}{2} \sum_{i=1}^{n} \sum_{j=1}^{n} \pi_i^T m_{ij}^I(q)\pi_j + U(q). \tag{5.32}$$

Hamilton's equations, expressed in terms of (q, π), describe the Hamiltonian flow on the cotangent bundle $T(S^2)^n$ according to:

$$\dot{q}_i = -S(q_i) \left\{ \sum_{j=1}^{n} m_{ij}^I(q)\pi_j \right\}, \qquad\qquad i = 1, \ldots, n, \tag{5.33}$$

$$\dot{\pi}_i = -\frac{1}{2} S(q_i) \frac{\partial}{\partial q_i} \sum_{j=1}^{n} \sum_{k=1}^{n} \pi_j^T m_{jk}^I(q)\pi_k$$

$$+ \left\{ \sum_{j=1}^{n} m_{ij}^I(q)\pi_j \right\} \times \pi_i - S(q_i)\frac{\partial U(q)}{\partial q_i}, \quad i = 1, \ldots, n. \tag{5.34}$$

If the Legendre transformation is globally invertible and the potential terms in (5.33) and (5.34) are globally defined on $(\mathbb{R}^3)^n$, then the domain of definition of (5.33) and (5.34) on $T^*(S^2)^n$ can be extended to $T^*(\mathbb{R}^3)^n$. This extension is natural in that it defines a Hamiltonian vector field on the cotangent bundle $T^*(\mathbb{R}^3)^n$. Alternatively, the manifold $T^*(S^2)^n$ is an invariant manifold of this Hamiltonian vector field on $T^*(\mathbb{R}^3)^n$ and its restriction to this invariant manifold describes the Hamiltonian flow of (5.33) and (5.34) on $T^*(S^2)^n$.

5.5 Linear Approximations of Dynamics on $(S^2)^n$

Geometric forms of the Euler–Lagrange equations and Hamilton's equations on the configuration manifold $(S^2)^n$ have been presented. These equations of motion provide insight into the geometry of the global dynamics of the associated Lagrangian vector field on $T(S^2)^n$ or the Hamiltonian vector field on $T^*(S^2)^n$.

Let $(q_e, 0) \in T(S^2)^n$ be an equilibrium solution of the Lagrangian vector field. It is possible to develop a linear vector field that approximates the Lagrangian vector field, at least locally in an open subset of $T(S^2)^n$.

A common approach in much of the literature on dynamical systems on $(S^2)^n$ introduces $2n$ local coordinates, such as angle coordinates, to describe the configuration. These descriptions involve complicated trigonometric or transcendental expressions and introduce additional complexity in the analysis and computations.

Although the main emphasis throughout this book is on global methods, we make use of local coordinates as a way of describing a linear vector field that approximates a nonlinear vector field on a manifold, at least in the neighborhood of an equilibrium solution. This approach is used subsequently in the chapter to study the local flow properties near an equilibrium. As further background for the subsequent development, Appendix B summarizes a linearization procedure for a Lagrangian vector field defined on TS^2.

5.6 Dynamics on $(S^2)^n$

In this section, several Lagrangian and Hamiltonian systems with configuration manifolds given by $(S^2)^n$ are introduced. In each case, the physical description and assumptions are made clear. The Euler–Lagrange equations are expressed in two different forms and Hamilton's equations are obtained in two different forms. These follow directly from the expression for the Lagrangian function in each example, and the general form of the equations of motion for dynamics on $(S^2)^n$ developed earlier in this chapter. Special features of these equations are described.

5.6.1 Dynamics of a Spherical Pendulum

A spherical pendulum consists of a rigid link connected to a frictionless, two degree of freedom ideal pivot; the mass of the spherical pendulum is assumed to be concentrated at the end of the massless link. The spherical pendulum acts under uniform, constant gravity.

Introduce an inertial Euclidean frame in three dimensions; the first two axes of the frame are horizontal and the third axis is vertical; the origin of the inertial frame is located at the pivot of the spherical pendulum. Most conventional treatments of the spherical pendulum define the attitude configuration in terms of two angles. Although this choice is natural for small angle motions of the spherical pendulum, it is problematic for large rotational motions of the spherical pendulum.

Here we introduce a globally defined notion of the attitude of the spherical pendulum. The attitude vector $q \in \mathsf{S}^2$ is a unit vector from the pivot point to the concentrated pendulum mass defined with respect to the inertial Euclidean frame. The configuration manifold is the sphere S^2, viewed as an embedded manifold in $\mathbb{R}^3$. Thus, the spherical pendulum has two degrees of freedom. A schematic of a spherical pendulum is shown in Figure 5.1.

Fig. 5.1 Spherical pendulum

Uniform gravity acts on the concentrated mass of the pendulum; g is the constant acceleration of gravity. The distance from the pivot to the pendulum mass is L and m is the mass of the pendulum. No forces, other than gravity, act on the spherical pendulum.

To illustrate the several possible formulations of the equations of motion, we present two equivalent versions of the Euler–Lagrange equations and two equivalent versions of Hamilton's equations.

5.6.1.1 Euler–Lagrange Equations in Terms of $(q, \dot{q})$

The Lagrangian function $L : \mathsf{T}\mathsf{S}^2 \to \mathbb{R}^1$ for a spherical pendulum can be written as the difference of the kinetic energy function and the gravitational potential energy function:

$$L(q, \dot{q}) = \frac{1}{2}mL^2 \|\dot{q}\|^2 - mgLe_3^T q. \tag{5.35}$$

The Euler–Lagrange equation for the spherical pendulum is obtained from (5.10) as

$$mL^2\ddot{q} + mL^2 \|\dot{q}\|^2 q + mgL(I_{3\times3} - qq^T)e_3 = 0. \tag{5.36}$$

This equation describes the Lagrangian dynamics of the spherical pendulum in terms of $(q, \dot{q}) \in \mathsf{T}\mathsf{S}^2$ on the tangent bundle of the configuration manifold S^2.

5.6.1.2 Euler–Lagrange Equations in Terms of (q, ω)

A convenient form of the Euler–Lagrange equations can also be obtained in terms of the angular velocity vector of the spherical pendulum $\omega \in T_q S^2$ which satisfies the rotational kinematics

$$\dot{q} = S(\omega)q. \tag{5.37}$$

Recall that the angular velocity vector satisfies $\omega^T q = 0$ and $S(\omega)$ is the 3×3 skew-symmetric matrix function.

The modified Lagrangian function can be expressed in terms of the angular velocity vector of the spherical pendulum as

$$\tilde{L}(q, \omega) = \frac{1}{2} m L^2 \|\omega\|^2 - mgLe_3^T q.$$

The Euler–Lagrange equation, expressed in terms of the angular velocity vector, consists of the rotational kinematics (5.37) and the equation obtained from (5.17), which can be written as

$$mL^2 \dot{\omega} + mgLS(q)e_3 = 0. \tag{5.38}$$

These equations (5.37) and (5.38) describe the Lagrangian dynamics of the spherical pendulum in terms of $(q, \omega) \in TS^2$ on the tangent bundle of S^2.

5.6.1.3 Hamilton's Equations in Terms of (q, μ)

Define the conjugate momentum

$$\mu = (I_{3\times 3} - qq^T) \frac{\partial L(q, \dot{q})}{\partial \dot{q}} = mL^2 \dot{q},$$

using the Legendre transformation. The Hamiltonian function can be expressed as

$$H(q, \mu) = \frac{1}{2mL^2} \|\mu\|^2 + mgLe_3^T q.$$

Then, Hamilton's equations of motion for the spherical pendulum are given by (5.25) and (5.26), which can be written in this case as

$$\dot{q} = \frac{\mu}{mL^2}, \tag{5.39}$$

$$\dot{\mu} = -\frac{\|\mu\|^2}{mL^2} q - mgL(I_{3\times 3} - qq^T)e_3. \tag{5.40}$$

Hamilton's equations (5.39) and (5.40) describe the Hamiltonian dynamics of the spherical pendulum $(q, \mu) \in \mathsf{T}^*S^2$ on the cotangent bundle of the configuration manifold.

5.6.1.4 Hamilton's Equations in Terms of (q, π)

Define the angular momentum

$$\pi = (I_{3 \times 3} - qq^T) \frac{\partial \tilde{L}(q, \omega)}{\partial \omega} = mL^2 \omega,$$

using the Legendre transformation. The modified Hamiltonian function is

$$\tilde{H}(q, \pi) = \frac{1}{2mL^2} \|\pi\|^2 + mgLe_3^T q.$$

Hamilton's equations of motion for the spherical pendulum, obtained from (5.33) and (5.34), are

$$\dot{q} = \frac{1}{mL^2} S(\pi) q, \tag{5.41}$$

$$\dot{\pi} = mgLS(e_3)q. \tag{5.42}$$

Thus, the Hamiltonian flow of the spherical pendulum is described by (5.41) and (5.42) in terms of the evolution of $(q, \pi) \in \mathsf{T}^*S^2$ on the cotangent bundle of S^2.

5.6.1.5 Conservation Properties

The Hamiltonian, which coincides with the total energy E in this case, is conserved, that is

$$H = \frac{1}{2} mL^2 \|\dot{q}\|^2 + mgLe_3^T q$$

is constant along each solution of the dynamical flow of the spherical pendulum.

It is also easy to show that the vertical component of the angular momentum, namely

$$e_3^T \pi = mL^2 e_3^T \omega$$

is constant along each solution of the dynamical flow of the spherical pendulum. This arises as a consequence of Noether's theorem, due to the invariance of the Lagrangian with respect to the lifted action of rotations about the gravity direction.

5.6.1.6 Equilibrium Properties

An important feature of the dynamics of the spherical pendulum is its equilibrium configurations in S^2. The conditions for an equilibrium are that the time derivative of the configuration vector, or equivalently the angular velocity vector or the momentum, is zero and the equilibrium configuration satisfies:

$$(I_{3\times3} - qq^T)e_3 = 0,$$

which implies that the time derivatives of the angular velocity or momentum both vanish as well. Hence, there are two equilibrium solutions given by $(-e_3, 0) \in \mathsf{TS}^2$ and by $(e_3, 0) \in \mathsf{TS}^2$. A configuration is an equilibrium if and only if it is collinear with the direction of gravity. The equilibrium $(-e_3, 0) \in \mathsf{TS}^2$ is referred to as the hanging equilibrium of the spherical pendulum; the equilibrium $(e_3, 0) \in \mathsf{TS}^2$ is the inverted equilibrium of the spherical pendulum.

The stability properties of each equilibrium are studied in turn using (5.47). We first linearize (5.36) about the inverted equilibrium $(e_3, 0) \in \mathsf{TS}^2$ to obtain

$$mL^2\ddot{\xi}_1 - mgL\xi_1 = 0,$$
$$mL^2\ddot{\xi}_2 - mgL\xi_2 = 0,$$

defined on the tangent space of TS^2 at $(e_3, 0) \in \mathsf{TS}^2$. These linear differential equations approximate the local dynamics of the spherical pendulum in a neighborhood of the inverted equilibrium $(e_3, 0) \in \mathsf{TS}^2$. The eigenvalues are easily determined to be $+\sqrt{\frac{g}{L}}, -\sqrt{\frac{g}{L}}, +\sqrt{\frac{g}{L}}, -\sqrt{\frac{g}{L}}$. Since there are positive eigenvalues, the inverted equilibrium $(e_3, 0) \in \mathsf{TS}^2$ is unstable.

We now linearize (5.36) about the hanging equilibrium $(-e_3, 0) \in \mathsf{TS}^2$. This gives the resulting linearized differential equations

$$mL^2\ddot{\xi}_1 + mgL\xi_1 = 0,$$
$$mL^2\ddot{\xi}_2 + mgL\xi_2 = 0,$$

defined on the tangent space of TS^2 at $(-e_3, 0) \in \mathsf{TS}^2$. These linear differential equations approximate the local dynamics of the spherical pendulum in a neighborhood of the hanging equilibrium $(-e_3, 0) \in \mathsf{TS}^2$. The eigenvalues are easily determined to be $+j\sqrt{\frac{g}{L}}, -j\sqrt{\frac{g}{L}}, +j\sqrt{\frac{g}{L}}, -j\sqrt{\frac{g}{L}}$. Since the eigenvalues are purely imaginary no conclusion can be drawn about the stability of the hanging equilibrium.

We now describe a Lyapunov method to demonstrate the stability of the hanging equilibrium. The total energy of the spherical pendulum can be shown to have a strict local minimum at the hanging equilibrium on TS^2, and the sublevel sets in TS^2 in a neighborhood of the hanging equilibrium are compact. Further, the time derivative of the total energy along the flow given by (5.36) is zero, that is the total energy does not increase along the flow. According to standard results, this guarantees that the hanging equilibrium $(-e_3, 0) \in \mathsf{TS}^2$ is stable.

These stability results confirm the physically intuitive conclusions about the local flow properties of these two equilibrium solutions.

5.6.2 Dynamics of a Particle Constrained to a Sphere That Rotates with Constant Angular Velocity

A particle, with mass m, is constrained to move, without friction, on a rigid spherical surface of radius $L > 0$. The sphere rotates about an inertially fixed diameter with a constant angular speed $\Omega \in \mathbb{R}^1$. This example can be compared with the prior example in Chapter 4 of a particle constrained to move on a rotating circular hoop. This example can be interpreted as a spherical pendulum whose pivot rotates at a constant angular velocity. Since spherical pendulum models usually include gravity effect, and the particle on a rotating sphere does not include gravity effects in the current formulation, the subsequent results are interpreted in terms of a particle constrained to a sphere that rotates with a constant angular velocity.

Introduce a rotating sphere-fixed Euclidean frame and an inertial Euclidean frame in three dimensions; the origins of the two frames are located at the fixed center of the rotating sphere. The third axes of the two frames are in the fixed direction of the angular velocity vector of the rotating sphere. The first two axes of the two frames are selected to form a right-hand Euclidean frame.

We use a globally defined notion of the configuration of the particle on a sphere: the attitude vector $q = (q_1, q_2, q_3) \in \mathsf{S}^2$ is a unit vector from the center of the sphere to the location of the particle defined with respect to the rotating sphere-fixed Euclidean frame. The configuration manifold is S^2, viewed as an embedded manifold in $\mathbb{R}^3$. Thus, the particle on a rotating sphere has two degrees of freedom. The position vector of the particle with respect to the sphere-fixed frame is $Lq \in \mathbb{R}^3$ and the position vector of the particle with respect to the inertial frame is $x \in \mathbb{R}^3$. A schematic of a particle on a rotating sphere is shown in Figure 5.2.

A conventional treatment of particle motion on a sphere would define the configuration in terms of latitude and longitude angles. Although this choice is natural for small angle motions, it is problematic for large motions through or near the poles.

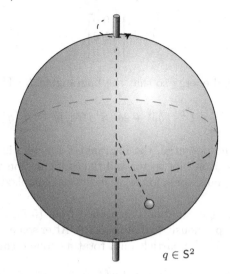

$q \in \mathsf{S}^2$

Fig. 5.2 Particle on a rotating sphere

5.6.2.1 Euler–Lagrange Equations in Terms of $(q, \dot{q})$

Assume that, initially, the first two axes of the sphere-fixed frame are aligned with the first two axes of the inertial frame. The position vector of the particle in the three-dimensional inertial frame can be expressed in terms of the configuration vector of the particle by

$$x = L \begin{bmatrix} \cos \Omega t & -\sin \Omega t & 0 \\ \sin \Omega t & \cos \Omega t & 0 \\ 0 & 0 & 1 \end{bmatrix} q,$$

where the transformation matrix is the rotation matrix from the sphere-fixed frame to the inertial frame. Thus, it follows that

$$\dot{x} = L \begin{bmatrix} \cos \Omega t & -\sin \Omega t & 0 \\ \sin \Omega t & \cos \Omega t & 0 \\ 0 & 0 & 1 \end{bmatrix} (S(\Omega e_3)q + \dot{q}),$$

since $\Omega e_3 \in \mathbb{R}^3$ is the angular velocity vector of the sphere in the sphere-fixed frame.

The kinetic energy of the particle on a rotating sphere can be expressed as

$$T(q, \dot{q}) = \frac{1}{2}m \left\| \dot{x} \right\|^2 = \frac{1}{2}mL^2 \left\{ \left\| \dot{q} \right\|^2 + 2q^T S^T(\Omega e_3)\dot{q} + \Omega^2 \left(q^T C q \right) \right\},$$

where

$$C = \begin{bmatrix} 1 & 0 & 0 \\ 0 & 1 & 0 \\ 0 & 0 & 0 \end{bmatrix}.$$

There is no potential energy so that the Lagrangian $L : \mathsf{TS}^2 \to \mathbb{R}^1$ is

$$L(q, \dot{q}) = \frac{1}{2}mL^2 \left\{ \|\dot{q}\|^2 + 2q^T S^T (\Omega e_3)\dot{q} + \Omega^2 \left(q^T C q \right) \right\}.$$

Although there is no potential, the Lagrangian function is a quadratic function in the configuration rate vector and the configuration vector; the second term in the Lagrangian expression adds additional complexity to the resulting dynamics.

The Euler–Lagrange equation is obtained from (5.7), following the same approach indicated previously to obtain (5.10). After some algebra, the Euler–Lagrange equation for the particle on a rotating sphere can be written as

$$mL^2\ddot{q} + mL^2 \|\dot{q}\|^2 q + 2mL^2(I_{3\times3} - qq^T)S(\Omega e_3)\dot{q}$$
$$- mL^2\Omega^2(I_{3\times3} - qq^T)Cq = 0. \tag{5.43}$$

This equation describes the Lagrangian dynamics of the particle on a rotating sphere in terms of $(q, \dot{q}) \in \mathsf{TS}^2$ on the tangent bundle of the configuration manifold S^2. The dynamics can also be used to describe the motion of the particle with respect to the inertial frame using the above relationships.

5.6.2.2 Euler–Lagrange Equations in Terms of (q, ω)

The Euler–Lagrange equations can also be expressed in terms of the angular velocity vector, which satisfies the rotational kinematics

$$\dot{q} = S(\omega)q, \tag{5.44}$$

where the angular velocity vector satisfies $\omega^T q = 0$ and $S(\omega)$ is the 3×3 skew-symmetric matrix function.

The modified Lagrangian function can be expressed in terms of the angular velocity vector as

$$\tilde{L}(q, \omega) = \frac{1}{2}mL^2 \left\{ \|\omega\|^2 + 2q^T S(\Omega e_3)S(q)\omega + \Omega^2 \left(q^T C q \right) \right\}.$$

The Euler–Lagrange equation, expressed in terms of the angular velocity vector ω, consists of the rotational kinematics (5.44) and an equation that can be obtained from (5.14); this is a lengthy equation, not given here, suggesting that it is not so convenient in this case to describe the Lagrangian dynamics in terms of the angular velocity vector.

5.6.2.3 Hamilton's Equations in Terms of (q, μ)

Define the momentum

$$\mu = (I_{3\times 3} - qq^T)\frac{\partial L(q, \dot{q})}{\partial \dot{q}} = mL^2\{\dot{q} + S(\Omega e_3)q\},$$

using the Legendre transformation. Thus, $\mu \in T_q^*S^2$ is conjugate to $\dot{q} \in T_qS^2$. The Hamiltonian function is

$$H(q, \mu) = \mu^T \dot{q} - \frac{1}{2}mL^2\{\|\dot{q}\|^2 - 2q^T S(\Omega e_3)\dot{q} + \Omega^2 (q^T Cq)\}$$

$$= \frac{1}{2}mL^2\{\|\dot{q}\|^2 - \Omega^2 (q^T Cq)\},$$

which can be expressed as

$$H(q, \mu) = \frac{1}{2mL^2}\|\mu - mL^2 S(\Omega e_3)q\|^2 - \frac{1}{2}mL^2\Omega^2 (q^T Cq).$$

Then, Hamilton's equations of motion for the particle on a rotating sphere are given by (5.20) and (5.21), which can be written in this case as

$$\dot{q} = \frac{1}{mL^2}(\mu - mL^2 S(\Omega e_3)q), \tag{5.45}$$

$$\dot{\mu} = -(I_{3\times 3} - qq^T)S(\Omega e_3)(\mu - mL^2 S(\Omega e_3)q)$$

$$+ \frac{1}{mL^2}(\mu - mL^2 S(\Omega e_3)q) \times (\mu \times q) + mL^2\Omega^2(I_{3\times 3} - qq^T)Cq. \tag{5.46}$$

Hamilton's equations (5.45) and (5.46) describe the Hamiltonian dynamics of the particle on a rotating sphere in terms of $(q, \mu) \in T^*S^2$ on the cotangent bundle of the configuration manifold.

5.6.2.4 Hamilton's Equations in Terms of (q, π)

Define the angular momentum

$$\pi = (I_{3\times 3} - qq^T)\frac{\partial \tilde{L}(q, \omega)}{\partial \omega} = mL^2\{\omega + S(q)S(\Omega e_3)q\},$$

using the Legendre transformation. Thus, $\pi \in T_q^*S^2$ is conjugate to the angular velocity $\omega \in T_qS^2$. The modified Hamiltonian function is

$$\tilde{H}(q, \pi) = \pi^T \omega - \frac{1}{2}mL^2\left\{\|\omega\|^2 + 2q^T S(\Omega e_3)S(q)\omega + \Omega^2 (q^T Cq)\right\}$$

$$= \frac{1}{2}mL^2\left\{\|\omega\|^2 - \Omega^2(q^T Cq)\right\},$$

which can be expressed as

$$\tilde{H}(q, \pi) = \frac{1}{2mL^2} \left\| \pi - mL^2 S(q) S(\Omega e_3) q \right\|^2 - \frac{1}{2} mL^2 \Omega^2 (q^T C q).$$

Hamilton's equations, expressed in terms of the conjugate momentum π, can be obtained from (5.28) and (5.29). This is a lengthy equation, not given here, suggesting that it is not so convenient in this case to describe the Lagrangian dynamics in terms of the angular velocity vector.

5.6.2.5 Conservation Properties

Equation (5.43) suggests that the dynamics of the particle on a rotating sphere do not depend on the mass value of the particle or the radius of the sphere but do depend on the constant angular velocity Ω of the rotating sphere.

It is easy to show that the Hamiltonian

$$H = \frac{1}{2} mL^2 \left\{ \|\dot{q}\|^2 + 2q^T S^T (\Omega e_3) \dot{q} + \Omega^2 \left(q^T C q \right) \right\}$$

which coincides with the total energy E in this case is constant along each solution of the dynamical flow of a particle on a rotating sphere.

5.6.2.6 Equilibrium Properties

We now determine the equilibrium configurations of the particle on a rotating sphere in S^2. The conditions for an equilibrium are that the time derivative of the configuration vector, or equivalently the angular velocity vector or the momentum, is zero and the equilibrium configuration satisfies:

$$-(I_{3\times3} - qq^T)Cq = \begin{bmatrix} -q_1 q_3^2 \\ -q_2 q_3^2 \\ q_3(1 - q_3^2) \end{bmatrix} = \begin{bmatrix} 0 \\ 0 \\ 0 \end{bmatrix},$$

which implies that the time derivatives of the angular velocity and momentum both vanish as well. Hence, an equilibrium vector in TS^2 is either a *polar* equilibrium: $(e_3, 0) \in TS^2$ or $(-e_3, 0) \in TS^2$ or an *equatorial* equilibrium $(q, 0) \in TS^2$ that satisfies

$$q_1^2 + q_2^2 = 1,$$
$$q_3 = 0.$$

A configuration vector is an equilibrium configuration if and only if it is either collinear or orthogonal to the angular velocity vector of the rotating sphere.

The stability properties of each equilibrium are studied in turn using (5.43). We first linearize (5.43) about the polar equilibrium $(e_3, 0) \in \mathsf{TS}^2$ to obtain

$$\ddot{\xi}_1 - \Omega^2 \xi_1 = 0,$$
$$\ddot{\xi}_2 - \Omega^2 \xi_2 = 0,$$

restricted to the tangent space of TS^2 at $(e_3, 0) \in \mathsf{TS}^2$. These linear differential equations approximate the local dynamics of the particle on a rotating sphere in a neighborhood of the equilibrium $(e_3, 0) \in \mathsf{TS}^2$. The eigenvalues are easily determined to be $+\Omega, +\Omega, -\Omega, -\Omega$. Since there are positive eigenvalues, the polar equilibrium $(e_3, 0) \in \mathsf{TS}^2$ is unstable.

It can be shown that the local dynamics of the particle on a rotating sphere in a neighborhood of the polar equilibrium $(-e_3, 0) \in \mathsf{TS}^2$ are also given by

$$\ddot{\xi}_1 - \Omega^2 \xi_1 = 0,$$
$$\ddot{\xi}_2 - \Omega^2 \xi_2 = 0,$$

so that this equilibrium is also unstable.

We now linearize (5.43) about the equatorial equilibrium $(e_1, 0) \in \mathsf{TS}^2$. This gives the resulting linearized differential equations

$$\ddot{\xi}_2 = 0,$$
$$\ddot{\xi}_3 + \Omega^2 \xi_3 = 0,$$

restricted to the tangent space of TS^2 at $(e_1, 0) \in \mathsf{TS}^2$. These linear differential equations approximate the local dynamics of the particle on a rotating sphere in a neighborhood of the equilibrium $(e_1, 0) \in \mathsf{TS}^2$. The eigenvalues are easily determined to be $j\Omega, -j\Omega, 0, 0$. Since the eigenvalues have zero real parts no conclusion can be drawn about the stability of this equatorial equilibrium.

It can be shown that the local dynamics of the particle on a rotating sphere in a neighborhood of any equatorial equilibrium have a similar form with eigenvalues $j\Omega, -j\Omega, 0, 0$.

5.6.3 Dynamics of a Spherical Pendulum Connected to Three Elastic Strings

An example of a spherical pendulum connected to three elastic strings is studied in detail in [90]. One end of a spherical pendulum, viewed as a thin, rigid link or rod, is connected to a spherical pivot joint at a fixed inertial position while three elastic strings, in tension, connect the other end of the pendulum link to fixed inertial positions. The pivot of the spherical pendulum and the fixed ends of the three elastic strings are assumed to lie in a fixed horizontal plane. Gravity acts on the rigid rod.

This example of a spherical pendulum connected to three elastic strings is used to demonstrate basic properties of a class of structures referred to as tensegrity structures in [90]. Our subsequent development of the tensegrity example has considerable overlap with the results presented in [90].

Introduce an inertial Euclidean frame in three dimensions; the first two axes of the frame are horizontal and the third axis is vertical; the origin of the inertial frame is located at the pivot of the pendulum link.

We introduce a globally defined notion of the attitude of the pendulum link. Let $q \in S^2$ be a unit vector from the pivot point in the direction of the center of mass of the pendulum link; it is defined with respect to the inertial Euclidean frame and referred to as the attitude vector of the link. The configuration manifold is the sphere S^2, viewed as an embedded manifold in $\mathbb{R}^3$. Thus, the pendulum link has two degrees of freedom.

Uniform gravity acts on the center of mass of the pendulum link; g is the constant acceleration of gravity. The length of the pendulum link is L and m is the mass of the pendulum link; the rod is assumed to be uniform so that its center of mass is located at the midpoint of the rod. The scalar moment of inertia of the pendulum link about its pivot is denoted by J.

Additional forces due to the three elastically connected strings act on the pendulum link. Each elastic string is connected to the free end of the pendulum link; the other end of the elastic string is connected to a fixed inertial support. The inertial supports are located at the vertices of an equilateral triangle within the horizontal plane containing the pivot of the pendulum link at the center of the equilateral triangle. These three support locations are given with respect to the inertial frame by

$$
z_1 = \begin{bmatrix} -\frac{L}{2} \\ -\frac{L}{2\sqrt{3}} \\ 0 \end{bmatrix}, \quad
z_2 = \begin{bmatrix} \frac{L}{2} \\ -\frac{L}{2\sqrt{3}} \\ 0 \end{bmatrix}, \quad
z_3 = \begin{bmatrix} 0 \\ \frac{L}{\sqrt{3}} \\ 0 \end{bmatrix}.
$$

Each elastic string is assumed to have elastic stiffness κ and un-stretched length L. A schematic of a spherical pendulum connected to three elastic strings is shown in Figure 5.3.

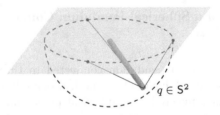

Fig. 5.3 Spherical pendulum connected to three elastic strings

To illustrate the several possible formulations of the equations of motion, we present two equivalent versions of the Euler–Lagrange equations and two equivalent versions of Hamilton's equations.

5.6.3.1 Euler–Lagrange Equations in Terms of $(q, \dot{q})$

The Lagrangian function $L : \mathsf{T}\mathsf{S}^2 \to \mathbb{R}^1$ for a pendulum link can be written as the difference of the kinetic energy and the potential energy. The kinetic energy function is

$$T(q, \dot{q}) = \frac{1}{2} J \|\dot{q}\|^2 .$$

The potential energy consists of the sum of the gravitational potential energy of the pendulum link and the elastic potential energy in the three strings. The gravitational potential energy function is

$$U_g(q) = \frac{mgL}{2} e_3^T q.$$

The elastic potential energy of the three strings is described as follows. Define the string vectors $s_i \in \mathbb{R}^3$ by

$$s_i = z_i - Lq, \quad i = 1, 2, 3.$$

These are geometric vectors from the ends of the three pendulum links to the attachment points of the three elastic strings in the inertial frame. The elastic potential energy of the three strings is

$$U_e(q) = \sum_{i=1}^{3} \frac{1}{2} \kappa(\|s_i\| - L)^2.$$

The tension forces that arise from the deformation of the three strings are

$$\begin{aligned}
F_i &= -\frac{\partial}{\partial(Lq)} \left(\frac{1}{2} \kappa(\|s_i\| - L)^2 \right) \\
&= -\kappa(\|s_i\| - L) \frac{\partial \|s_i\|}{\partial s_i} \frac{\partial s_i}{\partial(Lq)} \\
&= \kappa(\|s_i\| - L) \frac{s_i}{\|s_i\|}, \quad i = 1, 2, 3.
\end{aligned}$$

It is assumed throughout that all three strings remain in tension. This assumption requires that $\|s_i\| > L$, $i = 1, 2, 3$. If this condition is violated for a string, then the tension force on that string is necessarily zero; this could be incorporated into the above expressions. Thus, the Lagrangian function is

$$L(q, \dot{q}) = T(q, \dot{q}) - U_g(q) - U_e(q)$$

$$= \frac{1}{2} J \|\dot{q}\|^2 - \frac{mgL}{2} e_3^T q - \sum_{i=1}^{3} \frac{1}{2} \kappa(\|s_i\| - L)^2.$$

Following the prior development in Chapter 5, the Euler–Lagrange equation for the pendulum link with three elastically connected strings can be written as

$$J\ddot{q} + J \|\dot{q}\|^2 q + (I_{3\times 3} - qq^T) \left\{ \frac{mgL}{2} e_3 - L(F_1 + F_2 + F_3) \right\} = 0. \quad (5.47)$$

This equation describes the Lagrangian dynamics of the pendulum link and three elastically connected strings in terms of $(q, \dot{q}) \in \mathsf{TS}^2$ on the tangent bundle of the configuration manifold S^2.

5.6.3.2 Euler–Lagrange Equations in Terms of (q, ω)

A convenient form of the Euler–Lagrange equations can also be obtained in terms of the angular velocity vector of the pendulum link which satisfies the rotational kinematics

$$\dot{q} = S(\omega)q, \quad (5.48)$$

where the angular velocity vector satisfies $\omega^T q = 0$ and $S(\omega)$ is the 3×3 skew-symmetric matrix function.

The modified Lagrangian function can be expressed in terms of the angular velocity vector of the pendulum link as

$$\tilde{L}(q, \omega) = \frac{1}{2} J \|\omega\|^2 - \frac{mgL}{2} e_3^T q - \sum_{i=1}^{3} \frac{1}{2} \kappa(\|s_i\| - L)^2.$$

The Euler–Lagrange equation, expressed in terms of the angular velocity vector, consists of the rotational kinematics (5.48) and the equation

$$J\dot{\omega} + S(q) \left\{ \frac{mgL}{2} e_3 - L(F_1 + F_2 + F_3) \right\} = 0. \quad (5.49)$$

These equations (5.48) and (5.49) describe the Lagrangian dynamics of the spherical pendulum and three elastically connected strings in terms of $(q, \dot{q}) \in \mathsf{TS}^2$ on the tangent bundle of S^2.

5.6.3.3 Hamilton's Equations in Terms of (q, μ)

Define the momentum

$$\mu = (I_{3\times3} - qq^T)\frac{\partial L(q, \dot{q})}{\partial \dot{q}} = J\dot{q},$$

using the Legendre transformation. Thus, $\mu \in \mathsf{T}^*\mathsf{S}^2$ is conjugate to $\dot{q} \in \mathsf{TS}^2$. The Hamiltonian function can be expressed as

$$H(q, \mu) = \frac{1}{2J}\|\mu\|^2 + \frac{mgL}{2}e_3^T q + \sum_{i=1}^{3}\frac{1}{2}\kappa(\|s_i\| - L)^2.$$

Then, Hamilton's equations of motion for the spherical pendulum and three elastically connected strings are

$$\dot{q} = \frac{\mu}{J}, \tag{5.50}$$

$$\dot{\mu} = -\frac{\|\mu\|^2}{J}q - (I_{3\times3} - qq^T)\left\{\frac{mgL}{2}e_3 - L(F_1 + F_2 + F_3)\right\}. \tag{5.51}$$

Hamilton's equations (5.50) and (5.51) describe the Hamiltonian dynamics of the spherical pendulum and three elastically connected strings in terms of $(q, \mu) \in \mathsf{T}^*\mathsf{S}^2$ on the cotangent bundle of the configuration manifold.

5.6.3.4 Hamilton's Equations in Terms of (q, π)

Define the angular momentum

$$\pi = (I_{3\times3} - qq^T)\frac{\partial \tilde{L}(q, \omega)}{\partial \omega} = J\omega,$$

using the Legendre transformation. Thus, $\pi \in \mathsf{T}^*\mathsf{S}^2$ is conjugate to the angular velocity $\omega \in \mathsf{TS}^2$. The modified Hamiltonian function is

$$\tilde{H}(q, \pi) = \frac{1}{2J}\|\pi\|^2 + \frac{mgL}{2}e_3^T q + \sum_{i=1}^{3}\frac{1}{2}\kappa(\|s_i\| - L)^2.$$

Hamilton's equations of motion for the spherical pendulum and three elastically connected strings are

$$\dot{q} = \frac{1}{J}S(\pi)q, \tag{5.52}$$

$$\dot{\pi} = -S(q)\left\{\frac{mgL}{2}e_3 - L(F_1 + F_2 + F_3)\right\}. \tag{5.53}$$

Thus, the Hamiltonian flow of the spherical pendulum and three elastically connected strings is described by (5.52) and (5.53) in terms of the evolution of $(q, \pi) \in T^*S^2$ on the cotangent bundle of S^2.

5.6.3.5 Conservation Properties

It is easy to show that the Hamiltonian

$$H = \frac{1}{2} J \|\dot{q}\|^2 + \frac{mgL}{2} e_3^T q + \sum_{i=1}^{3} \frac{1}{2} \kappa (\|s_i\| - L)^2,$$

which coincides with the total energy E in this case, is constant along each solution of the dynamical flow of the spherical pendulum connected to three elastic strings.

5.6.3.6 Equilibrium Properties

An important feature of the dynamics of the spherical pendulum and three elastically connected strings is its equilibrium configurations lie in S^2. The conditions for an equilibrium are that the time derivative of the configuration vector, or equivalently the angular velocity vector or the momentum, is zero and the equilibrium configuration satisfies:

$$(I_{3 \times 3} - qq^T) \left\{ \frac{mgL}{2} e_3 - L(F_1 + F_2 + F_3) \right\} = 0,$$

which implies that the time derivatives of the angular velocity and momentum both vanish as well. This implies the condition: $\frac{mgL}{2} e_3 - L(F_1 + F_2 + F_3) \in \mathbb{R}^3$ and $q \in S^2$ are collinear. If $F_1 + F_2 + F_3 \in \mathbb{R}^3$ and $e_3 \in \mathbb{R}^3$ are collinear, then there are two equilibrium solutions given by $(-e_3, 0) \in TS^2$ and by $(e_3, 0) \in TS^2$. The equilibrium $(-e_3, 0) \in TS^2$ is referred to as the hanging equilibrium. At the hanging equilibrium, the string vectors are

$$s_1 = \begin{bmatrix} -\frac{L}{2} \\ -\frac{L}{2\sqrt{3}} \\ L \end{bmatrix}, \quad s_2 = \begin{bmatrix} \frac{L}{2} \\ -\frac{L}{2\sqrt{3}} \\ L \end{bmatrix}, \quad s_3 = \begin{bmatrix} 0 \\ \frac{L}{\sqrt{3}} \\ L \end{bmatrix}.$$

Thus, the tension force in each string can be computed and it can be shown that the net tension force, in the inertial frame, is

$$(F_1 + F_2 + F_3) = \frac{3(2 - \sqrt{3})}{2} \kappa L e_3.$$

The equilibrium $(e_3, 0) \in \mathsf{TS}^2$ is the inverted equilibrium. At the inverted equilibrium, the string vectors are

$$
s_1 = \begin{bmatrix} -\frac{L}{2} \\ -\frac{L}{2\sqrt{3}} \\ -L \end{bmatrix}, \quad
s_2 = \begin{bmatrix} \frac{L}{2} \\ -\frac{L}{2\sqrt{3}} \\ -L \end{bmatrix}, \quad
s_3 = \begin{bmatrix} 0 \\ \frac{L}{\sqrt{3}} \\ -L \end{bmatrix}.
$$

Thus, the tension force in each string can be computed and it can be shown that the net tension force, in the inertial frame, is

$$
(F_1 + F_2 + F_3) = -\frac{3(2 - \sqrt{3})}{2} \kappa L e_3.
$$

The local solution and stability properties of each equilibrium are studied in turn using (5.47). We first linearize (5.47) about the inverted equilibrium $(e_3, 0) \in \mathsf{TS}^2$ to obtain

$$
mL^2 \ddot{\xi}_1 + \left\{ \frac{3(2 - \sqrt{3})}{2} \kappa L^2 - \frac{mgL}{2} \right\} \xi_1 = 0,
$$

$$
mL^2 \ddot{\xi}_2 + \left\{ \frac{3(2 - \sqrt{3})}{2} \kappa L^2 - \frac{mgL}{2} \right\} \xi_2 = 0,
$$

restricted to the tangent space of TS^2 at $(e_3, 0) \in \mathsf{TS}^2$. These linear differential equations approximate the local dynamics of the spherical pendulum and three elastically connected strings in a neighborhood of the inverted equilibrium $(e_3, 0) \in \mathsf{TS}^2$. If $(3(2 - \sqrt{3})\kappa L^2 > mgL$, the eigenvalues are purely imaginary and, at least to first-order, the solutions in a neighborhood of the inverted equilibrium are oscillatory, but not necessarily periodic. If $3(2 - \sqrt{3})\kappa L^2 < mgL$, there is a positive eigenvalue and the inverted equilibrium $(e_3, 0) \in \mathsf{TS}^2$ is unstable.

We now linearize (5.47) about the hanging equilibrium $(-e_3, 0) \in \mathsf{TS}^2$. This gives the resulting linearized differential equations

$$
mL^2 \ddot{\xi}_1 + \left\{ \frac{3(2 - \sqrt{3})}{2} \kappa L^2 - \frac{mgL}{2} \right\} \xi_1 = 0,
$$

$$
mL^2 \ddot{\xi}_2 + \left\{ \frac{3(2 - \sqrt{3})}{2} \kappa L^2 - \frac{mgL}{2} \right\} \xi_2 = 0,
$$

restricted to the tangent space of TS^2 at $(-e_3, 0) \in \mathsf{TS}^2$. These linear differential equations approximate the local dynamics of the spherical pendulum and three elastically connected strings in a neighborhood of the hanging equilibrium $(-e_3, 0) \in \mathsf{TS}^2$. The eigenvalues are purely imaginary so, at least to first-order, the solutions in a neighborhood of the hanging equilibrium are oscillatory, but not necessarily periodic.

These local solution and stability results confirm the physically intuitive conclusions about the local flow properties of these two equilibrium solutions.

5.6.4 Dynamics of Two Elastically Connected Spherical Pendulums

Two identical pendulums are attached to a common inertially fixed support by a frictionless pivot. The origin of an inertial Euclidean frame is located at the common pivot. Each pendulum can rotate in three dimensions. Gravitational effects are ignored, but an elastic restoring force acts on the two pendulums. Each pendulum is assumed to be a thin rigid link of concentrated mass m located at distance L from the pivot. A schematic of the two elastically connected spherical pendulums is shown in Figure 5.4.

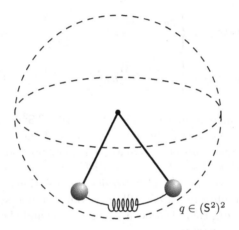

Fig. 5.4 Two elastically connected spherical pendulums

Let $q_i \in S^2$, $i = 1, 2$ denote the attitude of the i-th pendulum link in the inertial frame. Consequently, the configuration vector of the two pendulums is given by $q = (q_1, q_2) \in (S^2)^2$. Thus, the configuration manifold is $(S^2)^2$ and the dynamics of the elastically connected spherical pendulums have four degrees of freedom.

5.6.4.1 Euler–Lagrange Equations in Terms of $(q, \dot{q})$

The total kinetic energy is the sum of the kinetic energies of the two spherical pendulums

$$T(q, \dot{q}) = \frac{1}{2}mL^2 \|\dot{q}_1\|^2 + \frac{1}{2}mL^2 \|\dot{q}_2\|^2.$$

The elastic potential energy is assumed to be proportional to $(1 - \cos\theta)$, where θ is the angle between the two pendulum links. Since $\cos\theta = q_1^T q_2$, the elastic potential energy can be expressed in terms of the configuration as

$$U(q) = \kappa(1 - q_1^T q_2),$$

where κ is a positive elastic stiffness constant. Note that the elastic potential energy has zero gradient, and hence the force vanishes, when the two pendulums are collinear, that is the angle between the two pendulums is either 0 radians or π radians.

The Lagrangian function for the two elastically connected spherical pendulum is thus given by

$$L(q, \dot{q}) = \frac{1}{2}mL^2\|\dot{q}_1\|^2 + \frac{1}{2}mL^2\|\dot{q}_2\|^2 - \kappa(1 - q_1^T q_2).$$

The inertia matrix is constant so that the Euler–Lagrange equations, according to (5.10), are

$$mL^2\ddot{q}_1 + mL^2\|\dot{q}_1\|^2 q_1 - \kappa(I_{3\times3} - q_1 q_1^T)q_2 = 0, \tag{5.54}$$

$$mL^2\ddot{q}_2 + mL^2\|\dot{q}_2\|^2 q_2 - \kappa(I_{3\times3} - q_2 q_2^T)q_1 = 0. \tag{5.55}$$

This version of the Euler–Lagrange equations describes the Lagrangian dynamics of the two elastically connected spherical pendulums in terms of $(q, \dot{q}) \in T(S^2)^2$ on the tangent bundle of the configuration manifold $(S^2)^2$.

5.6.4.2 Euler–Lagrange Equations in Terms of (q, ω)

An alternative version of the Euler–Lagrange equations of motion for the two elastically connected spherical pendulums is expressed in terms of the angular velocity vector of the two links. The rotational kinematics are given by

$$\dot{q}_1 = S(\omega_1)q_1, \tag{5.56}$$

$$\dot{q}_2 = S(\omega_2)q_2, \tag{5.57}$$

where the angular velocity vector $\omega = (\omega_1, \omega_2) \in T_q(S^2)^2$.

The modified Lagrangian can be expressed in terms of the angular velocity vector as

$$\tilde{L}(q, \omega) = \frac{1}{2}mL^2\|\omega_1\|^2 + \frac{1}{2}mL^2\|\omega_2\|^2 - \kappa(1 - q_1^T q_2).$$

Following the prior results in (5.17), the Euler–Lagrange equations for the elastically connected spherical pendulums, expressed in terms of the angular velocity vector, are

$$mL^2\dot{\omega}_1 + \kappa S(q_2)q_1 = 0, \tag{5.58}$$

$$mL^2\dot{\omega}_2 + \kappa S(q_1)q_2 = 0. \tag{5.59}$$

Equations (5.56), (5.57), (5.58), and (5.59) also describe the Lagrangian dynamics of the two elastically connected spherical pendulums in terms of $(q, \omega) \in \mathsf{T}(S^2)^2$ on the tangent bundle of $(S^2)^2$.

5.6.4.3 Hamilton's Equations in Terms of (q, μ)

Hamilton's equations on the cotangent bundle $\mathsf{T}^*(S^2)^2$ are obtained by defining the conjugate momenta according to the Legendre transformation

$$\mu_1 = (I_{3\times 3} - q_1 q_1^T)\frac{\partial L(q, \dot{q})}{\partial \dot{q}_1} = mL^2 \dot{q}_1,$$

$$\mu_2 = (I_{3\times 3} - q_2 q_2^T)\frac{\partial L(q, \dot{q})}{\partial \dot{q}_2} = mL^2 \dot{q}_2,$$

where the momentum $\mu = (\mu_1, \mu_2) \in \mathsf{T}_q^*(S^2)^2$.

The Hamiltonian function is

$$H(q, \mu) = \frac{1}{2}\frac{\|\mu_1\|^2}{mL^2} + \frac{1}{2}\frac{\|\mu_2\|^2}{mL^2} + \kappa(1 - q_1^T q_2).$$

Hamilton's equations of motion, obtained from (5.25) and (5.26), are given by the kinematics equations

$$\dot{q}_1 = \frac{\mu_1}{mL^2}, \tag{5.60}$$

$$\dot{q}_2 = \frac{\mu_2}{mL^2}, \tag{5.61}$$

and the dynamics equations

$$\dot{\mu}_1 = -\frac{\|\mu_1\|^2}{mL^2}q_1 + \kappa(I_{3\times 3} - q_1 q_1^T)q_2, \tag{5.62}$$

$$\dot{\mu}_2 = -\frac{\|\mu_2\|^2}{mL^2}q_2 + \kappa(I_{3\times 3} - q_2 q_2^T)q_1. \tag{5.63}$$

This form of Hamilton's equations, given by (5.60), (5.61), (5.62), and (5.63), describes the Hamiltonian dynamics of the two elastically connected spherical pendulums in terms of $(q, \mu) \in \mathsf{T}^*(S^2)^2$ on the cotangent bundle of $(S^2)^n$.

5.6.4.4 Hamilton's Equations in Terms of (q, π)

A different form of Hamilton's equations on the cotangent bundle $T^*(S^2)^2$ can be obtained by defining the momentum according to the Legendre transformation $\pi = (\pi_1, \pi_2) \in T_q^*(S^2)^2$ that is conjugate to the angular velocity vector $\omega = (\omega_1, \omega_2) \in T_q(S^2)^2$. This gives

$$\pi_1 = (I_{3\times3} - q_1 q_1^T)\frac{\partial \tilde{L}(q, \omega)}{\partial \omega_1} = mL^2 \omega_1,$$

$$\pi_2 = (I_{3\times3} - q_2 q_2^T)\frac{\partial \tilde{L}(q, \omega)}{\partial \omega_2} = mL^2 \omega_2.$$

The modified Hamiltonian function is

$$\tilde{H}(q, \pi) = \frac{1}{2}\frac{\|\pi_1\|^2}{mL^2} + \frac{1}{2}\frac{\|\pi_2\|^2}{mL^2} + \kappa(1 - q_1^T q_2).$$

Hamilton's equations of motion for the elastically connected spherical pendulums, according to (5.33) and (5.34), are given by the kinematics equations

$$\dot{q}_1 = \frac{S(\pi_1)}{mL^2}q_1, \tag{5.64}$$

$$\dot{q}_2 = \frac{S(\pi_2)}{mL^2}q_2, \tag{5.65}$$

and the dynamics equations

$$\dot{\pi}_1 = -\kappa S(q_2)q_1, \tag{5.66}$$

$$\dot{\pi}_2 = -\kappa S(q_1)q_2. \tag{5.67}$$

The Hamiltonian flow of the elastically connected spherical pendulums is described by equations (5.64), (5.65), (5.66), and (5.67) in terms of the evolution of $(q, \pi) \in T^*(S^2)^2$ on the cotangent bundle of $(S^2)^2$.

5.6.4.5 Conservation Properties

It is easy to show that the Hamiltonian of the two elastically connected spherical pendulums

$$H = \frac{1}{2}mL^2\|\dot{q}_1\|^2 + \frac{1}{2}mL^2\|\dot{q}_2\|^2 + k(1 - q_1^T q_2),$$

which coincides with the total energy E in this case, is constant along each solution of the dynamical flow.

It is also easy to see that the total angular momentum of the two elastically connected spherical pendulums

$$\pi_1 + \pi_2 = mL^2(\omega_1 + \omega_2)$$

is constant along each solution of the dynamical flow. This arises as a consequence of Noether's theorem, due to the invariance of the Lagrangian with respect to the lifted action of rotations about the pivot.

5.6.4.6 Equilibrium Properties

The equilibrium solutions of the elastically connected spherical pendulums occur when the time derivative of the configuration vector, or equivalently the angular velocity vector or momenta, is zero, and the configuration satisfies:

$$(I_{3\times 3} - q_1 q_1^T)q_2 = 0,$$
$$(I_{3\times 3} - q_2 q_2^T)q_1 = 0,$$

which implies that the time derivatives of the angular velocity and momenta both vanish as well. This is equivalent to

$$q_1 \times q_2 = 0.$$

Consequently, equilibrium solutions occur when the pendulum links are stationary in an arbitrary direction but with the angle between them either 0 radians or π radians. There are two disjoint manifolds of equilibrium configurations given by

$$\left\{(q_1, q_2) \in (\mathsf{S}^2)^2 : q_1 = q_2\right\},$$
$$\left\{(q_1, q_2) \in (\mathsf{S}^2)^2 : q_1 = -q_2\right\}.$$

In the former case, the pendulum links are said to be *in phase*, while in the latter case they are said to be *out of phase*.

The stability properties of two typical equilibrium solutions are studied using (5.54) and (5.55). Without loss of generality, we first develop a linearization of (5.54) and (5.55) at the equilibrium $(e_3, -e_3, 0, 0) \in \mathsf{T}(\mathsf{S}^2)^2$; this equilibrium corresponds to the pendulum links out of phase. We use the notation $q_i = (q_{i1}, q_{i2}) \in \mathsf{S}^2$, $i = 1, 2$ so that the first index of each double subscript refers to the pendulum index.

First, consider the out of phase equilibrium solution $(e_3, -e_3, 0, 0) \in \mathsf{T}(\mathsf{S}^2)^2$. The linearized differential equations can be shown to be

$$mL^2\ddot{\xi}_{11} - \kappa(\xi_{21} + \xi_{11}) = 0,$$
$$mL^2\ddot{\xi}_{12} = 0,$$
$$mL^2\ddot{\xi}_{21} - \kappa(\xi_{11} + \xi_{21}) = 0,$$
$$mL^2\ddot{\xi}_{22} = 0,$$

restricted to the tangent space of $\mathsf{T}(\mathsf{S}^2)^2$ at $(e_3, -e_3, 0, 0) \in \mathsf{T}(\mathsf{S}^2)^2$. These linearized differential equations approximate the local dynamics of the elastically connected spherical pendulums in a neighborhood of the out of phase equilibrium $(e_3, -e_3, 0, 0) \in \mathsf{T}(\mathsf{S}^2)^2$. The eigenvalues of these linearized equations are $+\sqrt{\frac{2\kappa}{mL^2}}, -\sqrt{\frac{2\kappa}{mL^2}}, +\sqrt{\frac{2\kappa}{mL^2}}, -\sqrt{\frac{2\kappa}{mL^2}}, 0, 0, 0, 0$. Since there are positive eigenvalues, this equilibrium with out of phase pendulum links is unstable. Similarly, it can be shown that all out of phase equilibrium solutions are unstable.

Without loss of generality, we now develop a linearization of (5.54) and (5.55) at the equilibrium $(e_3, e_3, 0, 0) \in \mathsf{T}(\mathsf{S}^2)^2$; this equilibrium corresponds to the pendulum links in phase. The resulting linearized differential equations are

$$mL^2\ddot{\xi}_{11} - \kappa(\xi_{21} - \xi_{11}) = 0,$$

$$mL^2\ddot{\xi}_{12} = 0,$$

$$mL^2\ddot{\xi}_{21} - \kappa(\xi_{11} - \xi_{21}) = 0,$$

$$mL^2\ddot{\xi}_{22} = 0,$$

restricted to the tangent space of $\mathsf{T}(\mathsf{S}^2)^2$ at $(e_3, e_3, 0, 0) \in \mathsf{T}(\mathsf{S}^2)^2$. These linearized differential equations approximate the local dynamics of the elastically connected spherical pendulum links in a neighborhood of the in phase equilibrium $(e_3, e_3, 0, 0) \in \mathsf{T}(\mathsf{S}^2)^2$. The eigenvalues can be shown to be $+j\sqrt{\frac{2\kappa}{mL^2}}, -j\sqrt{\frac{2\kappa}{mL^2}}, +j\sqrt{\frac{2\kappa}{mL^2}}, -j\sqrt{\frac{2\kappa}{mL^2}}, 0, 0, 0, 0$. Since the eigenvalues are purely imaginary or zero, no conclusion can be made about the stability of this equilibrium on the basis of this analysis.

We mention that a Lyapunov approach, using the total energy of the elastically connected planar pendulums, does not provide a positive result in this case since the in phase equilibrium is not a strict local minimum of the total energy function on the tangent bundle $\mathsf{T}(\mathsf{S}^2)^2$.

Relative equilibrium solutions occur when the angular velocity vector of the two pendulums are identically equal and constant and the angle between them is either 0 radians or π radians. This corresponds to a constant rotation of the pendulum links about their common pivot, as a *rigid* system. An analysis of the stability of these relative equilibrium solutions could follow the developments in [70].

5.6.5 Dynamics of a Double Spherical Pendulum

A double spherical pendulum is defined by two rigid, massless links serially connected by a frictionless spherical joint with the first link connected to an inertially fixed frictionless spherical pivot. The mass of each link is concen-

trated at the outboard end of the link. The double spherical pendulum is
acted on by uniform, constant gravity.

The dynamics of a double spherical pendulum has been studied in [73]
using angle representations for the configuration; our approach here follows
the prior geometric development of this chapter allowing for global characterization of the dynamics.

An inertial Euclidean frame is selected with origin located at the fixed
pivot with the first two axes in a horizontal plane and the third axis vertical. The vector $q_1 \in S^2$ represents the attitude of the first link, from the
pivot to the first mass, and the vector $q_2 \in S^2$ represents the attitude of
the second link, from the first mass to the second mass, each vector being
defined with respect to the inertial Euclidean frame. Thus, the configuration
manifold is the product of two spheres, namely $(S^2)^2$ and the configuration vector $q = (q_1, q_2) \in (S^2)^2$. The time derivative of the configuration
$\dot{q} = (\dot{q}_1, \dot{q}_2) \in T_q(S^2)^2$. The double spherical pendulum has four degrees of
freedom. A schematic of a double spherical pendulum is shown in Figure 5.5.

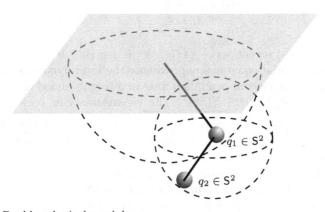

Fig. 5.5 Double spherical pendulum

Let the masses and the lengths of the two pendula be m_1, m_2 and L_1, L_2,
respectively; the constant acceleration of gravity is denoted by g.

5.6.5.1 Euler–Lagrange Equations in Terms of $(q, \dot{q})$

The kinetic energy is the sum of the kinetic energies for each link and is given
by

$$
\begin{aligned}
T(q, \dot{q}) &= \frac{1}{2} m_1 \left\| L_1 \dot{q}_1 \right\|^2 + \frac{1}{2} m_2 \left\| L_1 \dot{q}_1 + L_2 \dot{q}_2 \right\|^2 \\
&= \frac{1}{2}(m_1 + m_2) L_1^2 \left\| \dot{q}_1 \right\|^2 + m_2 L_1 L_2 (\dot{q}_1 \cdot \dot{q}_2) + \frac{1}{2} m_2 L_2^2 \left\| \dot{q}_2 \right\|^2
\end{aligned}
$$

$$= \frac{1}{2} \begin{bmatrix} \dot{q}_1 \\ \dot{q}_2 \end{bmatrix}^T \begin{bmatrix} (m_1 + m_2)L_1^2 I_{3\times3} & m_2 L_1 L_2 I_{3\times3} \\ m_2 L_1 L_2 I_{3\times3} & m_2 L_2^2 I_{3\times3} \end{bmatrix} \begin{bmatrix} \dot{q}_1 \\ \dot{q}_2 \end{bmatrix}.$$

The gravitational potential energy is

$$U(q) = m_1 g e_3^T (L_1 q_1) + m_2 g e_3^T (L_1 q_1 + L_2 q_2)$$
$$= (m_1 + m_2)g L_1 e_3^T q_1 + m_2 g L_2 e_3^T q_2.$$

Thus, the Lagrangian function $L : \mathsf{T}(S^2)^2 \to \mathbb{R}^1$ is

$$L(q, \dot{q}) = \frac{1}{2} \begin{bmatrix} \dot{q}_1 \\ \dot{q}_2 \end{bmatrix}^T \begin{bmatrix} (m_1 + m_2)L_1^2 I_{3\times3} & m_2 L_1 L_2 I_{3\times3} \\ m_2 L_1 L_2 I_{3\times3} & m_2 L_2^2 I_{3\times3} \end{bmatrix} \begin{bmatrix} \dot{q}_1 \\ \dot{q}_2 \end{bmatrix}$$
$$- (m_1 + m_2)g L_1 e_3^T q_1 - m_2 g L_2 e_3^T q_2.$$

The inertia matrix is constant so that the Euler–Lagrange equations, according to (5.10), are:

$$\begin{bmatrix} (m_1 + m_2)L_1^2 I_{3\times3} & m_2 L_1 L_2 (I_{3\times3} - q_1 q_1^T) \\ m_2 L_1 L_2 (I_{3\times3} - q_2 q_2^T) & m_2 L_2^2 I_{3\times3} \end{bmatrix} \begin{bmatrix} \ddot{q}_1 \\ \ddot{q}_2 \end{bmatrix}$$
$$+ \begin{bmatrix} (m_1 + m_2)L_1^2 \|\dot{q}_1\|^2 q_1 \\ m_2 L_2^2 \|\dot{q}_2\|^2 q_2 \end{bmatrix} + \begin{bmatrix} (m_1 + m_2)g L_1 (I_{3\times3} - q_1 q_1^T)e_3 \\ m_2 g L_2 (I_{3\times3} - q_2 q_2^T)e_3 \end{bmatrix} = \begin{bmatrix} 0 \\ 0 \end{bmatrix}.$$
$$(5.68)$$

This version of the Euler–Lagrange equations, given by (5.68), describes the Lagrangian dynamics of the double spherical pendulum in terms of $(q, \dot{q}) \in \mathsf{T}(S^2)^2$ on the tangent bundle of the configuration manifold.

5.6.5.2 Euler–Lagrange Equations in Terms of (q, ω)

An alternate version of the Euler–Lagrange equations for the double spherical pendulum is expressed in terms of the angular velocity vector of the two links. The rotational kinematics for each pendulum link are given by

$$\dot{q}_1 = S(\omega_1)q_1, \qquad (5.69)$$
$$\dot{q}_2 = S(\omega_2)q_2, \qquad (5.70)$$

where $\omega = (\omega_1, \omega_2) \in \mathsf{T}_q(S^2)^2$.

The modified Lagrangian $\tilde{L} : \mathsf{T}S^2 \times \mathbb{R}^3 \to \mathbb{R}^1$ can be expressed in terms of the angular velocity vector as

$$\tilde{L}(q, \omega) = \frac{1}{2} \begin{bmatrix} \omega_1 \\ \omega_2 \end{bmatrix}^T \begin{bmatrix} (m_1 + m_2)L_1^2 I_{3\times3} & m_2 L_1 L_2 S(q_1)^T S(q_2) \\ m_2 L_1 L_2 S(q_2)^T S(q_1) & m_2 L_2^2 I_{3\times3} \end{bmatrix} \begin{bmatrix} \omega_1 \\ \omega_2 \end{bmatrix}$$
$$- (m_1 + m_2)g L_1 e_3^T q_1 - m_2 g L_2 e_3^T q_2.$$

Following the prior results in (5.17), the Euler–Lagrange equations for the double spherical pendulum, expressed in terms of the angular velocity vector, are

$$
\begin{bmatrix} (m_1 + m_2)L_1^2 I_{3\times3} & m_2 L_1 L_2 S^T(q_1) S(q_2) \\ m_2 L_1 L_2 S^T(q_2) S(q_1) & m_2 L_2^2 I_{3\times3} \end{bmatrix} \begin{bmatrix} \dot{\omega}_1 \\ \dot{\omega}_2 \end{bmatrix}
$$
$$
+ \begin{bmatrix} m_2 L_1 L_2 \|\omega_2\|^2 S(q_2)q_1 \\ m_2 L_1 L_2 \|\omega_1\|^2 S(q_1)q_2 \end{bmatrix} + \begin{bmatrix} (m_1 + m_2)gL_1 S(q_1)e_3 \\ m_2 g L_2 S(q_2)e_3 \end{bmatrix} = \begin{bmatrix} 0 \\ 0 \end{bmatrix}. \quad (5.71)
$$

Equations (5.69), (5.70), and (5.71) describe the Lagrangian dynamics of the double spherical pendulum in terms of $(q, \omega) \in T(S^2)^2$ on the tangent bundle of $(S^2)^2$.

5.6.5.3 Hamilton's Equations in Terms of (q, μ)

Hamilton's equations on the cotangent bundle $T^*(S^2)^2$ are obtained by defining the conjugate momentum according to the Legendre transformation

$$
\begin{bmatrix} \mu_1 \\ \mu_2 \end{bmatrix} = \begin{bmatrix} (I_{3\times3} - q_1 q_1^T)\frac{\partial L(q, \dot{q})}{\partial \dot{q}_1} \\ (I_{3\times3} - q_2 q_2^T)\frac{\partial L(q, \dot{q})}{\partial \dot{q}_2} \end{bmatrix},
$$
$$
= \begin{bmatrix} (m_1 + m_2)L_1^2 I_{3\times3} & m_2 L_1 L_2 (I_{3\times3} - q_1 q_1^T) \\ m_2 L_1 L_2 (I_{3\times3} - q_2 q_2^T) & m_2 L_2^2 I_{3\times3} \end{bmatrix} \begin{bmatrix} \dot{q}_1 \\ \dot{q}_2 \end{bmatrix},
$$

so that the momentum $\mu = (\mu_1, \mu_2) \in T_q^*(S^2)^2$. This can be inverted in the form

$$
\begin{bmatrix} \dot{q}_1 \\ \dot{q}_2 \end{bmatrix} = \begin{bmatrix} (I_{3\times3} - q_1 q_1^T)\left(m_{11}^I(q)\mu_1 + m_{12}^I(q)\mu_2\right) \\ (I_{3\times3} - q_2 q_2^T)\left(m_{21}^I(q)\mu_1 + m_{22}^I(q)\mu_2\right) \end{bmatrix}, \quad (5.72)
$$

where

$$
\begin{bmatrix} m_{11}^I(q) & m_{12}^I(q) \\ m_{21}^I(q) & m_{22}^I(q) \end{bmatrix} = \begin{bmatrix} (m_1 + m_2)L_1^2 I_{3\times3} & m_2 L_1 L_2 (I_{3\times3} - q_1 q_1^T) \\ m_2 L_1 L_2 (I_{3\times3} - q_2 q_2^T) & m_2 L_2^2 I_{3\times3} \end{bmatrix}^{-1}
$$

denotes the 6×6 inverse matrix.

The Hamiltonian function $H : T^*(S^2)^2 \rightarrow \mathbb{R}^1$ is

$$
H(q, \mu) = \frac{1}{2} \begin{bmatrix} \mu_1 \\ \mu_2 \end{bmatrix}^T \begin{bmatrix} m_{11}^I(q) & m_{12}^I(q) \\ m_{21}^I(q) & m_{22}^I(q) \end{bmatrix} \begin{bmatrix} \mu_1 \\ \mu_2 \end{bmatrix}
$$
$$
+ (m_1 + m_2)gL_1 e_3^T q_1 + m_2 g L_2 e_3^T q_2.
$$

Hamilton's equations of motion, obtained from (5.25) and (5.26), are given by the kinematics equations (5.72) and the dynamics equations

$$\begin{bmatrix} \dot{\mu}_1 \\ \dot{\mu}_2 \end{bmatrix} = \begin{bmatrix} \sum_{j=1}^{2} \left(m_{1j}^I(q)\mu_j\right) \times (\mu_1 \times q_1) - (I_{3\times3} - q_1 q_1^T)\frac{1}{2}\frac{\partial}{\partial q_1} \sum_{j,k=1}^{2} \mu_j^T m_{jk}^I(q)\mu_k \\ \sum_{j=1}^{2} \left(m_{2j}^I(q)\mu_j\right) \times (\mu_2 \times q_2) - (I_{3\times3} - q_2 q_2^T)\frac{1}{2}\frac{\partial}{\partial q_2} \sum_{j,k=1}^{2} \mu_j^T m_{jk}^I(q)\mu_k \end{bmatrix}$$
$$- \begin{bmatrix} (m_1 + m_2)gL_1(I_{3\times3} - q_1 q_1^T)e_3 \\ m_2 g L_2(I_{3\times3} - q_2 q_2^T)e_3 \end{bmatrix}. \tag{5.73}$$

Hamilton's equations (5.72) and (5.73) describe the Hamiltonian dynamics of the double spherical pendulum in terms of $(q,\mu) \in \mathsf{T}^*(\mathsf{S}^2)^2$ on the cotangent bundle of $(\mathsf{S}^2)^2$.

5.6.5.4 Hamilton's Equations in Terms of (q, π)

A different form of Hamilton's equations on the cotangent bundle $\mathsf{T}^*(\mathsf{S}^2)^2$ can be obtained by defining the conjugate momentum according to the Legendre transformation

$$\begin{bmatrix} \pi_1 \\ \pi_2 \end{bmatrix} = \begin{bmatrix} (I_{3\times3} - q_1 q_1^T)\frac{\partial \tilde{L}(q,\omega)}{\partial \omega_1} \\ (I_{3\times3} - q_2 q_2^T)\frac{\partial \tilde{L}(q,\omega)}{\partial \omega_2} \end{bmatrix}$$
$$= \begin{bmatrix} (m_1 + m_2)L_1^2 I_{3\times3} & m_2 L_1 L_2 S^T(q_1)S(q_2) \\ m_2 L_1 L_2 S^T(q_2)S(q_1) & m_2 L_2^2 I_{3\times3} \end{bmatrix} \begin{bmatrix} \omega_1 \\ \omega_2 \end{bmatrix},$$

so that the momentum $\pi = (\pi_1, \pi_2) \in \mathsf{T}_q^*(\mathsf{S}^2)^2$ is conjugate to the angular velocity vector $\omega = (\omega_1, \omega_2) \in \mathsf{T}_q(\mathsf{S}^2)^2$. Thus, we obtain

$$\begin{bmatrix} \omega_1 \\ \omega_2 \end{bmatrix} = \begin{bmatrix} m_{11}^I(q) & m_{12}^I(q) \\ m_{21}^I(q) & m_{22}^I(q) \end{bmatrix} \begin{bmatrix} \pi_1 \\ \pi_2 \end{bmatrix},$$

where

$$\begin{bmatrix} m_{11}^I(q) & m_{12}^I(q) \\ m_{21}^I(q) & m_{22}^I(q) \end{bmatrix} = \begin{bmatrix} (m_1 + m_2)L_1^2 I_{3\times3} & m_2 L_1 L_2 S^T(q_1)S(q_2) \\ m_2 L_1 L_2 S^T(q_2)S(q_1) & m_2 L_2^2 I_{3\times3} \end{bmatrix}^{-1}$$

is the 6×6 inverse matrix.

The modified Hamiltonian function is

$$\tilde{H}(q, \pi) = \frac{1}{2}\begin{bmatrix} \pi_1 \\ \pi_2 \end{bmatrix}^T \begin{bmatrix} m_{11}^I(q) & m_{12}^I(q) \\ m_{21}^I(q) & m_{22}^I(q) \end{bmatrix} \begin{bmatrix} \pi_1 \\ \pi_2 \end{bmatrix}$$
$$+ (m_1 + m_2)gL_1 e_3^T q_1 + m_2 g L_2 e_3^T q_2.$$

Thus, Hamilton's equations of motion for the double spherical pendulum, according to (5.33) and (5.34), consist of the kinematics equations

$$\begin{bmatrix} \dot{q}_1 \\ \dot{q}_2 \end{bmatrix} = \begin{bmatrix} -S(q_1)m_{11}^I(q) & -S(q_1)m_{12}^I(q) \\ -S(q_2)m_{21}^I(q) & -S(q_2)m_{22}^I(q) \end{bmatrix} \begin{bmatrix} \pi_1 \\ \pi_2 \end{bmatrix}, \tag{5.74}$$

and the dynamical equations

$$\begin{bmatrix} \dot{\pi}_1 \\ \dot{\pi}_2 \end{bmatrix} = \begin{bmatrix} -S(q_1)\frac{1}{2}\frac{\partial}{\partial q_1}\sum_{j=1}^{2}\sum_{k=1}^{2}\pi_j^T m_{jk}^I(q)\pi_k \\ -S(q_2)\frac{1}{2}\frac{\partial}{\partial q_2}\sum_{j=1}^{2}\sum_{k=1}^{2}\pi_j^T m_{jk}^I(q)\pi_k \end{bmatrix}$$

$$+ \begin{bmatrix} (\sum_{j=1}^{2} m_{1j}^I(q)\pi_j) \times \pi_1 \\ (\sum_{j=1}^{2} m_{2j}^I(q)\pi_j) \times \pi_2 \end{bmatrix} + \begin{bmatrix} (m_1 + m_2)gL_1S(e_3)q_1 \\ m_2gL_2S(e_3)q_2 \end{bmatrix}. \tag{5.75}$$

The Hamiltonian flow of the double spherical pendulum is described by equations (5.74) and (5.75) in terms of the evolution of $(q, \pi) \in \mathsf{T}^*(S^2)^2$ on the cotangent bundle of $(S^2)^2$.

5.6.5.5 Conservation Properties

It is easy to show that the Hamiltonian of the double spherical pendulum

$$H = \frac{1}{2}\begin{bmatrix} \dot{q}_1 \\ \dot{q}_2 \end{bmatrix}^T \begin{bmatrix} (m_1 + m_2)L_1^2 I_{3\times 3} & m_2 L_1 L_2 I_{3\times 3} \\ m_2 L_1 L_2 I_{3\times 3} & m_2 L_2^2 I_{3\times 3} \end{bmatrix} \begin{bmatrix} \dot{q}_1 \\ \dot{q}_2 \end{bmatrix}$$

$$+ (m_1 + m_2)gL_1 e_3^T q_1 + m_2 g L_2 e_3^T q_2,$$

which coincides with the total energy E in this case, is constant along each solution of the Lagrangian flow.

In addition, the vertical component of angular momentum is preserved as a consequence of Noether's theorem, due to the invariance of the Lagrangian with respect to the lifted action of rotations about the gravity direction.

5.6.5.6 Equilibrium Properties

The equilibrium solutions of the double spherical pendulum occur when the time derivative of the configuration vector, or equivalently the angular velocity vector or momenta, is zero, and the direction vectors of the two pendulum links are collinear with the direction vector e_3, which implies that the time derivatives of the angular velocity and momenta both vanish as well. Thus, there are four distinct equilibrium solutions: $(-e_3, -e_3, 0, 0)$, $(-e_3, e_3, 0, 0)$, $(e_3, -e_3, 0, 0)$, and $(e_3, e_3, 0, 0)$ in $\mathsf{T}(S^2)^2$.

The total energy can be shown to have a strict local minimum at the first equilibrium $(-e_3, -e_3, 0, 0) \in \mathsf{T}(S^2)^2$ on the tangent bundle $\mathsf{T}(S^2)^2$. Since the time derivative of the total energy is zero along the flow of (5.68), it follows that this equilibrium is stable.

The other three equilibrium solutions are unstable. We demonstrate this fact only for the equilibrium solution $(e_3, e_3, 0, 0) \in \mathsf{T}(S^2)^2$ by analysis of the properties of the linearized differential equations at this equilibrium. The linearized equations, restricted to the eight-dimensional tangent space of $\mathsf{T}(S^2)^2$ at $(e_3, e_3, 0, 0) \in \mathsf{T}(S^2)^2$, are

$$\begin{bmatrix} (m_1 + m_2)L_1^2 I_{2\times2} & m_2 L_1 L_2 I_{2\times2} \\ m_2 L_1 L_2 I_{2\times2} & m_2 L_2^2 I_{2\times2} \end{bmatrix} \begin{bmatrix} \ddot{\xi}_1 \\ \ddot{\xi}_2 \end{bmatrix}$$
$$- \begin{bmatrix} (m_1 + m_2)g L_1 I_{2\times2} & 0 \\ 0 & m_2 g L_2 I_{2\times2} \end{bmatrix} \begin{bmatrix} \xi_1 \\ \xi_2 \end{bmatrix} = \begin{bmatrix} 0 \\ 0 \end{bmatrix}.$$

These linearized differential equations approximate the local dynamics of the double planar pendulum in a neighborhood of the equilibrium $(e_2, e_2, 0, 0) \in \mathsf{T}(S^1)^2$. The characteristic equation can be shown to be

$$\det \begin{bmatrix} (m_1 + m_2)(L_1^2 \lambda^2 - g L_1) I_{2\times2} & m_2 L_1 L_2 \lambda^2 I_{2\times2} \\ m_2 L_1 L_2 \lambda^2 I_{2\times2} & m_2 (L_2^2 \lambda^2 - g L_2) I_{2\times2} \end{bmatrix} = 0.$$

This characteristic equation is quadratic in λ^2 with sign changes in its coefficients. This guarantees that there are one or more eigenvalues with positive real parts. Consequently, this equilibrium solution $(e_3, e_3, 0, 0) \in \mathsf{T}(S^2)^2$ is unstable. A similar analysis shows that the other equilibrium solutions, namely $(e_3, -e_3, 0, 0) \in \mathsf{T}(S^2)^2$ and $(-e_3, e_3, 0, 0) \in \mathsf{T}(S^2)^2$, are also unstable.

5.7 Problems

5.1. Starting with the Euler–Lagrange equations (5.6), view the kinematics equations (5.1) as defining a change of variables $(q, \dot{q}) \in \mathsf{T}(S^2)^n \to (q, \omega) \in T(S^2)^n$. Show that this change of variables can be used to derive the Euler–Lagrange equations in the form (5.14) according to the following:

(a) Show that

$$\frac{\partial L(q, \dot{q})}{\partial \dot{q}_i} = -S(q_i) \frac{\partial \tilde{L}(q, \omega)}{\partial \omega_i}, \qquad i = 1, \dots, n,$$

$$\frac{\partial L(q, \dot{q})}{\partial q_i} = \frac{\partial \tilde{L}(q, \omega)}{\partial q_i} + S(\dot{q}_i) \frac{\partial \tilde{L}(q, \omega)}{\partial \omega_i}, \qquad i = 1, \dots, n.$$

(b) Substitute these expressions into (5.6) and simplify using several of the matrix identities in Chapter 1 to obtain the equations (5.14).

5.2. Show that the Euler–Lagrange equation (5.7) can be expressed as

$$\frac{d}{dt}\left(\frac{\partial L(q,\dot q)}{\partial \dot q_i}\right) - q_i \frac{d}{dt}\left(q_i^T \frac{\partial L(q,\dot q)}{\partial \dot q_i}\right)$$

$$- q_i \dot q_i^T \frac{\partial L(q,\dot q)}{\partial \dot q_i} - \left(I_{3\times 3} - q_i q_i^T\right)\frac{\partial L(q,\dot q)}{\partial q_i} = 0, \quad i = 1,\ldots,n.$$

5.3. In Chapter 4, the definition of the conjugate momentum $\pi_i \in T_{q_i}S^1$ given in (4.23) is the derivative of the modified Lagrangian with respect to ω_i. In this chapter, the definition of the conjugate momentum $\pi_i \in T_{q_i}S^2$ given in (5.27) is the gradient of the modified Lagrangian with respect to ω_i projected onto the tangent space $T_{q_i}S^2$. That is, no projection appears in the definition used in Chapter 4. Why are these two definitions, in fact, analogous even though there is no projection operator required in the results of Chapter 4? Give an argument that the apparent difference in definitions is due to the different geometries of S^1 in Chapter 4 and S^2 in Chapter 5.

5.4. Consider the dynamics of a particle, with mass m, constrained to move, without friction, on a sphere of radius R that rotates about a diameter with constant angular velocity. The origin of an inertial Euclidean frame is located at the center of the sphere. Assume uniform, constant gravity acts along the direction $-e_3$ of the inertial frame. The configuration vector is the direction of the particle in the inertial frame, so that the configuration manifold is S^2. The angular velocity vector of the sphere, in the inertial frame, is $\Omega e_3 \in \mathbb{R}^3$ where Ω is a real constant.

(a) Determine the Lagrangian function $L : TS^2 \to \mathbb{R}^1$ defined on the tangent bundle of the configuration manifold.
(b) What are the Euler–Lagrange equations for the particle constrained to a sphere that rotates with constant angular velocity under the action of uniform, constant gravity?
(c) Determine the Hamiltonian function $H : T^*S^2 \to \mathbb{R}^1$ defined on the cotangent bundle of the configuration manifold.
(d) What are Hamilton's equations for the particle constrained to a sphere that rotates with constant angular velocity under the action of uniform, constant gravity?
(e) What are conserved quantities for the dynamical flow on TS^2?
(f) What are the equilibrium solutions of the dynamical flow on TS^2?
(g) For one equilibrium solution, what are linearized equations that approximate the dynamical flow in a neighborhood of that equilibrium?

5.5. Consider the dynamics of a particle, with mass m, constrained to move, without friction, on a sphere of radius R that rotates about a diameter with constant angular velocity. The origin of an inertial Euclidean frame is located at the center of the sphere. Assume Newtonian gravity, in the sense that the gravity force on the particle always has constant magnitude with direction opposite to the radius vector from the center of the sphere to the location of the particle on the sphere. The configuration vector is the direction of the particle in the inertial frame, so that the configuration manifold is S^2. The angular velocity vector of the sphere, in the inertial frame, is $\Omega e_3 \in \mathbb{R}^3$ where Ω is a real constant.

(a) Determine the Lagrangian function $L : TS^2 \to \mathbb{R}^1$ defined on the tangent bundle of the configuration manifold.

(b) What are the Euler–Lagrange equations for the particle constrained to a sphere that rotates with constant angular velocity under the action of Newtonian gravity?

(c) Determine the Hamiltonian function $H : T^*S^2 \to \mathbb{R}^1$ defined on the cotangent bundle of the configuration manifold.

(d) What are Hamilton's equations for the particle constrained to a sphere that rotates with constant angular velocity under the action of Newtonian gravity?

(e) What are conserved quantities for the dynamical flow on TS^2?

(f) What are the equilibrium solutions of the dynamical flow on TS^2?

(g) For one equilibrium solution, what are linearized equations that approximate the dynamical flow in a neighborhood of that equilibrium?

5.6. Consider the dynamics of a *Foucault pendulum*, modeled as a particle of mass m, constrained to move without friction on a sphere of radius L about a center or pivot that is fixed to the surface of the Earth An inertial frame has its first two axes horizontal and third axis in the direction of the North pole (assumed to be the rotation axis of the Earth). The first axis of an Earth-fixed frame is in the direction of the local vertical. It is convenient to assume the origin of the inertial frame is located at the Earth's center and the origin of the Earth-fixed frame is located at the pivot on the Earth's surface. The pivot point is located at a latitude angle β. The Earth rotates about its North-South axis with constant angular velocity vector Ωe_3 in the inertial frame; the constant radius of the Earth is r. Assume uniform, constant gravity acts along the direction $-e_1$ of the Earth-fixed frame. The configuration vector of the Foucault pendulum is the direction of the particle in the Earth-fixed frame, so that the configuration manifold is S^2.

(a) Show that the Lagrangian function $L : TS^2 \to \mathbb{R}^1$ defined on the tangent bundle of the configuration manifold is given by

$$L(q, \dot{q}) = \frac{1}{2} m \, \|L\dot{q} + S(\Omega e_\beta)(re_1 + Lq)\|^2 - mgLe_1^T q,$$

where the axis of rotation of the Earth, expressed in the Earth-fixed frame, is given by the unit vector

$$e_\beta = \begin{bmatrix} \sin\beta \\ 0 \\ \cos\beta \end{bmatrix}.$$

(b) What are the Euler–Lagrange equations for the Foucault pendulum? Hint: Use care in the variational analysis noting that this Lagrangian is quadratic in $(q, \dot{q}) \in \mathsf{TS}^2$ with nontrivial product terms.

(c) Simplify these Euler–Lagrange equations for the Foucault pendulum assuming that $L \ll r$ and $r\Omega^2 \ll g$. Show that these simplified Euler–Lagrange equations also define a Lagrangian flow on the tangent bundle TS^2.

(d) What are conserved quantities for the dynamical flow on TS^2?

(e) Show that there are two equilibrium solutions of the dynamical flow on TS^2. Identify one as a hanging equilibrium and the other as an inverted equilibrium. Show that these equilibrium solutions are not necessarily exactly aligned with the local vertical direction of gravity.

(f) For the hanging equilibrium solution, what are linearized equations that approximate the dynamical flow in a neighborhood of that equilibrium? Obtain these linearized equations using the approximated form of the Euler–Lagrange equations.

5.7. Consider the dynamics of a spherical pendulum where the end of the pendulum link is connected to four elastic strings in tension. The mass of the pendulum is m, its moment of inertia about its fixed spherical pivot is J, and its length is L. Uniform gravity acts on the pendulum. The frictionless pivot of the spherical pendulum is located at the origin of an inertial Euclidean frame, where the third axis of the frame is vertical. One end of each of the four elastic strings is connected to the free end of the pendulum while the other end of the each string is connected to a fixed support; the locations of the four string supports are

$$z_1 = \begin{bmatrix} -\frac{\sqrt{2}L}{2} \\ -\frac{\sqrt{2}L}{2} \\ 0 \end{bmatrix}, \quad z_2 = \begin{bmatrix} \frac{\sqrt{2}L}{2} \\ -\frac{\sqrt{2}L}{2} \\ 0 \end{bmatrix}, \quad z_3 = \begin{bmatrix} -\frac{\sqrt{2}L}{2} \\ \frac{\sqrt{2}L}{2} \\ 0 \end{bmatrix}, \quad z_4 = \begin{bmatrix} \frac{\sqrt{2}L}{2} \\ \frac{\sqrt{2}L}{2} \\ 0 \end{bmatrix},$$

with respect to the inertial frame. Thus, the string connections are the vertices of a square of side $\sqrt{2}L$ in the ground plane, with the pivot of the spherical pendulum located at the center of the square. This can be viewed as another example of a tensegrity structure.

(a) Determine the Lagrangian function $L : \mathsf{TS}^2 \to \mathbb{R}^1$ defined on the tangent bundle of the configuration manifold.

(b) Determine expressions for the tension forces in the four strings.

(c) What are the Euler–Lagrange equations for the spherical pendulum connected to four strings in tension?

(d) Determine the Hamiltonian function $H : \mathsf{T}^*\mathsf{S}^2 \to \mathbb{R}^1$ defined on the cotangent bundle of the configuration manifold.

(e) What are Hamilton's equations for the spherical pendulum connected to four strings in tension?

(f) What are conserved quantities for the dynamical flow on TS^2?

(g) What are the equilibrium solutions of the dynamical flow on $\mathsf{T}\mathsf{S}^2$?
(h) For one equilibrium solution, what are linearized equations that approximate the dynamical flow in a neighborhood of that equilibrium?

5.8. Consider two identical particles, each of mass m, that translate in $\mathbb{R}^3$. Each particle is constrained to translate, without friction, on its own sphere embedded in $\mathbb{R}^3$. The two spheres have a common center; the radius of the first sphere is $R_1 > 0$ and the radius of the second sphere is $R_2 > R_1 > 0$. The position vector of the first particles in an inertial Euclidean frame is $x_1 = R_1 q_1 \in \mathbb{R}^3$ and the position vector of the second particle in the inertial Euclidean frame is $x_2 = R_2 q_2 \in \mathbb{R}^3$, where $q = (q_1, q_2) \in (\mathsf{S}^2)^2$. The configuration manifold is given by $(\mathsf{S}^2)^2$. The two particles act under the influence of a mutual potential function given by

$$U(x) = K \left\| x_1 - x_2 \right\|^2,$$

where $K > 0$ is constant.

(a) Determine the Lagrangian function $L : \mathsf{T}(\mathsf{S}^2)^2 \to \mathbb{R}^1$ defined on the tangent bundle $\mathsf{T}(\mathsf{S}^2)^2$ of the configuration manifold.
(b) What are the resulting Euler–Lagrange equations?
(c) Determine the Hamiltonian function $H : \mathsf{T}^*(\mathsf{S}^2)^2 \to \mathbb{R}^1$ defined on the cotangent bundle $\mathsf{T}^*(\mathsf{S}^2)^2$ of the configuration manifold.
(d) What are Hamilton's equations?
(e) What are conservation properties of the dynamical flow on $\mathsf{T}(\mathsf{S}^2)^2$?
(f) What are the equilibrium solutions of the dynamical flow on $\mathsf{T}(\mathsf{S}^2)^2$?

5.9. Consider the dynamics of two elastically connected spherical pendulums, with colocated pivots, under the influence of uniform, constant gravity. The pendulum masses are m_1 and m_2 and the pendulum lengths are L_1 and L_2. The attitudes of the two pendulum links in an inertial Euclidean frame with origin located at the pivot are $q_1 \in \mathsf{S}^2$ and $q_2 \in \mathsf{S}^2$. Thus, the configuration manifold is $(\mathsf{S}^2)^2$. The potential energy for the elastic connection between the two pendulum links is given by $U(q) = \kappa(1 - q_1^T q_2)$, where κ is an elastic stiffness constant.

(a) Determine the Lagrangian function $L : \mathsf{T}(\mathsf{S}^2)^2 \to \mathbb{R}^1$ defined on the tangent bundle of the configuration manifold.
(b) What are the Euler–Lagrange equations for the elastically connected pendulums?
(c) Determine the Hamiltonian function $H : \mathsf{T}^*(\mathsf{S}^2)^2 \to \mathbb{R}^1$ defined on the cotangent bundle of the configuration manifold.
(d) What are Hamilton's equations for the elastically connected pendulums?
(e) What are conserved quantities for the dynamical flow on $\mathsf{T}(\mathsf{S}^2)^2$?
(f) What are the equilibrium solutions of the dynamical flow on $\mathsf{T}(\mathsf{S}^2)^2$?
(g) For one equilibrium solution, what are linearized equations that approximate the dynamical flow in a neighborhood of that equilibrium?

5.10. Consider the dynamics of two elastically connected spherical pendulums, with colocated frictionless pivots. The pendulum masses are m_1 and m_2 and the pendulum lengths are L_1 and L_2. The attitudes of the two pendulum links in an inertial Euclidean frame with origin located at the colocated pivots are $q_1 \in S^2$ and $q_2 \in S^2$. Thus, the configuration manifold is $(S^2)^2$. The potential energy for the elastic connection is given by $U(q) = \kappa(1 - \sqrt{1 - (q_1^T q_2)^2})$, where κ is an elastic stiffness constant.

(a) Determine the Lagrangian function $L : \mathsf{T}(S^2)^2 \to \mathbb{R}^1$ defined on the tangent bundle of the configuration manifold.
(b) What are the Euler–Lagrange equations for the elastically connected pendulums?
(c) Determine the Hamiltonian function $H : \mathsf{T}^*(S^2)^2 \to \mathbb{R}^1$ defined on the cotangent bundle of the configuration manifold.
(d) What are Hamilton's equations for the elastically connected pendulums?
(e) What are conserved quantities for the dynamical flow on $\mathsf{T}(S^2)^2$?
(f) What are the equilibrium solutions of the dynamical flow on $\mathsf{T}(S^2)^2$?
(g) For one equilibrium solution, what are linearized equations that approximate the dynamical flow in a neighborhood of that equilibrium?

5.11. Consider the dynamics of a double spherical pendulum as previously studied in this chapter, but without gravity. The pendulum masses are m_1 and m_2 and the pendulum lengths are L_1 and L_2. The configuration manifold is $(S^2)^2$.

(a) Determine the Lagrangian function $L : \mathsf{T}(S^2)^2 \to \mathbb{R}^1$ defined on the tangent bundle of the configuration manifold.
(b) What are the Euler–Lagrange equations for the double spherical pendulum?
(c) Determine the Hamiltonian function $H : \mathsf{T}^*(S^2)^2 \to \mathbb{R}^1$ defined on the cotangent bundle of the configuration manifold.
(d) What are Hamilton's equations for the double spherical pendulum?
(e) What are conserved quantities for the dynamical flow on $\mathsf{T}(S^2)^2$?
(f) What are the equilibrium solutions of the dynamical flow on $\mathsf{T}(S^2)^2$?
(g) For one equilibrium solution, what are linearized equations that approximate the dynamical flow in a neighborhood of that equilibrium?

5.12. Consider the dynamics of a double spherical pendulum as previously studied in this chapter, without gravity but with an elastic joint between the two pendulum links that resists bending. The pendulum masses are m_1 and m_2 and the pendulum lengths are L_1 and L_2. The potential for the elastic connection is given by $U(q) = \kappa(1 - q_1^T q_2)$, where κ is a elastic stiffness constant. The configuration manifold is $(S^2)^2$.

(a) Determine the Lagrangian function $L : \mathsf{T}(S^2)^2 \to \mathbb{R}^1$ defined on the tangent bundle of the configuration manifold.
(b) What are the Euler–Lagrange equations for the double spherical pendulum?

(c) Determine the Hamiltonian function $H : T^*(S^2)^2 \to \mathbb{R}^1$ defined on the cotangent bundle of the configuration manifold.
(d) What are Hamilton's equations for the double spherical pendulum?
(e) What are conserved quantities for the dynamical flow on $T(S^2)^2$?
(f) What are the equilibrium solutions of the dynamical flow on $T(S^2)^2$?
(g) For one equilibrium solution, what are linearized equations that approximate the dynamical flow in a neighborhood of that equilibrium?

5.13. Consider the dynamics of a particle in three dimensions. The particle is constrained to move on an inertially fixed sphere, under a gravitational field given, in an inertial frame with origin located at the center of the sphere, by the gravitational field $G : \mathbb{R}^3 \to T\mathbb{R}^3$. That is, the gravitational force on a particle located at $x \in \mathbb{R}^3$ is given by $mG(x)$, where m denotes the constant mass of the particle. The configuration manifold is S^2.

(a) Determine the Lagrangian function $L : TS^2 \to \mathbb{R}^1$ of the particle defined on the tangent bundle of the configuration manifold.
(b) What are the Euler–Lagrange equations for the particle?
(c) Determine the Hamiltonian function $H : T^*S^2 \to \mathbb{R}^1$ of the particle defined on the cotangent bundle of the configuration manifold.
(d) What are Hamilton's equations for the particle?
(e) What are the conditions for an equilibrium solution of a particle in a gravitational field?
(f) Suppose that the gravitational field is constant. What are the conditions for an equilibrium solution of a particle?

5.14. Consider the dynamics of a charged particle, with mass m, in three dimensions. The particle is constrained to move on an inertially fixed sphere, under an electric field given, in an inertial frame with origin located at the center of the sphere, by an electric field $E : \mathbb{R}^3 \to T\mathbb{R}^3$ and a magnetic field given, in the inertial frame, by $B : \mathbb{R}^3 \to T\mathbb{R}^3$. The Lorentz force on a particle located at $x \in \mathbb{R}^3$ is given by $Q(E(x) + \dot{x} \times B(x))$, where Q denotes the constant charge on the particle. The configuration manifold is S^2.

(a) Determine the Lagrangian function $L : TS^2 \to \mathbb{R}^1$ of the particle defined on the tangent bundle of the configuration manifold.
(b) What are the Euler–Lagrange equations for the particle?
(c) Determine the Hamiltonian function $H : T^*S^2 \to \mathbb{R}^1$ of the particle defined on the cotangent bundle of the configuration manifold.
(d) What are Hamilton's equations for the particle?
(e) What are the conditions for an equilibrium solution of a particle in an electric field and a magnetic field?
(f) Suppose that the electric field and the magnetic field are constant. What are the conditions for an equilibrium solution of a particle?

5.15. Consider the problem of finding the curve(s) $[0, 1] \to S^2$ of shortest length that connect two fixed points in S^2. Such curves are referred to as geodesic curve(s) on S^2.

(a) If the curve is parameterized by $t \to q(t) \in S^2$, show that the incremental arc length of the curve is $ds = \sqrt{\|dq\|^2}$ so that the geodesic curve(s) minimize $\int_0^1 \sqrt{\|\dot{q}\|^2}\, dt$ among all curves on S^2 that connect the two fixed points.

(b) Show that the geodesic curve(s) necessarily satisfy the variational property $\delta \int_0^1 \|\dot{q}\|\, dt = 0$ for all smooth curves $t \to q(t) \in S^2$ that satisfy the boundary conditions $q(0) = q_0 \in S^2$, $q(1) = q_1 \in S^2$.

(c) What are the Euler–Lagrange equations and Hamilton's equations that geodesic curves in S^2 must satisfy?

(d) Suppose that $q_0 \in S^2$, $q_1 \in S^2$ do not lie on a common diameter of the sphere. Show that there is a unique geodesic curve in S^2. Describe the geodesic curve. Show that the geodesic curve is actually a minimum of $\int_0^1 \|\dot{q}\|\, dt$.

(e) Suppose that $q_0 \in S^2$, $q_1 \in S^2$ lie on a common diameter of the sphere. Show that there are two geodesic curves in S^2. Describe each geodesic curve. Show that each geodesic curve is actually a minimum of $\int_0^1 \|\dot{q}\|\, dt$.

5.16. A thin rigid rod, under uniform, constant gravity, is supported at its ends by two identical spherical pendulums. Each pendulum is attached to an inertially fixed support by a frictionless pivot; the two pivots are assumed to lie in a common horizontal plane. An inertial frame in three dimensions is selected; its origin is located on the midpoint of the line between the two pivots; the first axis is along this line in the horizontal plane; the second axis is orthogonal to the first axis in the horizontal plane and the third axis is vertical. The two fixed pivots are a distance L apart. Each pendulum is assumed to be a rigid, massless link with length L. The ends of the rigid rod are attached to the ends of the pendulums through spherical joints. The rod has length L and it has concentrated mass m located at each end of the rigid rod, so that the total mass of the rod is $2m$. Let $q_1 \in S^2$ denote the attitude of the first pendulum in the inertial frame; let $q_2 \in S^2$ denote the attitude of the second pendulum in the inertial frame. The rigidity of the rod imposes a holonomic constraint given by $\|q_1 - q_2\|^2 = 1$, so that the constraint manifold $M = \{(q_1, q_2) \in (S^2)^2 : \|q_1 - q_2\|^2 = 1\}$ is embedded in $\mathbb{R}^6$.

(a) Show that the augmented Lagrangian function $L^a : TM \times \mathbb{R}^1 \to \mathbb{R}^1$ is

$$L^a(q, \dot{q}) = \frac{1}{2}m\|L\dot{q}_1\|^2 + \frac{1}{2}m\|L\dot{q}_2\|^2 - mgLe_3^T q_1 - mgLe_3^T q_2$$
$$+ \lambda\left(\|q_1 - q_2\|^2 - 1\right).$$

(b) What are the Euler–Lagrange equations for the rigid rod?

(c) Determine the augmented Hamiltonian function $H^a : T^*M \times \mathbb{R}^1 \to \mathbb{R}^1$ of the rigid rod, including a Lagrange multiplier.
(d) What are Hamilton's equations for the rigid rod?
(e) What are conserved quantities for the rigid rod?
(f) What are the equilibrium solutions on the tangent bundle of the constraint manifold?

5.17. A four bar linkage in three dimensions consists of three rigid links with one end of each of two of the links constrained to rotate in three dimensions about its fixed pivot. The fourth link can rotate subject to the fact that its two ends are pinned to the ends of the opposite ends of the first two links. The pivots and joint connections are assumed to be frictionless and to allow rotation of the three links in three dimensions. The two links with an inertially fixed end are assumed to have length L with mass m concentrated at the mid points of the links. The connecting link is assumed to have length $2.5\,L$ with mass $\frac{m}{2}$ concentrated at its mid point. The distance between the two fixed pivots is $2.5\,L$. We assume there are no collisions between links or with the inertial base, and we ignore all gravity effects. Choose an inertially fixed three-dimensional Euclidean frame with origin located at the first fixed pivot point and with its first axis in the direction of the second fixed pivot point. With respect to the inertially fixed frame, let $q_i \in S^2$ denote the attitude vector of the i-th link with respect to the inertial frame, for $i = 1, 2, 3$. The configuration vector is $q = (q_1, q_2, q_3) \in (S^2)^3$.

(a) Show that Lagrangian function $L : T(S^2)^3 \to \mathbb{R}^1$, ignoring the holonomic constraint defined by the locations of the fixed pivots, can be written as

$$ L(q, \dot{q}) = \frac{1}{2} m \left\| \frac{L}{2} \dot{q}_1 \right\|^2 + \frac{1}{2} \left(\frac{m}{2} \right) \left\| \frac{5\,L}{2} \dot{q}_2 \right\|^2 + \frac{1}{2} m \left\| \frac{L}{2} \dot{q}_3 \right\|^2. $$

(b) Show that the vector holonomic constraint that arises from the locations of the fixed pivots can be written as

$$ Lq_1 + 2.5\,Lq_2 + Lq_3 - 2.5\,Le_1 = 0. $$

(c) Describe the constraint manifold M embedded in $(S^2)^3$.
(d) Determine the augmented Lagrangian function $L^a : M \times \mathbb{R}^3 \to \mathbb{R}^1$ on the tangent bundle of the constraint manifold, including Lagrange multipliers.
(e) What are the Euler–Lagrange equations for the four-bar linkage?
(f) Determine the augmented Hamiltonian function $H^a : T^*M \times \mathbb{R}^3 \to \mathbb{R}^1$ on the cotangent bundle of the constraint manifold, including Lagrange multipliers.
(g) What are Hamilton's equations for the four bar linkage?
(h) What are conserved quantities for the four bar linkage?
(i) What are the equilibrium solutions of the dynamical flow?
(j) For one equilibrium solution, what are linearized equations that approximate the dynamical flow in a neighborhood of that equilibrium?

5.18. Consider the dynamics of an ideal particle that evolves on a manifold M that is a smooth deformation of the embedded manifold S^2 in $\mathbb{R}^3$ in the sense that it is the image of a diffeomorphism $\phi : S^2 \to M$. Suppose that the Lagrangian function is $L : TM \to \mathbb{R}^1$. Use the form for the Lagrangian dynamics on S^2 and this diffeomorphism to show that the Lagrangian dynamics on M satisfy

$$(I_{3\times3} - qq^T)\left(\frac{\partial\phi(q)}{\partial q}\right)\left\{\frac{d}{dt}\left(\frac{\partial L(x,\dot{x})}{\partial \dot{x}}\right) - \frac{\partial L(x,\dot{x})}{\partial x}\right\} = 0,$$

where $q = \phi^{-1}(x)$ is the inverse of the diffeomorphism.

5.19. Consider an ideal particle that is constrained to move on the surface of an ellipsoid, identified as the configuration manifold $M = \{q \in \mathbb{R}^3 : \{\frac{q_1}{a}\}^2 + \{\frac{q_2}{b}\}^2 + \{\frac{q_3}{c}\}^2 - 1 = 0\}$ embedded in $\mathbb{R}^3$. Assume one axis of the ellipsoid is vertical and the other two axes of the ellipsoid are horizontal and uniform gravity acts on the particle.

(a) Determine the Lagrangian function $L : TM \to \mathbb{R}^1$ defined on the tangent bundle of the configuration manifold.
(b) Use the results of the previous problem to determine the Euler–Lagrange equations for the particle on the ellipsoid.
(c) Determine the Hamiltonian function $H : T^*M \to \mathbb{R}^1$ defined on the cotangent bundle of the configuration manifold.
(d) What are Hamilton's equations for the particle on the ellipsoid?
(e) What are conserved quantities of the dynamical flow on TM?
(f) What are the equilibrium solutions of the dynamical flow on TM?
(g) For one equilibrium solution, what are linearized equations that approximate the dynamical flow in a neighborhood of that equilibrium?

5.20. Suppose the configuration manifold is $(S^2)^n$ and the kinetic energy has the form of a general quadratic function in the time derivative of the configuration vector, so that the Lagrangian function $L : T(S^2)^n \to \mathbb{R}^1$ is given by

$$L(q,\dot{q}) = \frac{1}{2}\sum_{i=1}^{n}\sum_{j=1}^{n}m_{ij}(q)\dot{q}_i^T\dot{q}_j + \sum_{i=1}^{n}a_i(q)^T\dot{q}_i - U(q),$$

where $q = (q_1,\ldots,q_n) \in (S^2)^n$ and $m_{ij}(q) = m_{ji}(q) > 0$, $i = 1,\ldots,n$, $j = 1,\ldots,n$, $a_i(q)$, $i = 1,\ldots,n$ are vector functions and $U(q)$ is a real scalar function.

(a) What are the Euler–Lagrange equations for this Lagrangian?
(b) What are Hamilton's equations for the Hamiltonian associated with this Lagrangian?
(c) Determine a modified Lagrangian function expressed in terms of angular velocity vector.

(d) What are the Euler–Lagrange equations for this modified Lagrangian?
(e) Determine the modified Hamiltonian function expressed in terms of the momentum conjugate to the angular velocity vector.
(f) What are Hamilton's equations for this modified Hamiltonian function?

Chapter 6
Lagrangian and Hamiltonian Dynamics on $\mathsf{SO}(3)$

This chapter treats the Lagrangian dynamics and Hamiltonian dynamics of a rotating rigid body. A rigid body is a collection of mass particles whose relative positions do not change, that is the body does not deform when acted on by external forces. A rigid body is a useful idealization.

The most general form of rigid body motion consists of a combination of rotation and translation. In this chapter, we consider rotational motion only. Combined rotational and translational dynamics of a rigid body are studied in the subsequent chapter.

We begin by identifying the configurations of a rotating rigid body in three dimensions as elements of the Lie group $\mathsf{SO}(3)$. Equations of motion for the Lagrangian and Hamiltonian dynamics, expressed as Euler–Lagrange (or Euler) equations and Hamilton's equations, are developed for rigid body rotations in three dimensions. These results are illustrated by several examples of the rotational dynamics of a rigid body.

There are many books and research papers that treat rigid body kinematics and dynamics from both theoretical and applied perspectives. It is a common approach in the published literature to describe rigid body kinematics and dynamics in terms of rotation matrices, but not to fully exploit such geometric representations. For example, books such as [5, 10, 26, 30, 32, 70] introduce rotation matrices but make substantial use of local coordinates, such as Euler angles, in analysis and computations. The references [23, 40, 68, 77] are notable for their emphasis on rotation matrices as the primary representation for kinematics and dynamics of rigid body motion on $\mathsf{SO}(3)$ in applications to spacecraft and robotics. In the context of multi-body spacecraft control, [84] was one of the first publications formulating multi-body dynamics using the configuration manifold $(\mathsf{SO}(3))^n$.

© Springer International Publishing AG 2018   273
T. Lee et al., *Global Formulations of Lagrangian and Hamiltonian Dynamics on Manifolds*, Interaction of Mechanics and Mathematics, DOI 10.1007/978-3-319-56953-6_6

6.1 Configurations as Elements in the Lie Group SO(3)

Two Euclidean frames are introduced; these aid in defining the attitude configuration of a rotating rigid body. A reference Euclidean frame is arbitrarily
selected; it is often selected to be an inertial frame but this is not essential.
A Euclidean frame fixed to the rigid body is also introduced; this fixed frame
rotates as the rigid body rotates. The origin of this body-fixed frame can be
arbitrarily selected, but it is often convenient to locate it at the center of
mass of the rigid body.

As a manifold embedded in $\mathsf{GL}(3)$ or $\mathbb{R}^{3\times3}$, recall that

$$\mathsf{SO}(3) = \left\{ R \in \mathsf{GL}(3) : R^T R = RR^T = I_{3\times3}, \det(R) = +1 \right\},$$

has dimension three. The tangent space of $\mathsf{SO}(3)$ at $R \in \mathsf{SO}(3)$ is given by

$$\mathsf{T}_R\mathsf{SO}(3) = \left\{ R\xi \in \mathbb{R}^{3\times3} : \xi \in \mathfrak{so}(3) \right\},$$

and has dimension three. The tangent bundle of $\mathsf{SO}(3)$ is given by

$$\mathsf{TSO}(3) = \left\{ (R, R\xi) \in \mathsf{SO}(3) \times \mathbb{R}^{3\times3} : \xi \in \mathfrak{so}(3) \right\},$$

and has dimension six.

We can view $R \in \mathsf{SO}(3)$ as representing the attitude of the rigid body, so
that $\mathsf{SO}(3)$ is the configuration manifold for rigid body rotational motion. An
attitude matrix $R \in \mathsf{SO}(3)$ can be viewed as a linear transformation on $\mathbb{R}^3$ in
the sense that a representation of a vector in the body-fixed frame is transformed into a representation of the vector in the reference frame. Thus, the
transpose of an attitude matrix $R^T \in \mathsf{SO}(3)$ denotes a linear transformation
from a representation of a vector in the reference frame into a representation
of the vector in the body-fixed frame. These two important properties are
summarized as:

- If $b \in \mathbb{R}^3$ is a representation of a vector expressed in the body-fixed frame,
  then $Rb \in \mathbb{R}^3$ denotes the same vector in the reference frame.
- If $x \in \mathbb{R}^3$ is a representation of a vector expressed in the reference frame,
  then $R^T x \in \mathbb{R}^3$ denotes the same vector in the body-fixed frame.

These are important relationships that are used extensively in the subsequent
developments.

In addition, $R \in \mathsf{SO}(3)$ can be viewed as defining a rigid body rotation
on $\mathbb{R}^3$ according to the rules of matrix multiplication. In this interpretation,
$R \in \mathsf{SO}(3)$ is viewed as a rotation matrix that defines a linear transformation
that acts on rigid body attitudes. This makes $\mathsf{SO}(3)$ a Lie group manifold
using standard matrix multiplication as the group operation, as discussed
in Chapter 1. Since the dimension of $\mathsf{SO}(3)$ is three, rigid body rotational
motion has three degrees of freedom.

6.2 Kinematics on $\mathsf{SO(3)}$

The rotational kinematics of a rotating rigid body are described in terms of the time evolution of the attitude and attitude rate of the rigid body given by $(R, \dot{R}) \in \mathsf{TSO(3)}$. As in Chapter 2, the rotational kinematics equations for a rotating rigid body are given by

$$\dot{R} = R\xi,$$

where $\xi \in \mathfrak{so}(3)$.

We make use of the isomorphism between the Lie algebra $\mathfrak{so}(3)$ and $\mathbb{R}^3$ given by $\xi = S(\omega)$ with $\xi \in \mathfrak{so}(3)$, $\omega \in \mathbb{R}^3$. This perspective is utilized in the subsequent development. This leads to the expression for the attitude or rotational kinematics given by

$$\dot{R} = RS(\omega), \tag{6.1}$$

where $\omega \in \mathbb{R}^3$ is referred to as the angular velocity vector of the rigid body expressed in the body-fixed Euclidean frame.

It is sometimes convenient to partition the rigid body attitude or rotation $R \in \mathsf{SO(3)}$ as a 3×3 matrix into its rows. We use the notation $r_i \in \mathsf{S}^2$ to denote the i-th column of $R^T \in \mathsf{SO(3)}$ for $i = 1, 2, 3$. This is equivalent to the partition

$$R = \begin{bmatrix} r_1^T \\ r_2^T \\ r_3^T \end{bmatrix}.$$

Thus, the rotational kinematics of a rotating rigid body can also be described by the three vector differential equations

$$\dot{r}_i = -\xi r_i, \quad i = 1, 2, 3,$$

or equivalently by

$$\dot{r}_i = S(r_i)\omega, \quad i = 1, 2, 3.$$

We subsequently describe the attitude configuration of a rotating rigid body by the equivalent descriptions $R \in \mathsf{SO(3)}$ or $r_i \in \mathsf{S}^2$, $i = 1, 2, 3$, depending on whichever is the most convenient description.

6.3 Lagrangian Dynamics on SO(3)

A Lagrangian function is introduced. Euler–Lagrange equations are derived using Hamilton's principle that the infinitesimal variation of the action integral is zero. The Euler–Lagrange equations are first expressed for an arbitrary Lagrangian function; then Euler–Lagrange equations are obtained for the case that the kinetic energy term in the Lagrangian function is a quadratic function of the angular velocity vector.

6.3.1 Hamilton's Variational Principle

The Lagrangian function is defined on the tangent bundle of SO(3), that is $L : \mathsf{TSO}(3) \to \mathbb{R}^1$.

We identify the tangent bundle $\mathsf{TSO}(3)$ with $\mathsf{SO}(3) \times \mathfrak{so}(3)$ or with $\mathsf{SO}(3) \times \mathbb{R}^3$ using the isomorphism between $\mathfrak{so}(3)$ and $\mathbb{R}^3$. Thus, we can express the Lagrangian as a function $L(R, \dot{R}) = L(R, R\xi) = L(R, RS(\omega))$ defined on the tangent bundle $\mathsf{TSO}(3)$. We make use of the modified Lagrangian function $\tilde{L}(R, \omega) = L(R, RS(\omega))$, where we view $\tilde{L} : \mathsf{TSO}(3) \to \mathbb{R}^1$ according to the kinematics (6.1).

In studying the dynamics of a rotating rigid body, the Lagrangian function is the difference of a kinetic energy function and a potential energy function; thus the modified Lagrangian function is

$$\tilde{L}(R, \omega) = T(R, \omega) - U(R),$$

where the kinetic energy function $T(R, \omega)$ is viewed as being defined on the tangent bundle $\mathsf{TSO}(3)$ and the potential energy function $U(R)$ is defined on SO(3).

The subsequent development describes variations of functions with values in the special orthogonal group SO(3); rather than using the abstract Lie group formalism, we obtain the results explicitly for the rotation group. In particular, we introduce variations of a rotational motion $t \to R(t) \in \mathsf{SO}(3)$, denoted by $t \to R^\epsilon(t) \in \mathsf{SO}(3)$, by using the exponential map and the isomorphism between $\mathfrak{so}(3)$ and $\mathbb{R}^3$.

The variation of $R : [t_0, t_f] \to \mathsf{SO}(3)$ is a differentiable curve $R^\epsilon : (-c, c) \times [t_0, t_f] \to \mathsf{SO}(3)$ for $c > 0$ such that $R^0(t) = R(t)$, and $R^\epsilon(t_0) = R(t_0)$, $R^\epsilon(t_f) = R(t_f)$ for any $\epsilon \in (-c, c)$.

The variation of a rotational motion can be described using the exponential map as

$$R^\epsilon(t) = R(t)e^{\epsilon S(\eta(t))},$$

where $\epsilon \in (-c, c)$ and $\eta : [t_0, t_f] \to \mathbb{R}^3$ is a differentiable curve that vanishes at t_0 and t_f. Consequently, $S(\eta(t)) \in \mathfrak{so}(3)$ defines a differentiable curve with values in the Lie algebra of skew symmetric matrices that vanishes at t_0 and t_f, and $e^{\epsilon S(\eta(t))} \in SO(3)$ defines a differentiable curve that takes values in the Lie group of rotation matrices and is the identity matrix at t_0 and t_f. Thus, the time derivative of the variation of the rotational motion of a rigid body is

$$\dot{R}^\epsilon(t) = \dot{R}(t) e^{\epsilon S(\eta(t))} + \epsilon R(t) e^{\epsilon S(\eta(t))} S(\dot{\eta}(t)).$$

Suppressing the time dependence in the subsequent notation, the varied curve satisfies

$$\begin{aligned}
\xi^\epsilon &= (R^\epsilon)^T \dot{R}^\epsilon \\
&= e^{-\epsilon S(\eta)} \xi e^{\epsilon S(\eta)} + \epsilon S(\dot{\eta}) \\
&= \xi + \epsilon \left(S(\dot{\eta}) + \xi S(\eta) - S(\eta)\xi \right) + \mathcal{O}(\epsilon^2).
\end{aligned}$$

Define the variation of the angular velocity by $\xi^\epsilon = S(\omega^\epsilon)$ and use the fact that $\xi = S(\omega)$ to obtain

$$S(\omega^\epsilon) = S(\omega) + \epsilon(S(\dot{\eta}) + S(\omega)S(\eta) - S(\eta)S(\omega)) + \mathcal{O}(\epsilon^2).$$

We use a skew symmetric matrix identity to obtain

$$S(\omega^\epsilon) = S(\omega) + \epsilon(S(\dot{\eta}) + S(\omega \times \eta)) + \mathcal{O}(\epsilon^2),$$

or equivalently

$$S(\omega^\epsilon) = S(\omega + \epsilon(\dot{\eta} + \omega \times \eta)) + \mathcal{O}(\epsilon^2).$$

Thus, the variation of the angular velocity satisfies

$$\omega^\epsilon = \omega + \epsilon(\dot{\eta} + \omega \times \eta) + \mathcal{O}(\epsilon^2).$$

From these expressions, we determine the infinitesimal variations

$$\delta R = \left.\frac{d}{d\epsilon} R^\epsilon\right|_{\epsilon=0} = RS(\eta), \tag{6.2}$$

$$\delta\omega = \left.\frac{d}{d\epsilon} \omega^\epsilon\right|_{\epsilon=0} = \dot{\eta} + \omega \times \eta = \dot{\eta} + S(\omega)\eta. \tag{6.3}$$

This framework allows us to introduce the action integral and Hamilton's principle to obtain Euler–Lagrange equations that describe the rotational dynamics of a rigid body.

The action integral is the integral of the Lagrangian function, or equivalently the modified Lagrangian function, along a rotational motion of the rigid body over a fixed time period. The variations are taken over all differentiable curves with values in $SO(3)$ for which the initial and final values are fixed.

The action integral along a rotational motion of a rotating rigid body is

$$\mathfrak{G} = \int_{t_0}^{t_f} \tilde{L}(R, \omega) \, dt.$$

The action integral along a variation of a rotational motion of the rigid body is

$$\mathfrak{G}^\epsilon = \int_{t_0}^{t_f} \tilde{L}(R^\epsilon, \omega^\epsilon) \, dt.$$

The varied value of the action integral along a variation of a rotational motion of the rigid body can be expressed as a power series in ϵ as

$$\mathfrak{G}^\epsilon = \mathfrak{G} + \epsilon \delta \mathfrak{G} + \mathcal{O}(\epsilon^2),$$

where the infinitesimal variation of the action integral is

$$\delta \mathfrak{G} = \frac{d}{d\epsilon} \mathfrak{G}^\epsilon \Big|_{\epsilon=0}.$$

Hamilton's principle states that the infinitesimal variation of the action integral along any rotational motion of the rigid body is zero:

$$\delta \mathfrak{G} = \frac{d}{d\epsilon} \mathfrak{G}^\epsilon \Big|_{\epsilon=0} = 0, \tag{6.4}$$

for all possible infinitesimal variations $\eta : [t_0, t_f] \to \mathbb{R}^3$ satisfying $\eta(t_0) = \eta(t_f) = 0$.

6.3.2 Euler–Lagrange Equations: General Form

We first compute the infinitesimal variation of the action integral as

$$\frac{d}{d\epsilon} \mathfrak{G}^\epsilon \Big|_{\epsilon=0} = \int_{t_0}^{t_f} \left\{ \frac{\partial \tilde{L}(R, \omega)}{\partial \omega} \cdot \delta \omega + \frac{\partial \tilde{L}(R, \omega)}{\partial R} \cdot \delta R \right\} dt.$$

Examining the first term, we obtain

$$\int_{t_0}^{t_f} \frac{\partial \tilde{L}(R,\omega)}{\partial \omega} \cdot \delta \omega \, dt = \int_{t_0}^{t_f} \frac{\partial \tilde{L}(R,\omega)}{\partial \omega} \cdot (\dot{\eta} + \omega \times \eta) \, dt$$

$$= -\int_{t_0}^{t_f} \left\{ \frac{d}{dt} \left(\frac{\partial \tilde{L}(R,\omega)}{\partial \omega} \right) + S(\omega) \frac{\partial \tilde{L}(R,\omega)}{\partial \omega} \right\} \cdot \eta \, dt,$$

where the first term is integrated by parts, using the fact that $\eta(t_0) = \eta(t_f) = 0$, and the second term is rewritten using a cross product identity.

The second term above is now rewritten. We use the notation $r_i \in S^2$ and $\delta r_i \in T_{r_i} S^2$ to denote the i-th column of $R^T \in SO(3)$ and $\delta R^T \in T_R SO(3)$, respectively. This is equivalent to partitioning R and δR into row vectors as

$$R = \begin{bmatrix} r_1^T \\ r_2^T \\ r_3^T \end{bmatrix}, \quad \delta R = \begin{bmatrix} \delta r_1^T \\ \delta r_2^T \\ \delta r_3^T \end{bmatrix}.$$

We use the fact that $\delta r_i = S(r_i)\eta$ to obtain

$$\int_{t_0}^{t_f} \frac{\partial \tilde{L}(R,\omega)}{\partial R} \cdot \delta R \, dt = \int_{t_0}^{t_f} \sum_{i=1}^{3} \frac{\partial \tilde{L}(R,\omega)}{\partial r_i} \cdot \delta r_i \, dt$$

$$= \int_{t_0}^{t_f} \sum_{i=1}^{3} \frac{\partial \tilde{L}(R,\omega)}{\partial r_i} \cdot S(r_i)\eta \, dt$$

$$= -\int_{t_0}^{t_f} \sum_{i=1}^{3} \left(S(r_i) \frac{\partial \tilde{L}(R,\omega)}{\partial r_i} \right) \cdot \eta \, dt. \qquad (6.5)$$

Substituting, the expression for the infinitesimal variation of the action integral is obtained:

$$\left. \frac{d}{d\epsilon} \mathfrak{G}^\epsilon \right|_{\epsilon=0}$$

$$= -\int_{t_0}^{t_f} \left\{ \frac{d}{dt} \left(\frac{\partial \tilde{L}(R,\omega)}{\partial \omega} \right) + S(\omega) \frac{\partial \tilde{L}(R,\omega)}{\partial \omega} + \sum_{i=1}^{3} S(r_i) \frac{\partial \tilde{L}(R,\omega)}{\partial r_i} \right\} \cdot \eta \, dt.$$

From Hamilton's principle, the above expression for the infinitesimal variation of the action integral should be zero for all differentiable variations $\eta : [t_0, t_f] \to \mathbb{R}^3$ with fixed endpoints. The fundamental lemma of the calculus of variations leads to the Euler–Lagrange equations.

Proposition 6.1 *The Euler–Lagrange equations for a modified Lagrangian function $\tilde{L} : TSO(3) \to \mathbb{R}^1$ are*

$$\frac{d}{dt} \left(\frac{\partial \tilde{L}(R,\omega)}{\partial \omega} \right) + \omega \times \frac{\partial \tilde{L}(R,\omega)}{\partial \omega} + \sum_{i=1}^{3} r_i \times \frac{\partial \tilde{L}(R,\omega)}{\partial r_i} = 0. \qquad (6.6)$$

This form of the Euler–Lagrange equations, together with the rotational kinematics equations (6.1), describe the Lagrangian flow of a rotating rigid body
on the tangent bundle TSO(3) in terms of $(R, \omega) \in$ TSO(3).

6.3.3 Euler–Lagrange Equations: Quadratic Kinetic Energy

We now determine a more explicit expression for the kinetic energy of a
rotating rigid body. This expression is used to obtain a standard form of the
Euler–Lagrange equations. For simplicity, the reference frame is assumed to
be an inertial frame, and the origin of the body-fixed frame is assumed to be
located at the center of mass of the rigid body.

Let $\rho \in \mathcal{B} \subset \mathbb{R}^3$ be a vector from the origin of the body-fixed frame to
a mass element of the rigid body expressed in the body-fixed frame. Here
$\mathcal{B}$ denotes the set of material points that constitute the rigid body in the
body-fixed frame. Thus, $\dot{R}\rho$ is the velocity vector of this mass element in the
inertial frame. The kinetic energy of the rotating rigid body can be expressed
as the body integral

$$T(R, \omega) = \frac{1}{2} \int_{\mathcal{B}} \|\dot{R}\rho\|^2 \, dm(\rho)$$

$$= \frac{1}{2} \int_{\mathcal{B}} \|RS(\rho)\omega\|^2 \, dm(\rho)$$

$$= \frac{1}{2}\omega^T \left(\int_{\mathcal{B}} S(\rho)^T S(\rho) \, dm(\rho) \right) \omega,$$

where $dm(\rho)$ denotes the mass of the incremental element located at $\rho \in \mathcal{B}$.
Thus, we can express the kinetic energy as a quadratic function of the angular
velocity vector

$$T(R, \omega) = \frac{1}{2}\omega^T J \omega,$$

where

$$J = \int_{\mathcal{B}} S(\rho)^T S(\rho) \, dm(\rho),$$

is the 3×3 standard inertia matrix of the rigid body that characterizes the
rotational inertia of the rigid body in the body-fixed frame.

The inertia matrix can be shown to be a symmetric and positive-definite
matrix. It has three positive eigenvalues and three eigenvectors that form an
orthonormal basis for $\mathbb{R}^3$. This special basis defines the principal axes of the
rigid body and it is sometimes convenient to select the body-fixed frame to

be aligned with the principal axes of the body. In this case, the inertia matrix is diagonal.

Consequently, the modified Lagrangian function has the special form

$$\tilde{L}(R,\omega) = \frac{1}{2}\omega^T J\omega - U(R). \tag{6.7}$$

This gives the standard form of the equations for a rotating rigid body, often referred to as the Euler equations for rigid body rotational dynamics, as

$$J\dot{\omega} + S(\omega)J\omega - \sum_{i=1}^{3} S(r_i)\frac{\partial U(R)}{\partial r_i} = 0. \tag{6.8}$$

These Euler equations (6.8), together with the rotational kinematics (6.1), describe the Lagrangian flow of a rotating rigid body in terms of the evolution of $(R,\omega) \in \mathsf{TSO}(3)$ on the tangent bundle $\mathsf{TSO}(3)$.

If the potential energy terms in (6.8) are globally defined on $\mathbb{R}^{3\times 3}$, then the domain of definition of the rotational kinematics (6.1) and the Euler equations (6.8) on $\mathsf{TSO}(3)$ can be extended to $\mathsf{T}\mathbb{R}^{3\times 3}$. This extension is natural and useful in that it defines a Lagrangian vector field on the tangent bundle $\mathsf{T}\mathbb{R}^{3\times 3}$ Alternatively, the manifold $\mathsf{TSO}(3)$ is an invariant manifold of this Lagrangian vector field on $\mathsf{T}\mathbb{R}^{3\times 3}$ and its restriction to this invariant manifold describes the Lagrangian flow of (6.1) and (6.8) on $\mathsf{TSO}(3)$.

6.4 Hamiltonian Dynamics on SO(3)

We introduce the Legendre transformation to obtain the angular momentum and the Hamiltonian function. We make use of Hamilton's phase space variational principle to derive Hamilton's equations for a rotating rigid body.

6.4.1 Hamilton's Phase Space Variational Principle

As in the prior section, we begin with a modified Lagrangian function $\tilde{L}$: $\mathsf{TSO}(3) \to \mathbb{R}^1$, which is a real-valued function defined on the tangent bundle of the configuration manifold $\mathsf{SO}(3)$; we assume that the modified Lagrangian function

$$\tilde{L}(R,\omega) = T(R,\omega) - U(R),$$

is given by the difference between a kinetic energy function $T(R,\omega)$ defined on the tangent bundle and a configuration dependent potential energy function $U(R)$.

The angular momentum of the rotating rigid body in the body-fixed frame is defined by the Legendre transformation

$$\Pi = \frac{\partial \tilde{L}(R, \omega)}{\partial \omega}, \qquad (6.9)$$

where we assume the Lagrangian has the property that the map $\omega \in \mathfrak{so}(3) \to \Pi \in \mathfrak{so}(3)^*$ is invertible. The angular momentum is viewed as being conjugate to the angular velocity vector.

The modified Hamiltonian function $\tilde{H} : T^*SO(3) \to \mathbb{R}^1$ is defined on the cotangent bundle of $SO(3)$ by

$$\tilde{H}(R, \Pi) = \Pi \cdot \omega - \tilde{L}(R, \omega),$$

using the Legendre transformation.

Consider the modified action integral of the form,

$$\tilde{\mathfrak{G}} = \int_{t_0}^{t_f} \left\{ \Pi \cdot \omega - \tilde{H}(R, \Pi) \right\} dt.$$

The infinitesimal variation of the action integral is given by

$$\delta\tilde{\mathfrak{G}} = \int_{t_0}^{t_f} \left\{ \Pi \cdot \delta\omega - \frac{\partial \tilde{H}(R, \Pi)}{\partial R} \cdot \delta R + \delta\Pi \cdot \left(\omega - \frac{\partial \tilde{H}(R, \Pi)}{\partial \Pi} \right) \right\} dt.$$

Recall from (6.2) and (6.3) that the infinitesimal variations can be written as

$$\delta R = R S(\eta),$$
$$\delta\omega = \dot{\eta} + S(\omega)\eta,$$

for differentiable curves $\eta : [t_0, t_f] \to \mathbb{R}^3$. Following the arguments used to obtain (6.5),

$$\int_{t_0}^{t_f} \frac{\partial \tilde{H}(R, \Pi)}{\partial R} \cdot \delta R \, dt = -\int_{t_0}^{t_f} \sum_{i=1}^{3} \left(S(r_i) \frac{\partial \tilde{H}(R, \Pi)}{\partial r_i} \right) \cdot \eta \, dt.$$

6.4.2 Hamilton's Equations: General Form

We now derive Hamilton's equations. Substitute the preceding expressions into the expression for the infinitesimal variation of the modified action integral and integrate by parts to obtain

$$\delta\tilde{\mathfrak{G}} = \int_{t_0}^{t_f} \Pi \cdot (\dot{\eta} + S(\omega)\eta) + \sum_{i=1}^{3} \left(S(r_i) \frac{\partial \tilde{H}(R, \Pi)}{\partial r_i} \right) \cdot \eta$$

$$+ \delta\Pi \cdot \left(\omega - \frac{\partial \tilde{H}(R, \Pi)}{\partial \Pi} \right) dt$$

$$= \int_{t_0}^{t_f} \left\{ -\dot{\Pi} - S(\omega)\Pi + \sum_{i=1}^{3} \left(S(r_i) \frac{\partial \tilde{H}(R, \Pi)}{\partial r_i} \right) \right\} \cdot \eta$$

$$+ \delta\Pi \cdot \left(\omega - \frac{\partial \tilde{H}(R, \Pi)}{\partial \Pi} \right) dt.$$

Invoke Hamilton's phase space variational principle that $\delta\tilde{\mathfrak{G}} = 0$ for all possible functions $\eta : [t_0, t_f] \rightarrow \mathbb{R}^3$ and $\delta\Pi : [t_0, t_f] \rightarrow \mathbb{R}^3$ that satisfy $\eta(t_0) = \eta(t_f) = 0$. This implies that the expression in each of the braces of the above equation should be zero. We thus obtain Hamilton's equations, expressed in terms of (R, Π).

Proposition 6.2 *Hamilton's equations for a modified Hamiltonian function $\tilde{H} : \mathsf{T}^*\mathsf{SO}(3) \rightarrow \mathbb{R}^1$ are*

$$\dot{r}_i = r_i \times \frac{\partial \tilde{H}(R, \Pi)}{\partial \Pi}, \quad i = 1, 2, 3, \tag{6.10}$$

$$\dot{\Pi} = \Pi \times \frac{\partial \tilde{H}(R, \Pi)}{\partial \Pi} + \sum_{i=1}^{3} r_i \times \frac{\partial \tilde{H}(R, \Pi)}{\partial r_i}. \tag{6.11}$$

Thus, equations (6.10) and (6.11) define Hamilton's equations of motion for the dynamics of the Hamiltonian flow in terms of the evolution of $(R, \Pi) \in \mathsf{T}^*\mathsf{SO}(3)$ on the cotangent bundle $\mathsf{TSO}(3)$.

The following property follows directly from the above formulation of Hamilton's equations on $\mathsf{SO}(3)$:

$$\frac{d\tilde{H}(R, \Pi)}{dt} = \sum_{i=1}^{3} \frac{\partial \tilde{H}(R, \Pi)}{\partial r_i} \cdot \dot{r}_i + \frac{\partial \tilde{H}(R, \Pi)}{\partial \Pi} \cdot \dot{\Pi}$$

$$= \frac{\partial \tilde{H}(R, \Pi)}{\partial \Pi} \cdot S(\Pi) \frac{\partial \tilde{H}(R, \Pi)}{\partial \Pi}$$

$$= 0.$$

The modified Hamiltonian function is constant along each solution of Hamilton's equation. This property does not hold if the modified Hamiltonian function has a nontrivial explicit dependence on time.

6.4.3 Hamilton's Equations: Quadratic Kinetic Energy

Suppose the kinetic energy is a quadratic in the angular velocity vector

$$\tilde{L}(R,\omega) = \frac{1}{2}\omega^T J\omega - U(R).$$

The Legendre transformation gives

$$\Pi = J\omega,$$

and the modified Hamiltonian function can be expressed as

$$\tilde{H}(R,\Pi) = \frac{1}{2}\Pi^T J^{-1}\Pi + U(R). \qquad (6.12)$$

Hamilton's equations for a rotating rigid body are described on the cotangent bundle T*SO(3) as:

$$\dot{r}_i = r_i \times J^{-1}\Pi, \quad i = 1,2,3, \qquad (6.13)$$

$$\dot{\Pi} = \Pi \times J^{-1}\Pi + \sum_{i=1}^{3} r_i \times \frac{\partial U(R)}{\partial r_i}. \qquad (6.14)$$

Equations (6.13) and (6.14) define Hamilton's equations of motion for rigid body dynamics and they describe the Hamiltonian flow in terms of the evolution of $(R,\Pi) \in$ T*SO(3) on the cotangent bundle T*SO(3).

If the potential energy terms in (6.14) are globally defined on $\mathbb{R}^{3\times3}$, then the domain of definition of (6.13) and (6.14) on T*SO(3) can be extended to T*$\mathbb{R}^{3\times3}$. This extension is natural and useful in that it defines a Hamiltonian vector field on the cotangent bundle T*$\mathbb{R}^{3\times3}$ Alternatively, the manifold T*SO(3) is an invariant manifold of this Hamiltonian vector field on T*$\mathbb{R}^{3\times3}$ and its restriction to this invariant manifold describes the Hamiltonian flow of (6.13) and (6.14) on T*SO(3).

6.5 Linear Approximations of Dynamics on SO(3)

Geometric forms of the Euler–Lagrange equations and Hamilton's equations on the configuration manifold SO(3) have been presented. This yields equations of motion that provide insight into the geometry of the global dynamics on SO(3).

A linear vector field can be determined that approximates the Lagrangian vector field on TSO(3), at least locally in an open subset of TSO(3). Such linear approximations allow a straightforward analysis of local dynamics properties.

A common approach in the literature on the dynamics of rotating rigid bodies involves introducing local coordinates in the form of three angle coordinates; the most common local coordinates are Euler angles, but exponential local coordinates have some advantages as described in Appendix B. These descriptions often involve complicated trigonometric or transcendental expressions and introduce complexity in the analysis and computations.

Although our main emphasis is on global methods, we make use of local coordinates as a way of describing a linear vector field that approximates a vector field on TSO(3), at least in the neighborhood of an equilibrium solution. This approach is used subsequently in this chapter to study the local flow properties near an equilibrium. As further background, linearized equations are developed in local coordinates for SO(3) in Appendix B.

6.6 Dynamics on SO(3)

We study several physical examples of a rotating rigid body in three dimensions. In each, the configuration manifold is SO(3); consequently each of the dynamics has three degrees of freedom. Lagrangian and Hamiltonian formulations of the equations of motion are presented; a few simple flow properties are identified.

6.6.1 Dynamics of a Freely Rotating Rigid Body

We consider a freely rotating rigid body, also referred to as the free rigid body, in the sense that no moments act on the body. In this case, the prior development holds with zero potential energy $U(R) = 0$. This is the simplest case of a rotating rigid body in three dimensions.

An inertial Euclidean frame is selected arbitrarily. The origin of the body-fixed Euclidean frame is assumed to be located at the center of mass of the rigid body which is assumed to be fixed in the inertial frame. A schematic of a freely rotating rigid body is shown in Figure 6.1.

6.6.1.1 Euler–Lagrange Equations

The attitude kinematics equation for the free rigid body is described by

$$\dot{R} = RS(\omega). \tag{6.15}$$

The modified Lagrangian function $\tilde{L} : \mathsf{TSO}(3) \to \mathbb{R}^1$ is

$$\tilde{L}(R, \omega) = \frac{1}{2}\omega^T J\omega.$$

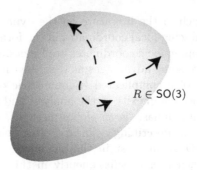

$R \in$ SO(3)

Fig. 6.1 Freely rotating rigid body

Following the results in (6.8) with zero potential energy, the Euler–Lagrange equations of motion for the free rigid body, referred to as the Euler equations, are given by

$$J\dot{\omega} + \omega \times J\omega = 0, \tag{6.16}$$

where $J = \int_B S(\rho)^T S(\rho)dm(\rho)$ is the standard 3×3 inertia matrix of the rigid body in the body-fixed frame. These equations of motion (6.15) and (6.16) define the Lagrangian flow for the free rigid body dynamics described by the evolution of $(R, \omega) \in$ TSO(3) on the tangent bundle of SO(3).

6.6.1.2 Hamilton's Equations

Using the Legendre transformation, let

$$\Pi = \frac{\partial \tilde{L}(R, \omega)}{\partial \omega} = J\omega$$

be the angular momentum of the free rigid body expressed in the body-fixed frame. The modified Hamiltonian is

$$\tilde{H}(R, \Pi) = \frac{1}{2}\Pi^T J^{-1}\Pi.$$

The rotational kinematics equation can be written as

$$\dot{R} = RS(J^{-1}\Pi). \tag{6.17}$$

Using (6.16), Hamilton's equations are given by

$$\dot{\Pi} = \Pi \times J^{-1}\Pi, \tag{6.18}$$

Thus, Hamilton's equations of motion (6.17) and (6.18) describe the Hamiltonian dynamics of the free rigid body as $(R, \Pi) \in T^*SO(3)$ as they evolve on the cotangent bundle of $SO(3)$.

6.6.1.3 Conservation Properties

There are two conserved quantities, or integrals of motion, for the rotational dynamics of a free rigid body. First, the Hamiltonian, which is the rotational kinetic energy and coincides with the total energy E in this case, is conserved; that is

$$H = \frac{1}{2} \omega^T J \omega$$

is constant along each solution of the dynamical flow of the free rigid body.

In addition, there is a rotational symmetry: the Lagrangian is invariant with respect to the tangent lift of arbitrary rigid body rotations. This symmetry leads to conservation of the angular momentum in the inertial frame; that is

$$R\Pi = RJ\omega$$

is constant along each solution of the dynamical flow of the free rigid body. Consequently the magnitude of the angular momentum in the body-fixed frame is also conserved, that is

$$\|J\omega\|^2$$

is constant along each solution of the dynamical flow of the free rigid body. These results are well known for the free rigid body and they guarantee that the free rigid body is integrable [10].

There are additional conservation properties if the distribution of mass in the rigid body has a symmetry. There are many published results for such cases.

6.6.1.4 Equilibrium Properties

The equilibria or constant solutions are easily identified. The free rigid body is in equilibrium at any attitude in $SO(3)$ if the angular velocity vector is zero.

To illustrate the linearization of the dynamics of a rotating rigid body, consider the equilibrium solution $(I_{3\times 3}, 0) \in TSO(3)$. According to Appendix B, $\theta = (\theta_1, \theta_2, \theta_3) \in \mathbb{R}^3$ are exponential local coordinates for $SO(3)$ in a neighborhood of $I_{3\times 3} \in SO(3)$. Following the results in Appendix B, the linearized differential equations defined on the six-dimensional tangent space of $TSO(3)$

at $(I_{3\times3}, 0) \in \mathsf{TSO}(3)$ are given by

$$J\ddot{\xi} = 0.$$

These linearized differential equations approximate the rotational dynamics of the rigid body in a neighborhood of $(I_{3\times3}, 0) \in \mathsf{TSO}(3)$. These simple linear dynamics are accurate to first-order in the perturbations expressed in local coordinates. Higher-order coupling effects are important for large perturbations of the angular velocity vector of the rigid body from equilibrium.

Solutions for which the angular velocity vector are constant can also be identified; these are referred to as relative equilibrium solutions and they necessarily satisfy

$$\omega \times J\omega = 0.$$

Thus, the relative equilibrium solutions occur when the angular velocity vector is collinear with an eigenvector of the inertia matrix J. A comprehensive treatment of relative equilibria of the free rigid body is given in [36].

6.6.2 Dynamics of a Three-Dimensional Pendulum

A three-dimensional pendulum is a rigid body supported by a fixed, frictionless pivot, acted on by uniform, constant gravity. The terminology *three-dimensional pendulum* refers to the fact that the pendulum is a rigid body, with three rotational degrees of freedom, that rotates under uniform, constant gravity. The formulation of a three-dimensional pendulum seems first to have been introduced in [87] and its dynamics developed further in [18, 20, 21, 58]. The development that follows is based on these sources.

An inertial Euclidean frame is selected so that the first two axes lie in a horizontal plane and the third axis is vertical. The origin of the inertial Euclidean frame is selected to be the location of the pendulum pivot. The body-fixed frame is selected so that its origin is located at the center of mass of the rigid body. Let m be the mass of the three-dimensional pendulum and let $\rho_0 \in \mathbb{R}^3$ be the nonzero vector from the center of mass of the body to the pivot, expressed in the body-fixed frame. Let J be the constant 3×3 inertia matrix of the rigid body described subsequently. As before, g denotes the constant acceleration of gravity. A schematic of a three-dimensional pendulum is shown in Figure 6.2.

The attitude of the rigid body is $R \in \mathsf{SO}(3)$ and $\omega \in \mathbb{R}^3$ is the angular velocity vector of the rigid body. The attitude kinematics equation for the three-dimensional pendulum is

$$\dot{R} = RS(\omega). \tag{6.19}$$

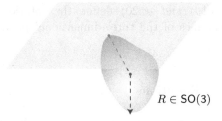

$R \in \mathsf{SO}(3)$

Fig. 6.2 Three-dimensional pendulum

6.6.2.1 Euler–Lagrange Equations

Let $\rho \in \mathbb{R}^3$ be a vector from the origin of the body-fixed frame to a mass element of the rigid body expressed in the body-fixed frame. Thus, $\dot{R}(-\rho_0 + \rho)$ is the velocity vector of this mass element in the inertial frame. The kinetic energy of the rotating rigid body can be expressed as the body integral

$$T(R, \omega) = \frac{1}{2} \int_{\mathcal{B}} \left\| \dot{R}(-\rho_0 + \rho) \right\|^2 dm(\rho)$$
$$= \frac{1}{2} \int_{\mathcal{B}} \left\| RS(-\rho_0 + \rho)\omega \right\|^2 dm(\rho)$$
$$= \frac{1}{2} \omega^T J \omega,$$

where the moment of inertia matrix is

$$J = \int_{\mathcal{B}} S(\rho)^T S(\rho) dm(\rho) + m S^T(\rho_0) S(\rho_0).$$

The gravitational potential energy of the three-dimensional pendulum arises from the gravitational field acting on each material particle in the pendulum body. This can be expressed as

$$U(R) = -\int_{\mathcal{B}} g e_3^T R \rho \, dm(\rho) = -m g e_3^T R \rho_0.$$

The modified Lagrangian function of the three-dimensional pendulum can be expressed as:

$$\tilde{L}(R, \omega) = \frac{1}{2} \omega^T J \omega + m g e_3^T R \rho_0.$$

The Euler–Lagrange equations for the three-dimensional pendulum are given by

$$J \dot{\omega} + \omega \times J \omega - m g \rho_0 \times R^T e_3 = 0. \tag{6.20}$$

These equations (6.19) and (6.20) define the rotational kinematics and the Lagrangian dynamics of the three-dimensional pendulum described by $(R, \omega) \in \mathsf{TSO}(3)$.

6.6.2.2 Hamilton's Equations

Hamilton's equations of motion are easily obtained. According to the Legendre transformation,

$$\Pi = \frac{\partial \tilde{L}(R, \omega)}{\partial \omega} = J\omega$$

is the angular momentum of the three-dimensional pendulum expressed in the body-fixed frame. Thus, the modified Hamiltonian is

$$\tilde{H}(R, \Pi) = \frac{1}{2}\Pi^T J^{-1}\Pi - mgS(\rho_0)R^T e_3.$$

Hamilton's equations of motion are given by the rotational kinematics

$$\dot{R} = RS(J^{-1}\Pi). \tag{6.21}$$

and

$$\dot{\Pi} = \Pi \times J^{-1}\Pi + mg\rho_0 \times R^T e_3, \tag{6.22}$$

Thus, the Hamiltonian dynamics of the three-dimensional pendulum, described by equations (6.21) and (6.22), characterize the evolution of (R, Π) on the cotangent bundle $\mathsf{T^*SO}(3)$.

6.6.2.3 Conservation Properties

There are two conserved quantities, or integrals of motion, for the three-dimensional pendulum. First, the Hamiltonian, which coincides with the total energy E in this case, is conserved, that is

$$H = \frac{1}{2}\omega^T J\omega - mg\rho_0^T R^T e_3,$$

and it is constant along each solution of the dynamical flow of the three-dimensional pendulum.

In addition, the modified Lagrangian is invariant with respect to the lifted action of rotations about the vertical or gravity direction. By Noether's theorem, this symmetry leads to conservation of the component of angular momentum about the vertical or gravity direction; that is

$$h = \omega^T J R^T e_3,$$

and it is constant along each solution of the dynamical flow of the three-dimensional pendulum.

6.6.2.4 Equilibrium Properties

The equilibrium or constant solutions of the three-dimensional pendulum are easily obtained. The conditions for an equilibrium solution are:

$$\omega \times J\omega - mg\rho_0 \times R^T e_3 = 0,$$
$$RS(\omega) = 0.$$

Since $R \in SO(3)$ is non-singular, it follows that the angular velocity vector $\omega = 0$. Thus, an equilibrium attitude satisfies

$$\rho_0 \times R^T e_3 = 0,$$

which implies that

$$R^T e_3 = \frac{\rho_0}{\|\rho_0\|},$$

or

$$R^T e_3 = -\frac{\rho_0}{\|\rho_0\|}.$$

An attitude R is an equilibrium attitude if and only if the vertical direction or equivalently the gravity direction $R^T e_3$, resolved in the body-fixed frame, is collinear with the body-fixed vector ρ_0 from the center of mass of the rigid body to the pivot. If $R^T e_3$ is in the opposite direction to the vector ρ_0, then $(R, 0) \in TSO(3)$ is an *inverted* equilibrium of the three-dimensional pendulum; if $R^T e_3$ is in the same direction to the vector ρ_0, then $(R, 0)$ is a *hanging* equilibrium of the three-dimensional pendulum.

Without loss of generality, it is convenient to assume that the constant center of mass vector, in the body-fixed frame, satisfies

$$\frac{\rho_0}{\|\rho_0\|} = -e_3.$$

Consequently, if $R \in SO(3)$ defines an equilibrium attitude for the three-dimensional pendulum, then an arbitrary rotation of the three-dimensional pendulum about the vertical is also an equilibrium attitude. In summary, there are two disjoint equilibrium manifolds for the three-dimensional pendulum.

The manifold

$$\left\{ R \in \mathsf{SO}(3) : R^T e_3 = \frac{\rho_0}{\|\rho_0\|} \right\},$$

is referred to as the inverted equilibrium manifold, since the center of mass is directly above the pivot.

We now obtain linearized equations at the inverted equilibrium $(I_{3\times3}, 0) \in$ $\mathsf{TSO}(3)$. According to Appendix B, $\theta = (\theta_1, \theta_2, \theta_3) \in \mathbb{R}^3$ are exponential local coordinates for $\mathsf{SO}(3)$ in a neighborhood of $I_{3\times3} \in \mathsf{SO}(3)$. Following the results in Appendix B, the linearized differential equations for the three-dimensional pendulum are defined on the six-dimensional tangent space of $\mathsf{TSO}(3)$ at $(I_{3\times3}, 0) \in \mathsf{TSO}(3)$ and are given by

$$J\ddot{\xi} - mg\|\rho_0\| \begin{bmatrix} 1 & 0 & 0 \\ 0 & 1 & 0 \\ 0 & 0 & 0 \end{bmatrix} \xi = 0.$$

These linearized differential equations approximate the rotational dynamics of a rotating rigid body in a neighborhood of $(I_{3\times3}, 0) \in \mathsf{TSO}(3)$. These linear dynamics are accurate to first-order in the perturbations expressed in local coordinates.

The eigenvalues of the linearized equations can be shown to have the following pattern: two pairs of eigenvalues that are real with equal magnitudes and opposite signs and one pair of eigenvalues at the origin. Since there is a positive eigenvalue, this inverted equilibrium solution is unstable.

Next, the manifold

$$\left\{ R \in \mathsf{SO}(3) : R^T e_3 = -\frac{\rho_0}{\|\rho_0\|} \right\},$$

is referred to as the hanging equilibrium manifold, since the center of mass is directly below the pivot.

We obtain linearized differential equations at the hanging equilibrium $(-I_{3\times3}, 0) \in \mathsf{TSO}(3)$. According to Appendix B, $\theta = (\theta_1, \theta_2, \theta_3) \in \mathbb{R}^3$ are exponential local coordinates for $\mathsf{SO}(3)$ in a neighborhood of $-I_{3\times3} \in \mathsf{SO}(3)$. The linearized differential equations for the three-dimensional pendulum are defined on the six-dimensional tangent space of $\mathsf{TSO}(3)$ at $(-I_{3\times3}, 0) \in \mathsf{TSO}(3)$ and are given by

$$J\ddot{\xi} + mg\|\rho_0\| \begin{bmatrix} 1 & 0 & 0 \\ 0 & 1 & 0 \\ 0 & 0 & 0 \end{bmatrix} \xi = 0.$$

These linearized differential equations approximate the rotational dynamics of a rotating rigid body in a neighborhood of the hanging equilibrium $(-I_{3\times3}, 0) \in \mathsf{TSO}(3)$. These linear dynamics, with two pairs of purely imaginary eigenvalues and one pair of zero eigenvalues, are accurate to first-order in the perturbations expressed in local coordinates.

Solutions for which the angular velocity vector are constant can also be identified; these are relative equilibrium solutions and they necessarily satisfy

$$\omega \times J\omega - mg\rho_0 \times R^T e_3 = 0.$$

Thus, the relative equilibrium solutions occur when the angular velocity vector is collinear with an eigenvector of the inertia matrix J, and the direction of this angular velocity vector, in the inertial frame, is collinear with the gravity direction.

6.6.3 Dynamics of a Rotating Rigid Body in Orbit

Consider the rotational motion of a rigid body in a circular orbit about a large central body. A Newtonian gravity model is used, which gives rise to a differential gravity force on each mass element of the rigid body; this gravity gradient moment is included in our subsequent analysis. The subsequent development follows the presentations in [50, 51].

Three Euclidean frames are introduced: an inertial frame whose origin is at the center of the central body, a body-fixed frame whose origin is located at the center of mass of the orbiting rigid body, and a so-called local vertical, local horizontal (LVLH) frame, whose first axis is tangent to the circular orbit, the second axis is perpendicular to the plane of the orbit, and the third axis is along the orbit radius vector. The origin of the LVLH frame is located at the center of mass of the rigid body and remains on the circular orbit, so that the LVLH frame necessarily rotates at the orbital rate. The LVLH frame is not an inertial frame, but it does have physical significance; it is used to describe the gravity gradient moment. A schematic of a rotating rigid body in a circular orbit is shown in Figure 6.3.

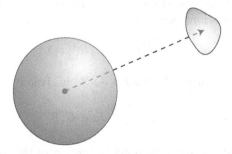

Fig. 6.3 Rotating rigid body in a circular orbit

We define three sets of rotation matrices in SO(3): $R^{bi} \in$ SO(3) denotes a rotation matrix from the body-fixed frame to the inertial frame, $R^{li} \in$ SO(3) denotes a rotation matrix from the LVLH frame to the inertial frame, and $R^{bl} \in$ SO(3) denotes a rotation matrix from the body-fixed frame to the LVLH frame. Thus, the three rotation matrices satisfy $R^{bl} = (R^{li})^T R^{bi}$. We show that the dynamics of a rotating rigid body in a circular orbit can be expressed in terms of the rotation matrix $R^{bi} \in$ SO(3), so that SO(3) is the configuration manifold.

Let $\omega \in \mathbb{R}^3$ be the angular velocity of the rigid body expressed in the body-fixed frame. The 3×3 constant matrix J is the standard inertia matrix of the rigid body in the body-fixed frame. The scalar orbital angular velocity is $\omega_0 = \sqrt{\frac{GM}{r_0^3}}$, where M denotes the mass of the central body, G is the universal gravitational constant, and r_0 is the constant radius of the circular orbit. The inertial frame is selected so that the orbital plane is orthogonal to the second inertial axis; hence the orbital angular velocity vector is $\omega_0 e_2$ in the inertial frame. The LVLH frame is selected so that the orbit radius vector of the body is $r_0 e_3$ in the LVLH frame.

6.6.3.1 Euler–Lagrange Equations

Based on the prior developments in this chapter, the on-orbit rigid body rotational kinematics equations are given as follows. The attitude of the body-fixed frame with respect to the inertial frame is described by the rotational kinematics

$$\dot{R}^{bi} = R^{bi} S(\omega),$$

the attitude of the LVLH frame with respect to the inertial frame is described by the rotational kinematics

$$\dot{R}^{li} = R^{li} S(\omega_0 e_2),$$

and the attitude of the body-fixed frame with respect to the LVLH frame is described by the rotational kinematics

$$\dot{R}^{bl} = R^{bl} S(\omega - \omega_0 R^{bl^T} e_2).$$

The modified Lagrangian $\tilde{L} :$ TSO(3) $\rightarrow \mathbb{R}^1$ is given by

$$\tilde{L}(R^{bi}, \omega) = \frac{1}{2}\omega^T J\omega - U(R^{bi}),$$

where $U(R^{bi})$ is the gravitational potential energy of the rigid body in orbit. Thus, the Euler–Lagrange equations of motion are given by

$$J\dot{\omega} + \omega \times J\omega = M^g,$$

where

$$M^g = \sum_{i=1}^{3} r_i \times \frac{\partial U(R^{bi})}{\partial r_i},$$

is the gravity gradient moment on the rigid body due to the gravity potential $U(R^{bi})$. In the gravity gradient moment expression, r_1, r_2, r_3 denote the
column partitions of $(R^{bi})^T \in \mathsf{SO}(3)$.

Since the orbital angular velocity ω_0 is constant, the rotational kinematics
equation for $R^{li} \in \mathsf{SO}(3)$ can be explicitly solved to obtain

$$R^{li}(t) = R^{li}(t_0)e^{S(\omega_0 e_2)(t-t_0)}.$$

This describes the rotation of the LVLH frame with respect to the inertial
frame.

The gravity potential for the full orbiting rigid body is obtained by integrating the gravity potential for each element in the body over the body; this
leads to

$$U(R^{bi}) = -\int_{\mathcal{B}} \frac{GM}{\|x + R^{bi}\rho\|}\, dm(\rho),$$

where $x \in \mathbb{R}^3$ is the position of the center of mass of the orbiting rigid body
in the inertial frame, and $\rho \in \mathbb{R}^3$ is a vector from the center of mass of the
rigid body to the body element with mass $dm(\rho)$ in the body-fixed frame.

We now derive a closed form approximation for the gravitational moment
M^g using the fact that the rigid body is in a circular orbit so that the norm
of x is constant. The size of the rigid body is assumed to be much smaller
than the orbital radius.

Since the rigid body position vector in the LVLH frame is $r_0 e_3$, the position
vector of the rigid body in the inertial frame is given by $x = r_0 R^{li} e_3$. Using
this expression, the matrix of derivatives of the gravitational potential energy
is

$$\frac{\partial U(R^{bi})}{\partial R^{bi}} = \int_{\mathcal{B}} \frac{GM r_0 R^{li} e_3 \rho^T}{\| r_0 e_3 + R^{bl}\rho \|^3}\, dm(\rho)$$

$$= \frac{GM}{r_0} \int_{\mathcal{B}} \frac{\left(R^{li} e_3 \hat{\rho}^T\right)\frac{\|\rho\|}{r_0}}{\left[1 + 2\left(e_3^T R^{bl}\hat{\rho}\right)\frac{\|\rho\|}{r_0} + \frac{\|\rho\|^2}{r_0^2}\right]^{\frac{3}{2}}}\, dm(\rho),$$

where $\hat{\rho} = \frac{\rho}{\|\rho\|} \in \mathbb{R}^3$ is the unit vector along the direction of ρ. Since the size
of the rigid body is significantly smaller than the orbital radius, it follows
that $\frac{\|\rho\|}{r_0} \ll 1$. Using a Taylor series expansion, we obtain the second-order
approximation:

$$\frac{\partial U(R^{bi})}{\partial R^{bi}} = \frac{GM}{r_0} \int_{\mathcal{B}} R^{li} e_3 \hat{\rho}^T \left\{ \frac{\|\rho\|}{r_0} - 3e_3^T R^{bl} \hat{\rho} \frac{\|\rho\|^2}{r_0^2} \right\} dm(\rho).$$

Since the body-fixed frame is located at the center of mass of the rigid body, $\int_{\mathcal{B}} \rho \, dm(\rho) = 0$. Therefore, the first term in the above equation vanishes. Since $e_3^T R^{bl} \rho$ is a scalar, it can be shown that the above partial derivative matrix can be written as

$$\frac{\partial U(R^{bi})}{\partial R^{bi}} = -3\omega_0^2 R^{li} e_3 e_3^T R^{bl} \left(\frac{1}{2} \text{tr}[J] I_{3\times3} - J \right).$$

This can be used to obtain an expression for the gravity gradient moment on the full rigid body:

$$M^g = \sum_{i=1}^{3} r_i \times \frac{\partial U(R^{bi})}{\partial r^i} = 3\omega_0^2 R^{bl^T} e_3 \times J R^{bl^T} e_3.$$

In summary, the Euler equations can be written as

$$J\dot{\omega} + \omega \times J\omega = 3\omega_0^2 R^{bl^T} e_3 \times J R^{bl^T} e_3, \tag{6.23}$$

and the attitude kinematics equation with respect to the LVLH frame is

$$\dot{R}^{bl} = R^{bl} S(\omega - \omega_0 R^{bl^T} e_2). \tag{6.24}$$

These rotational equations of motion (6.23) and (6.24) define the Lagrangian flow of an on-orbit rigid body as the dynamics described by $(R^{bl}, \omega) \in \text{TSO}(3)$ evolve on the tangent bundle of $\text{SO}(3)$. Rotational dynamics that describe the attitude of the rigid body in the inertial frame or in the body-fixed frame can be obtained from the above development.

6.6.3.2 Hamilton's Equations

Hamilton's equations are easily obtained by defining the angular momentum

$$\Pi = \frac{\partial \tilde{L}(R, \omega)}{\partial \omega} = J\omega.$$

Thus, the modified Hamiltonian function is

$$\tilde{H}(R^{bi}, \Pi) = \frac{1}{2} \Pi^T J^{-1} \Pi + U(R^{bi}).$$

Hamilton's equations of motion for the on-orbit rigid body can be written as the attitude kinematics equation with respect to the LVLH frame, namely

$$\dot{R}^{bl} = R^{bl} S(J^{-1}\Pi - \omega_0 R^{bl^T} e_2), \tag{6.25}$$

and the Euler equations

$$\dot{\Pi} = \Pi \times J^{-1}\Pi + 3\omega_0^2 R^{bl^T} e_3 \times J R^{bl^T} e_3. \tag{6.26}$$

These equations (6.25) and (6.26) define the Hamiltonian flow of the rotational dynamics of an on-orbit rigid body as described by $(R^{bl}, \Pi) \in T^*SO(3)$ on the cotangent bundle of SO(3). Rotational dynamics that describe the attitude of the rigid body in the inertial frame or in the body-fixed frame can be obtained from the above development.

6.6.3.3 Conservation Properties

The Hamiltonian, which coincides with the total energy E in this case, is

$$H = \frac{1}{2}\omega^T J \omega + U(R^{bi});$$

the Hamiltonian is constant along each solution of the dynamical flow.

6.6.3.4 Equilibrium Properties

The orbiting rigid body is in a relative equilibrium when the attitude of the body with respect to the LVLH frame is constant. The relative equilibria can be obtained by assuming that (R^{bl}, ω) are constant in (6.23) and (6.24). This leads to the requirement that the constant angular velocity of the orbiting body is

$$\omega = \omega_0 R^{bl^T} e_3,$$

and the attitude of the rigid body in the LVLH frame is such that the gravity moment on the rigid body is zero, namely

$$R^{bl^T} e_3 \times J R^{bl^T} e_3 = 0.$$

Thus, an attitude $R^{bl} \in SO(3)$ is a relative equilibrium of the orbiting rigid body if $R^{bl^T} e_3 \in \mathbb{R}^3$ is an eigenvector of the inertia matrix J.

6.6.4 Dynamics of a Rigid Body Planar Pendulum

A rigid body planar pendulum is a rigid body that is constrained to rotate about an inertially fixed revolute joint under the influence of uniform,
constant gravity. Since the revolute joint allows one degree of freedom rotation about its axis, each material point in the rigid body necessarily rotates
along a circular arc, centered at the closest point on the axis, in a fixed
two-dimensional plane. This motivates the designation of rigid body planar
pendulum. This is a generalization of the lumped mass planar pendulum example that was introduced in Chapter 4 using the configuration manifold
S^1.

As usual we consider an inertial Euclidean frame in $\mathbb{R}^3$ and we select
a body-fixed frame. The inertial frame is selected so that the third axis is
vertical. For convenience, the origin of the inertial frame is located on the
axis of the revolute joint at the point on the axis that is closest to the center
of mass of the rigid body; the origin of the body-fixed frame coincides with
the center of mass of the rigid body. We denote the direction vector of the
axis of the revolute joint, in the inertial frame, by $a \in S^2$ and we denote the
vector from the center of mass of the rigid body to the origin of the inertial
frame, expressed in the body-fixed frame, by $\rho_0 \in \mathbb{R}^3$. The mass of the rigid
body is m and the inertia matrix of the rigid body, computed subsequently,
is denoted by J. A schematic of a rigid body planar pendulum is shown in
Figure 6.4.

Fig. 6.4 Rigid body planar pendulum

It is an important observation that rotations of the rigid body about the
axis leave material points in the rigid body located on the axis unchanged. If
$R \in SO(3)$ denotes the attitude of the rigid body, then it follows that $Ra = a$
expresses the fact that the direction of the revolute joint axis is unchanged
under rotations about that axis. Thus, the configuration manifold for the
rigid body planar pendulum is

$$M = \{R \in SO(3) : Ra = a\}.$$

This is a differentiable submanifold of SO(3) with dimension one. Consequently, the rigid body planar pendulum has one degree of freedom.

6.6.4.1 Kinematics and Variations

Since the configuration manifold is a submanifold of SO(3), the kinematics and the expressions for the infinitesimal variations must be suitably modified from the prior development in this chapter.

The angular velocity vector of the rigid body $\Omega \in \mathbb{R}^3$ is introduced according to the usual rigid body kinematics

$$\dot{R} = RS(\Omega).$$

We first see that the constraint $Ra = a$ implies that $\dot{R}a = 0$; thus $S(\Omega)a = 0$, that is $\Omega \times a = 0$. This implies that Ω is collinear with a, that is there is $\omega : [t_0, t_f] \to \mathbb{R}^1$ such that

$$\Omega = \omega a,$$

where ω is the scalar angular velocity of the rigid body about its rotation axis. Thus, the rigid body angular velocity vector, in the body-fixed frame, has magnitude given by the scalar angular velocity in the direction of the axis of rotation. Thus, the rotational kinematics of the rigid body can be expressed as

$$\dot{R} = RS(\omega a). \tag{6.27}$$

From the prior analysis in this chapter, it follows that the infinitesimal variation of the rigid body attitude is

$$\delta R = RS(\eta),$$

where $\eta : [t_0, t_f] \to \mathbb{R}^3$ is a differentiable curve that vanishes at its endpoints. Since $Ra = a$, it follows that

$$\delta Ra = 0.$$

This constraint is satisfied if $S(a)\eta = 0$, or equivalently $\eta = \beta a$, where $\beta : [t_0, t_f] \to \mathbb{R}$ is a differentiable curve that vanishes at its endpoints. Thus,

$$\delta R = RS(\beta a).$$

Further, the infinitesimal variation of the angular velocity vector is

$$\delta\Omega = \dot{\eta} + S(\omega a)\eta$$
$$= \dot{\beta}a + \omega S(a)\beta a$$
$$= \dot{\beta}a,$$

since $S(a)a = 0$. Thus,

$$\delta\omega = a^T \delta\Omega = \dot{\beta}.$$

6.6.4.2 Euler–Lagrange Equations

We now derive Euler–Lagrange equations for the rigid body planar pendulum. The above expressions for the infinitesimal variations play a key role.

The inertial position of a material point located in the rigid body at $\rho \in \mathcal{B}$ is given by $R(-\rho_0 + \rho) \in \mathbb{R}^3$. The kinetic energy of the rigid body is

$$T = \frac{1}{2}\int_{\mathcal{B}}\left\|\dot{R}(-\rho_0 + \rho)\right\|^2 dm(\rho)$$
$$= \frac{1}{2}\int_{\mathcal{B}}\|RS(\Omega)(-\rho_0 + \rho)\|^2 dm(\rho)$$
$$= \frac{1}{2}\Omega^T J\Omega,$$

where the rigid body moment of inertia matrix is

$$J = \int_{\mathcal{B}} S^T(\rho)S(\rho)\,dm(\rho) + mS^T(\rho_0)S(\rho_0).$$

The gravitational potential energy of the rigid body is

$$U(R) = \int_{\mathcal{B}} ge_3^T R(-\rho_0 + \rho)\,dm(\rho)$$
$$= -mge_3^T R\rho_0.$$

The modified Lagrangian function is

$$\tilde{L}(R,\Omega) = \frac{1}{2}\Omega^T J\Omega + mge_3^T R\rho_0,$$

or equivalently

$$\tilde{L}(R,\omega) = \frac{1}{2}a^T Ja\,\omega^2 + mge_3^T R\rho_0.$$

The infinitesimal variation of the action integral is

$$\frac{d}{d\epsilon}\mathfrak{G}^\epsilon\bigg|_{\epsilon=0} = \int_{t_0}^{t_f} a^T Ja\,\omega\,\delta\omega + mg\rho_0^T\,\delta R^T e_3\,dt.$$

Use the expression

$$\delta R^T e_3 = -S(\beta a)R^T e_3 = \beta S(R^T e_3)a$$

to obtain the infinitesimal variation of the action integral:

$$\frac{d}{d\epsilon}\mathfrak{G}^\epsilon\bigg|_{\epsilon=0} = \int_{t_0}^{t_f} a^T Ja\,\omega\dot\beta + mg\rho_0^T S(R^T e_3)a\beta\,dt.$$

Integrating by parts and using the fact that the variations vanish at the endpoints, we obtain

$$\frac{d}{d\epsilon}\mathfrak{G}^\epsilon\bigg|_{\epsilon=0} = -\int_{t_0}^{t_f}\left\{a^T Ja\dot\omega - mg\rho_0^T S(R^T e_3)a\right\}\cdot\beta\,dt.$$

Hamilton's principle and the fundamental lemma of the calculus of variations give the Euler–Lagrange equation

$$a^T Ja\,\dot\omega - mga^T(\rho_0 \times R^T e_3) = 0. \tag{6.28}$$

The equations (6.27) and (6.28) describe the dynamical flow of the rigid body planar pendulum on the tangent bundle $\mathsf{T}M$.

6.6.4.3 Hamilton's Equations

According to the Legendre transformation,

$$\pi = \frac{\partial\tilde{L}(R,\omega)}{\partial\omega} = a^T Ja\,\omega$$

is the scalar angular momentum of the rigid body pendulum about its axis of rotation. Thus, the modified Hamiltonian is

$$\tilde{H}(R,\pi) = \frac{1}{2}\frac{\pi^2}{a^T Ja} - mge_3^T R\rho_0.$$

Hamilton's equations of motion are given by the rotational kinematics

$$\dot R = RS\left(\frac{\pi a}{a^T Ja}\right), \tag{6.29}$$

and

$$\dot\pi = mga^T(\rho_0 \times R^T e_3). \tag{6.30}$$

The Hamiltonian dynamics of the rigid body planar pendulum, characterized by equations (6.29) and (6.30), are described by the evolution of (R, π) on the cotangent bundle T^*M.

6.6.4.4 Reduced Equations for the Rigid Body Planar Pendulum

As we have shown, each material point in the rigid body rotates along a planar circular arc about a center on the axis of the revolute joint. In particular, the center of mass vector ρ_0 rotates along a planar circular arc, with center at the origin of the inertial frame. The two-dimensional plane containing each such circular arc is inertially fixed and orthogonal to the axis $a \in S^2$. This suggests that it should be possible to describe such rotations in terms of planar rotations in S^1 as discussed previously in Chapter 4. This connection is clarified in the following development, where the rigid body planar pendulum equations are used to obtain reduced equations that evolve on S^1.

To this end, define the direction of the position vector of the center of mass of the rigid body, expressed in the inertial frame:

$$\zeta = \frac{R\rho_0}{\|R\rho_0\|} = \frac{R\rho_0}{\|\rho_0\|},$$

which follows since $\|R\rho_0\| = \|\rho_0\|$. Thus, $\zeta \in S^2$.

It is easy to see that the rotational kinematics (6.27) can be used to obtain

$$\begin{aligned}
\dot{\zeta} &= \dot{R}R^T\zeta \\
&= RS(\omega a)R^T\zeta \\
&= S(R\omega a)\zeta \\
&= S(\omega a)\zeta,
\end{aligned}$$

where we have used a matrix identity and the fact that $a = Ra$.

We now construct a Euclidean orthonormal basis for the inertial frame given by the ordered triple in S^2:

$$a_1, a_2, a.$$

Since $a^T\zeta = 0$, we can express

$$\zeta = q_1 a_1 + q_2 a_2,$$

where $q = (q_1, q_2) \in S^1$. Substituting this into the above rotational kinematics, we obtain

$$\begin{aligned}
\dot{q}_1 a_1 + \dot{q}_2 a_2 &= \omega S(a)\{q_1 a_1 + q_2 a_2\} \\
&= \omega \{q_1 a_2 - q_2 a_1\}.
\end{aligned}$$

Consequently,

$$\dot{q}_1 = -\omega q_2,$$
$$\dot{q}_2 = \omega q_1.$$

In vector form, this can be written as

$$\dot{q} = \omega S q, \tag{6.31}$$

where S is the constant 2×2 skew-symmetric matrix used in Chapter 4.

We now express the Euler–Lagrange equation (6.28) in a different form. Consider the expression

$$
\begin{aligned}
-mga^T(\rho_0 \times R^T e_3) &= mga^T S(R^T e_3)\rho_0 \\
&= mga^T R^T S(e_3) R\rho_0 \\
&= mg\,\|\rho_0\|\, a^T S(e_3)\zeta \\
&= mg\,\|\rho_0\| \left\{ a^T S(e_3)a_1 q_1 + a^T S(e_3)a_2 q_2 \right\},
\end{aligned}
$$

where we have used a matrix identity and the fact that $Ra = a$. The Euler–Lagrange equation can be expressed as

$$a^T Ja\,\dot{\omega} + mg\,\|\rho_0\| \left\{ a^T S(e_3)a_1 q_1 + a^T S(e_3)a_2 q_2 \right\} = 0. \tag{6.32}$$

Thus, the rotational kinematics (6.31) and the Euler–Lagrange equation (6.32) describe the dynamics of the rigid body planar pendulum in terms of $(q, \omega) \in \mathsf{TS}^1$. These are referred to as reduced equations since they describe only the dynamics of the position vector of the center of mass of the rigid body in the inertial frame.

Following a similar development, a reduced form for Hamilton's equations can be obtained that describe the evolution on the cotangent bundle T^*M. These results require introduction of the reduced Lagrangian, expressed on the tangent bundle TS^1, definition of the conjugate momentum using the Legendre transformation, and derivation of the reduced Hamilton's equations on $\mathsf{T}^*\mathsf{S}^1$. These details are not given here.

6.6.4.5 Conservation Properties

The Hamiltonian, which coincides with the total energy E in this case, is conserved. This can be expressed as

$$H = \frac{1}{2}a^T Ja\,\omega^2 - mge_3^T R\rho_0,$$

which is constant along each solution of the dynamical flow of the rigid body planar pendulum.

6.6.4.6 Equilibrium Properties

The equilibrium or constant solutions of the rigid body planar pendulum occur when the angular velocity $\omega = 0$ and the rigid body attitude satisfies the algebraic equations on the configuration manifold M:

$$mga^T \left(\rho_0 \times R^T e_3 \right) = 0,$$

which implies that the time derivative of the angular velocity vanishes. This requires that the direction of gravity, expressed in the body-fixed frame, and the center of mass vector ρ_0 be collinear.

6.7 Problems

6.1. In this problem, we derive an alternative expression of the moment caused by an attitude-dependent potential, summarized in Proposition 6.1.

(a) Consider two matrices $A, B \in \mathbb{R}^{3 \times 3}$. Let $a_i, b_i \in \mathbb{R}^3$ be the i-th column of A^T and B^T for $i \in \{1, 2, 3\}$, respectively, such that the matrices A and B are partitioned into

$$A = \begin{bmatrix} a_1^T \\ a_2^T \\ a_3^T \end{bmatrix}, \quad B = \begin{bmatrix} b_1^T \\ b_2^T \\ b_3^T \end{bmatrix}.$$

Show that

$$B^T A - A^T B = \sum_{i=1}^{3} b_i a_i^T - a_i b_i^T = \sum_{i=1}^{3} S(a_i \times b_i).$$

(b) Using the above identify, show that the moment caused by an attitude-dependent potential can be rewritten as

$$-\sum_{i=1}^{3} r_i \times \frac{\partial \tilde{L}(R, \omega)}{\partial r_i} = \left(R^T \frac{\partial \tilde{L}(R, \omega)}{\partial R} - \frac{\partial \tilde{L}(R, \omega)}{\partial R}^T R \right)^{\vee},$$

where $\frac{\partial \tilde{L}(R, \omega)}{\partial R} \in \mathbb{R}^{3 \times 3}$ is defined such that its i, j-th element corresponds to the derivative of $L(R, \omega)$ with respect to the i, j-th element of R for $i, j \in \{1, 2, 3\}$.

6.2. Consider the attitude dynamics of a rigid body described in Section 6.3.3. Here, we rederive the Euler–Lagrange equation given in (6.8) to include the effects of an external moment according to the Lagrange–d'Alembert principle.

Suppose that there exists an external moment $M \in \mathbb{R}^3$ acting on the rigid body. Assume it is resolved in the body-fixed frame.

(a) Let $\rho \in \mathbb{R}^3$ be the vector from the mass center of the rigid body to a mass element $dm(\rho)$. Let $dF(\rho) \in \mathbb{R}^3$ be the force acting on $dm(\rho)$. Assume that both of ρ and $dF(\rho)$ are expressed in the body-fixed frame. As there is no external force, $\int_\mathcal{B} dF(\rho) = 0$. Due to the external moment, we have $\int_\mathcal{B} \rho \times dF(\rho) = M$. Show that the virtual work due to the external moment is given by

$$\delta W = \int_\mathcal{B} R dF(\rho) \cdot \delta R \rho = \int_\mathcal{B} \eta \cdot (\rho \times dF(\rho)) = \eta \cdot M,$$

where $\delta R = R\hat{\eta}$ for $\eta \in \mathbb{R}^3$.
(b) From the Lagrange–d'Alembert principle, show that the Euler–Lagrange equation is given by

$$J\dot{\omega} + S(\omega)J\omega - \sum_{i=1}^3 S(r_i)\frac{\partial U(R)}{\partial r_i} = M.$$

6.3. Consider the dynamics of a rotating rigid body that is constrained to planar rotational motion in $\mathbb{R}^2$. That is, the configuration manifold is taken as the Lie group of 2×2 orthogonal matrices with determinant $+1$, namely $\mathsf{SO}(2)$. The rotational kinematics, expressed in terms of the rotational motion $t \to R \in \mathsf{SO}(2)$, are given by

$$\dot{R} = RS\omega,$$

for some scalar angular velocity $t \to \omega \in \mathbb{R}^1$; as before, S is the standard 2×2 skew-symmetric matrix. The modified Lagrangian function is given by

$$\tilde{L}(R,\omega) = \frac{1}{2}J\omega^2 - U(R),$$

where J is the scalar moment of inertia of the rigid body and $U(R)$ is the configuration dependent potential energy function.

(a) What are expressions for the infinitesimal variations of $R \in \mathsf{SO}(2)$, $\dot{R} \in T_R\mathsf{SO}(2)$, and $\omega \in \mathbb{R}^1$?
(b) Use Hamilton's principle to derive the Euler equations for the planar rotations of the rigid body.
(c) Use the Legendre transformation to derive Hamilton's equations for the planar rotations of the rigid body.
(d) What are conserved quantities of the dynamical flow on $\mathsf{TSO}(2)$?
(e) What are conditions that define equilibrium solutions of the dynamical flow on $\mathsf{TSO}(2)$?

6.4. Consider a planar pendulum, with scalar moment of inertia J, under constant, uniform gravity. Assume the configuration manifold of the planar pendulum is taken to be the Lie group SO(2). Use the results in the prior problem for the following.

(a) What are the Euler equations for the planar pendulum on the tangent bundle TSO(2)?
(b) What are Hamilton's equations for the planar pendulum on the cotangent bundle T*SO(2)?
(c) What are the conserved quantities of the dynamical flow on TSO(2)?
(d) What are the equilibrium solutions of the dynamical flow on TSO(2)?

6.5. Consider a double planar pendulum under constant, uniform gravity. The first link rotates about an inertially fixed one degree of freedom revolute joint. The two links are connected by another revolute joint fixed in the two links, constraining the two links to rotate in a common vertical plane. The scalar moments of inertia of the two pendulums are J_1 and J_2 about the two joint axes. Assume the configuration manifold of the planar pendulum is taken to be the Lie group product $(SO(2))^2$. Use the results in the prior problem for the following.

(a) What are the Euler–Lagrange equations for the double planar pendulum on the tangent bundle $T(SO(2))^2$?
(b) What are Hamilton's equations for the double planar pendulum on the cotangent bundle $T^*(SO(2))^2$?
(c) What are the conserved quantities of the dynamical flow on $T(SO(2))^2$?
(d) What are the equilibrium solutions of the dynamical flow on $T(SO(2))^2$?

6.6. Consider the rigid body planar pendulum considered in subsection 6.6.4. The configuration manifold is $M = \{R \in SO(3) : Ra = a\}$.

(a) Show that the configuration manifold M, which is a submanifold of the Lie group SO(3), is a one-dimensional matrix Lie group.
(b) Show that the configuration manifold M is diffeomorphic to SO(2).

6.7. Consider the free rotational motion of a symmetric rigid body in $\mathbb{R}^3$. Assume the moment of inertia in the body-fixed frame is $J = J_s I_{3 \times 3}$, where $J_s > 0$ is a scalar.

(a) What are the Euler equations for the free rotational motion of a symmetric rigid body?
(b) Given initial conditions $w(t_0) = w_0 \in \mathbb{R}^3$, $R(t_0) = R_0 \in SO(3)$, determine analytical expressions for the angular velocity and for the rigid body attitude, the latter described using exponential matrices.

6.8. Consider the free rotational motion of an asymmetric rigid body in $\mathbb{R}^3$. Assume the body-fixed frame is selected so that the moment of inertia is $J = \text{diag}(J_1, J_2, J_3)$, where $J_i > 0$, $i = 1, 2, 3$, are distinct.

(a) What are the Euler equations for the free rotational motion of an asymmetric rigid body?

(b) What are the equilibrium solutions for the dynamical flow defined by the Euler equations? These equilibrium solutions of the Euler equations can be viewed as relative equilibrium solutions for the complete rotational dynamics of the asymmetric rigid body.

(c) For each equilibrium solution of the Euler equations, describe the time dependence of the resulting rigid body attitude in SO(3).

6.9. Consider the free rotational motion of a rigid body, with an axis of symmetry, in $\mathbb{R}^3$. Assume the body-fixed frame is selected so that the moment of inertia is $J = \text{diag}(J_1, J_1, J_2)$, where $J_i > 0$, $i = 1, 2$, are distinct.

(a) What are the Euler equations for the free rotational motion of a rigid body with an axis of symmetry?

(b) What are the equilibrium solutions for dynamical flow defined by the Euler equations? These equilibrium solutions of the Euler equations can be viewed as relative equilibrium solutions for the complete rotational dynamics of the rigid body with an axis of symmetry.

(c) For each equilibrium solution of the Euler equations, describe the time dependence of the resulting rigid body attitude in SO(3).

6.10. Consider the rotational motion of a rigid body in $\mathbb{R}^3$. Let $b \in \mathcal{B} \subset \mathbb{R}^3$ denote the location of a material point in the body, expressed in the body-fixed frame.

(a) Assume an external force $F \in \mathbb{R}^3$, expressed in the inertial frame, acts on the rigid body at the single point in the rigid body denoted by $b \in \mathcal{B}$. Show that the component of the force $R^T F \in \mathbb{R}^3$ in the direction $b \in \mathcal{B}$ does not influence the rotational dynamics of the rigid body.

(b) What are the Euler equations for the rotational motion of a rigid body, expressed in terms of the external force acting on the rigid body in the inertial frame?

(c) Assume an external force $F \in \mathbb{R}^3$, expressed in the body-fixed frame, acts on the rigid body at the single point in the rigid body denoted by $b \in \mathcal{B}$. Show that the component of the force $F \in \mathbb{R}^3$ in the direction $b \in \mathcal{B}$ does not influence the rotational dynamics of the rigid body.

(d) What are the Euler equations for the rotational motion of a rigid body, expressed in terms of the external force acting on the rigid body in the body-fixed frame?

6.11. Consider the dynamics of a rigid body, consisting of material points denoted by $\mathcal{B}$ in a body-fixed frame, under the influence of a gravitational field. The configuration $R \in \text{SO}(3)$ denotes the attitude of the rigid body. Assume the origin of the body-fixed frame is located at the center of mass of the rigid body. A gravitational force acts on each material point in the rigid body. The net moment of all of the gravity forces is obtained by integrating the gravity moment for each mass increment of the body over the whole body. The gravitational field, expressed in the inertial frame, is given by $G : \mathbb{R}^3 \to T\mathbb{R}^3$. The incremental gravitational moment vector on a mass increment $dm(\rho)$ of the rigid body, located at $\rho \in \mathcal{B}$ in the body-fixed frame, is given in the inertial frame by $R\rho \times dm(\rho)G(R\rho)$ or, equivalently in the body-fixed frame, by $\rho \times dm(\rho)R^T G(R\rho)$. Thus, the net gravity moment, in the body-fixed frame, is $\int_{\mathcal{B}} \rho \times R^T G(R\rho)\, dm(\rho)$.

(a) What are the Euler equations for the rotational dynamics of the rigid body in the gravitational field?
(b) What are Hamilton's equations for the rotational dynamics of the rigid body in the gravitational field?
(c) What are the conditions for an equilibrium solution of a rotating rigid body in the gravitational field?
(d) Suppose the gravitational field $G(x) = -ge_3$ is constant. What are the simplified Euler equations for the rotational dynamics of the rigid body? What are the conditions for an equilibrium solution of a rotating rigid body in a constant gravitational field?

6.12. Consider the dynamics of a charged rigid body, consisting of material points denoted by $\mathcal{B}$ in a body-fixed frame, under the influence of an electric field and a magnetic field. The configuration $R \in SO(3)$ denotes the attitude of the rigid body. Assume the origin of the body-fixed frame is located at the center of mass of the rigid body. An electric force and a magnetic force act on each material point in the rigid body. The net moment of all of the electric and magnetic forces is obtained by integrating the incremental electric and magnetic moments for each volume increment of the body over the whole body. The electric field, expressed in the inertial frame, is given by $E : \mathbb{R}^3 \to T\mathbb{R}^3$; the magnetic field, expressed in the inertial frame, is given by $B : \mathbb{R}^3 \to T\mathbb{R}^3$ The incremental electric and magnetic moment vector on a volume increment with charge dQ, located at $\rho \in \mathcal{B}$ in the body-fixed frame, is given in the inertial frame by $R\rho \times dQ(E(R\rho) + \dot{R}\rho \times B(R\rho))$ or, equivalently in the body-fixed frame, by $\rho \times dQR^T(E(R\rho) + \dot{R}\rho \times B(R\rho))$. Thus, the total electric and magnetic moment, in the body-fixed frame, is $\int_{\mathcal{B}} \rho \times R^T(E(R\rho) + \dot{R}\rho \times B(R\rho))\, dQ$.

(a) What are the Euler equations for the rotational dynamics of the rigid body in the electric field and the magnetic field?
(b) What are Hamilton's equations for the rotational dynamics of the rigid body in the electric field and the magnetic field?
(c) What are the conditions for an equilibrium solution of a rotating rigid body in the electric and the magnetic field?
(d) Suppose the electric field $E(x) = -Ee_3$ and the magnetic field $B(x) = Be_2$ are constant, where E and B are scalar constants. What are the simplified Euler equations for the rotational dynamics of the rigid body?

What are the conditions for an equilibrium solution of a rotating rigid body in this constant electric and magnetic field?

6.13. Consider the rotational motion of a rigid body in $\mathbb{R}^3$ acted on by a force $F \in \mathbb{R}^3$. The Euler equations are

$$J\dot{\omega} + \omega \times J\omega = r \times F.$$

In the body-fixed frame, $r = \sum_{i=1}^{3} a_i e_i$ is a constant vector and $F = \sum_{i=1}^{3} f_i R^T e_i$ is the force. These are expressed in terms of scalar constants $a_i, f_i, i = 1, 2, 3$.

(a) Show that the moment vector is constant in the inertial frame.
(b) What are conditions on the constants $a_i, f_i, i = 1, 2, 3$ that guarantee that the Euler equations have an equilibrium solution?
(c) What are conditions on the constants $a_i, f_i, i = 1, 2, 3$ that guarantee that the rigid body dynamical flow on $\mathsf{TSO}(3)$ has an equilibrium solution $(R, \omega) = (I_{3\times3}, 0) \in \mathsf{TSO}(3)$? Are there other equilibrium solutions in this case? What are they?

6.14. Consider the rotational motion of a rigid body in $\mathbb{R}^3$ acted on by a moment vector that is constant in the body-fixed frame. The Euler equations are

$$J\dot{\omega} + \omega \times J\omega = M,$$

where $M = \sum_{i=1}^{3} a_i e_i$ is the constant moment vector for scalar constants $a_i, i = 1, 2, 3$.

(a) Confirm that the moment vector is constant in the body-fixed frame.
(b) Assume the rigid body is asymmetric so that the moment of inertia matrix $J = \mathsf{diag}(J_1, J_2, J_3)$ with distinct entries. Obtain algebraic equations that characterize when the Euler equations have relative equilibrium solutions; that is, the angular velocity vector is constant.

6.15. Consider two concentric rigid spherical shells with common inertially fixed centers. The shells, viewed as rigid bodies, are free to rotate subject to a potential that depends only on the relative attitude of the two spherical shells. The configuration manifold is $(\mathsf{SO}(3))^2$ and the modified Lagrangian function $\tilde{L} : \mathsf{T}(\mathsf{SO}(3))^2 \to \mathbb{R}^1$ is given by

$$\tilde{L}(R_1, R_1, \omega_1, \omega_2) = \frac{1}{2}\omega_1^T J_1 \omega_1 + \frac{1}{2}\omega_2^T J_2 \omega_2 - K\mathsf{trace}(R_1^T R_2 - I_{3\times3}),$$

where $(R_i, \omega_i), i = 1, 2$, denote the attitudes and angular velocity vectors of the two spherical shells and J_1, J_2 are 3×3 inertia matrices of the two spherical shells and K is an elastic constant.

(a) What are the Euler–Lagrange equations for the two concentric shells on the tangent bundle $\mathsf{T}(\mathsf{SO}(3))^2$?
(b) What are Hamilton's equations for the two concentric shells on the cotangent bundle $\mathsf{T}^*(\mathsf{SO}(3))^2$?
(c) What are the conserved quantities of the dynamical flow on $\mathsf{T}(\mathsf{SO}(3))^2$?
(d) What are the equilibrium solutions of the dynamical flow on $\mathsf{T}(\mathsf{SO}(3))^2$?
(e) Determine the linearization that approximates the dynamical flow in a neighborhood of a selected equilibrium solution.

6.16. It can be shown that the problem of finding the geodesic curves on the Lie group $\mathsf{SO}(3)$ is equivalent to the problem of finding smooth curves on $\mathsf{SO}(3)$ that minimize $\int_0^1 \|\omega\|^2 \, dt$ and connect two fixed points in $\mathsf{SO}(3)$.

(a) Using curves described on the interval $[0, 1]$ by $t \to R(t) \in \mathsf{SO}(3)$, show that geodesic curves satisfy the variational property $\delta \int_0^1 \|\omega\|^2 \, dt = 0$ for all smooth curves $t \to R(t) \in \mathsf{SO}(3)$ that satisfy the boundary conditions $R(0) = R_0 \in \mathsf{SO}(3)$, $R(1) = R_1 \in \mathsf{SO}(3)$.
(b) What are the Euler–Lagrange equations and Hamilton's equations that geodesic curves on $\mathsf{SO}(3)$ must satisfy?
(c) Use the equations and boundary conditions for the geodesic curves to describe the geodesic curves on $\mathsf{SO}(3)$.

6.17. Consider the problem of finding the geodesic curves on the Lie group $\mathsf{SO}(3)$ that minimize $\int_0^1 \omega^T J \omega \, dt$ and connect two fixed points in $\mathsf{SO}(3)$. Here J is a symmetric, positive-definite 3×3 matrix that is not a scalar multiple of the identity $I_{3 \times 3}$.

(a) Using curves described on the interval $[0, 1]$ by $t \to R(t) \in \mathsf{SO}(3)$, show that geodesic curves satisfy the variational property $\delta \int_0^1 \omega^T J \omega \, dt = 0$ for all smooth curves $t \to R(t) \in \mathsf{SO}(3)$ that satisfy the boundary conditions $R(0) = R_0 \in \mathsf{SO}(3)$, $R(1) = R_1 \in \mathsf{SO}(3)$.
(b) What are the Euler–Lagrange equations and Hamilton's equations that such geodesic curves on $\mathsf{SO}(3)$ must satisfy?
(c) Describe the impediments in obtaining an analytical expression for such geodesics on $\mathsf{SO}(3)$.

6.18. Consider n rotating rigid bodies that are coupled through the potential energy; the configuration manifold is $(\mathsf{SO}(3))^n$. With respect to a common inertial Euclidean frame, the attitudes of the rigid bodies are given by $R_i \in \mathsf{SO}(3)$, $i = 1, \ldots, n$, and we use the notation $R = (R_1, \ldots, R_n) \in (\mathsf{SO}(3))^n$; similarly, $\omega = (\omega_1, \ldots, \omega_n) \in (\mathbb{R}^3)^n$. Suppose the kinetic energy of the rigid bodies is a quadratic function in the angular velocities of the bodies, so that the modified Lagrangian function $\tilde{L} : \mathsf{T}(\mathsf{SO}(3))^n \to \mathbb{R}^1$ is given by

$$\tilde{L}(R, \omega) = \frac{1}{2} \sum_{i=1}^n \omega_i^T J_i \omega_i + \sum_{i=1}^n a_i^T \omega_i - U(R),$$

where J_i are 3×3 symmetric and positive-definite matrices, $i = 1, \ldots, n$, $a_i \in \mathbb{R}^3$, $i = 1, \ldots, n$, and $U : (\mathsf{SO(3)})^n \to \mathbb{R}^1$ is the potential energy that characterizes the coupling of the rigid bodies.

(a) What are the Euler–Lagrange equations for this modified Lagrangian for n coupled rigid bodies?
(b) What are Hamilton's equations for the modified Hamiltonian associated with this modified Lagrangian for n coupled rigid bodies?

Chapter 7
Lagrangian and Hamiltonian Dynamics on $\mathsf{SE}(3)$

We now study the Lagrangian and Hamiltonian dynamics of a rotating and translating rigid body. A rigid body that is simultaneously translating and rotating is said to undergo *Euclidean motion*. We begin by identifying the configuration of a translating and rotating rigid body in three dimensions as an element of the Lie group $\mathsf{SE}(3)$. Lagrangian and Hamiltonian dynamics for such general rigid body motion in three dimensions, expressed as Euler equations and Hamilton's equations, are obtained. Several specific illustrations of Lagrangian dynamics and Hamiltonian dynamics of a rotating and translating rigid body are studied.

Publications that treat Euclidean rigid body motion are numerous, but many of these make use of local coordinates to describe rigid body rotational motion. Two publications that treat Euclidean motion in a unified way using the geometry of $\mathsf{SE}(3)$ are [5, 77].

7.1 Configurations as Elements in the Lie Group $\mathsf{SE}(3)$

As in the prior chapter, two Euclidean frames are introduced; these aid in defining the configuration of a rotating and translating rigid body. An inertial Euclidean frame is arbitrarily selected. A Euclidean frame fixed in the rigid body is also introduced. The origin of this body-fixed frame can be arbitrarily selected, but it is often convenient to locate it at the center of mass of the rigid body. The body-fixed frame translates and rotates with the rigid body.

As a manifold, recall that

$$\mathsf{SE}(3) = \left\{ (R, x) \in \mathsf{GL}(3) \times \mathbb{R}^3 : R^T R = R R^T = I_{3\times 3}, \det(R) = +1 \right\},$$

© Springer International Publishing AG 2018
T. Lee et al., *Global Formulations of Lagrangian and Hamiltonian Dynamics on Manifolds*, Interaction of Mechanics and Mathematics,
DOI 10.1007/978-3-319-56953-6_7

has dimension six. The tangent space of SE(3) at $(R, x) \in$ SE(3) is given by

$$\mathsf{T}_{(R,x)}\mathsf{SE}(3) = \left\{ (R\xi, \zeta) \in \mathbb{R}^{3 \times 3} \times \mathbb{R}^3 : \xi \in \mathfrak{so}(3) \right\},$$

and has dimension six. The tangent bundle of SE(3) is given by

$$\mathsf{TSE}(3) = \left\{ (R, x, R\xi, \zeta) \in \mathsf{GL}(3) \times \mathbb{R}^3 \times \mathbb{R}^{3 \times 3} \times \mathbb{R}^3 : R \in \mathsf{SO}(3), \xi \in \mathfrak{so}(3) \right\},$$

and has dimension twelve.

By partitioning the elements $(R, x) \in$ SE(3) into a 4×4 homogenous matrix

$$G = \begin{bmatrix} R & x \\ 0 & 1 \end{bmatrix}, \tag{7.1}$$

as demonstrated in Chapter 1, SE(3) can be viewed as a Lie group manifold embedded in GL(4) or $\mathbb{R}^{4 \times 4}$, where matrix multiplication is the group operation.

Thus, we can view $(R, x) \in$ SE(3) as representing a configuration in the sense that $R \in$ SO(3) is the attitude of a rigid body and $x \in \mathbb{R}^3$ is the location of the origin of the body-fixed frame in the inertial frame. Consequently, SE(3) can be viewed as the configuration manifold for a rotating and translating rigid body.

In addition, the pair $(R, x) \in$ SE(3), represented as a homogeneous matrix, can also be viewed as defining a rigid body transformation that describes a Euclidean motion (rotation and translation) on $\mathbb{R}^3$ according to the rules of matrix multiplication of homogeneous matrices. In this interpretation, $R \in$ SO(3) is viewed as a rotational transformation that acts on the attitude of the rigid body and $x \in \mathbb{R}^3$ is viewed as a translational transformation of the rigid body. Consequently, a rigid body transformation, consisting of both rotation and translation, acts on a rigid body configuration to give a transformed configuration according to matrix multiplication of homogeneous matrices. This makes SE(3) a Lie group manifold, as discussed in Chapter 1. Since the dimension of SE(3) is six, Euclidean motion of a rigid body has six degrees of freedom.

7.2 Kinematics on SE(3)

The rotational and translational kinematics of a rigid body are described in terms of the Euclidean motion given by $t \to (R, x, \dot{R}, \dot{x}) \in$ TSE(3), where $R \in$ SO(3) is the attitude of the rigid body and $x \in \mathbb{R}^3$ is the position vector of the origin of the body-fixed frame, expressed in the inertial frame.

As in the prior chapter, the rotational kinematics equation can be described as

$$\dot{R} = RS(\omega), \tag{7.2}$$

where $\omega \in \mathbb{R}^3$ is the angular velocity vector in the body-fixed frame.

The translational velocity vector of the origin of the body-fixed frame, sometimes referred to as the translational velocity vector of the rigid body, is the time derivative of the inertial position vector from the origin of the inertial frame to the origin of the body-fixed frame. In the inertial frame, the translational velocity vector $\dot{x} \in \mathbb{R}^3$ of the rigid body is

$$\dot{x} = Rv, \tag{7.3}$$

where $v \in \mathbb{R}^3$ denotes the translational velocity vector of the rigid body, expressed in the body-fixed frame. These are referred to as the translational kinematics of a rigid body.

Equations (7.2) and (7.3) can be rewritten as

$$\begin{bmatrix} \dot{R} & \dot{x} \\ 0 & 0 \end{bmatrix} = \begin{bmatrix} R & x \\ 0 & 1 \end{bmatrix} \begin{bmatrix} S(\omega) & v \\ 0 & 0 \end{bmatrix}, \tag{7.4}$$

or equivalently,

$$\dot{G} = GV, \tag{7.5}$$

where the 4×4 matrix V is an element of the Lie algebra $\mathfrak{se}(3)$, defined as

$$\mathfrak{se}(3) = \left\{ \begin{bmatrix} S(\omega) & v \\ 0 & 0 \end{bmatrix} \in \mathbb{R}^{4\times4} : \omega, v \in \mathbb{R}^3 \right\}.$$

From the above definition, it is straightforward to see that the Lie algebra $\mathfrak{se}(3)$ can be identified with $\mathbb{R}^6$, and therefore, the tangent bundle TSE(3) can be identified with SE(3) $\times \mathbb{R}^6$. The isomorphism between $\mathbb{R}^6$ and $\mathfrak{se}(3)$ is denoted by $\mathcal{S} : \mathbb{R}^3 \times \mathbb{R}^3 \to \mathfrak{se}(3)$, defined as

$$V = \mathcal{S}\left(\begin{bmatrix} \omega \\ v \end{bmatrix}\right) = \mathcal{S}(\omega, v) = \begin{bmatrix} S(\omega) & v \\ 0 & 0 \end{bmatrix}.$$

These representations are used interchangeably. This induces an inner product on $\mathfrak{se}(3)$ from the standard inner product on $\mathbb{R}^3$ as

$$\mathcal{S}(\omega_1, v_1) \cdot \mathcal{S}(\omega_2, v_2) = \omega_1 \cdot \omega_2 + v_1 \cdot v_2,$$

for any $\omega_1, v_1, \omega_2, v_2 \in \mathbb{R}^3$. This inner product notation is extensively used in the subsequent variational analysis. This relationship implies that the dual space $\mathfrak{se}(3)^*$ can be identified with $\mathfrak{se}(3)$.

7.3 Lagrangian Dynamics on SE(3)

We now consider the dynamics of a rotating and translating rigid body as it evolves on the configuration manifold SE(3). The Lagrangian function is defined on the tangent bundle of SE(3), that is $L : \mathsf{TSE}(3) \to \mathbb{R}^1$. Using the associations: TSE(3) with SE(3) × $\mathfrak{se}(3)$, SE(3) with SO(3) × $\mathbb{R}^3$, and $\mathfrak{se}(3)$ with $\mathfrak{so}(3)$ × $\mathbb{R}^3$, the Lagrangian function can be viewed as $L : \mathsf{SO}(3) \times \mathbb{R}^3 \times \mathfrak{so}(3) \times \mathbb{R}^3 \to \mathbb{R}^1$.

7.3.1 Hamilton's Variational Principle

We can express the Lagrangian as a function $L(G, \dot{G}) = L(G, GV)$ defined on the tangent bundle TSE(3). We make use of the modified Lagrangian function $\tilde{L}(G, V) = L(G, GV)$. This is a traditional point of view in formulating Euclidean rigid body dynamics and it is followed in the subsequent development.

In studying the dynamics of a rotating and translating rigid body, the modified Lagrangian function is the difference of a kinetic energy function and a potential energy function

$$\tilde{L}(G, V) = T(G, V) - U(G),$$

where the kinetic energy function $T(G, V)$ is defined on the tangent bundle TSE(3) and the potential energy function $U(G)$ is defined on SE(3).

The variations of a Euclidean motion $G : [t_0, t_f] \to \mathsf{SE}(3)$ are differentiable curves $G^\epsilon : (-c, c) \times [t_0, t_f] \to \mathsf{SE}(3)$, for $c > 0$, such that $G^0(t) = G(t)$, and $G^\epsilon(t_0) = G(t_0)$, $G^\epsilon(t_f) = G(t_f)$ for any $\epsilon \in (-c, c)$. It can be described using the exponential map as

$$G^\epsilon(t) = G(t) e^{\epsilon \Gamma(t)},$$

where $\Gamma : [t_0, t_f] \to \mathfrak{se}(3)$ denotes a differentiable curve, with values in $\mathfrak{se}(3)$ that vanishes at t_0 and t_f, of the form

$$\Gamma = \mathcal{S}(\eta, \chi) = \begin{bmatrix} S(\eta) & \chi \\ 0 & 0 \end{bmatrix},$$

where $\eta : [t_0, t_f] \to \mathbb{R}^3$ and $\chi : [t_0, t_f] \to \mathbb{R}^3$ denote differentiable curves that vanish at t_0 and t_f. From this, the infinitesimal variation of a Euclidean motion is given by

$$\delta G = \left. \frac{d}{d\epsilon} G^\epsilon \right|_{\epsilon=0} = G\Gamma, \tag{7.6}$$

or equivalently

$$\begin{bmatrix} \delta R & \delta x \\ 0 & 0 \end{bmatrix} = \begin{bmatrix} R & x \\ 0 & 1 \end{bmatrix} \begin{bmatrix} S(\eta) & \chi \\ 0 & 0 \end{bmatrix} = \begin{bmatrix} RS(\eta) & R\chi \\ 0 & 0 \end{bmatrix}. \tag{7.7}$$

Next, we find the variation of $V \in \mathfrak{se}(3)$. Taking the time derivative of (7.6), we obtain

$$\delta \dot{G} = \dot{G}\Gamma + G\dot{\Gamma} = GV\Gamma + G\dot{\Gamma}.$$

Also, taking the variation of the kinematics equation (7.5),

$$\delta \dot{G} = \delta GV + G\delta V = G\Gamma V + G\delta V.$$

By combining these, the variation of V is seen to be given by

$$\delta V = \dot{\Gamma} + V\Gamma - \Gamma V. \tag{7.8}$$

The last two terms of (7.8) can be rewritten as

$$\begin{aligned} V\Gamma - \Gamma V &= \begin{bmatrix} S(\omega) & v \\ 0 & 0 \end{bmatrix} \begin{bmatrix} S(\eta) & \chi \\ 0 & 0 \end{bmatrix} - \begin{bmatrix} S(\eta) & \chi \\ 0 & 0 \end{bmatrix} \begin{bmatrix} S(\omega) & v \\ 0 & 0 \end{bmatrix} \\ &= \begin{bmatrix} S(\omega)S(\eta) - S(\eta)S(\omega) & S(\omega)\chi - S(\eta)v \\ 0 & 0 \end{bmatrix} \\ &= \begin{bmatrix} S(\omega \times \eta) & S(\omega)\chi - S(\eta)v \\ 0 & 0 \end{bmatrix} \\ &= S(\omega \times \eta, \omega \times \chi - \eta \times v) \\ &= S\left(\begin{bmatrix} S(\omega) & 0 \\ S(v) & S(\omega) \end{bmatrix} \begin{bmatrix} \eta \\ \chi \end{bmatrix} \right). \end{aligned}$$

For any $\eta_0, \chi_0 \in \mathbb{R}^3$, we have the following property:

$$\begin{aligned} (V\Gamma - \Gamma V) \cdot S(\eta_0, \chi_0) &= \eta_0^T (\omega \times \eta) + \chi_0^T (\omega \times \chi - \eta \times v) \\ &= \eta^T (\eta_0 \times \omega + \chi_0 \times v) + \chi^T (\chi_0 \times \omega) \\ &= \begin{bmatrix} \eta \\ \chi \end{bmatrix}^T \begin{bmatrix} -S(\omega) & -S(v) \\ 0 & -S(\omega) \end{bmatrix} \begin{bmatrix} \eta_0 \\ \chi_0 \end{bmatrix} \\ &= \Gamma \cdot S\left(\begin{bmatrix} -S(\omega) & -S(v) \\ 0 & -S(\omega) \end{bmatrix} \begin{bmatrix} \eta_0 \\ \chi_0 \end{bmatrix} \right). \tag{7.9} \end{aligned}$$

This framework allows us to introduce the action integral and Hamilton's principle to obtain equations for the Euclidean dynamics of a rigid body.

The action integral is the integral of the Lagrangian function, or equivalently the modified Lagrangian function, along a Euclidean motion of the rigid body over a fixed time period. The variations are taken over all differentiable curves with values in SE(3) for which the initial and final values are fixed. The action integral along a Euclidean motion of a rigid body is

$$\mathfrak{G} = \int_{t_0}^{t_f} \tilde{L}(G, V) \, dt.$$

The action integral along a variation of a Euclidean motion of the rigid body is

$$\mathfrak{G}^\epsilon = \int_{t_0}^{t_f} \tilde{L}(G^\epsilon, V^\epsilon) \, dt.$$

The varied value of the action integral along a variation of a rotational and translational motion of the rigid body can be expressed as a power series in ϵ as

$$\mathfrak{G}^\epsilon = \mathfrak{G} + \epsilon \delta \mathfrak{G} + \mathcal{O}(\epsilon^2),$$

where the infinitesimal variation of the action integral is

$$\delta \mathfrak{G} = \frac{d}{d\epsilon} \mathfrak{G}^\epsilon \bigg|_{\epsilon=0}.$$

Hamilton's principle states that the infinitesimal variation of the action integral along any Euclidean motion of the rigid body is zero:

$$\delta \mathfrak{G} = \frac{d}{d\epsilon} \mathfrak{G}^\epsilon \bigg|_{\epsilon=0} = 0, \tag{7.10}$$

for all possible differentiable variations in SE(3), that is for all possible infinitesimal variations $\Gamma : [t_0, t_f] \rightarrow \mathfrak{se}(3)$, satisfying $\Gamma(t_0) = \Gamma(t_f) = 0$.

7.3.2 Euler–Lagrange Equations: General Form

The infinitesimal variation of the action integral is

$$\frac{d}{d\epsilon} \mathfrak{G}^\epsilon \bigg|_{\epsilon=0} = \int_{t_0}^{t_f} \left\{ \frac{\partial \tilde{L}(G, V)}{\partial V} \cdot \delta V + \frac{\partial \tilde{L}(G, V)}{\partial G} \cdot \delta G \right\} dt.$$

Examining the first term, and using (7.8), we obtain

$$\int_{t_0}^{t_f} \frac{\partial \tilde{L}(G,V)}{\partial V} \cdot \delta V \, dt$$

$$= \int_{t_0}^{t_f} \mathcal{S}\left(\frac{\partial \tilde{L}(G,V)}{\partial \omega}, \frac{\partial \tilde{L}(G,V)}{\partial v} \right) \cdot (\dot{\Gamma} + V\Gamma - \Gamma V) \, dt$$

$$= \int_{t_0}^{t_f} \mathcal{S}\left(-\frac{d}{dt} \begin{bmatrix} \frac{\partial \tilde{L}(G,V)}{\partial \omega} \\ \frac{\partial \tilde{L}(G,V)}{\partial v} \end{bmatrix} + \begin{bmatrix} -S(\omega) & -S(v) \\ 0 & -S(\omega) \end{bmatrix} \begin{bmatrix} \frac{\partial \tilde{L}(G,V)}{\partial \omega} \\ \frac{\partial \tilde{L}(G,V)}{\partial v} \end{bmatrix} \right) \cdot \Gamma \, dt,$$

where the first term is integrated by parts, using the fact that $\Gamma(t_0) = \Gamma(t_f) = 0$, and the second term is obtained using (7.9).

The second term above is given by

$$\int_{t_0}^{t_f} \frac{\partial \tilde{L}(G,V)}{\partial G} \cdot \delta G \, dt = \int_{t_0}^{t_f} \frac{\partial \tilde{L}(G,V)}{\partial R} \cdot \delta R + \frac{\partial \tilde{L}(G,V)}{\partial x} \cdot \delta x \, dt.$$

By using (7.6) and (7.7), it can be rewritten as

$$\int_{t_0}^{t_f} \frac{\partial \tilde{L}(G,V)}{\partial G} \cdot \delta G \, dt = \int_{t_0}^{t_f} -\sum_{i=1}^{3} \left(S(r_i) \frac{\partial \tilde{L}(G,V)}{\partial r_i} \right) \cdot \eta + \frac{\partial \tilde{L}(G,V)}{\partial x} \cdot R\chi \, dt$$

$$= \int_{t_0}^{t_f} \mathcal{S}\left(\begin{bmatrix} -\sum_{i=1}^{3} S(r_i) \frac{\partial \tilde{L}(G,V)}{\partial r_i} \\ R^T \frac{\partial \tilde{L}(G,V)}{\partial x} \end{bmatrix} \right) \cdot \Gamma \, dt.$$

Substituting these two results, the expression for the infinitesimal variation of the action integral is given by

$$\frac{d}{d\epsilon} \mathfrak{G}^\epsilon \Big|_{\epsilon=0} = \int_{t_0}^{t_f} \mathcal{S}\left(-\frac{d}{dt} \begin{bmatrix} \frac{\partial \tilde{L}(G,V)}{\partial \omega} \\ \frac{\partial \tilde{L}(G,V)}{\partial v} \end{bmatrix} + \begin{bmatrix} -S(\omega) & -S(v) \\ 0 & -S(\omega) \end{bmatrix} \begin{bmatrix} \frac{\partial \tilde{L}(G,V)}{\partial \omega} \\ \frac{\partial \tilde{L}(G,V)}{\partial v} \end{bmatrix} \right.$$

$$\left. + \begin{bmatrix} -\sum_{i=1}^{3} S(r_i) \frac{\partial \tilde{L}(G,V)}{\partial r_i} \\ R^T \frac{\partial \tilde{L}(G,V)}{\partial x} \end{bmatrix} \right) \cdot \Gamma \, dt.$$

From Hamilton's principle, the above expression for the infinitesimal variation of the action integral should be zero for all infinitesimal variations $\Gamma : [t_0, t_f] \to \mathfrak{se}(3)$ that vanish at the endpoints. The fundamental lemma of the calculus of variations gives the Euler–Lagrange equations

$$\frac{d}{dt} \begin{bmatrix} \frac{\partial \tilde{L}(G,V)}{\partial \omega} \\ \frac{\partial \tilde{L}(G,V)}{\partial v} \end{bmatrix} + \begin{bmatrix} S(\omega) & S(v) \\ 0 & S(\omega) \end{bmatrix} \begin{bmatrix} \frac{\partial \tilde{L}(G,V)}{\partial \omega} \\ \frac{\partial \tilde{L}(G,V)}{\partial v} \end{bmatrix} - \begin{bmatrix} -\sum_{i=1}^{3} S(r_i) \frac{\partial \tilde{L}(G,V)}{\partial r_i} \\ R^T \frac{\partial \tilde{L}(G,V)}{\partial x} \end{bmatrix} = 0.$$

This can be written as follows.

Proposition 7.1 *The Euler–Lagrange equations for a modified Lagrangian function $\tilde{L} : \mathsf{TSE}(3) \rightarrow \mathbb{R}^1$ are*

$$\frac{d}{dt}\left(\frac{\partial \tilde{L}(G,V)}{\partial \omega}\right) + \omega \times \frac{\partial \tilde{L}(G,V)}{\partial \omega}$$

$$+v \times \frac{\partial \tilde{L}(G,V)}{\partial v} + \sum_{i=1}^{3} r_i \times \frac{\partial \tilde{L}(G,V)}{\partial r_i} = 0, \tag{7.11}$$

$$\frac{d}{dt}\left(\frac{\partial \tilde{L}(G,V)}{\partial v}\right) + \omega \times \frac{\partial \tilde{L}(G,V)}{\partial v} - R^T \frac{\partial \tilde{L}(G,V)}{\partial x} = 0. \tag{7.12}$$

Thus, (7.11) and (7.12), together with the rotational kinematics (7.2) and the translational kinematics (7.3), describe the Lagrangian flow of the rotational and translational dynamics of a rigid body in terms of the evolution of $(R, x, \omega, v) \in \mathsf{TSE}(3)$ on the tangent bundle $\mathsf{TSE}(3)$.

7.3.3 Euler–Lagrange Equations: Quadratic Kinetic Energy

We now determine an expression for the kinetic energy of a rotating and translating rigid body. This expression is used to obtain a standard form of the Euler–Lagrange equations.

Let $\rho \in \mathbb{R}^3$ be a vector from the origin of the body-fixed frame to a mass element, with mass $dm(\rho)$, of the rigid body expressed in the body-fixed frame. The position vector of this mass element, in the inertial frame, is $x + R\rho \in \mathbb{R}^3$, and the velocity vector of this mass element, in the inertial frame, is $\dot{x} + \dot{R}\rho \in \mathbb{R}^3$. The kinetic energy of the rigid body is obtained by integrating the kinetic energy of each mass element of the body over the material points in the body denoted by $\mathcal{B}$:

$$T(R, x, \omega, \dot{x}) = \frac{1}{2} \int_{\mathcal{B}} \|\dot{x} + \dot{R}\rho\|^2 \, dm(\rho)$$

$$= \frac{1}{2} \int_{\mathcal{B}} \|\dot{x} + RS(\omega)\rho\|^2 \, dm(\rho)$$

$$= \frac{1}{2} m \|\dot{x}\|^2 - m \dot{x}^T RS(\rho_{cm})\omega + \frac{1}{2}\omega^T J\omega,$$

where $J = \int_{\mathcal{B}} S(\rho)^T S(\rho) dm(\rho)$ is the 3×3 standard moment of inertia of the rigid body and ρ_{cm} is the location of the center of mass of the body in the body-fixed frame defined by

$$\rho_{cm} = \frac{\int_{\mathcal{B}} \rho \, dm(\rho)}{\int_{\mathcal{B}} dm(\rho)}.$$

The total mass of the rigid body is $m = \int_{\mathcal{B}} dm(\rho)$.

The kinetic energy can also be written as a quadratic function in terms of the translational velocity vector of the origin of the body-fixed frame $v = R^T \dot{x} \in \mathbb{R}^3$ and the angular velocity vector $\omega \in \mathbb{R}^3$:

$$T(R, x, \omega, v) = \frac{1}{2} m \|v\|^2 - m v^T S(\rho_{cm}) \omega + \frac{1}{2} \omega^T J \omega.$$

The potential energy of the rigid body can be obtained by integrating the potential energy of each body element over the body. The potential energy of the translating and rotating rigid body is assumed to depend on the configuration $G = (R, x) \in \mathsf{SE}(3)$. Consequently, the modified Lagrangian function can be expressed in the equivalent forms:

$$\tilde{L}(R, x, \omega, \dot{x}) = \frac{1}{2} \omega^T J \omega + m \omega^T S(\rho_{cm}) R^T \dot{x} + \frac{1}{2} m \|\dot{x}\|^2 - U(R, x). \quad (7.13)$$

and

$$\tilde{L}(R, x, \omega, v) = \frac{1}{2} \omega^T J \omega + m \omega^T S(\rho_{cm}) v + \frac{1}{2} m \|v\|^2 - U(R, x). \quad (7.14)$$

Equations (7.11) and (7.12) can be simplified to obtain a standard form of the Euler–Lagrange equations for the Euclidean motion of a rigid body as it evolves on the tangent bundle $\mathsf{TSE}(3)$:

$$J\dot{\omega} + m\rho_{cm} \times \dot{v} + \omega \times (J\omega + mS(\rho_{cm})v)$$
$$-mS(v)S(\rho_{cm})\omega - \sum_{i=1}^{3} r_i \times \frac{\partial U(G)}{\partial r_i} = 0, \quad (7.15)$$

$$m\dot{v} - m\rho_{cm} \times \dot{\omega} + m\omega \times (v - \rho_{cm} \times \omega) + R^T \frac{\partial U(G)}{\partial x} = 0. \quad (7.16)$$

These equations (7.15) and (7.16), together with the rotational kinematics (7.2) and the translational kinematics (7.3), describe the dynamical flow of a rotating and translating rigid body in terms of the evolution of $(R, x, \omega, v) \in \mathsf{TSE}(3)$ on the tangent bundle of $\mathsf{SE}(3)$.

It is sometimes convenient to express the equations of motion in terms of the translational velocity vector in the inertial frame. The Euler–Lagrange equations can also be expressed as

$$J\dot{\omega} + m\rho_{cm} \times R^T \ddot{x} + \omega \times J\omega - \sum_{i=1}^{3} r_i \times \frac{\partial U(G)}{\partial r_i} = 0, \quad (7.17)$$

$$m\ddot{x} - mR\rho_{cm} \times \dot{\omega} - mRS(\omega)S(\rho_{cm})\omega + \frac{\partial U(G)}{\partial x} = 0. \quad (7.18)$$

These equations (7.17) and (7.18), together with the rotational kinematics (7.2) and the translational kinematics (7.3), describe the dynamical flow of a rotating and translating rigid body in terms of the evolution of $(R, x, \omega, \dot{x}) \in \mathsf{TSE}(3)$ on the tangent bundle of $\mathsf{SE}(3)$.

In the special case that the origin of the body-fixed frame is located at the center of mass of the rigid body, $\rho_{cm} = 0$, we obtain a simplified form of the Euler–Lagrange equations

$$J\dot{\omega} + \omega \times J\omega - \sum_{i=1}^{3} r_i \times \frac{\partial U(G)}{\partial r_i} = 0, \tag{7.19}$$

$$m\dot{v} + m\omega \times v + R^T \frac{\partial U(G)}{\partial x} = 0. \tag{7.20}$$

The Euler–Lagrange equations when the origin of the body-fixed frame is located at the center of mass of the rigid body can also be expressed in terms of the translational velocity vector in the inertial frame as:

$$J\dot{\omega} + \omega \times J\omega - \sum_{i=1}^{3} r_i \times \frac{\partial U(G)}{\partial r_i} = 0, \tag{7.21}$$

$$m\ddot{x} + \frac{\partial U(G)}{\partial x} = 0. \tag{7.22}$$

In this case, the coupling between the rotational dynamics and the translational dynamics occurs only through the potential energy.

If the potential terms in (7.15) and (7.16) are globally defined on $\mathbb{R}^{3\times3} \times \mathbb{R}^3$, the domain of definition of the rotational kinematics (7.2) and the Euler–Lagrange equations (7.15) and (7.16) on TSE(3) can be extended to $\mathsf{T}(\mathbb{R}^{3\times3} \times \mathbb{R}^3)$. This extension is natural in that it defines a Lagrangian vector field on the tangent bundle $\mathsf{T}(\mathbb{R}^{3\times3} \times \mathbb{R}^3)$. Alternatively, the manifold TSE(3) is an invariant manifold of this Lagrangian vector field on $\mathsf{T}(\mathbb{R}^{3\times3} \times \mathbb{R}^3)$ and its restriction to this invariant manifold describes the Lagrangian flow of (7.2), (7.15), and (7.16) on TSE(3). Such extensions can also be made in the other Lagrangian formulations above as well.

7.4 Hamiltonian Dynamics on SE(3)

We introduce the Legendre transformation to obtain the angular momentum and the Hamiltonian function. We make use of Hamilton's phase space variational principle to derive Hamilton's equations for a rotating and translating rigid body.

7.4.1 Hamilton's Phase Space Variational Principle

The Legendre transformation is introduced for a rotating and translating rigid body. The Legendre transformation gives the conjugate momentum as

$$P = \frac{\partial \tilde{L}(G,V)}{\partial V} = \mathcal{S}(\Pi, p) = \begin{bmatrix} S(\Pi) & p \\ 0 & 0 \end{bmatrix},$$

where the Legendre transformation $V \in \mathfrak{se}(3) \rightarrow P \in \mathfrak{se}(3)^*$ is assumed to be invertible. This shows that the angular momentum in the body-fixed frame and the translational momentum in the body-fixed frame are

$$\Pi = \frac{\partial \tilde{L}(G,V)}{\partial \omega}, \tag{7.23}$$

$$p = \frac{\partial \tilde{L}(G,V)}{\partial v}. \tag{7.24}$$

The modified Hamiltonian function $\tilde{H} : \mathsf{T}^*\mathsf{SE}(3) \rightarrow \mathbb{R}^1$, defined on the cotangent bundle of $\mathsf{T}^*\mathsf{SE}(3)$, is

$$\tilde{H}(G,P) = P \cdot V - \tilde{L}(G,V),$$

using the Legendre transformation.

7.4.2 Hamilton's Equations: General Form

Consider the modified action integral

$$\tilde{\mathfrak{G}} = \int_{t_0}^{t_f} \left\{ P \cdot V - \tilde{H}(G,P) \right\} dt.$$

The infinitesimal variation of the action integral is given by

$$\delta \tilde{\mathfrak{G}} = \int_{t_0}^{t_f} \left\{ P \cdot \delta V - \frac{\partial \tilde{H}(G,P)}{\partial G} \cdot \delta G + \delta P \cdot \left(V - \frac{\partial \tilde{H}(G,P)}{\partial P} \right) \right\} dt. \tag{7.25}$$

Recall from (7.8) that the infinitesimal variations δV can be written as

$$\delta V = \dot{\Gamma} + V\Gamma - \Gamma V,$$

for differentiable curves $\Gamma : [t_0, t_f] \rightarrow \mathfrak{se}(3)$. Therefore, the first term of (7.25) can be written as

$$\int_{t_0}^{t_f} P \cdot \delta V \, dt = \int_{t_0}^{t_f} \mathcal{S}(\Pi, p) \cdot (\dot{\Gamma} + V\Gamma - \Gamma V) \, dt$$

$$= \int_{t_0}^{t_f} \mathcal{S} \left(-\begin{bmatrix} \dot{\Pi} \\ \dot{p} \end{bmatrix} + \begin{bmatrix} -S(\omega) & -S(v) \\ 0 & -S(\omega) \end{bmatrix} \begin{bmatrix} \Pi \\ p \end{bmatrix} \right) \cdot \Gamma \, dt,$$

where the first term is integrated by parts, using the fact that $\Gamma(t_0) = \Gamma(t_f) = 0$, and the second term is obtained by using (7.9).

The second term of (7.25) is given by

$$\int_{t_0}^{t_f} \frac{\partial \tilde{H}(G,P)}{\partial G} \cdot \delta G \, dt = \int_{t_0}^{t_f} \frac{\partial \tilde{H}(G,P)}{\partial R} \cdot \delta R + \frac{\partial \tilde{H}(G,P)}{\partial x} \cdot \delta x \, dt.$$

By using (6.5) and (7.7), this can be rewritten as

$$\int_{t_0}^{t_f} \frac{\partial \tilde{H}(G,P)}{\partial G} \cdot \delta G \, dt = \int_{t_0}^{t_f} \left[-\sum_{i=1}^{3} \left(S(r_i) \frac{\partial \tilde{H}(G,P)}{\partial r_i} \right) \cdot \eta \right.$$

$$\left. + \frac{\partial \tilde{H}(G,P)}{\partial x} \cdot R\chi \right] dt$$

$$= \int_{t_0}^{t_f} S\left(\begin{bmatrix} -\sum_{i=1}^{3} S(r_i) \frac{\partial \tilde{H}(G,P)}{\partial r_i} \\ R^T \frac{\partial \tilde{H}(G,P)}{\partial x} \end{bmatrix} \right) \cdot \Gamma \, dt.$$

Substituting these two results, the expression for the infinitesimal variation of the action integral is obtained:

$$\frac{d}{d\epsilon} \mathfrak{G}^{\epsilon} \Big|_{\epsilon=0} = \int_{t_0}^{t_f} S\left(-\begin{bmatrix} \dot{\Pi} \\ \dot{p} \end{bmatrix} + \begin{bmatrix} -S(\omega) & -S(v) \\ 0 & -S(\omega) \end{bmatrix} \begin{bmatrix} \Pi \\ p \end{bmatrix} \right.$$

$$\left. - \begin{bmatrix} -\sum_{i=1}^{3} S(r_i) \frac{\partial \tilde{H}(G,P)}{\partial r_i} \\ R^T \frac{\partial \tilde{H}(G,P)}{\partial x} \end{bmatrix} \right) \cdot \Gamma + \delta P \cdot \left(V - \frac{\partial \tilde{H}(G,P)}{\partial P} \right) dt.$$

From Hamilton's principle, the above expression for the infinitesimal variation of the action integral should be zero for all infinitesimal variations $\Gamma : [t_0, t_f] \to \mathfrak{se}(3)$ and $\delta P : [t_f, t_f] \to \mathfrak{se}(3)^*$ that satisfy $\Gamma(t_0) = \Gamma(t_f) = 0$. The fundamental lemma of the calculus of variations gives Hamilton's equations,

$$V = \frac{\partial \tilde{H}(G,P)}{\partial P},$$

$$\begin{bmatrix} \dot{\Pi} \\ \dot{p} \end{bmatrix} = -\begin{bmatrix} S(\omega) & S(v) \\ 0 & S(\omega) \end{bmatrix} \begin{bmatrix} \Pi \\ p \end{bmatrix} - \begin{bmatrix} -\sum_{i=1}^{3} S(r_i) \frac{\partial \tilde{H}(G,P)}{\partial r_i} \\ R^T \frac{\partial \tilde{H}(G,P)}{\partial x} \end{bmatrix}.$$

Thus, we obtain the general form of Hamilton's equations for a rotating and translating rigid body.

Proposition 7.2 *Hamilton's equations for a modified Hamiltonian function* $\tilde{H} : \mathsf{T}^*\mathsf{SE}(3) \to \mathbb{R}^1$ *are*

$$\dot{r}_i = r_i \times \frac{\partial \tilde{H}(G,P)}{\partial \Pi}, \quad i = 1, 2, 3, \tag{7.26}$$

$$\dot{x} = R\frac{\partial \tilde{H}(G, P)}{\partial p}, \tag{7.27}$$

$$\dot{\Pi} = \Pi \times \frac{\partial \tilde{H}(G, P)}{\partial \Pi} + p \times \frac{\partial \tilde{H}(G, P)}{\partial p} + \sum_{i=1}^{3} r_i \times \frac{\partial \tilde{H}(G, P)}{\partial r_i}, \tag{7.28}$$

$$\dot{p} = p \times \frac{\partial \tilde{H}(G, P)}{\partial \Pi} - R^T\frac{\partial \tilde{H}(G, P)}{\partial x}. \tag{7.29}$$

Equations (7.26), (7.27), (7.28), and (7.29) define Hamilton's equations for the Euclidean dynamics of a rigid body in terms of the evolution of $(R, x, \Pi, p) \in T^*SE(3)$ on the cotangent bundle $T^*SE(3)$.

The following property follows directly from the above formulation of Hamilton's equations on $SE(3)$:

$$\begin{aligned}
\frac{d\tilde{H}(G, P)}{dt} &= \sum_{i=1}^{3} \frac{\partial \tilde{H}(G, P)}{\partial r_i} \cdot \dot{r}_i + \frac{\partial \tilde{H}(G, P)}{\partial \Pi} \cdot \dot{\Pi} \\
&\quad + \frac{\partial \tilde{H}(G, P)}{\partial x} \cdot \dot{x} + \frac{\partial \tilde{H}(G, P)}{\partial p} \cdot \dot{p}, \\
&= 0.
\end{aligned}$$

The modified Hamiltonian function is constant along each solution of Hamilton's equation. This property does not hold if the modified Hamiltonian function has a nontrivial explicit dependence on time.

7.4.3 Hamilton's Equations: Quadratic Kinetic Energy

If the kinetic energy function is quadratic in $(\omega, \dot{x})$, then the Legendre transformation is

$$\begin{bmatrix} \Pi \\ p \end{bmatrix} = \frac{\partial \tilde{L}(G, P)}{\partial P} = \begin{bmatrix} J & mS(\rho_{cm})R^T \\ mRS(\rho_{cm})^T & mI_{3\times 3} \end{bmatrix} \begin{bmatrix} \omega \\ \dot{x} \end{bmatrix},$$

and we assume that this can be inverted to give

$$\begin{bmatrix} \omega \\ \dot{x} \end{bmatrix} = \begin{bmatrix} M_{11}^I & M_{12}^I \\ M_{21}^I & M_{22}^I \end{bmatrix} \begin{bmatrix} \Pi \\ p \end{bmatrix},$$

where

$$\begin{bmatrix} M_{11}^I & M_{12}^I \\ M_{21}^I & M_{22}^I \end{bmatrix} = \begin{bmatrix} J & mS(\rho_{cm})R^T \\ mRS(\rho_{cm})^T & mI_{3\times 3} \end{bmatrix}^{-1}.$$

The modified Hamiltonian function is

$$\tilde{H}(G, P) = \frac{1}{2} \begin{bmatrix} \Pi \\ p \end{bmatrix}^T \begin{bmatrix} M_{11}^I & M_{12}^I \\ M_{21}^I & M_{22}^I \end{bmatrix} \begin{bmatrix} \Pi \\ p \end{bmatrix} + U(G). \tag{7.30}$$

Hamilton's equations for the Euclidean dynamics of a rigid body as they evolve on the cotangent bundle T*SE(3) are:

$$\dot{R} = RS(M_{11}^I \Pi + M_{12}^I p), \tag{7.31}$$

$$\dot{x} = R(M_{21}^I \Pi + M_{22}^I p), \tag{7.32}$$

$$\dot{\Pi} = \Pi \times (M_{11}^I \Pi + M_{12}^I p) + p \times (M_{21}^I \Pi + M_{22}^I p)$$

$$+ \sum_{i=1}^{3} r_i \times \frac{\partial U(G)}{\partial r_i}, \tag{7.33}$$

$$\dot{p} = p \times (M_{11}^I \Pi + M_{12}^I p) - R^T \frac{\partial U(G)}{\partial x}. \tag{7.34}$$

Hamilton's equations describe the Hamiltonian flow of the Euclidean dynamics of a rigid body in terms of the evolution of $(R, x, \Pi, p) \in$ T*SE(3) on the cotangent bundle T*SE(3).

In the special case that the origin of the body-fixed frame is located at the center of mass of the rigid body, $S(\rho_{cm}) = 0$, and we obtain a simplification of Hamilton's equations

$$\dot{R} = RS(J^{-1}\Pi), \tag{7.35}$$

$$\dot{x} = \frac{1}{m} Rp, \tag{7.36}$$

$$\dot{\Pi} = \Pi \times J^{-1}\Pi + \sum_{i=1}^{3} r_i \times \frac{\partial U(G)}{\partial r_i}, \tag{7.37}$$

$$\dot{p} = p \times J^{-1}\Pi - R^T \frac{\partial U(G)}{\partial x}. \tag{7.38}$$

The relative simplicity of this form of Hamilton's equations for Euclidean motion of a rigid body motivates the choice of a body-fixed Euclidean frame whose origin is located at the center of mass of the rigid body.

If the potential terms in (7.33) and (7.34) are globally defined on $\mathbb{R}^{3\times3} \times \mathbb{R}^3$, the domain of definition of (7.31), (7.32), (7.33), and (7.34) on T*SE(3) can be extended to T*$(\mathbb{R}^{3\times3} \times \mathbb{R}^3)$. This extension is natural and useful in that it defines a Hamiltonian vector field on the cotangent bundle T*$(\mathbb{R}^{3\times3} \times \mathbb{R}^3)$. The manifold T*SE(3) is an invariant manifold of this Hamiltonian vector field on T*$(\mathbb{R}^{3\times3} \times \mathbb{R}^3)$ and its restriction to this invariant manifold describes the Hamiltonian flow of (7.31), (7.32), (7.33), and (7.34) on T*SE(3).

7.5 Linear Approximations of Dynamics on **SE(3)**

Geometric forms of the Euler–Lagrange equations and Hamilton's equations on SE(3) have been presented that provide insight into the geometry of the global dynamics on the configuration manifold SE(3). It is possible to determine a linear vector field that approximates the Lagrangian vector field on TSE(3), at least locally in an open subset of TSE(3).

7.6 Dynamics on **SE(3)**

In this section, we present examples of a rigid body undergoing Euclidean motion in three dimensions. In each case, the configuration manifold is SE(3); consequently, each of the illustrated dynamics has six degrees of freedom. Lagrangian and Hamiltonian formulations of the equations of motion are presented; a few simple flow properties are identified for each illustration.

7.6.1 Dynamics of a Rotating and Translating Rigid Body

A rigid body can rotate and translate in three dimensions under the action of uniform, constant gravity. We also consider the special case that the gravity force is zero.

The dynamics of such a rigid body are described using an inertial Euclidean frame and a body-fixed Euclidean frame. The first two axes of the inertial frame are selected to lie in a horizontal plane and the third axis is selected to be vertical. The origin of the body-fixed frame is located at the center of mass of the rigid body. The mass of the rigid body is m and the standard inertia matrix in the body-fixed frame is J.

Let $R \in \mathsf{SO}(3)$ define the attitude of the rigid body as the linear transformation from the body-fixed frame to the inertial frame. Let $x \in \mathbb{R}^3$ denote the position vector from the center of the inertial frame to the center of mass of the rigid body, in the inertial frame. Thus, the configuration of the rotating and translating rigid body is $(R, x) \in \mathsf{SE}(3)$ and SE(3) is the configuration manifold. The rigid body has three translational degrees of freedom and three rotational degrees of freedom.

7.6.1.1 Euler–Lagrange Equations

The modified Lagrangian function is

$$\tilde{L}(R, x, \omega, \dot{x}) = \frac{1}{2}\omega^T J \omega + \frac{1}{2}m \left\| \dot{x} \right\|^2 - mge_3^T x.$$

Following the prior results in (7.21) and (7.22), the Euler–Lagrange equations of motion for the Euclidean motion of a rigid body under the action of constant gravity are given by

$$J\dot{\omega} + \omega \times J\omega = 0, \tag{7.39}$$

$$m\ddot{x} + mge_3 = 0. \tag{7.40}$$

The moment due to the gravity potential is zero since the gravity potential is attitude independent; the gravity force on the rigid body can be viewed as acting on the center of mass of the rigid body.

The rotational kinematics equation for the rigid body are

$$\dot{R} = RS(\omega). \tag{7.41}$$

These equations (7.39), (7.40), and (7.41) describe the Lagrangian dynamics of the rigid body in terms of $(R, x, \omega, \dot{x}) \in \mathsf{TSE}(3)$ on the tangent bundle of $\mathsf{SE}(3)$.

Even in the presence of constant gravity, the translational dynamics and the rotational dynamics of a rigid body are decoupled. Thus, the rotational dynamics of the rigid body are not influenced by the translational motion and the translational dynamics are not influenced by the rotational motion. The translational motion and the rotational motion evolve independently.

7.6.1.2 Hamilton's Equations

Hamilton's equations of motion are obtained by introducing the rotational and translational momenta defined according to the Legendre transformation by $\Pi = J\omega$, $p = m\dot{x}$. Thus, the modified Hamiltonian is

$$\tilde{H}(R, x, \Pi, p) = \frac{1}{2}\Pi^T J^{-1}\Pi + \frac{1}{2m}\|p\|^2 + mge_3^T x.$$

Following the results in (7.35), (7.36), (7.37), and (7.38), Hamilton's equations consist of the rotational kinematics equations

$$\dot{R} = RS(J^{-1}\Pi), \tag{7.42}$$

the translational kinematics equations

$$\dot{x} = \frac{p}{m}, \tag{7.43}$$

the rotational dynamics equations

$$\dot{\Pi} = \Pi \times J^{-1}\Pi, \tag{7.44}$$

and the translational dynamics equations

$$\dot{p} = -mge_3. \tag{7.45}$$

These equations (7.44), (7.45), (7.42), and (7.43) describe the Hamiltonian dynamics of the rigid body under gravity in terms of $(R, x, \Pi, p) \in T^*SE(3)$ on the cotangent bundle of $SE(3)$.

7.6.1.3 Conservation Properties

Conserved quantities, or integrals of motion, exist for the rotational and translational dynamics of a rigid body acting under constant gravity. The Hamiltonian, which coincides with the energy in this case, is given by

$$H = \frac{1}{2}\omega^T J\omega + \frac{1}{2}m\|\dot{x}\|^2 + mge_3^T x,$$

which is constant along each solution of the dynamical flow.

In addition, there is a rotational symmetry: the Lagrangian is invariant with respect to the tangent lift of any rigid body rotation about the center of mass. This symmetry leads to conservation of the angular momentum in the inertial frame; that is

$$R\Pi = RJ\omega$$

is constant along each solution of the dynamical flow of the rigid body under the action of constant gravity. From this, it follows that the square of the magnitude of the angular momentum

$$\|J\omega\|^2,$$

is also constant along each solution of the dynamical flow.

Finally, there is a translational symmetry: the Lagrangian is invariant with respect to the tangent lift of translations in the horizontal plane. This symmetry leads to conservation of the two horizontal components of the translational momentum in the inertial frame; that is the horizontal projections of the translational momentum in the inertial frame

$$me_i^T \dot{x}, \quad i = 1, 2,$$

are constant along each solution of the dynamical flow for the Euclidean motion of a rigid body under constant gravity.

7.6.1.4 Equilibrium Properties

If there is a constant nonzero gravity force on the rigid body, there are no equilibrium solutions.

7.6.1.5 Conservation and Equilibrium Properties in Zero Gravity

The prior results hold if gravity is absent, that is if $g = 0$. This case is referred to as the dynamics of a free rigid body. In this special case, there are additional conservation and equilibrium properties that are now identified.

The free rigid body has several conservation properties due to symmetries in the Lagrangian function. There is a rotational symmetry due to the fact that the Lagrangian is invariant with respect to the tangent lift of any rotation of the rigid body. This symmetry leads to conservation of the angular momentum in the inertial frame as indicated above. There is a translational symmetry: the Lagrangian is invariant with respect to the tangent lift of any rigid body translation. This symmetry leads to conservation of the translational momentum in the inertial frame; that is the translational velocity $\dot{x}$ and the translational momentum p are constant along each solution of the dynamical flow of the free rigid body.

The equilibria or constant solutions of a freely rotating and translating rigid body are easily identified. The free rigid body is in equilibrium if the configuration in SE(3) is constant, that is the angular velocity vector $\omega = 0$ and the translational velocity vector $\dot{x} = 0$; the rigid body can be in equilibrium for any configuration in SE(3).

To illustrate the linearization of the dynamics of a freely rotating and translating rigid body, we consider the equilibrium solution $(I_{3\times3}, 0, 0, 0) \in$ TSE(3). The linearized differential equations on the twelve-dimensional tangent space of TSE(3) at $(I_{3\times3}, 0, 0, 0) \in$ TSE(3) are given by

$$J\ddot{\xi} = 0,$$
$$m\ddot{\chi} = 0.$$

These linearized differential equations approximate the rotational and translational dynamics of the rigid body in a neighborhood of $(I_{3\times3}, 0, 0, 0) \in$ TSE(3).

Solutions for which the angular velocity vector and the translational velocity vector are constant can also be identified; these are referred to as relative equilibrium solutions and they necessarily satisfy

$$\omega \times J\omega = 0,$$
$$\ddot{x} = 0.$$

Thus, the relative equilibrium solutions occur when the angular velocity vector is collinear with an eigenvector of the inertia matrix J and the translational velocity vector is constant.

In particular, consider the initial-value problem for the freely rotating and translating rigid body given by $R(t_0) = R_0$, $\omega(t_0) = \omega_0$, $\dot{x}(t_0) = v_0$, where we assume that $\omega_0 \in \mathbb{R}^3$ is an eigenvector of the inertia matrix J. The resulting rotational motion of the free rigid body can be expressed as

$$
\begin{aligned}
R(t) &= R_0 e^{S(\omega_0)(t-t_0)}, \\
x(t) &= x_0 + v_0(t - t_0), \\
\omega(t) &= \omega_0, \\
\dot{x}(t) &= v_0,
\end{aligned}
$$

which verifies that the rotational and translational velocity vectors are constant.

7.6.2 Dynamics of an Elastically Supported Rigid Body

Consider a rigid body that is supported by multiple elastic connections. The rigid body is acted on by uniform constant gravity. The elastic connections are characterized by n elastic springs: one end of each spring is attached to a fixed point on the rigid body while the other end of the spring is attached to a fixed inertial support. Each spring is assumed to be massless with a known elastic stiffness, and the spring always remains straight. Each spring can be in either compression or tension, and the restoring force in each spring is proportional to the axial deflection of the spring in the direction of the vector between its two attachment points.

An inertial three-dimensional Euclidean frame is constructed so that the first two axes lie in a fixed horizontal and the third axis is vertical, opposite to the direction of gravity. A Euclidean frame is fixed in the rigid body, with its origin at the center of mass of the body.

Let m denote the mass of the rigid body and J denote the standard moment of inertia matrix of the rigid body with respect to its body-fixed frame. The springs are ordered, so that κ_i denotes the linear elastic stiffness and L_i denotes the natural length of the i-th spring, for $i = 1, \ldots, n$. The i-th spring is connected to the rigid body at location $\rho_i \in \mathbb{R}^3$ in the body-fixed frame and to an inertial support at location $z_i \in \mathbb{R}^3$ in the inertial frame, for $i = 1, \ldots, n$.

Let $R \in \mathsf{SO}(3)$ denote the attitude of the rigid body and let $x \in \mathbb{R}^3$ denote the position vector of the center of mass of the rigid body, expressed in the inertial frame. The configuration is $(R, x) \in \mathsf{SE}(3)$ and the configuration manifold of an elastically supported rigid body is $\mathsf{SE}(3)$. This system has six degrees of freedom.

As a special case, consider the case $n = 1$ of a rigid body with a single elastic support. For simplicity, we locate the origin of the inertial frame at the inertial support point of the spring. A schematic of a rigid body with a single elastic support is shown in Figure 7.1.

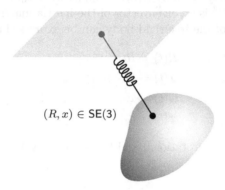

$(R, x) \in \mathsf{SE}(3)$

Fig. 7.1 Rigid body with a single elastic support

7.6.2.1 Euler–Lagrange Equations

The rotational kinematics of the rigid body are given by

$$\dot{R} = RS(\omega), \tag{7.46}$$

where $\omega \in \mathbb{R}^3$ is the angular velocity vector of the rigid body in the body-fixed frame.

We use the Lagrange–d'Alembert principle to obtain the equations of motion, viewing the elastic forces as external forces and the gravitational force as arising from a potential.

The modified Lagrangian function of a rotating and translating rigid body $\tilde{L} : \mathsf{TSE}(3) \to \mathbb{R}^1$ is the sum of the rotational kinetic energy and the translational kinetic energy minus the gravitational potential energy:

$$\tilde{L}(R, x, \omega, \dot{x}) = \frac{1}{2}m \left\| \dot{x} \right\|^2 + \frac{1}{2}\omega^T J\omega - mge_3^T x.$$

The Euler–Lagrange equations can be obtained using the prior results in (7.21) and (7.22) to obtain

$$J\dot{\omega} + \omega \times J\omega = \sum_{i=1}^{n} M_i,$$

$$m\ddot{x} + mge_3^T x = \sum_{i=1}^{n} F_i.$$

Here, $M_i \in \mathbb{R}^3$ is the moment vector on the rigid body produced by the i-th elastic spring, expressed in the body-fixed frame. And $F_i \in \mathbb{R}^3$ is the force on the rigid body produced by the i-th elastic spring, for $i = 1, \ldots, n$, expressed in the inertial frame.

Based on the assumptions, the elastic force of the i-th spring on the rigid body is given by

$$F_i = \kappa_i \{\|s_i\| - L_i\} \frac{s_i}{\|s_i\|}, \quad i = 1, \ldots, n,$$

where the vectors in the inertial frame, defined by the spring attachment points, are

$$s_i = z_i - (x + R\rho_i), \quad i = 1, \ldots, n.$$

Thus, if $\|s_i\| > L_i$, it follows that the i-th elastic spring is in tension and $F_i \in \mathbb{R}^3$ is proportional to the extension in the direction $s_i \in \mathbb{R}^3$; if $\|s_i\| < L_i$, it follows that the i-th elastic spring is in compression and $F_i \in \mathbb{R}^3$ is proportional to the compression in the direction $-s_i \in \mathbb{R}^3$; if $\|s_i\| = L_i$, then $F_i = 0$.

The elastic forces, expressed in the body-fixed frame, are $R^T F_i$, $i = 1, \ldots, n$. Thus, the elastic moments on the rigid body, expressed in the body-fixed frame, are given by

$$M_i = \rho_i \times R^T F_i, \quad i = 1, \ldots, n.$$

These expressions for the elastic forces and moments are the same expressions used for elastic cables or spring in tension; see [90] where these same elastic relations are used to describe tensegrity structures.

In summary, the Euler–Lagrange equations for an elastically supported rigid body are

$$J\dot{\omega} + \omega \times J\omega + \sum_{i=1}^{n} \kappa_i \rho_i \times \{\|z_i - x - R\rho_i\| - L_i\} \frac{R^T(x - z_i) + \rho_i}{\|z_i - x - R\rho_i\|} = 0,$$
(7.47)

$$m\ddot{x} + mge_3^T x + \sum_{i=1}^{n} \kappa_i \{\|z_i - x - R\rho_i\| - L_i\} \frac{(x - z_i) + R\rho_i}{\|z_i - x - R\rho_i\|} = 0. \quad (7.48)$$

These equations assume that the vectors defined by the spring attachment points are never zero. We assume that there are no collisions between the rigid body, the inertially fixed spring attachment points, or the springs. This assumption limits the global validity of the equations of motion on the tangent bundle. With this qualification, the dynamics of an elastically supported rigid body are described by the rotational kinematics (7.46) and the Euler–Lagrange equations (7.47) and (7.48) in terms of the evolution of $(R, x, \omega, \dot{x})$ on the tangent bundle TSE(3).

In the case of a rigid body with a single elastic support, the Euler–Lagrange equations are

$$J\dot{\omega} + \omega \times J\omega + \kappa S(\rho)\{\|x + R\rho\| - L\}\frac{R^T x + \rho}{\|x + R\rho\|} = 0, \qquad (7.49)$$

$$m\ddot{x} + mge_3^T x + \kappa\{\|x + R\rho\| - L\}\frac{x + R\rho}{\|x + R\rho\|} = 0. \qquad (7.50)$$

7.6.2.2 Hamilton's Equations

Hamilton's equations are determined by introducing the Legendre transformation

$$p = m\dot{x},$$
$$\Pi = J\omega,$$

where $(p, \Pi) \in \mathfrak{se}(3)^*$ is conjugate to $(\dot{x}, \omega) \in \mathfrak{se}(3)$. Thus, Hamilton's equations consist of the kinematics equations

$$\dot{x} = \frac{p}{m}, \qquad (7.51)$$

$$\dot{R} = RS(J^{-1}\Pi), \qquad (7.52)$$

and the dynamics equations

$$\dot{\Pi} = \Pi \times J^{-1}\Pi - \sum_{i=1}^{n} \kappa_i \rho_i \times \{\|z_i - x - R\rho_i\| - L_i\}\frac{R^T(x - z_i) + \rho_i}{\|z_i - x - R\rho_i\|}, \qquad (7.53)$$

$$\dot{p} = -mge_3^T x - \sum_{i=1}^{n} \kappa_i\{\|z_i - x - R\rho_i\| - L_i\}\frac{(x - z_i) + R\rho_i}{\|z_i - x - R\rho_i\|}. \qquad (7.54)$$

These equations (7.51), (7.52), (7.53), and (7.54) describe the Hamiltonian flow of an elastically supported rigid body in terms of the evolution of $(R, x, \Pi, p) \in \mathsf{T}^*\mathsf{SE}(3)$ on the cotangent bundle of the configuration manifold.

In the case of a rigid body with a single elastic support, Hamilton's equations are

$$\dot{x} = \frac{p}{m}, \qquad (7.55)$$

$$\dot{R} = RS(J^{-1}\Pi), \qquad (7.56)$$

and

$$\dot{\Pi} = \Pi \times J^{-1}\Pi - \kappa\rho \times \{\|x + R\rho\| - L\}\frac{R^T x + \rho}{\|x + R\rho\|}, \qquad (7.57)$$

$$\dot{p} = -mge_3^T x - \kappa\{\|x + R\rho\| - L\}\frac{x + R\rho}{\|x + R\rho\|}. \qquad (7.58)$$

7.6.2.3 Conservation Properties

The Hamiltonian, which coincides with the total energy E in this case, is conserved along the dynamical flow. Since the expression for the elastic energy is complicated, we do not present an expression for the Hamiltonian.

7.6.2.4 Equilibrium Properties

The equilibrium solutions of an elastically supported rigid body occur when the rigid body is stationary, that is the translational and rotational velocity vectors and their time derivatives are zero. The equilibrium configurations satisfy the algebraic equations

$$0 = \sum_{i=1}^{n} \kappa_i \rho_i \times \{\|z_i - x - R\rho_i\| - L_i\}\frac{R^T(x - z_i) + \rho_i}{\|z_i - x - R\rho_i\|}, \qquad (7.59)$$

$$0 = \sum_{i=1}^{n} \kappa_i \{\|z_i - x - R\rho_i\| - L_i\}\frac{(x - z_i) + R\rho_i}{\|z_i - x - R\rho_i\|} + mge_3^T x. \qquad (7.60)$$

The first equilibrium condition implies that the net elastic moment on the rigid body is zero. The second equilibrium condition implies that the net elastic force on the rigid body balances the weight of the body.

It is possible to obtain linear differential equations that approximate the dynamical flow in a neighborhood of any equilibrium solution; this analysis would follow the methods described in Appendix B. Such developments are left to the reader.

The equilibrium solutions for a rigid body with a single elastic support occur when the rigid body is stationary, so that the following algebraic conditions holds:

$$0 = \rho \times R^T x,$$

$$0 = \kappa\{\|x + R\rho\| - L\}\frac{x + R\rho}{\|x + R\rho\|} + mge_3^T x.$$

The first condition implies that the position vector of the center of mass of the rigid body and the vector to the point at which the spring is attached to the rigid body are collinear. The second condition implies that the point at which the elastic support is attached to the rigid body is collinear with the direction of gravity, and the elastic extension (or compression) balances the gravitational force on the rigid body. Consequently, there are four distinct manifolds that describe the equilibrium configurations, namely

$$\{(R,x) \in \mathsf{SE}(3) : \frac{R^T \rho}{\|R\rho\|} = -e_3, \, x = -\left(L + \frac{mg}{\kappa}\right)e_3\},$$

$$\{(R,x) \in \mathsf{SE}(3) : \frac{R^T \rho}{\|R\rho\|} = -e_3, \, x = \left(L - \frac{mg}{\kappa}\right)e_3\},$$

$$\{(R,x) \in \mathsf{SE}(3) : \frac{R^T \rho}{\|R\rho\|} = e_3, \, x = -\left(L + \frac{mg}{\kappa}\right)e_3\},$$

$$\{(R,x) \in \mathsf{SE}(3) : \frac{R^T \rho}{\|R\rho\|} = e_3, \, x = \left(L - \frac{mg}{\kappa}\right)e_3\}.$$

The first and third equilibrium manifolds correspond to extension of the spring; the second and fourth equilibrium manifolds correspond to compression of the spring. The first and second equilibrium manifolds correspond to the point of attachment of the spring below the center of mass of the rigid body; the third and fourth equilibrium manifolds correspond to the point of attachment of the spring above the center of mass of the rigid body.

Note that if $(R,x) \in \mathsf{SE}(3)$ is an equilibrium configuration then any rotation of the rigid body about the vertical direction is also an equilibrium configuration.

7.6.3 Dynamics of a Rotating and Translating Rigid Dumbbell Satellite in Orbit

A dumbbell satellite, consisting of two identical rigid spherical bodies connected by a massless rigid link, can translate and rotate in three dimensions acted on by gravity forces that arises from an inertially fixed central body. Assuming the mass of the central body M is much larger than the total mass m of the dumbbell satellite, this model is referred to as a restricted full two-body model for the dumbbell satellite. It has the important feature that the rotational dynamics and the translational dynamics are coupled through the gravitational potential that depends on the position and attitude of the orbiting rigid body.

We introduce an inertial Euclidean frame and a body-fixed Euclidean frame. The origin of the inertially fixed frame is located at the center of the large central spherical body. The body-fixed frame is aligned with the principal axes of the body, with the third body-fixed axis along the line between the centers of mass of the two rigid spherical bodies. Thus, the standard inertial matrix J of the dumbbell satellite is diagonal. The center of the body-fixed frame is located at the center of mass of the dumbbell at the midpoint of the link between the two spherical bodies.

We use $R \in \mathsf{SO}(3)$ to denote the attitude of the rigid body as a rotation matrix from the body-fixed frame to the inertial frame and we use $x \in \mathbb{R}^3$ to denote the position vector from the center of the central body to the center of mass of the rigid body in the inertial frame. This defines the configuration (R, x) in the configuration manifold $\mathsf{SE}(3)$. The orbiting dumbbell satellite has three translational degrees of freedom and three rotational degrees of freedom. A schematic of a rotating and translating rigid dumbbell satellite in orbit is shown in Figure 7.2.

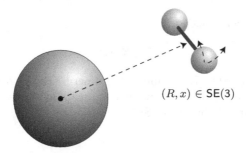

$(R, x) \in \mathsf{SE}(3)$

Fig. 7.2 Rotating and translating rigid dumbbell satellite in orbit

The subsequent development follows the presentations in [50, 51], where full body dynamics are treated in detail.

7.6.3.1 Euler–Lagrange Equations

The modified Lagrangian function is

$$\tilde{L}(R, x, \omega, \dot{x}) = \frac{1}{2}\omega^T J \omega + \frac{1}{2}m \|\dot{x}\|^2 - U(R, x).$$

The Newtonian gravity potential is the sum of the gravitational potentials of the two spherical bodies that define the dumbbell satellite; it can be shown to be

$$U(R, x) = -\frac{GMm}{2} \sum_{i=1}^{2} \|x + R\rho_i\|^{-1},$$

where G is the universal gravitational constant, and $\rho_i \in \mathbb{R}^3$ is a vector from the origin of the body-fixed frame to the center of the i-th sphere of the dumbbell satellite in the body-fixed frame for $i = 1, 2$. From the assumption that the third body-fixed axis is along the line between the centers of mass of the two spherical bodies that define the dumbbell satellite, we have $\rho_1 = \frac{L}{2}e_3$, $\rho_2 = -\frac{L}{2}e_3$ with L being the axial length of the dumbbell satellite.

The Euler–Lagrange equations for the dumbbell satellite can be obtained from (7.21) and (7.22) as

$$J\dot{\omega} + \omega \times J\omega - \frac{GMm}{2}\sum_{i=1}^{2}\frac{\rho_i \times (x + R\rho_i)}{\|x + R\rho_i\|^3} = 0, \tag{7.61}$$

$$m\ddot{x} + \frac{GMm}{2}\sum_{i=1}^{2}\frac{x + R\rho_i}{\|x + R\rho_i\|^2} = 0. \tag{7.62}$$

The rotational kinematics are

$$\dot{R} = RS(\omega). \tag{7.63}$$

These equations (7.61), (7.62), and (7.63) describe the Lagrangian dynamics of a rigid body acted on by Newtonian gravity in terms of $(R, x, \omega, \dot{x}) \in$ TSE(3) on the tangent bundle of SE(3).

7.6.3.2 Hamilton's Equations

Hamilton's equations of motion are obtained using the Legendre transformation to introduce the rotational and translational momenta defined by $\Pi = J\omega$, $p = m\dot{x}$. The modified Hamiltonian is

$$\tilde{H}(R, x, \Pi, p) = \frac{1}{2}\Pi^T J^{-1}\Pi + \frac{1}{2m}\|p\|^2 - \frac{GMm}{2}\sum_{i=1}^{2}\|x + R\rho_i\|^{-1}.$$

Hamilton's equations for the Euclidean motion of a rigid body in Newtonian gravity are obtained from (7.35), (7.36), (7.37), and (7.38); they consist of the rotational kinematics equations

$$\dot{R} = RS(J^{-1}\Pi), \tag{7.64}$$

the translational kinematics equations

$$\dot{x} = \frac{p}{m}, \tag{7.65}$$

and the dynamics equations

$$\dot{\Pi} = \Pi \times J^{-1}\Pi + \frac{GMm}{2} \sum_{i=1}^{2} \frac{\rho_i \times (x + R\rho_i)}{\|x + R\rho_i\|^3}, \tag{7.66}$$

$$\dot{p} = -\frac{GMm}{2} \sum_{i=1}^{2} \frac{x + R\rho_i}{\|x + R\rho_i\|^2}. \tag{7.67}$$

These differential equations (7.64), (7.65), (7.66), and (7.67) describe the Hamiltonian dynamics for the dumbbell satellite in a Newtonian gravity potential in terms of $(R, x, \Pi, p) \in T^*SE(3)$ on the cotangent bundle of $SE(3)$.

The equations of motion illustrate that the translational, or orbital, dynamics and the rotational dynamics of the dumbbell satellite are coupled through the Newtonian gravity. This coupling can give rise to complicated full body dynamics.

7.6.3.3 Conservation Properties

The Hamiltonian of the dumbbell satellite, which coincides with the total energy E in this case, is conserved; that is

$$H = \frac{1}{2}\omega^T J\omega + \frac{1}{2}m\|\dot{x}\|^2 - \frac{GMm}{2} \sum_{i=1}^{2} \|x + R\rho_i\|^{-1}$$

is constant along each solution of the dynamical flow of the dumbbell satellite.

7.6.3.4 Equilibrium Properties

There are no equilibrium solutions, but there are important relative equilibrium solutions. For example, these equations can be used to determine solutions for which the dumbbell satellite is in a circular orbit in a fixed orbital plane, that is the center of each spherical body of the satellite (and its center of mass) follow a circular path in $\mathbb{R}^3$ in a fixed orbital plane. Such solutions require the angular velocity of the dumbbell satellite to be identical to the orbital angular velocity, and the body-fixed axes of the dumbbell satellite to be properly aligned with the orbital position vector. Such relative equilibrium solutions involve a complicated interplay between the rotational motion and the translational motion of the dumbbell satellite.

7.7 Problems

7.1. Consider a rotating and translating rigid body, with the configuration $(R, x) \in SE(3)$ defined with respect to an inertial frame and a body-fixed

frame whose origin is located at the center of mass of the rigid body. The
rigid body has mass m and moment of inertia matrix J. Assume the modified
Lagrangian is separable in the sense that

$$\tilde{L}(R, x, \omega, \dot{x}) = \frac{1}{2}\omega^T J\omega + \frac{1}{2}m \left\| \dot{x} \right\|^2 - U_r(R) - U_t(x).$$

(a) What are the Euler–Lagrange equations for the rigid body?
(b) What is the modified Hamiltonian function defined on the cotangent bun-
    dle of the configuration manifold?
(c) What are Hamilton's equations for the rigid body?
(d) Confirm from these equations of motion that the rotational dynamics and
    the translational dynamics of the rigid body are completely decoupled in
    the sense that the rotational dynamics do not depend on the translational
    motion and the translational dynamics do not depend on the rotational
    motion.
(e) What are conserved quantities for the rigid body rotational and transla-
    tional dynamics?

7.2. Consider the dynamics of a rigid body that is constrained to planar
rotational and translational motion in $\mathbb{R}^2$. This is Euclidean motion of a
rigid body where each material point in the body is constrained to move
in an inertially fixed two-dimensional plane. Assume the configuration is
$(R, x) \in$ SE(2), where $R \in$ SO(2) is a 2×2 attitude matrix and $x \in \mathbb{R}^2$
is the position vector of the center of mass of the rigid body. The configuration is defined with respect to a two-dimensional inertial Euclidean frame
and a two-dimensional body-fixed frame. The modified Lagrangian function
is

$$\tilde{L}(R, x, \omega, \dot{x}) = \frac{1}{2}J\omega^2 + \frac{1}{2}m \left\| \dot{x} \right\|^2 - U(R, x),$$

where J is the scalar moment of inertia of the rigid body and m is the mass
of the rigid body.

(a) What are expressions for the infinitesimal variations of $(R, \omega) \in$ TSO(2)
    and $(x, \dot{x}) \in$ T$\mathbb{R}^2$?
(b) What are the Euler–Lagrange equations for planar rotation and transla-
    tion of a rigid body on the tangent bundle of the configuration manifold?
(c) What is the modified Hamiltonian function defined on the cotangent bun-
    dle of the configuration manifold?
(d) What are Hamilton's equations for planar rotation and translation of a
    rigid body on the cotangent bundle of the configuration manifold?
(e) What are conserved quantities of the dynamical flow on TSE(2)?
(f) What are the conditions for an equilibrium solution of the dynamical flow
    on TSE(2)?

7.3. Consider a rigid body, constrained to move within a fixed vertical plane,
under constant, uniform gravity. Use the formulation and results of the prior

problem. The mass of the rigid body is m and the scalar moment of inertia of the rigid body is J. Assume the configuration manifold of the rigid body, assuming planar rotation and translation, is taken to be the Lie group $\mathsf{SE}(2)$.

(a) What are the Euler–Lagrange equations for planar rotation and translation of a rigid body?
(b) What are Hamilton's equations for the planar rotation and translation of a rigid body?
(c) What are the conserved quantities of the dynamical flow on $\mathsf{TSE}(2)$?

7.4. Consider the Euclidean dynamics of a rigid body, consisting of material points denoted by $\mathcal{B}$ in a body-fixed frame, under a gravitational field. The configuration $(R, x) \in \mathsf{SE}(3)$. Assume the origin of the body-fixed frame is located at the center of mass of the rigid body. A gravitational force acts on each material point in the rigid body. The net force and net moment of all the gravity forces is obtained by integrating the gravity force and moment for each mass increment of the body over the whole body. The gravitational field, expressed in the inertial frame, is given by $G : \mathbb{R}^3 \to \mathsf{T}\mathbb{R}^3$. The incremental gravitational force on a mass increment $dm(\rho)$ of the rigid body, located at $\rho \in \mathcal{B}$ in the body-fixed frame, is given in the inertially fixed frame by $dm(\rho)G(R\rho)$. Thus, the net gravity force, in the inertial frame, is $\int_{\mathcal{B}} G(R\rho)\,dm(\rho)$. The incremental gravitational moment vector on a mass increment $dm(\rho)$ of the rigid body, located at $\rho \in \mathcal{B}$ in the body-fixed frame, is given in the inertially fixed frame by $R\rho \times dm(\rho)G(R\rho)$ or, equivalently in the body-fixed frame, by $\rho \times dm(\rho)R^T G(R\rho)$. Thus, the net gravity moment, in the body-fixed frame, is $\int_{\mathcal{B}} \rho \times R^T G(R\rho)\,dm(\rho)$.

(a) What are the Euler–Lagrange equations for the Euclidean dynamics of the rigid body in the gravitational field?
(b) What are Hamilton's equations for the Euclidean dynamics of the rigid body in the gravitational field?
(c) What are the conditions for an equilibrium solution of the Euclidean motion of a rigid body in the gravitational field?
(d) Suppose the gravitational field $G(x) = -ge_3$ is constant. What are the simplified Euler–Lagrange equations for the Euclidean dynamics of the rigid body? What are the conditions for an equilibrium solution of a rigid body in a constant gravitational field?

7.5. Consider the Euclidean dynamics of a charged rigid body, consisting of material points denoted by $\mathcal{B}$ in a body-fixed frame, under an electric field and a magnetic field. The configuration is denoted by $(R, x) \in \mathsf{SE}(3)$. Assume the origin of a body-fixed frame is located at the center of mass of the rigid body. The mass of the rigid body is m and J is the standard inertia matrix of the rigid body. An electric force and a magnetic force act on each material point in the rigid body. The net force and the net moment of all of the electric and magnetic forces is obtained by integrating the incremental electric and magnetic forces and moments for each volume increment of the body over the whole body. The electric field, expressed in the inertial frame, is given by $E : \mathbb{R}^3 \to T\mathbb{R}^3$; the magnetic field, expressed in the inertial frame, is given by $B : \mathbb{R}^3 \to T\mathbb{R}^3$. The incremental electric and magnetic force on a volume increment with charge dQ, located at $\rho \in \mathcal{B}$ in the body-fixed frame, is given in the inertially fixed frame by $dQ(E(R\rho) + \dot{R}\rho \times B(R\rho))$. Thus, the net electric and magnetic force, in the inertial frame, is $\int_{\mathcal{B}}(E(R\rho) + \dot{R}\rho \times B(R\rho))\,dQ$. The incremental electric and magnetic moment vector on a volume increment with charge dQ, located at $\rho \in \mathcal{B}$ in the body-fixed frame, is given in the body-fixed frame by $\rho \times dQ R^T(E(R\rho) + \dot{R}\rho \times B(R\rho))$. Thus, the total electric and magnetic moment, in the body-fixed frame, is $\int_{\mathcal{B}} \rho \times R^T(E(R\rho) + \dot{R}\rho \times B(R\rho))\,dQ$.

(a) What are the Euler–Lagrange equations for the Euclidean dynamics of the rigid body in the electric field and the magnetic field?
(b) What are Hamilton's equations for the Euclidean dynamics of the rigid body in the electric field and the magnetic field?
(c) What are the conditions for an equilibrium solution of the Euclidean motion of a rigid body in the electric field and the magnetic field?
(d) Suppose the electric field $E(x) = -Ee_3$ and the magnetic field $B(x) = Be_2$ are constant, where E and B are scalar constants. What are the simplified Euler–Lagrange equations for the Euclidean dynamics of the rigid body? What are the conditions for an equilibrium solution of a rigid body in a constant electric field and a constant magnetic field?

7.6. Consider the Euclidean motion of a rigid body in $\mathbb{R}^3$, with configuration $(R, x) \in \mathsf{SE}(3)$, acted on by a force that is constant in the inertial frame. Assume the origin of the body-fixed frame is located at the center of mass of the rigid body. The Euler–Lagrange equations are

$$J\dot{\omega} + \omega \times J\omega = r \times F,$$
$$m\ddot{x} = F.$$

where $r = \sum_{i=1}^{3} a_i e_i$ is a constant vector in the body-fixed frame and $F = \sum_{i=1}^{3} f_i R^T e_i$ is an external force. Here, a_i, f_i, $i = 1, 2, 3$ are scalar constants.

(a) What are conditions on the constants a_i, f_i, $i = 1, 2, 3$, that guarantee the rigid body has an equilibrium solution? What are the resulting equilibrium solutions?
(b) What are conditions on the constants a_i, f_i, $i = 1, 2, 3$, that guarantee the rigid body has a relative equilibrium solution, i.e., the rotational and translational velocity vectors defining the configuration are constant? What are the resulting relative equilibrium solutions?

7.7. Consider the Euclidean motion of a rigid body in $\mathbb{R}^3$, with configuration $(R, x) \in \mathsf{SE}(3)$, acted on by a force that is constant in the body-fixed frame. Assume the origin of the body-fixed frame is located at the center of mass of the rigid body. The Euler–Lagrange equations are

$$Jw + w \times Jw = r \times F,$$
$$m\dot{v} + mw \times v = F,$$

where $r = \sum_{i=1}^{3} a_i e_i$ is a constant vector in the body-fixed frame and $F = \sum_{i=1}^{3} f_i e_i$ is an external force. Here a_i, f_i, $i = 1, 2, 3$, are scalar constants. As previously, $v = R^T \dot{x} \in \mathbb{R}^3$ denotes the translational velocity vector of the center of mass of the rigid body in the body-fixed frame.

(a) What are conditions on the constants a_i, f_i, $i = 1, 2, 3$, that guarantee the rigid body has an equilibrium solution? What are the resulting equilibrium solutions?

(b) What are conditions on the constants a_i, f_i, $i = 1, 2, 3$, that guarantee the rigid body has a relative equilibrium solution, i.e., the rotational and translational velocity vectors defining the configuration are constant? What are the resulting relative equilibrium solutions?

7.8. A uniform rigid cube, acted on by uniform, constant gravity, is supported by an elastic foundation. The rigid cube can rotate and translate so that its configuration is $(R, x) \in \mathsf{SE}(3)$. An inertial frame is constructed with its third axis opposite to the direction of gravity. The origin of the body-fixed frame for the rigid cube is located at the center of mass of the cube and its axes are principle axes of the cube. The length of each edge of the cube is L, the mass of the cube is m and the scalar matrix $JI_{3\times3}$, where the scalar $J > 0$, denotes the standard inertia matrix of the rigid body. The elastic foundation that supports the cube is described as follows. Four elastic springs connect the four bottom corners of the cube to four fixed inertial supports. The four connection points of the elastic springs to the cube are given, in the body-fixed frame, by:

$$b_1 = \frac{L}{2} e_1 + \frac{L}{2} e_2 - \frac{L}{2} e_3,$$
$$b_2 = \frac{L}{2} e_1 - \frac{L}{2} e_2 - \frac{L}{2} e_3,$$
$$b_3 = -\frac{L}{2} e_1 + \frac{L}{2} e_2 - \frac{L}{2} e_3,$$
$$b_4 = -\frac{L}{2} e_1 - \frac{L}{2} e_2 - \frac{L}{2} e_3.$$

The connection points of the elastic springs to the inertial supports are given, in the inertial frame, by:

$$c_1 = \frac{L}{2}e_1 + \frac{L}{2}e_2,$$

$$c_2 = \frac{L}{2}e_1 - \frac{L}{2}e_2,$$

$$c_3 = -\frac{L}{2}e_1 + \frac{L}{2}e_2,$$

$$c_4 = -\frac{L}{2}e_1 - \frac{L}{2}e_2.$$

These four pairs of points are connected to the four vertices on one side of the cube by linearly elastic springs. These springs can deform axially with identical elastic stiffness κ; they have zero elastic axial forces when the scalar axial deflection is D. Assume these elastic springs satisfy the inequality $4\kappa D > mg$. In addition, each of the four elastic springs that support the cube is constructed so that there is an elastic vertical restoring moment proportional to the angle between the axis of the spring and the vertical; the elastic stiffness K for this moment is the same for the four springs.

(a) Show that the four elastic spring vectors from the inertial attachment points to their respective attachment points on the cube are given, in the inertial frame, by

$$s_i = x + Rb_i - c_i, \quad i = 1, 2, 3, 4.$$

(b) Show that the four elastic axial forces $F_i \in \mathbb{R}^3$, in the inertial frame, that act on the cube are

$$F_i = \kappa(\|s_i\| - D)\frac{s_i}{\|s_i\|}, \quad i = 1, 2, 3, 4.$$

(c) Show that the four elastic moment vectors $M_i \in \mathbb{R}^3$ on the cube, due to these axial forces, are

$$M_i = \kappa(\|s_i\| - D)b_i \times R^T \frac{s_i}{\|s_i\|}, \quad i = 1, 2, 3, 4,$$

expressed in the body-fixed frame.

(d) Show that the restoring moment vector $M \in \mathbb{R}^3$ on the cube, due to the vertical restoring moments on the four elastic springs, is given in the body-fixed frame by

$$M = \sum_{i=1}^{4} Kb_i \times R^T e_3.$$

(e) Show that the Euler–Lagrange equations, including the forces and moments due to gravity and the elastic foundation, are given by

$$J\dot{\omega} + \omega \times J\omega = \sum_{i=1}^{4} \kappa(\|s_i\| - D)b_i \times R^T \frac{s_i}{\|s_i\|} + \sum_{i=1}^{4} Kb_i \times R^T e_3,$$

$$m\ddot{x} + mge_3 = \sum_{i=1}^{4} \kappa(\|s_i\| - D)\frac{s_i}{\|s_i\|},$$

for the rotational and translational dynamics of a cube on an elastic foundation.

(f) Show that $(R, \omega, x, \dot{x}) = (I_{3\times3}, 0, \frac{mge_3}{4\kappa}, 0) \in \mathsf{TSE}(3)$ defines an equilibrium of the cube on an elastic foundation.

(g) What are linearized dynamics that approximate the dynamical flow of the cube on an elastic foundation in a neighborhood of the above equilibrium?

7.9. It is often convenient to consider flight of a fixed-wing aircraft in $\mathbb{R}^3$ by viewing it as a rigid body so that the configuration manifold is $\mathsf{SE}(3)$. The inertial frame is chosen so that the third axis is vertical; the body-fixed frame is chosen so that the first axis points along the nose of the aircraft and the first and third axes define a plane of mass symmetry of the aircraft. Suppose the origin of the body-fixed frame is located at the center of mass of the aircraft. Without going into detail, the important forces that act on the aircraft in flight arise from aerodynamics, thrust, and gravity. Using the notation introduced in this chapter, the aerodynamics force, including both lift and drag effects, is most conveniently described as a vector function with values in the body-fixed frame that depends on the aircraft rotational and translational velocity vectors in the body-fixed frame, that is $F_a : \mathbb{R}^3 \times \mathbb{R}^3 \to \mathsf{T}^*\mathbb{R}^3$; further $F_a(v, \omega)$ lies in the plane of mass symmetry of the aircraft. The aerodynamics moment vector, including both lift and drag effects, is described as a vector function with values in the body-fixed frame that depends on the aircraft rotational and translational velocity vectors in the body-fixed frame, that is $M_a : \mathbb{R}^3 \times \mathbb{R}^3 \to \mathsf{T}^*\mathbb{R}^3$. The thrust force is most conveniently described as a vector function with values in the body-fixed frame that depends on the aircraft translational velocity vector in the body-fixed frame, that is $T : \mathbb{R}^3 \to \mathsf{T}^*\mathbb{R}^3$; further $T(v)$ has a fixed direction in the plane of mass symmetry of the aircraft. A constant gravity force also acts on the aircraft in flight.

(a) What are the rotational kinematics? What are the translational kinematics?

(b) Suppose that the aerodynamics force and the thrust force have nontrivial moment arms. Introduce suitable notation to describe the rotational dynamics of the aircraft in terms of the configuration $(R, x, \omega, v) \in \mathsf{TSE}(3)$.

(c) Describe the translational dynamics of the aircraft in terms of the configuration variables $(R, x, \omega, v) \in \mathsf{TSE}(3)$.

(d) Describe the translational dynamics of the aircraft in terms of the configuration variables $(R, \omega, \dot{x}) \in \mathsf{TSE}(3)$.

(e) What algebraic conditions characterize the relative equilibrium solutions, i.e., solutions for which the attitude is constant, the angular velocity vector is zero, and the aircraft translational velocity vector is constant?

(f) What complications arise if the origin of the body-fixed frame is not located at the center of mass of the aircraft? What modifications need to be made in the rotational and translations equations of motion?

7.10. Consider the rigid body maneuver problem of finding curve(s) on the Lie group $\mathsf{SE}(3)$ that connects two fixed points in $\mathsf{SE}(3)$ and minimizes $\int_0^1 \{\|\omega\|^2 + \|\dot{x}\|^2\} dt$. The origin of the body-fixed frame is located at the center of mass of the rigid body.

(a) If the curve is described on the interval $[0,1]$ by $t \to (R(t), x(t)) \in \mathsf{SE}(3)$, show that the minimizing curves satisfy the variational property $\delta \int_0^1 \{\|\omega\|^2 + \|\dot{x}\|^2\} dt = 0$ for all smooth curves $t \to (R(t), x(t)) \in \mathsf{SE}(3)$ that satisfy the boundary conditions $(R(0), x(0)) = (R_0, x_0) \in \mathsf{SE}(3)$, $(R(1), x(1)) = (R_1, x_1) \in \mathsf{SE}(3)$.

(b) What are the Euler–Lagrange equations and Hamilton's equations that the minimizing curves must satisfy?

(c) Use the above equations and boundary conditions to determine analytical expressions for the minimizing curves.

7.11. Consider n rotating and translating rigid bodies that are coupled through the potential energy; the configuration manifold is $(\mathsf{SE}(3))^n$. With respect to a common inertial Euclidean frame, the attitudes of the rigid bodies are given by $R_i \in \mathsf{SO}(3)$, $i = 1, \ldots, n$, the position vectors of the centers of mass of the rigid bodies are $x_i \in \mathbb{R}^3$, $i = 1, \ldots, n$, and we use the notation $R = (R_1, \ldots, R_n) \in (\mathsf{SO}(3))^n$ and $x = (x_1, \ldots, x_n) \in (\mathbb{R}^3)^n$; similarly, $\omega = (\omega_1, \ldots, \omega_n) \in (\mathfrak{so}(3))^n$. Suppose the kinetic energy of the rigid bodies is a quadratic function in the angular velocity vectors of the bodies and the translational velocity vectors of the bodies, so that the modified Lagrangian function $\tilde{L} : \mathsf{T}(\mathsf{SE}(3))^n \to \mathbb{R}^1$ is given by

$$\tilde{L}(R, x, \omega, \dot{x}) = \frac{1}{2} \sum_{i=1}^{n} \omega_i^T J_i \omega_i + \sum_{i=1}^{n} a_i^T \omega_i + \frac{1}{2} \sum_{i=1}^{n} \dot{x}_i^T m_i \dot{x}_i$$

$$+ \sum_{i=1}^{n} b_i^T \dot{x}_i - U(R, x),$$

where J_i are 3×3 symmetric and positive-definite matrices, $i = 1, \ldots, n$, $a_i \in \mathbb{R}^3$, $i = 1, \ldots, n$, $m_i > 0$, $i = 1, \ldots, n$, are scalars, $b_i \in \mathbb{R}^3$, $i = 1, \ldots, n$, and $U : (\mathsf{SE}(3))^n \to \mathbb{R}^1$ is the potential energy that characterizes the coupling of the rigid bodies.

(a) What are the Euler–Lagrange equations for this modified Lagrangian?

(b) What are Hamilton's equations for the modified Hamiltonian associated with this modified Lagrangian?

Chapter 8
Lagrangian and Hamiltonian Dynamics on Manifolds

In Chapters 3 through 7, we developed Lagrangian and Hamiltonian dynamics for systems that evolve on specific configuration manifolds, namely $\mathbb{R}^n$, $(\mathsf{S}^1)^n$, $(\mathsf{S}^2)^n$, $\mathsf{SO}(3)$, $\mathsf{SE}(3)$. These developments were direct in that they made use of variational calculus and the geometric features of these particular configuration manifolds. The observant reader should recognize a common pattern in those variational developments and should expect that all of these developments can be viewed as special cases of a more abstract result. In this chapter, such results are obtained. The development is built around the concept of a configuration manifold M, embedded in $\mathbb{R}^n$, and a Lagrangian function $L : \mathsf{T}M \to \mathbb{R}^1$ defined on the tangent bundle of the configuration manifold.

We obtain two different forms of variational conditions that characterize curves that lie in a configuration manifold and connect two fixed points in the manifold for which the action integral is stationary; such curves are said to be extremal. Both sets of variational conditions are expressed in terms of an orthogonal projection operator: for each $x \in M$ the orthogonal projection operator maps $\mathbb{R}^n$ onto the tangent space $\mathsf{T}_x M$ [62]. Euler–Lagrange equations are derived that are expressed in terms of the Lagrangian function and the orthogonal projection operator. Hamilton's equations are obtained that are expressed in terms of a Hamiltonian function, obtained from the Legendre transformation, and the orthogonal projection operator.

We next consider special cases where the configuration manifold has a product structure: a product of linear manifolds embedded in $\mathbb{R}^3$, a product of copies of the manifold S^1 embedded in $\mathbb{R}^2$, and a product of copies of the manifold S^2 embedded in $\mathbb{R}^3$ [63]. We obtain Euler–Lagrange equations and Hamilton's equations that evolve on such configuration manifolds.

Then, we develop abstract variational results for a configuration manifold that is a Lie group, and we obtain Euler–Lagrange equations and Hamilton's equations. Additionally, abstract variational results are developed for

© Springer International Publishing AG 2018　　　　　　　　　347
T. Lee et al., *Global Formulations of Lagrangian and Hamiltonian Dynamics on Manifolds*, Interaction of Mechanics and Mathematics,
DOI 10.1007/978-3-319-56953-6_8

a configuration manifold that is a homogeneous manifold, and we obtain Euler–Lagrange equations and Hamilton's equations for this case.

Geometric mechanics, as presented in this text, emphasizes the development of Lagrangian and Hamiltonian equations of motion, expressed in terms of configuration vectors in a configuration manifold. These variational results are based on the geometric properties of the embedded configuration manifold and the Lagrangian function defined on the tangent bundle of the configuration manifold or the Hamiltonian function defined on the cotangent bundle of the configuration manifold. In many cases, the variational equations can be globally extended in the sense they can be naturally extended from defining flows on the embedded configuration manifold to flows on the embedding vector space.

The classical approach to Lagrangian and Hamiltonian dynamics on manifolds is obtained by imposing manifold evolution via algebraic constraints. There is a long history of such developments; see [94, 95, 96, 98]; these and other classical developments are primarily analytical and do not make use of differential geometric features of the manifolds.

Alternatively, Lagrangian and Hamiltonian dynamics on manifolds have been investigated using methods of geometric mechanics as in [5, 10, 11, 16, 37, 38, 39, 69, 72, 92]. These references use geometric concepts in the analysis of solutions and flow properties, but not in the formulation of the basic equations of motion as is emphasized here. This text is primarily concerned with the formulation of Lagrangian and Hamiltonian dynamics in a form that is consistent with fact that the configuration manifold is embedded in a vector space. The text does not delve deeply into the use of differential geometry in analysis of solution properties: a comprehensive treatment that combines the role of differential geometry in both the formulation and advanced analysis of Lagrangian dynamics and Hamiltonian dynamics remains to be developed.

8.1 Lagrangian Dynamics on a Manifold

An abstract version of the Euler–Lagrange equations that evolve on the tangent bundle TM is given in [5]. This result, expressed in terms of infinitesimal variations, is derived. It is then used, together with an orthogonal projection operator, to obtain several different but equivalent forms of the Euler–Lagrange equations.

8.1.1 Variations on the Tangent Bundle TM

Assume M is a differentiable manifold embedded in $\mathbb{R}^n$; we denote the tangent space at $x \in M$ by $T_x M$ and we denote the tangent bundle of M by TM.

The subsequent development describes variations of functions with values in the manifold M. Let $x : [t_0, t_f] \to M$ be a differentiable curve. The family of variations of x is defined by differentiable mappings $x^\epsilon : (-c, c) \times [t_0, t_f] \to M$ for $c > 0$; since M is an embedded manifold, each member of the family can be shown to have the power series

$$x^\epsilon(t) = x(t) + \epsilon \delta x(t) + \mathcal{O}(\epsilon^2), \quad t_0 \le t \le t_f,$$

for a differentiable curve $\delta x : [t_0, t_f] \to \mathbb{R}^n$ that satisfies $\delta x(t) \in T_{x(t)}M$, $t_0 \le t \le t_f$, and $\delta x(t_0) = \delta x(t_f) = 0$. It is easy to see that the variations satisfy $x^0(t) = x(t)$, $t_0 \le t \le t_f$, and $x^\epsilon(t_0) = x(t_0)$ and $x^\epsilon(t_f) = x(t_f)$ for any $\epsilon \in (-c, c)$.

Suppose that $(x, \dot{x}) : [t_0, t_f] \to TM$ is a differentiable curve on the tangent bundle. This construction also allows us to define variations of $(x, \dot{x})$ by the family of differentiable mappings $(x^\epsilon, \dot{x}^\epsilon) : (-c, c) \times [t_0, t_f] \to TM$ in the natural way.

8.1.2 Lagrangian Variational Conditions

Consider a Lagrangian $L : TM \to \mathbb{R}^1$. We now derive variational conditions according to Hamilton's principle, namely that solution curves correspond to extremal curves of the action integral, i.e., the infinitesimal variation of the action integral is zero.

Define the action integral along a motion that evolves on the manifold M as

$$\mathfrak{G} = \int_{t_0}^{t_f} L(x, \dot{x}) \, dt.$$

Hamilton's principle is that a solution curve $(x, \dot{x}) : [t_0, t_f] \to TM$ satisfies the variational condition

$$\delta \mathfrak{G} = \frac{d}{d\epsilon} \int_{t_0}^{t_f} L(x^\epsilon, \dot{x}^\epsilon) \, dt \Big|_{\epsilon=0} = 0.$$

The infinitesimal variation of the action integral can be written as

$$\delta \mathfrak{G} = \int_{t_0}^{t_f} \left\{ \frac{\partial L(x, \dot{x})}{\partial \dot{x}} \cdot \delta \dot{x} + \frac{\partial L(x, \dot{x})}{\partial x} \cdot \delta x \right\} dt.$$

Integrating the first term on the right by parts, and using the fact that δx vanishes at $t = t_0$ and t_f, the infinitesimal variation of the action integral is given by

$$\delta \mathfrak{G} = \int_{t_0}^{t_f} \left\{ -\frac{d}{dt} \left(\frac{\partial L(x, \dot{x})}{\partial \dot{x}} \right) + \frac{\partial L(x, \dot{x})}{\partial x} \right\} \cdot \delta x \, dt.$$

The fundamental lemma of the calculus of variations on a configuration manifold M implies the following:

Proposition 8.1 *Consider a differentiable manifold M embedded in $\mathbb{R}^n$ and assume $(x, \dot{x}) : [t_0, t_f] \to TM$. Then, $(x, \dot{x}) : [t_0, t_f] \to TM$ is an extremal curve of the action integral if and only if*

$$\left\{ \frac{d}{dt}\left(\frac{\partial L(x, \dot{x})}{\partial \dot{x}} \right) - \frac{\partial L(x, \dot{x})}{\partial x} \right\} \cdot \delta x = 0, \quad \delta x \in T_x M. \tag{8.1}$$

This is an abstract condition, essentially in the form given by [5], that characterizes the Lagrangian flow that evolves on the tangent bundle of the configuration manifold. In geometric terms, the condition can also be stated as requiring that for each $(x, \dot{x}) \in TM$, the Euler–Lagrange expression

$$\left\{ \frac{d}{dt}\left(\frac{\partial L(x, \dot{x})}{\partial \dot{x}} \right) - \frac{\partial L(x, \dot{x})}{\partial x} \right\}$$

always lies in the orthogonal complement of the tangent space $T_x M$.

We now introduce the orthogonal projection matrix for each $x \in M$ as $P(x) : \mathbb{R}^n \to T_x M$ which satisfies the orthogonality condition: for each $y \in \mathbb{R}^n$

$$(y - P(x)y) \cdot z = 0, \text{ for all } z \in T_x M. \tag{8.2}$$

We assume the orthogonal projection, viewed as an operator from M to $\mathbb{R}^{n \times n}$, is a differentiable matrix-valued function. The projection matrix $P(x)$ is symmetric as follows. From the orthogonality, we have

$$0 = (y - P(x)y) \cdot P(x)z = y^T(I_{n \times n} - P(x)^T)P(x)z$$
$$= y^T(P(x) - P(x)^T P(x))z.$$

Since this is satisfied for any $y, z \in \mathbb{R}^n$, we have $P(x) = P(x)^T P(x)$, so that $P^T(x) = (P(x)^T P(x))^T = P(x)^T P(x) = P(x)$. The transpose of the orthogonal projection can be viewed as $P(x)^T : \mathbb{R}^n \to T_x^* M$, and this form is often used to clarify that the result is a covector.

Thus, Proposition 8.1 holds for all infinitesimal variations at $x \in M$ given by $\delta x = P(x)y$, $y \in \mathbb{R}^n$. This implies that Proposition 8.1 can also be expressed in terms of the orthogonal projection operator as follows.

Proposition 8.2 *Consider a differentiable manifold M embedded in $\mathbb{R}^n$ and assume $(x, \dot{x}) : [t_0, t_f] \to TM$. Then, $(x, \dot{x}) : [t_0, t_f] \to TM$ is an extremal curve of the action integral if and only if*

$$P(x)^T \left\{ \frac{d}{dt}\left(\frac{\partial L(x, \dot{x})}{\partial \dot{x}} \right) - \frac{\partial L(x, \dot{x})}{\partial x} \right\} = 0. \tag{8.3}$$

This is an abstract condition that characterizes the Lagrangian flow that evolves on the tangent bundle of the configuration manifold. Propositions 8.1 and 8.2 provide different but equivalent geometric characterizations.

8.1.3 Euler–Lagrange Equations on $\mathsf{T}M$

We now derive Euler–Lagrange equations that describe solution curves on the tangent bundle $\mathsf{T}M$. These Euler–Lagrange equations are expressed in terms of the Lagrangian function $L : \mathsf{T}M \to \mathbb{R}^1$ and the orthogonal projection operator $P(x) : \mathbb{R}^n \to \mathsf{T}_x M$ defined for each $x \in M$. We show that these Euler–Lagrange equations define a smooth vector field, called the Lagrangian vector field, on the tangent bundle $\mathsf{T}M$.

To this end, we can write

$$\frac{d}{dt}\left(\frac{\partial L(x,\dot{x})}{\partial \dot{x}}\right) = \frac{\partial^2 L(x,\dot{x})}{\partial \dot{x}^2}\ddot{x} + \frac{\partial}{\partial x}\left(\frac{\partial L(x,\dot{x})}{\partial \dot{x}}\right)\dot{x},$$

so that (8.3) is rewritten as

$$P(x)^T\left\{\frac{\partial^2 L(x,\dot{x})}{\partial \dot{x}^2}\ddot{x} + \frac{\partial}{\partial x}\left(\frac{\partial L(x,\dot{x})}{\partial \dot{x}}\right)\dot{x} - \frac{\partial L(x,\dot{x})}{\partial x}\right\} = 0. \qquad (8.4)$$

However, $\ddot{x}$ is not uniquely determined by (8.4) since the projection operator $P(x)$ is not invertible.

To determine $\ddot{x}$, we require an additional constraint on $\ddot{x}$. Since $\dot{x} \in \mathsf{T}_x M$, it follows that $(I_{n\times n} - P(x))\dot{x} = 0$ from the definition of the orthogonal projection operator. Taking the time-derivative, we obtain

$$(I_{n\times n} - P(x))\ddot{x} - \frac{\partial(P(x)\dot{x})}{\partial x}\dot{x} = 0, \qquad (8.5)$$

which provides a constraint on the component of $\ddot{x}$ that is normal to $\mathsf{T}_x M$.

We now solve (8.4) and (8.5) together. These two expressions can be rearranged into the following matrix equation:

$$\begin{bmatrix} P(x)^T \frac{\partial^2 L(x,\dot{x})}{\partial \dot{x}^2} \\ I_{n\times n} - P(x) \end{bmatrix}\ddot{x} = \begin{bmatrix} -P(x)^T\left(\frac{\partial}{\partial x}\left(\frac{\partial L(x,\dot{x})}{\partial \dot{x}}\right)\dot{x} - \frac{\partial L(x,\dot{x})}{\partial x}\right) \\ \frac{\partial(P(x)\dot{x})}{\partial x}\dot{x} \end{bmatrix}. \qquad (8.6)$$

The Lagrangian is hyperregular so that the Hessian of the Lagrangian is positive-definite; consequently, the $2n \times n$ matrix function on the left-hand side of equation (8.6) has full column rank (see Problem 8.1). Multiply (8.6) by the transpose of this $2n\times n$ matrix to obtain the Euler–Lagrange equations.

Proposition 8.3 *Consider a differentiable manifold M embedded in $\mathbb{R}^n$ and a hyperregular Lagrangian $L : \mathsf{T}M \to \mathbb{R}^1$. Assume $(x, \dot{x}) : [t_0, t_f] \to \mathsf{T}M$. Then, $(x, \dot{x}) : [t_0, t_f] \to \mathsf{T}M$ is an extremal curve of the action integral if and only if*

$$\left\{ I_{n \times n} - P(x) + \frac{\partial^2 L(x, \dot{x})}{\partial \dot{x}^2} P(x) \frac{\partial^2 L(x, \dot{x})}{\partial \dot{x}^2} \right\} \ddot{x}$$

$$+ \frac{\partial^2 L(x, \dot{x})}{\partial \dot{x}^2} P(x)^T \left(\frac{\partial}{\partial x} \left(\frac{\partial L(x, \dot{x})}{\partial \dot{x}} \right) \dot{x} - \frac{\partial L(x, \dot{x})}{\partial x} \right)$$

$$- (I_{n \times n} - P(x)) \frac{\partial (P(x) \dot{x})}{\partial x} \dot{x} = 0. \tag{8.7}$$

This abstract version of the Euler–Lagrange equations (8.7) characterizes the Lagrangian flow on the tangent bundle of the configuration manifold. Since the $n \times n$ matrix function that multiplies $\ddot{x}$ in (8.7) is necessarily invertible for all $x \in M$, it follows that the Euler–Lagrange equations (8.7) determines a unique vector field on $\mathsf{T}M$; that is, for each $(x, \dot{x}) \in \mathsf{T}M$ there is a unique $(x, \dot{x}, \ddot{x}) \in \mathsf{T}(\mathsf{T}M)$. This map from $\mathsf{T}M$ to $\mathsf{T}(\mathsf{T}M)$ defines a smooth vector field on $\mathsf{T}M$ that is denoted by $F|_{\mathsf{T}M}$.

8.1.4 Extension of the Lagrangian Vector Field from $\mathsf{T}M$ to $\mathsf{T}\mathbb{R}^n$

Assume the Lagrangian function can be defined everywhere on the embedding tangent bundle $\mathsf{T}\mathbb{R}^n$ and assume the orthogonal projection operator $P(x)$ can be globally extended so that it is defined for each $x \in \mathbb{R}^n$. These extensions are made without changing the Lagrangian on $\mathsf{T}M$ and without changing the orthogonal projection property on M.

Then, the extended differential equation (8.7) can be viewed as defining a smooth Lagrangian vector field $F : \mathsf{T}\mathbb{R}^n \to \mathsf{T}(\mathsf{T}\mathbb{R}^n)$. This globally defined vector field necessarily has an invariant manifold $\mathsf{T}M$ and the restriction of the global Lagrangian vector field F to $\mathsf{T}M$ is in fact the Lagrangian vector field $F|_{\mathsf{T}M}$ on $\mathsf{T}M$ that is of primary interest. This vector field F on $\mathsf{T}\mathbb{R}^n$ can be viewed as an extension of the Lagrangian vector field $F|_{\mathsf{T}M}$ on $\mathsf{T}M$.

It should be emphasized that Propositions 8.1 and 8.2, under the stated assumptions, only guarantee the existence of a Lagrangian vector field on the tangent bundle $\mathsf{T}M$. The additional assumptions that the Lagrangian function and the orthogonal projection operator are globally defined on $\mathsf{T}\mathbb{R}^n$ are required to guarantee that the Lagrangian vector field on $\mathsf{T}M$ can be extended to a vector field on $\mathsf{T}\mathbb{R}^n$.

Global extension of the Lagrangian vector field is possible for many physical systems such as multi-body connections of mass particles and rigid bodies. The implications are important in that many classical analytical and computational methods can be utilized for the extended Lagrangian vector field on $\mathsf{T}\mathbb{R}^n$, while maintaining the understanding that the results are only of interest on the tangent bundle $\mathsf{T}M$. As a practical matter, even if a global extension to the entire embedding vector space is not possible, a local extension to a tubular neighborhood of the embedded configuration manifold can still be useful.

8.2 Hamiltonian Dynamics on a Manifold

An abstract version of Hamilton's equations that evolve on the cotangent bundle T^*M is obtained by introducing the Legendre transformation and the Hamiltonian function. Hamilton's equations are expressed in terms of the Hamiltonian function and the orthogonal projection operator.

8.2.1 Legendre Transformation and the Hamiltonian

For a hyperregular Lagrangian function $L : \mathsf{T}M \to \mathbb{R}^1$, the Legendre transformation $\mathbb{F}L : \mathsf{T}M \to \mathsf{T}^*M$ is defined as

$$\mathbb{F}L(x, \dot{x}) = (x, \mu), \tag{8.8}$$

where $\mu \in (\mathsf{T}_x M)^*$, referred to as the momentum conjugate to $\dot{x} \in \mathsf{T}_x M$, is given by

$$\mu = P(x)^T \frac{\partial L(x, \dot{x})}{\partial \dot{x}}. \tag{8.9}$$

Since the Lagrangian is hyperregular on $\mathsf{T}M$ for each $x \in M$, the Legendre transformation $\dot{x} \to \mu$ is a diffeomorphism from $\mathsf{T}_x M$ to $(\mathsf{T}_x M)^*$.

It is convenient to introduce the Hamiltonian function $H : \mathsf{T}^*M \to \mathbb{R}^1$ as

$$H(x, \mu) = \mu \cdot \dot{x} - L(x, \dot{x}),$$

using the Legendre transformation (8.9).

8.2.2 Variations on the Cotangent Bundle T^*M

Suppose that $(x, \dot{x}) : [t_0, t_f] \to \mathsf{T}M$ is a differentiable curve on the tangent bundle. Using the Legendre transformation, there is a corresponding differentiable curve $(x, \mu) : [t_0, t_f] \to \mathsf{T}^*M$ on the cotangent bundle. The construction of the family of variations $(x^\epsilon, \dot{x}^\epsilon) : (-c, c) \times [t_0, t_f] \to \mathsf{T}M$ on the tangent bundle also allows us to define the family of variations $(x^\epsilon, \mu^\epsilon) : (-c, c) \times [t_0, t_f] \to \mathsf{T}^*M$ on the cotangent bundle.

8.2.3 Hamilton's Phase Space Variational Principle

Hamilton's equations on a manifold can be most directly obtained from a variational principle on T^*M. *Hamilton's phase space variational principle* states that solution curves are extremal curves of the action integral, where the Lagrangian in the action integral is expressed in terms of the Hamiltonian. That is to say that the infinitesimal variation of the action integral

$$\mathfrak{G} = \int_{t_0}^{t_f} \{ \mu \cdot \dot{x} - H(x, \mu) \} \, dt$$

along any solution curve with fixed end points is zero. Equivalently, a solution curve $(x, \mu) : [t_0, t_f] \to \mathsf{T}^*M$ satisfies the variational condition

$$\delta\mathfrak{G} = \frac{d}{d\epsilon} \int_{t_0}^{t_f} \{ \mu^\epsilon \cdot \dot{x}^\epsilon - H(x^\epsilon, \mu^\epsilon) \} \, dt \Big|_{\epsilon=0} = 0.$$

8.2.4 Hamilton's Equations on T^*M

We now derive Hamilton's equations that describe solution curves on the cotangent bundle T^*M. Hamilton's equations are expressed in terms of the Hamiltonian function $H : \mathsf{T}^*M \to \mathbb{R}^1$ and the orthogonal projection operator $P(x)^T : \mathbb{R}^n \to (\mathsf{T}_x M)^*$ defined for each $x \in M$. We show that these Hamilton's equations define a smooth vector field, called the Hamiltonian vector field, on the cotangent bundle T^*M.

To this end, the infinitesimal variation of the action integral can be written as

$$\delta\mathfrak{G} = \int_{t_0}^{t_f} \left\{ \mu \cdot \delta\dot{x} - \frac{\partial H(x, \mu)}{\partial x} \cdot \delta x + \delta\mu \cdot \left(\dot{x} - \frac{\partial H(x, \mu)}{\partial \mu} \right) \right\} \, dt.$$

Integrating the first term on the right by parts, the infinitesimal variation of the action integral is given by

$$\delta\mathfrak{G} = \mu \cdot \delta x \Big|_{t_0}^{t_f} + \int_{t_0}^{t_f} \left\{ \left(-\dot{\mu} - \frac{\partial H(x, \mu)}{\partial x} \right) \cdot \delta x + \delta\mu \cdot \left(\dot{x} - \frac{\partial H(x, \mu)}{\partial \mu} \right) \right\} \, dt.$$

Using the definition (8.9) of μ, $P(x)^T \mu = \mu$, or equivalently, $(I_{n \times n} - P(x)^T)\mu = 0$. Thus, the infinitesimal variations satisfy

$$-\left(\frac{\partial P(x)^T \mu}{\partial x} \right) \delta x + (I_{n \times n} - P(x)^T)\delta\mu = 0.$$

The first term of the above equation can be viewed as a weighted linear combination of matrices acting on μ. This term can also be written as

$$\left(\sum_{i=1}^{n} \frac{\partial P(x)^T}{\partial x_i} \delta x_i \right) \mu = \left[\frac{\partial P(x)^T}{\partial x_1} \mu \, \middle| \, \cdots \, \middle| \, \frac{\partial P(x)^T}{\partial x_n} \mu \right] \delta x$$

$$= \left(\frac{\partial P(x)^T \mu}{\partial x} \right) \delta x, \tag{8.10}$$

where $\left(\frac{\partial P(x)^T \mu}{\partial x} \right)$ is a $n \times n$ matrix that is the Jacobian of the vector-valued function $P(x)^T \mu$. Equivalently, the first term is obtained by concatenating column vectors that are given by matrix-vector products of the matrices $\frac{\partial P(x)^T}{\partial x_i}$ and the vector μ for $i = 1, \ldots, n$.

To impose this constraint on the infinitesimal variations, we decompose the variation $\delta \mu$ into the sum of two orthogonal components: one component in $T_x^* M$, namely $\delta \mu^M = P(x)^T \delta \mu$, and the other component orthogonal to $T_x^* M$, namely $\delta \mu^C = (I_{n \times n} - P(x)^T) \delta \mu$. Then, satisfying the constrained variation implies that $\delta \mu^C = (I_{n \times n} - P(x)^T) \delta \mu = \left(\frac{\partial P(x)^T \mu}{\partial x} \right) \delta x$.

From Hamilton's phase space variational principle, $(x, \mu) : [t_0, t_f] \to T^* M$ is a solution curve if and only if $\delta \mathfrak{G} = 0$ for all continuous infinitesimal variations $\delta x : [t_0, t_f] \to \mathbb{R}^n$, $\delta \mu : [t_0, t_f] \to \mathbb{R}^n$, that satisfy $(\delta x(t), \delta \mu(t)) \in T_{(x(t), \mu(t))} T^* M$ and $\delta x(t_0) = \delta x(t_f) = 0$. The vanishing infinitesimal variations δx at the endpoints mean that the boundary terms vanish, and by decomposing the variation $\delta \mu$, we obtain,

$$\delta \mathfrak{G} = \int_{t_0}^{t_f} \left(-\dot{\mu} - \frac{\partial H(x, \mu)}{\partial x} \right) \cdot \delta x + \delta \mu^M \cdot \left(\dot{x} - \frac{\partial H(x, \mu)}{\partial \mu} \right)$$

$$+ \delta \mu^C \cdot \left(\dot{x} - \frac{\partial H(x, \mu)}{\partial \mu} \right) dt$$

$$= \int_{t_0}^{t_f} \left(-\dot{\mu} - \frac{\partial H(x, \mu)}{\partial x} \right) \cdot \delta x + (P(x)^T \delta \mu) \cdot \left(\dot{x} - \frac{\partial H(x, \mu)}{\partial \mu} \right)$$

$$+ \left(\dot{x} - \frac{\partial H(x, \mu)}{\partial \mu} \right) \cdot \left(\left(\frac{\partial P(x)^T \mu}{\partial x} \right) \delta x \right) dt$$

$$= \int_{t_0}^{t_f} \left(-\dot{\mu} - \frac{\partial H(x, \mu)}{\partial x} + \left(\frac{\partial P(x)^T \mu}{\partial x} \right)^T \left(\dot{x} - \frac{\partial H(x, \mu)}{\partial \mu} \right) \right) \cdot \delta x$$

$$+ \delta \mu \cdot P(x) \left(\dot{x} - \frac{\partial H(x, \mu)}{\partial \mu} \right) dt.$$

Then, the fundamental lemma of the calculus of variations on a configuration manifold M implies that

$$P(x)^T \left(-\dot{\mu} - \frac{\partial H(x,\mu)}{\partial x} + \left(\frac{\partial P(x)^T \mu}{\partial x} \right)^T \left(\dot{x} - \frac{\partial H(x,\mu)}{\partial \mu} \right) \right) = 0, \quad (8.11)$$

$$P(x) \left(\dot{x} - \frac{\partial H(x,\mu)}{\partial \mu} \right) = 0. \quad (8.12)$$

The projections $P(x)$ and $P(x)^T$ in the equations above arise from the fact that we only require that the variation of the action vanishes for infinitesimal variations that satisfy $(\delta x, \delta \mu) \in \mathsf{T}_{(x,\mu)} \mathsf{T}^* M$. Since $\dot{x} \in T_x M$, it follows that $P(x)\dot{x} = \dot{x}$, and we can rewrite (8.12) as

$$\dot{x} = P(x) \frac{\partial H(x,\mu)}{\partial \mu}. \quad (8.13)$$

Equation (8.11) is incomplete since it only determines the component of $\dot{\mu}$ projected onto $T_x^* M$. The remaining component is determined by taking the time derivative of $(I_{n \times n} - P(x)^T)\mu = 0$ to obtain

$$- \left(\frac{\partial P(x)^T \mu}{\partial x} \right) \dot{x} + (I_{n \times n} - P(x)^T) \dot{\mu} = 0,$$

which yields

$$(I_{n \times n} - P(x)^T)\dot{\mu} = \left(\frac{\partial P(x)^T \mu}{\partial x} \right) \dot{x} = \left(\frac{\partial P(x)^T \mu}{\partial x} \right) P(x) \frac{\partial H(x,\mu)}{\partial \mu}. \quad (8.14)$$

Combining (8.11) with (8.14), and substituting (8.13), we obtain

$$\dot{\mu} = P(x)^T \dot{\mu} + (I_{n \times n} - P(x)^T)\dot{\mu}$$

$$= -P(x)^T \frac{\partial H(x,\mu)}{\partial x} + P(x)^T \left(\frac{\partial P(x)^T \mu}{\partial x} \right)^T \left(\dot{x} - \frac{\partial H(x,\mu)}{\partial \mu} \right)$$

$$\quad + \left(\frac{\partial P(x)^T \mu}{\partial x} \right) \dot{x}$$

$$= -P(x)^T \frac{\partial H(x,\mu)}{\partial x} - P(x)^T \left(\frac{\partial P(x)^T \mu}{\partial x} \right)^T \frac{\partial H(x,\mu)}{\partial \mu}$$

$$\quad + \left(P(x)^T \left(\frac{\partial P(x)^T \mu}{\partial x} \right)^T + \left(\frac{\partial P(x)^T \mu}{\partial x} \right) \right) P(x) \frac{\partial H(x,\mu)}{\partial \mu}$$

$$= -P(x)^T \frac{\partial H(x,\mu)}{\partial x} + \left\{ P(x)^T \left(\frac{\partial P(x)^T \mu}{\partial x} \right)^T P(x) \right.$$

$$\quad + \left. \left(\frac{\partial P(x)^T \mu}{\partial x} \right) P(x) - P(x)^T \left(\frac{\partial P(x)^T \mu}{\partial x} \right)^T \right\} \frac{\partial H(x,\mu)}{\partial \mu}.$$

In summary, Hamilton's equations are given in the following.

Proposition 8.4 *Consider a differentiable manifold M embedded in $\mathbb{R}^n$ and a Hamiltonian $H : \mathsf{T}^*M \to \mathbb{R}^1$. Then, $(x, \mu) : [t_0, t_f] \to \mathsf{T}^*M$ is an extremal curve of the action integral if and only if*

$$\dot{x} = P(x)\frac{\partial H(x, \mu)}{\partial \mu}, \tag{8.15}$$

$$\dot{\mu} = -P(x)^T \frac{\partial H(x, \mu)}{\partial x} + \left\{ P(x)^T \left(\frac{\partial P(x)^T \mu}{\partial x} \right)^T P(x) \right.$$

$$\left. + \left(\frac{\partial P(x)^T \mu}{\partial x} \right) P(x) - P(x)^T \left(\frac{\partial P(x)^T \mu}{\partial x} \right)^T \right\} \frac{\partial H(x, \mu)}{\partial \mu}. \tag{8.16}$$

This abstract version of Hamilton's equations (8.15) and (8.16) characterizes the Hamiltonian flow that evolves on the cotangent bundle of the configuration manifold. Equations (8.15) and (8.16) determine a unique vector field on T^*M; that is, for each $(x, \mu) \in \mathsf{T}^*M$ there is a unique $(x, \mu, \dot{x}, \dot{\mu}) \in \mathsf{T}(\mathsf{T}^*M)$. This map from T^*M to $\mathsf{T}(\mathsf{T}^*M)$ defines a smooth vector field on T^*M that is denoted by $F^*|_{\mathsf{T}^*M}$.

8.2.5 Invariance of the Hamiltonian

The following property follows directly from Hamilton's equation on T^*M:

$$\frac{dH(x, \mu)}{dt} = \frac{\partial H(x, \mu)}{\partial x} \cdot \dot{x} + \frac{\partial H(x, \mu)}{\partial \mu} \cdot \dot{\mu}$$

$$= \frac{\partial H(x, \mu)}{\partial x} \cdot \dot{x} + P(x)^T \frac{\partial H(x, \mu)}{\partial \mu} \cdot P(x)^T \dot{\mu}$$

$$+ (I - P(x)^T)\frac{\partial H(x, \mu)}{\partial \mu} \cdot (I - P(x)^T)\dot{\mu},$$

where $\dot{\mu}$ is decomposed into two parts using the orthogonality of the projection. Substituting (8.11), (8.14), and (8.15),

$$\frac{dH(x, \mu)}{dt} = \frac{\partial H(x, \mu)}{\partial x} \cdot P(x)^T \frac{\partial H(x, \mu)}{\partial \mu} - P(x)^T \frac{\partial H(x, \mu)}{\partial \mu} \cdot P(x)^T \frac{\partial H(x, \mu)}{\partial x}$$

$$+ P(x)^T \frac{\partial H(x, \mu)}{\partial \mu} \cdot P(x)^T \left(\frac{\partial P(x)^T \mu}{\partial x} \right)^T \left(\dot{x} - \frac{\partial H(x, \mu)}{\partial \mu} \right)$$

$$+ (I - P(x)^T)\frac{\partial H(x, \mu)}{\partial \mu} \cdot \left(\frac{\partial P(x)^T \mu}{\partial x} \right) P(x)\frac{\partial H(x, \mu)}{\partial \mu}.$$

Since $P(x)P(x)^T = P(x)^T$, the first two terms of the above equation add to zero. Also, substituting (8.15), and rearranging with $P(x)P(x)^T = P(x)^T = P(x)$, the result reduces to

$$\frac{dH(x,\mu)}{dt} = P(x)\frac{\partial H(x,\mu)}{\partial \mu} \cdot \left(\frac{\partial P(x)^T \mu}{\partial x}\right)^T (P(x) - I)\frac{\partial H(x,\mu)}{\partial \mu}$$
$$+ (I - P(x)^T)\frac{\partial H(x,\mu)}{\partial \mu} \cdot \left(\frac{\partial P(x)^T \mu}{\partial x}\right) P(x)\frac{\partial H(x,\mu)}{\partial \mu}.$$

Rearranging the first term and using the symmetry of the projection yields

$$\frac{dH(x,\mu)}{dt} = 0.$$

Therefore, the Hamiltonian function is constant along each solution of Hamilton's equations. This property does not hold if the Hamiltonian function has a nontrivial explicit dependence on time.

8.2.6 Extension of the Hamiltonian Vector Field from T^*M to $\mathsf{T}^*\mathbb{R}^n$

Assume the Hamiltonian function can be defined everywhere on the embedding cotangent bundle $\mathsf{T}^*\mathbb{R}^n$ and assume the orthogonal projection operator $P(x)$ can be globally extended so that it is defined for each $x \in \mathbb{R}^n$. These extensions are made without changing the Hamiltonian on T^*M and without changing the orthogonal projection property on M.

The extended differential equations (8.15) and (8.16) can be viewed as defining a smooth Hamiltonian vector field $F : \mathsf{T}^*\mathbb{R}^n \to \mathsf{T}(\mathsf{T}^*\mathbb{R}^n)$. This globally defined vector field necessarily has an invariant manifold T^*M and the restriction of the global Hamiltonian vector field F^* on $\mathsf{T}^*\mathbb{R}^n$ is in fact the Hamiltonian vector field $F^*|_{\mathsf{T}^*M}$ on T^*M that is of primary interest. This vector field F^* defined on $\mathsf{T}^*\mathbb{R}^n$ can be viewed as an extension of the Hamiltonian vector field $F^*|_{\mathsf{T}^*M}$ on T^*M.

It should be emphasized that Hamilton's equations, under the stated assumptions, only guarantee the existence of a Hamiltonian vector field on the cotangent bundle T^*M. Additional assumptions that the Hamiltonian function and the constraint functions that define the configuration manifold are globally defined on $\mathsf{T}^*\mathbb{R}^n$ are required to guarantee that the Hamiltonian vector field on T^*M can be extended to a vector field on $\mathsf{T}^*\mathbb{R}^n$.

Global extension of the Hamiltonian vector field is possible for many physical systems, such as multi-body systems. The implications are important in that many classical analytical and computational methods can be utilized for the extended Hamiltonian vector field on $\mathsf{T}^*\mathbb{R}^n$, while maintaining the understanding that the only results of interest are for the Hamiltonian vector field restricted to T^*M. As a practical matter, even if a global extension to the entire embedding vector space is not possible, a local extension to a tubular neighborhood of the embedded configuration manifold can still be useful.

8.3 Lagrangian and Hamiltonian Dynamics on Products of Manifolds

In many multi-body systems, it is convenient to view the configuration manifold as a product of embedded manifolds. Let $M_1, \ldots, M_n$ denote differentiable manifolds and suppose that the configuration manifold is given by $M = M_1 \times \cdots \times M_n$. That is, the configuration vector is $x = (x_1, \ldots, x_n) \in M$ where $x_i \in M_i$, $i = 1, \ldots, n$. Suppose that the corresponding orthogonal projection operators are denoted by $P_i(x_i)$ for $x_i \in M_i$ with $i = 1, \ldots, n$. The Lagrangian function is given by $L : \mathsf{T}M \to \mathbb{R}^1$.

The development used to obtain Proposition 8.2 can be easily followed to obtain an Euler–Lagrange condition for extremals on a configuration manifold with this product structure.

Proposition 8.5 *Consider a differentiable manifold $M = M_1 \times \cdots \times M_n$ and a hyperregular Lagrangian $L : \mathsf{T}M \to \mathbb{R}^1$. Then, $(x, \dot{x}) : [t_0, t_f] \to \mathsf{T}M$ is an extremal curve of the action integral if and only if*

$$P_i(x_i)^T \left\{ \frac{d}{dt} \left(\frac{\partial L(x, \dot{x})}{\partial \dot{x}_i} \right) - \frac{\partial L(x, \dot{x})}{\partial x_i} \right\} = 0, \quad i = 1, \ldots, n. \tag{8.17}$$

Similarly, the development used to obtain Proposition 8.4 can be easily followed to obtain Hamilton's equations for extremal curves on a configuration manifold with this product structure. The Legendre transformation is

$$\mu_i = P_i(x_i)^T \frac{\partial L(x, \dot{x})}{\partial \dot{x}_i}, \quad i = 1, \ldots, n, \tag{8.18}$$

and the Hamiltonian function $H : \mathsf{T}^*M \to \mathbb{R}^1$ is

$$H(x, \mu) = \sum_{i=1}^{n} \mu_i \cdot \dot{x}_i - L(x, \dot{x}), \tag{8.19}$$

using the Legendre transformation (8.18).

Proposition 8.6 *Consider a differentiable manifold $M = M_1 \times \cdots \times M_n$ and a Hamiltonian $H : \mathsf{T}^*M \to \mathbb{R}^1$. Then, $(x, \mu) : [t_0, t_f] \to \mathsf{T}^*M$ is an extremal curve of the action integral if and only if*

$$\dot{x}_i = P_i(x_i)\frac{\partial H(x,\mu)}{\partial \mu_i}, \quad i = 1,\ldots,n, \tag{8.20}$$

$$\dot{\mu}_i = -P_i(x_i)^T\frac{\partial H(x,\mu)}{\partial x_i} + \left\{ P_i(x_i)^T\left(\frac{\partial P_i(x_i)^T\mu_i}{\partial x_i}\right)^T P_i(x_i) \right.$$

$$\left. + \left(\frac{\partial P_i(x_i)^T\mu_i}{\partial x_i}\right)P(x_i) - P_i(x_i)^T\left(\frac{\partial P_i(x_i)^T\mu_i}{\partial x_i}\right)^T \right\}\frac{\partial H(x,\mu)}{\partial \mu_i},$$

$$i = 1,\ldots,n. \tag{8.21}$$

These results are now used to obtain the Euler–Lagrange equations and Hamilton's equations for several categories of configuration manifolds with a product structure. These are not the most general possible results, but they clearly demonstrate the power of the prior abstract developments.

8.3.1 Lagrangian and Hamiltonian Dynamics on a Product of Linear Manifolds

Suppose that A_i, $i = 1,\ldots,n$ are full rank matrices with three columns and either one or two rows, and suppose b_i, $i = 1,\ldots,n$, are compatible vectors. Consider n linear manifolds, each embedded in $\mathbb{R}^3$, defined by

$$M_i = \left\{ x_i \in \mathbb{R}^3 : A_i x_i = b_i \right\}, \quad i = 1,\ldots,n.$$

The configuration manifold $M = M_1 \times \cdots \times M_n$ is a product of linear manifolds embedded in $\mathbb{R}^{3n}$.

The Lagrangian function $L : \mathsf{T}M \to \mathbb{R}^1$ is given by

$$L(x,\dot{x}) = \frac{1}{2}\sum_{i=1}^{n}\sum_{j=1}^{n} m_{ij}\dot{x}_i^T\dot{x}_j - U(x).$$

Here, $m_{ij} \in \mathbb{R}^1$, $i,j = 1,\ldots,n$, and viewed as an $n \times n$ array they form a symmetric, positive-definite matrix. The potential function $U : \mathbb{R}^{3n} \to \mathbb{R}^1$.

Since the tangent space of the linear manifold M_i can be identified with the null space $\mathcal{N}(A_i)$, the orthogonal projection matrices $P_i : \mathbb{R}^3 \to \mathcal{N}(A_i)$ are given by the 3×3 constant matrices

$$P_i = I_{3\times3} - A_i^T(A_iA_i^T)^{-1}A_i, \quad i = 1,\ldots,n.$$

Thus,

$$P_i(x)\frac{d}{dt}\left(\frac{\partial L(x,\dot{x})}{\partial \dot{x}_i}\right) = \left(I_{3\times3} - A_i^T(A_iA_i^T)^{-1}A_i\right)\sum_{j=1}^{n}m_{ij}\ddot{x}_j$$

$$= m_{ii}\ddot{x}_i + \left(I_{3\times3} - A_i^T(A_iA_i^T)^{-1}A_i\right)\sum_{j\neq i}^{n}m_{ij}\ddot{x}_j,$$

where we have used the fact that $A_i\ddot{x}_i = 0$, $i = 1,\ldots,n$. According to Proposition 8.4, the Euler–Lagrange equation (8.17) can be written as:

$$m_{ii}\ddot{x}_i + \left(I_{3\times3} - A_i^T(A_iA_i^T)^{-1}A_i,\right)\sum_{j\neq i}^{n}m_{ij}\ddot{x}_j$$

$$+ \left(I_{3\times3} - A_i^T(A_iA_i^T)^{-1}A_i,\right)\frac{\partial U(x)}{\partial x_i} = 0, \quad i = 1,\ldots,n. \qquad (8.22)$$

Equation (8.22) is defined only on the tangent bundle $\mathsf{T}M$. Since the potential function is viewed as being defined on $\mathbb{R}^{3n}$, equation (8.22) can be viewed as defining a Lagrangian vector field on $\mathsf{T}\mathbb{R}^{3n}$. In this sense, the Lagrangian vector field on $\mathsf{T}M$ is extended to a Lagrangian vector field on $\mathsf{T}\mathbb{R}^{3n}$.

The Legendre transformation $\mathbb{F}L : \mathsf{T}M \to \mathsf{T}M^*$, according to (8.18), is

$$\mu_i = m_{ii}\dot{x}_i + \left(I_{3\times3} - A_i^T(A_iA_i^T)^{-1}A_i\right)\sum_{j\neq i}^{n}m_{ij}\dot{x}_j, \quad i = 1,\ldots,n.$$

This can be inverted as

$$\dot{x}_i = \sum_{j=1}^{n}m_{ij}^I\mu_j, \quad i = 1,\ldots,n,$$

where the $3n \times 3n$ partitioned matrix

$$\begin{bmatrix} m_{11}^I & \cdots & m_{1n}^I \\ \vdots & \ddots & \vdots \\ m_{n1}^I & \cdots & m_{nn}^I \end{bmatrix}$$

$$= \begin{bmatrix} m_{11}I_{3\times3} & \cdots & m_{1n}(I_{3\times3} - A_1^T(A_1^TA_1)^{-1}A_1) \\ \vdots & \ddots & \vdots \\ m_{n1}(I_{3\times3} - A_n^T(A_n^TA_n)^{-1}A_n) & \cdots & m_{nn}I_{3\times3} \end{bmatrix}^{-1}.$$

The Hamiltonian function $H : \mathsf{T}M^* \to \mathbb{R}^1$, according to (8.19), is

$$H(x,\mu) = \frac{1}{2}\sum_{i=1}^{n}\sum_{j=1}^{n}\mu_i^T m_{ij}^I\mu_j + U(x).$$

Thus, Hamilton's equations, following equations (8.20) and (8.21), can be written as

$$\dot{x}_i = \sum_{j=1}^{n} m_{ij}^{I} \mu_j, \qquad\qquad i = 1, \ldots, n, \qquad (8.23)$$

$$\dot{\mu}_i = -\left(I_{3\times3} - A_i^T (A_i A_i^T)^{-1} A_i\right) \frac{\partial U(x)}{\partial x_i}, \quad i = 1, \ldots, n. \qquad (8.24)$$

Equations (8.23) and (8.24) are defined on the cotangent bundle T^*M. Since the potential function is defined on $\mathbb{R}^{3n}$, equations (8.23) and (8.24) can be viewed as defining a Hamiltonian vector field on $T^*\mathbb{R}^{3n}$. In this sense, the Hamiltonian vector field on T^*M is extended to a Hamiltonian vector field on $T^*\mathbb{R}^{3n}$.

8.3.2 Lagrangian and Hamiltonian Dynamics on $(S^1)^n$

The configuration manifold $M = S^1 \times \cdots \times S^1$, denoted by $M = (S^1)^n$, is embedded in $\mathbb{R}^{2n}$. The Lagrangian function $L : T(S^1)^n \to \mathbb{R}^1$ is given by

$$L(x, \dot{x}) = \frac{1}{2} \sum_{i=1}^{n} \sum_{j=1}^{n} m_{ij} \dot{x}_i^T \dot{x}_j - U(x).$$

Here, $m_{ij} \in \mathbb{R}^1$, $i, j = 1, \ldots, n$, and when viewed as an $n \times n$ array form a symmetric, positive-definite matrix. The potential function $U : \mathbb{R}^{2n} \to \mathbb{R}^1$.

The orthogonal projections are given by the 2×2 matrix functions

$$P_i(x_i) = I_{2\times2} - x_i x_i^T, \quad i = 1, \ldots, n.$$

Differentiate $x_i^T x_i = 1$ twice to obtain $x_i^T \ddot{x}_i = -\|\dot{x}_i\|^2$; then we obtain

$$P_i(x) \frac{d}{dt}\left(\frac{\partial L(x, \dot{x})}{\partial \dot{x}_i}\right) = \left(I_{2\times2} - x_i x_i^T\right) \sum_{j=1}^{n} m_{ij} \ddot{x}_j$$

$$= m_{ii}\left(\ddot{x}_i - x_i x_i^T \ddot{x}_i\right) + \left(I_{2\times2} - x_i x_i^T\right) \sum_{j\neq i}^{n} m_{ij} \ddot{x}_j$$

$$= m_{ii}\left(\ddot{x}_i + \|\dot{x}_i\|^2 x_i\right) + \left(I_{2\times2} - x_i x_i^T\right) \sum_{j\neq i}^{n} m_{ij} \ddot{x}_j.$$

According to Proposition 8.4, the Euler–Lagrange equation (8.17) can be shown to be:

$$m_{ii}\ddot{x}_i + \left(I_{2\times 2} - x_i x_i^T\right) \sum_{j \neq i}^{n} m_{ij}\ddot{x}_j$$

$$+ m_{ii}\left\|\dot{x}_i\right\|^2 x_i + \left(I_{2\times 2} - x_i x_i^T\right) \frac{\partial U(x)}{\partial x_i} = 0, \quad i = 1, \ldots, n. \quad (8.25)$$

Equation (8.25) is defined on the tangent bundle $\mathsf{T}(\mathsf{S}^1)^n$. Since the potential function is viewed as being defined on $\mathbb{R}^{2n}$, equation (8.25) can be viewed as defining a Lagrangian vector field on $\mathsf{T}\mathbb{R}^{2n}$. In this sense, the Lagrangian vector field on $\mathsf{T}(\mathsf{S}^1)^n$ is extended to a Lagrangian vector field on $\mathsf{T}\mathbb{R}^{2n}$. Equation (8.25) is a special case of the Euler–Lagrange equation (4.9) derived in Chapter 4.

Using (8.18), the Legendre transformation $\mathbb{F}L : \mathsf{T}M \to \mathsf{T}M^*$ is introduced to define the conjugate momenta

$$\mu_i = m_{ii}\dot{x}_i + \left(I_{2\times 2} - x_i x_i^T\right) \sum_{j \neq i}^{n} m_{ij}\dot{x}_j, \quad i = 1, \ldots, n.$$

This can be inverted as

$$\dot{x}_i = \sum_{j=1}^{n} m_{ij}^I \mu_j, \quad i = 1, \ldots, n,$$

where the $2n \times 2n$ partitioned matrix function

$$\begin{bmatrix} m_{11}^I & \cdots & m_{1n}^I \\ \vdots & \ddots & \vdots \\ m_{n1}^I & \cdots & m_{nn}^I \end{bmatrix} = \begin{bmatrix} m_{11}I_{2\times 2} & \cdots & m_{1n}(I_{2\times 2} - x_1 x_1^T) \\ \vdots & \ddots & \vdots \\ m_{n1}(I_{2\times 2} - x_n x_n^T) & \cdots & m_{nn}I_{2\times 2} \end{bmatrix}^{-1}.$$

The Hamiltonian function $H : \mathsf{T}^*(\mathsf{S}^1)^n \to \mathbb{R}^1$, according to (8.19), is

$$H(x, \mu) = \frac{1}{2} \sum_{i=1}^{n} \sum_{j=1}^{n} \mu_i^T m_{ij}^I \mu_j + U(x).$$

We now simplify (8.21) using the orthogonal projection. The derivative of the projection operator term in (8.21) is

$$\left(\frac{\partial P^T(x_i)\mu_i}{\partial x_i}\right) = \frac{\partial (I_{2\times 2} - x_i x_i^T)\mu_i}{\partial x_i}$$

$$= -(x_i^T \mu_i)I_{2\times 2} - x_i \mu_i^T$$

$$= -x_i \mu_i^T,$$

where we have used the fact that $x_i^T \mu_i = 0$. Substituting into the expression in the braces of (8.21), we obtain

$$P^T(x_i) \left(\frac{\partial P^T(x_i)\mu_i}{\partial x_i} \right)^T P(x_i) + \left(\frac{\partial P^T(x_i)\mu_i}{\partial x_i} \right)^T P(x_i)$$

$$- P^T(x_i) \left(\frac{\partial P^T(x_i)\mu_i}{\partial x_i} \right)^T$$

$$= -(I_{3\times 3} - x_i x_i^T)\mu_i x_i^T (I_{3\times 3} - x_i x_i^T)$$

$$- x_i \mu_i^T (I_{2\times 2} - x_i x_i^T) + (I_{3\times 3} - x_i x_i^T)\mu_i x_i^T$$

$$= -\mu_i x_i^T + \mu_i x_i^T x_i x_i^T - x_i \mu_i^T + \mu_i x_i^T$$

$$= \mu_i x_i^T - x_i \mu_i^T,$$

where we have used the facts that $\mu_i^T x_i = 0$ and $x_i^T x_i = 1$. Thus, Hamilton's equations, following equations (8.20) and (8.21), can be written as

$$\dot{x}_i = \sum_{j=1}^{n} m_{ij}^I \mu_j, \quad i = 1, \ldots, n, \tag{8.26}$$

$$\dot{\mu}_i = (\mu_i x_i^T - x_i \mu_i^T) \sum_{j=1}^{n} m_{ij}^I \mu_j - (I_{2\times 2} - x_i x_i^T)) \frac{\partial U(x)}{\partial x_i}, \quad i = 1, \ldots, n. \tag{8.27}$$

Equations (8.26) and (8.27) are defined on the cotangent bundle $T^*(S^1)^n$. Since the potential function is defined on $\mathbb{R}^{2n}$, equations (8.26) and (8.27) can be viewed as defining a Hamiltonian vector field on $T^*\mathbb{R}^{2n}$. In this sense, the Hamiltonian vector field on $T^*(S^1)^n$ is extended to a Hamiltonian vector field on $T^*\mathbb{R}^{2n}$. Hamilton's equations (8.26) and (8.27) are a special case of Hamilton's equations (4.21) and (4.22) derived in Chapter 4.

8.3.3 Lagrangian and Hamiltonian Dynamics on $(S^2)^n$

The configuration manifold $M = S^2 \times \cdots \times S^2$, denoted by $M = (S^2)^n$, is embedded in $\mathbb{R}^{3n}$. The Lagrangian function $L : T(S^2)^n \to \mathbb{R}^1$ is given by

$$L(x, \dot{x}) = \frac{1}{2} \sum_{i=1}^{n} \sum_{j=1}^{n} m_{ij} \dot{x}_i^T \dot{x}_j - U(x).$$

Here, $m_{ij} \in \mathbb{R}^1$, $i, j = 1, \ldots, n$, and viewed as an $n \times n$ array form a symmetric, positive-definite matrix. The potential function $U : \mathbb{R}^{3n} \to \mathbb{R}^1$.

The orthogonal projections are given by the 3×3 matrix functions

$$P_i(x_i) = I_{3\times 3} - x_i x_i^T, \quad i = 1, \ldots, n.$$

Differentiate $x_i^T x_i = 1$ twice to obtain $x_i^T \ddot{x}_i = - \|\dot{x}_i\|^2$; then we obtain

$$P_i(x)\frac{d}{dt}\left(\frac{\partial L(x,\dot{x})}{\partial \dot{x}_i}\right) = \left(I_{3\times3} - x_i x_i^T\right)\sum_{j=1}^{n} m_{ij}\ddot{x}_j$$

$$= m_{ii}\left(\ddot{x}_i - x_i x_i^T \ddot{x}_i\right) + \left(I_{3\times3} - x_i x_i^T\right)\sum_{j\neq i}^{n} m_{ij}\ddot{x}_j$$

$$= m_{ii}\left(\ddot{x}_i + \|\dot{x}_i\|^2 x_i\right) + \left(I_{3\times3} - x_i x_i^T\right)\sum_{j\neq i}^{n} m_{ij}\ddot{x}_j.$$

According to Proposition 8.4, the Euler–Lagrange equation (8.17) can be shown to be:

$$m_{ii}\ddot{x}_i + \left(I_{3\times3} - x_i x_i^T\right)\sum_{j\neq i}^{n} m_{ij}\ddot{x}_j$$

$$+ m_{ii}\|\dot{x}_i\|^2 x_i + \left(I_{3\times3} - x_i x_i^T\right)\frac{\partial U(x)}{\partial x_i} = 0, \quad i = 1,\dots,n. \quad (8.28)$$

Equation (8.28) is defined on the tangent bundle $\mathsf{T}(\mathsf{S}^2)^n$. Since the potential function is viewed as being defined on $\mathbb{R}^{3n}$, equation (8.28) can be viewed as defining a Lagrangian vector field on $\mathsf{T}\mathbb{R}^{3n}$. In this sense, the Lagrangian vector field on $\mathsf{T}(\mathsf{S}^2)^n$ is extended to a Lagrangian vector field on $\mathsf{T}\mathbb{R}^{3n}$. Equation (8.28) is a special case of the Euler–Lagrange equation (5.10) derived in Chapter 5.

According to (8.18), the Legendre transformation $\mathbb{FL} : \mathsf{T}M \to \mathsf{T}M^*$ is introduced to define the conjugate momenta

$$\mu_i = m_{ii}\dot{x}_i + \left(I_{3\times3} - x_i x_i^T\right)\sum_{j\neq i}^{n} m_{ij}\dot{x}_j, \quad i = 1,\dots,n.$$

This can be inverted as

$$\dot{x}_i = \sum_{j=1}^{n} m_{ij}^I \mu_j, \quad i = 1,\dots,n,$$

where the $3n \times 3n$ partitioned matrix function

$$\begin{bmatrix} m_{11}^I & \cdots & m_{1n}^I \\ \vdots & \ddots & \vdots \\ m_{n1}^I & \cdots & m_{nn}^I \end{bmatrix} = \begin{bmatrix} m_{11}I_{3\times3} & \cdots & m_{1n}(I_{3\times3} - x_1 x_1^T) \\ \vdots & \ddots & \vdots \\ m_{n1}(I_{3\times3} - x_n x_n^T) & \cdots & m_{nn}I_{3\times3} \end{bmatrix}^{-1}.$$

Using (8.19), the Hamiltonian function $H : \mathsf{T}^*(\mathsf{S}^2)^n \to \mathbb{R}^1$ is

$$H(x,\mu) = \frac{1}{2}\sum_{i=1}^{n}\sum_{j=1}^{n} \mu_i^T m_{ij}^I \mu_j + U(x).$$

Equation (8.21) is simplified using the orthogonal projection. The derivative of the projection operator term in (8.21) is

$$
\left(\frac{\partial P^T(x_i)\mu_i}{\partial x_i} \right) = \frac{\partial (I_{3\times3} - x_i x_i^T)\mu_i}{\partial x_i}
$$
$$
= -(x_i^T \mu_i) I_{3\times3} - x_i \mu_i^T
$$
$$
= -x_i \mu_i^T,
$$

where we have used the fact that $x_i^T \mu_i = 0$. Substituting this into the expression in the braces of (8.21), we obtain

$$
P^T(x_i) \left(\frac{\partial P^T(x_i)\mu_i}{\partial x_i} \right)^T P(x_i) + \left(\frac{\partial P^T(x_i)\mu_i}{\partial x_i} \right) P(x_i)
$$
$$
- P^T(x_i) \left(\frac{\partial P^T(x_i)\mu_i}{\partial x_i} \right)^T
$$
$$
= -(I_{3\times3} - x_i x_i^T)\mu_i x_i^T (I_{3\times3} - x_i x_i^T)
$$
$$
- x_i \mu_i^T (I_{3\times3} - x_i x_i^T) + (I_{3\times3} - x_i x_i^T)\mu_i x_i^T
$$
$$
= -\mu_i x_i^T + \mu_i x_i^T x_i x_i^T - x_i \mu_i^T + \mu_i x_i^T
$$
$$
= \mu_i x_i^T - x_i \mu_i^T,
$$

where we have used the facts that $\mu_i^T x_i = 0$ and $x_i^T x_i = 1$. Thus, we can write

$$
\left\{ P_i(x_i)^T \left(\frac{\partial P_i(x_i)^T \mu_i}{\partial x_i} \right)^T P_i(x_i) \right.
$$
$$
\left. + \left(\frac{\partial P_i(x_i)^T \mu_i}{\partial x_i} \right) P(x_i) - P_i(x_i)^T \left(\frac{\partial P_i(x_i)^T \mu_i}{\partial x_i} \right)^T \right\} \frac{\partial H(x,\mu)}{\partial \mu_i}
$$
$$
= \left(\mu_i x_i^T - x_i \mu_i^T \right) \frac{\partial H(x,\mu)}{\partial \mu_i},
$$
$$
= \frac{\partial H(x,\mu)}{\partial \mu_i} \times (\mu_i \times x_i).
$$

Hamilton's equations, following equations (8.20) and (8.21), can be written as

$$
\dot{x}_i = \sum_{j=1}^{n} m_{ij}^I \mu_j, \qquad\qquad\qquad i = 1,\dots,n,
$$

$$(8.29)$$

$$
\dot{\mu}_i = \left(\sum_{j=1}^{n} m_{ij}^I \mu_j \right) \times (\mu_i \times x_i) - (I_{3\times3} - x_i x_i^T) \frac{\partial U(x)}{\partial x_i}, \quad i = 1,\dots,n.
$$

$$(8.30)$$

Equations (8.29) and (8.30) are defined on the cotangent bundle $\mathsf{T}^*(\mathsf{S}^2)^n$. Since the potential function is defined on $\mathbb{R}^{3n}$, equations (8.29) and (8.30) can be viewed as defining a Hamiltonian vector field on $\mathsf{T}^*\mathbb{R}^{3n}$. In this sense, the Hamiltonian vector field on $\mathsf{T}^*(\mathsf{S}^2)^n$ is extended to a Hamiltonian vector field on $\mathsf{T}^*\mathbb{R}^{3n}$. Hamilton's equations (8.29) and (8.30) are a special case of Hamilton's equations (5.25) and (5.26) derived in Chapter 5.

8.4 Lagrangian and Hamiltonian Dynamics Using Lagrange Multipliers

We revisit the classical Euler–Lagrange equations and Hamilton's equations, with holonomic constraints, that are expressed using Lagrange multipliers [30, 32]. We show that Proposition 8.2 in the prior section can be used to obtain these classical results.

Recall, the configuration manifold M embedded in $\mathbb{R}^n$ can be described by algebraic equations expressed in terms of the configuration as

$$f_i(x) = 0, \quad i = 1, \ldots, m, \tag{8.31}$$

where $f_i : \mathbb{R}^n \to \mathbb{R}^1$, $i = 1, \ldots, m$, are continuously differentiable functions with linearly independent gradient functions for all $x \in M$.

According to Proposition 8.2, equation (8.3) can be rewritten as

$$\frac{d}{dt}\left(\frac{\partial L(x,\dot{x})}{\partial \dot{x}}\right) - \frac{\partial L(x,\dot{x})}{\partial x} + \sum_{i=1}^{m} \lambda_i \frac{\partial f_i(x)}{\partial x} = 0. \tag{8.32}$$

The Euler–Lagrange differential equations (8.32) and the algebraic equations (8.31) have been shown to have index two and thus define a continuous vector field on the tangent bundle of the constraint manifold $\mathsf{T}M$.

As in Chapter 3, these equations can be expressed in terms of an augmented Lagrangian function $L^a : \mathsf{T}^*M \times \mathbb{R}^m \to \mathbb{R}^1$ given by

$$L^a(x,\dot{x},\lambda) = L(x,\dot{x}) + \sum_{i=1}^{m} \lambda_i f_i(x).$$

The Euler–Lagrange equations can be expressed as

$$\frac{d}{dt}\left(\frac{\partial L^a(x,\dot{x},\lambda)}{\partial \dot{x}}\right) - \frac{\partial L^a(x,\dot{x},\lambda)}{\partial x} = 0. \tag{8.33}$$

The augmented Euler–Lagrange equations (8.33), together with the m algebraic constraint equations given in (8.31), can be shown to be index two differential-algebraic equations. They guarantee that the constrained Lagrangian dynamics described by $(x, \dot{x}) \in \mathsf{T}M$ evolve on the tangent bundle of the constraint manifold $\mathsf{T}M$.

If we use the classical definition of the Legendre transformation

$$\mu = \frac{\partial L(x, \dot{x})}{\partial \dot{x}},$$

and the classical definition of the Hamiltonian function

$$H(x, \mu) = \mu \cdot \dot{x} - L(x, \dot{x}),$$

then Hamilton's equations can be described by differential-algebraic equations that depend on Lagrange multipliers as

$$\dot{x} = \frac{\partial H(x, \mu)}{\partial \mu}, \tag{8.34}$$

$$\dot{\mu} = -\frac{\partial H(x, \mu)}{\partial x} - \sum_{i=1}^{m} \lambda_i \frac{\partial f_i(x)}{\partial x}, \tag{8.35}$$

together with the algebraic equations (8.31). These differential-algebraic equations can be shown to have index two and thus define a smooth vector field on the cotangent bundle of the constraint manifold T^*M.

As in Chapter 3 these equations can be expressed in terms of the augmented Hamiltonian function $H^a : \mathsf{T}^*M \times \mathbb{R}^m \to \mathbb{R}^1$ as

$$H^a(x, \mu, \lambda) = \mu \cdot \dot{x} - L^a(x, \dot{x}, \lambda),$$

so that we obtain Hamilton's equations

$$\dot{x} = \frac{\partial H^a(x, \mu, \lambda)}{\partial \mu}, \tag{8.36}$$

$$\dot{\mu} = -\frac{\partial H^a(x, \mu, \lambda)}{\partial x}. \tag{8.37}$$

Hamilton's equations (8.36) and (8.37), together with the m algebraic constraint equations (8.31), are also index two differential-algebraic equations that guarantee the constrained Hamiltonian dynamics evolve on the cotangent bundle of the constraint manifold T^*M.

These classical variational results, expressed in terms of Lagrange multipliers, have been previously introduced in Chapter 3. Here, we derived these classical variational results from the abstract variational results earlier in this chapter.

8.5 Lagrangian and Hamiltonian Dynamics on SO(3)

In the preceding sections of this chapter, the configuration manifold has been assumed to be embedded in $\mathbb{R}^n$ and we have derived Lagrangian and Hamiltonian equations of motion using the orthogonal projection operator. These same ideas can be applied to a configuration manifold that is viewed as an embedding in a vector space of matrices. This is best illustrated in the case where the configuration manifold is SO(3).

Instead of viewing SO(3) as embedded in $\mathbb{R}^9$ as in Proposition 8.2, we assume that the embedding space of SO(3) is $\mathbb{R}^{3\times 3}$. This avoids transforming a 3×3 attitude or rotation matrix $R \in$ SO(3) into the corresponding 9×1 vector; the resulting Euler–Lagrange equation is written in terms of the matrix R directly.

Suppose the configuration manifold is the special orthogonal group SO(3) $= \{R \in \mathbb{R}^{3\times 3} : R^T R = I_{3\times 3}, \det[R] = 1\}$. Recall the tangent space at R is given by T_RSO(3) $= \{V \in \mathbb{R}^{3\times 3} : R^T V + V^T R = 0\}$, and for any $V \in \mathsf{T}_R$SO(3), there exists $v \in \mathbb{R}^3$ such that $V = RS(v)$.

We recall the definitions of the inner product and the projection operator on $\mathbb{R}^{3\times 3}$ as follows. The inner-product is defined as an element-wise operation. More explicitly, for $V, W \in \mathsf{T}_R$SO(3),

$$V \cdot W = \operatorname{tr}\left[V^T W\right] = \sum_{i=1}^{3}\sum_{j=1}^{3} V_{ij} W_{ij}.$$

The projection operator acts on $Y \in \mathbb{R}^{3\times 3}$ at T_RSO(3). That is, $P(R, Y)$: SO(3) $\times \mathbb{R}^{3\times 3} \to \mathsf{T}_R$SO(3) can be written as

$$P(R, Y) = \frac{1}{2}R(R^T Y - Y^T R) = \frac{1}{2}(Y - RY^T R). \tag{8.38}$$

The projection is orthogonal since

$$(Y - P(R,Y)) \cdot RS(z) = \frac{1}{2}\operatorname{tr}\left[(Y + RY^T R)S(z)R^T\right]$$

$$= \frac{1}{2}\operatorname{tr}\left[(R^T Y + Y^T R)S(z)\right]$$

$$= 0,$$

for any $z \in \mathbb{R}^3$, where we have used the facts that the trace is invariant under transpose and cyclic permutation, and the trace of the product of any symmetric matrix and any compatible skew-symmetric matrix is zero. The above equation states that the difference between Y and its projection $P(R, Y)$ is normal to the tangent space T_RSO(3), and hence, it implies orthogonality of the projection. The projection is also symmetric since for any $Z \in \mathbb{R}^{3\times 3}$,

$$Z \cdot P(R, Y) = \frac{1}{2}\text{tr}\left[(Y - RY^T R)Z^T\right]$$
$$= \frac{1}{2}\text{tr}\left[Y^T (Z - RZ^T R)\right]$$
$$= Y \cdot P(R, Z).$$

To illustrate this, consider the attitude dynamics of a freely rotating rigid body described in Section 6.6.1. The Lagrangian is given by

$$\tilde{L}(R, \omega) = \frac{1}{2}\omega^T J \omega,$$

where $J \in \mathbb{R}^{3 \times 3}$ is the standard inertia matrix of the rigid body, and $\omega = S^{-1}(R^T \dot{R}) \in \mathbb{R}^3$ corresponds to the angular velocity vector. To rewrite the Lagrangian in terms of $(R, \dot{R})$, define a nonstandard inertia matrix $J_d = \frac{1}{2}\text{tr}[J]\, I_{3\times 3} - J$ as in [50]. The Lagrangian of the rigid body corresponding to the rotational kinetic energy can be rewritten in terms of $(R, \dot{R})$ as

$$L(R, \dot{R}) = \frac{1}{2}\text{tr}\left[\dot{R} J_d \dot{R}^T\right] = \frac{1}{2}\dot{R} \cdot \dot{R} J_d.$$

Thus,

$$\frac{\partial L}{\partial \dot{R}} = \dot{R} J_d.$$

From (8.3), and (8.38), we obtain

$$\frac{1}{2}(\ddot{R} J_d - R J_d \ddot{R}^T R) = 0,$$

which can be rearranged into

$$R^T \ddot{R} J_d - J_d \ddot{R}^T R = 0. \tag{8.39}$$

Next, we show that this is equivalent to (6.16) when it is written in terms of the standard inertia matrix J and the angular velocity vector. Since $\dot{R} = RS(\omega)$, we have $\ddot{R} = RS(\omega)^2 + RS(\dot{\omega})$. Therefore, $R^T \ddot{R} = S(\omega)^2 + S(\dot{\omega})$. Substituting and rearranging, we obtain

$$S(\dot{\omega})J_d + J_d S(\dot{\omega}) + S(\omega)^2 J_d - J_d S(\omega)^2 = 0.$$

Use the identities of the skew-symmetric matrix function $S(\cdot)$; we have $S(\dot{\omega})J_d + J_d S(\dot{\omega}) = S((\text{tr}[J_d]\, I_{3\times 3} - J_d)\dot{\omega}) = S(J\dot{\omega})$, and $S(\omega)^2 J_d - J_d S(\omega)^2 = S(J_d\omega)S(\omega) - S(\omega)S(J_d\omega) = -S(J\omega)S(\omega) + S(\omega)S(J\omega) = S(\omega \times J\omega)$. Therefore, the above equation reduces to

$$J\dot{\omega} + \omega \times J\omega = 0,$$

which is equivalent to (6.16), that was derived in Section 6.6.1; this describes the Lagrangian flow on $\mathsf{TSO}(3)$. This example illustrates that Proposition 8.2 can also be applied to $\mathsf{SO}(3)$, viewing the embedding space as $\mathbb{R}^{3 \times 3}$.

Hamilton's equations for the Hamiltonian flow on $\mathsf{T}^*\mathsf{SO}(3)$ can be easily obtained using the Legendre transformation and the definition of the Hamiltonian function. The details are omitted here.

This development provides a natural transition to the general case that the configuration manifold is a matrix Lie group. This is the topic of the next section.

8.6 Lagrangian and Hamiltonian Dynamics on a Lie Group

We now present an abstract geometric formulation for Lagrangian and Hamiltonian systems where the configuration manifold is a Lie group G embedded in $\mathbb{R}^{n \times n}$. The results in the prior sections can be utilized since a Lie group is a manifold; however, we provide a complete development, following the same line of arguments, to emphasize the special structure that arises from the Lie group assumption. See [22] for a different treatment of variational methods on Lie groups.

The action integral is defined in terms of a Lagrangian, and Hamilton's principle is used to derive an abstract version of the Euler–Lagrange equations on the Lie group G using variational calculus. The Legendre transformation is introduced, Hamilton's function is defined, and Hamilton's equations are derived.

8.6.1 Additional Material on Lie Groups and Lie Algebras

A Lie group is a group that is also a manifold, where the group operations are smooth maps with respect to the manifold structure. Thus, the elements in a Lie group satisfy both group properties and manifold properties. A Lie group, as a manifold, supports the definition of tangent bundle and cotangent bundle, but they have additional properties in this case. An element in the Lie group and an element of its tangent space, viewed as a pair, lie in the tangent bundle.

Let G denote a Lie group, for which the group operation is differentiable. A Lie algebra $\mathfrak{g}$ is a vector space associated with this Lie group. This Lie algebra $\mathfrak{g}$ is the tangent space of the Lie group G at the identity element $e \in \mathsf{G}$, with a *Lie bracket* $[\cdot, \cdot] : \mathfrak{g} \times \mathfrak{g} \to \mathfrak{g}$ that is bilinear, skew symmetric, and satisfies the Jacobi identity.

For $g, h \in G$, the *left translation map* $L_h : G \to G$ is defined as $L_h g = hg$. Similarly, the *right translation map* $R_h : G \to G$ is defined as $R_h g = gh$. Given $\xi \in \mathfrak{g}$, define a vector field $X_\xi : G \to TG$ such that $X_\xi(g) = T_e L_g \cdot \xi$, and let the corresponding unique integral curve passing through the identity e at $t = 0$ be denoted by $\gamma_\xi(t)$.

We can represent an element in the Lie group G in terms of the exponential map of an element in the associated Lie algebra $\mathfrak{g}$. The *exponential map* is denoted by $\exp : \mathfrak{g} \to G$ or by the common notation $e^\xi = \gamma_\xi(1)$. The exponential map is a local diffeomorphism from a neighborhood of zero in $\mathfrak{g}$ onto a neighborhood of the identity e in G.

Define the *inner automorphism* $i_g : G \to G$ as $i_g(h) = ghg^{-1}$. The *adjoint operator* $\mathrm{Ad}_g : \mathfrak{g} \to \mathfrak{g}$ is the differential of $i_g(h)$ with respect to h at $h = e$ along the direction $\eta \in \mathfrak{g}$, i.e., $\mathrm{Ad}_g \eta = T_e i_g \cdot \eta$. The *adjoint operator* $\mathrm{ad}_\xi : \mathfrak{g} \to \mathfrak{g}$ is obtained by differentiating $\mathrm{Ad}_g \eta$ with respect to g at e in the direction ξ, i.e., $\mathrm{ad}_\xi \eta = T_e(\mathrm{Ad}_g \eta) \cdot \xi$. In terms of the Lie bracket, $\mathrm{ad}_\xi \eta = [\xi, \eta]$.

Let $\langle \cdot, \cdot \rangle$ be the pairing between a tangent vector and a cotangent vector. The *coadjoint operator* $\mathrm{Ad}_g^* : G \times \mathfrak{g}^* \to \mathfrak{g}^*$ is defined by $\langle \mathrm{Ad}_g^* \alpha, \xi \rangle = \langle \alpha, \mathrm{Ad}_g \xi \rangle$ for $\alpha \in \mathfrak{g}^*$. The *coadjoint operator* $\mathrm{ad}^* : \mathfrak{g} \times \mathfrak{g}^* \to \mathfrak{g}^*$ is defined by $\langle \mathrm{ad}_\eta^* \alpha, \eta \rangle = \langle \alpha, \mathrm{ad}_\eta \xi \rangle$ for $\alpha \in \mathfrak{g}^*$.

As we will see below, one important consequence of the structure of a Lie group is we can identify the tangent bundle TG of a Lie group G with $G \times \mathfrak{g}$ by left trivialization. Therefore, by choosing a basis for the Lie algebra $\mathfrak{g}$, we can explicitly parameterize the tangent space at any point of the Lie group and thereby avoid the need for projections.

Accessible expanded introductions to Lie groups can be found in Chapter 9 of [70] or Chapter 1 of [78], and more in-depth expositions of the mathematical foundations can be found in [28, 35, 97].

8.6.2 Variations on a Lie Group

The configuration manifold is a Lie group G. We identify the tangent bundle TG with $G \times \mathfrak{g}$ by left trivialization. A tangent vector $(g, \dot{g}) \in T_g G$ is expressed as

$$\dot{g} = T_e L_g \cdot \xi = g\xi, \tag{8.40}$$

for $\xi \in \mathfrak{g}$. These can be thought of as the kinematics equation that relates the *velocity* variable $\xi \in \mathfrak{g}$ in the Lie algebra to the configuration variables $g \in G$ in the Lie group. Thus, any differentiable motion $g : [t_0, t_f] \to G$ with values in the Lie group G necessarily satisfies the kinematics equation (8.40) for some function $\xi : [t_0, t_f] \to \mathfrak{g}$ with values in the Lie algebra $\mathfrak{g}$.

The subsequent development describes variations of functions with values in the Lie group G. Let $g : [t_0, t_f] \to G$ be a differentiable curve. The family of variations of g is defined by a differentiable mapping $g^\epsilon : (-c, c) \times [t_0, t_f] \to \mathsf{G}$, for $c > 0$, defined using the exponential map as

$$g^\epsilon(t) = g e^{\epsilon \eta(t)},$$

for a curve $\eta : [t_0, t_f] \to \mathfrak{g}$. It is easy to show that the family of curves g^ϵ is well defined for some constant c as the exponential map is a local diffeomorphism between $\mathfrak{g}$ and G. Furthermore, the family of curves satisfies $g^\epsilon(t_0) = g(t_0)$, $g^\epsilon(t_f) = g(t_f)$ for any $\epsilon \in (-c, c)$, provided $\eta(t_0) = \eta(t_f) = 0$. It is also guaranteed that the family of varied curves have values in G.

The infinitesimal variation of g is defined by

$$\delta g(t) = \frac{d}{d\epsilon} g^\epsilon(t) \Big|_{\epsilon=0} = \mathsf{T}_e \mathsf{L}_{g(t)} \cdot \frac{d}{d\epsilon} e^{\epsilon \eta(t)} \Big|_{\epsilon=0}$$
$$= g(t) \eta(t). \tag{8.41}$$

For each $t \in [t_0, t_f]$, the infinitesimal variation $\delta g(t)$ lies in the tangent space $\mathsf{T}_{g(t)} \mathsf{G}$. Using this expression, the family of variations of $\xi(t) \in \mathfrak{g}$ is

$$\xi^\epsilon(t) = \xi(t) + \epsilon \delta \xi(t) + \mathcal{O}(\epsilon^2),$$

where the infinitesimal variation of $\xi : [t_0, t_f] \to \mathfrak{g}$ is

$$\delta \xi(t) = \dot{\eta}(t) + \mathrm{ad}_{\xi(t)} \eta(t). \tag{8.42}$$

Equations (8.41) and (8.42) define the variations of $(g(t), \xi(t)) : [t_0, t_f] \to \mathsf{G} \times \mathfrak{g}$.

8.6.3 Euler–Lagrange Equations

The Lagrangian function $L : \mathsf{G} \times \mathfrak{g} \to \mathbb{R}^1$ is the difference of the kinetic energy $T : \mathsf{G} \times \mathfrak{g} \to \mathbb{R}^1$ and the potential energy $U : \mathsf{G} \to \mathbb{R}^1$:

$$L(g, \xi) = T(g, \xi) - U(g).$$

Define the action integral along a motion that evolves on the Lie group configurations G as

$$\mathfrak{G} = \int_{t_0}^{t_f} L(g, \xi) \, dt.$$

The action integral along a variation of the motion is

$$\mathfrak{G}^\epsilon = \int_{t_0}^{t_f} L(g^\epsilon, \xi^\epsilon) \, dt.$$

The varied value of the action integral can be expressed as a power series in ϵ as

$$\mathfrak{G}^\epsilon = \mathfrak{G} + \epsilon \delta \mathfrak{G} + \mathcal{O}(\epsilon^2),$$

where the infinitesimal variation of the action integral is

$$\delta \mathfrak{G} = \frac{d}{d\epsilon} \mathfrak{G}^\epsilon \bigg|_{\epsilon=0}.$$

Hamilton's principle states that the infinitesimal variation of the action integral along any motion is zero, i.e.,

$$\delta \mathfrak{G} = \frac{d}{d\epsilon} \mathfrak{G}^\epsilon \bigg|_{\epsilon=0} = 0, \tag{8.43}$$

for all possible differentiable infinitesimal variations in $\mathfrak{g}$, i.e., for all possible differentiable infinitesimal variations $\eta : [t_0, t_f] \to \mathfrak{g}$ satisfying $\eta(t_0) = \eta(t_f) = 0$.

Since variation and integration commute, the variation of the action integral can be written as

$$\delta \mathfrak{G} = \frac{d}{d\epsilon} \mathfrak{G}^\epsilon \bigg|_{\epsilon=0} = \int_{t_0}^{t_f} \left\{ \left(\frac{\partial L(g,\xi)}{\partial g} \cdot \delta g \right) + \left(\frac{\partial L(g,\xi)}{\partial \xi} \cdot \delta \xi \right) \right\} dt,$$

where $\frac{\partial L(g,\xi)}{\partial g} \in \mathsf{T}^*\mathsf{G}$ denotes the derivative of the Lagrangian with respect to g, given by

$$\frac{d}{d\epsilon} L(g^\epsilon, \xi) \bigg|_{\epsilon=0} = \left(\frac{\partial L(g,\xi)}{\partial g} \cdot \delta g \right),$$

and $\frac{\partial L(g,\xi)}{\partial \xi} \in \mathfrak{g}^*$ denotes the derivative of the Lagrangian with respect to ξ, given by

$$\frac{d}{d\epsilon} L(g, \xi^\epsilon) \bigg|_{\epsilon=0} = \left(\frac{\partial L(g,\xi)}{\partial \xi} \cdot \delta \xi \right).$$

Since $\mathsf{T}(\mathsf{L}_g \circ \mathsf{L}_{g^{-1}}) = \mathsf{TL}_g \circ \mathsf{TL}_{g^{-1}}$ is equal to the identity map on TG, this can be written as

$$\delta \mathfrak{G} = \int_{t_0}^{t_f} \left\{ \left\langle \frac{\partial L(g,\xi)}{\partial g}, \delta g \right\rangle + \left\langle \frac{\partial L(g,\xi)}{\partial \xi}, \delta \xi \right\rangle \right\} dt$$

$$= \int_{t_0}^{t_f} \left\{ \left\langle \frac{\partial L(g,\xi)}{\partial g}, (\mathsf{T}_e \mathsf{L}_g \circ \mathsf{T}_g \mathsf{L}_{g^{-1}}) \cdot \delta g \right\rangle + \left\langle \frac{\partial L(g,\xi)}{\partial \xi}, \delta \xi \right\rangle \right\} dt.$$

Substituting (8.41) and (8.42), we obtain

$$\delta\mathfrak{G} = \int_{t_0}^{t_f} \left\{ \left\langle \frac{\partial L(g,\xi)}{\partial g}, \ \mathsf{T}_e\mathsf{L}_g \cdot \eta \right\rangle + \left\langle \frac{\partial L(g,\xi)}{\partial \xi}, \ \dot{\eta} + \mathrm{ad}_\xi\eta \right\rangle \right\} dt$$

$$= \int_{t_0}^{t_f} \left\{ \left\langle \mathsf{T}_e^*\mathsf{L}_g \cdot \frac{\partial L(g,\xi)}{\partial g} + \mathrm{ad}_\xi^* \cdot \frac{\partial L(g,\xi)}{\partial \xi}, \ \eta \right\rangle + \left\langle \frac{\partial L(g,\xi)}{\partial \xi}, \ \dot{\eta} \right\rangle \right\} dt.$$

$$(8.44)$$

Integrating by parts, the infinitesimal variation of the action integral is given by

$$\delta\mathfrak{G} = \int_{t_0}^{t_f} \left\{ \left\langle \mathsf{T}_e^*\mathsf{L}_g \cdot \frac{\partial L(g,\xi)}{\partial g} + \mathrm{ad}_\xi^* \cdot \frac{\partial L(g,\xi)}{\partial \xi}, \ \eta \right\rangle + \left\langle \frac{\partial L(g,\xi)}{\partial \xi}, \ \dot{\eta} \right\rangle \right\} dt$$

$$= \left\langle \frac{\partial L(g,\xi)}{\partial \xi}, \ \eta \right\rangle \Bigg|_{t_0}^{t_f} + \int_{t_0}^{t_f} \left\langle \mathsf{T}_e^*\mathsf{L}_g \cdot \frac{\partial L(g,\xi)}{\partial g} + \mathrm{ad}_\xi^* \cdot \frac{\partial L(g,\xi)}{\partial \xi}, \ \eta \right\rangle$$

$$- \left\langle \frac{d}{dt}\frac{\partial L(g,\xi)}{\partial \xi}, \ \eta \right\rangle dt.$$

Since $\eta(t_0) = 0$ and $\eta(t_f) = 0$, the first term of the above equation vanishes. Thus, we obtain

$$\delta\mathfrak{G} = \int_{t_0}^{t_f} \left\langle \mathsf{T}_e^*\mathsf{L}_g \cdot \frac{\partial L(g,\xi)}{\partial g} + \mathrm{ad}_\xi^* \cdot \frac{\partial L(g,\xi)}{\partial \xi}, \ \eta \right\rangle - \left\langle \frac{d}{dt}\frac{\partial L(g,\xi)}{\partial \xi}, \ \eta \right\rangle dt.$$

From Hamilton's principle, $\delta\mathfrak{G} = 0$ for all $\eta : [t_0, t_f] \to \mathfrak{g}$. According to the fundamental lemma of the calculus of variations, as described in Appendix A for a configuration manifold that is a Lie group, it follows that

$$\left\langle \frac{d}{dt}\left(\frac{\partial L(g,\xi)}{\partial \xi}\right) - \mathrm{ad}_\xi^* \cdot \frac{\partial L(g,\xi)}{\partial \xi} - \mathsf{T}_e^*\mathsf{L}_g \cdot \frac{\partial L(g,\xi)}{\partial g}, \ \eta \right\rangle = 0, \quad \eta \in \mathfrak{g}.$$

This implies that the first term in the above pairing is necessarily the zero linear functional in $\mathfrak{g}^*$. Recognizing this fact, the Euler–Lagrange equations on the Lie group configuration manifold G can be written as an equation with values in $\mathfrak{g}^*$:

$$\frac{d}{dt}\left(\frac{\partial L(g,\xi)}{\partial \xi}\right) - \mathrm{ad}_\xi^* \cdot \frac{\partial L(g,\xi)}{\partial \xi} - \mathsf{T}_e^*\mathsf{L}_g \cdot \frac{\partial L(g,\xi)}{\partial g} = 0, \qquad (8.45)$$

and the kinematics equation with values in $\mathfrak{g}$:

$$\dot{g} = g\xi. \qquad (8.46)$$

Thus, the Euler–Lagrange equations (8.45) and the kinematics equations (8.46) define the Lagrangian flow on the tangent bundle described by $(g, \xi) \in \mathsf{TG}$.

The essential idea of this development is expressing the variation of a curve in G using the exponential map. The expression for the variation is carefully chosen such that the varied curve lies on the configuration manifold G. The use of the exponential map $\exp : \mathfrak{g} \to \mathsf{G}$ is desirable in two aspects: (i) since the variation is obtained by a group operation, it is guaranteed to lie on G, and (ii) the variation is parameterized by a curve in a linear vector space $\mathfrak{g}$.

These equations are obtained using the left trivialization. Therefore, the velocity ξ may be considered as a quantity expressed in the body-fixed frame. We can develop similar equations using the right trivialization to obtain the equations of motion expressed in the fixed or inertial frame. This is summarized by the following statement.

The tangent bundle TG is identified with $\mathsf{G} \times \mathfrak{g}$ by right trivialization. Suppose that the Lagrangian is defined as $L(g, \varsigma) : \mathsf{G} \times \mathfrak{g} \to \mathbb{R}$. The corresponding Euler–Lagrange equations are given by

$$\dot{g} = \varsigma g, \tag{8.47}$$

$$\frac{d}{dt}\left(\frac{\partial L(g,\varsigma)}{\partial \varsigma}\right) + \mathrm{ad}^*_\varsigma \cdot \frac{\partial L(g,\varsigma)}{\partial \varsigma} - \mathsf{T}^*_e \mathsf{R}_g \cdot \frac{\partial L(g,\varsigma)}{\partial g} = 0. \tag{8.48}$$

Thus, the kinematics equations (8.47) and the Euler–Lagrange equations (8.48) define the Lagrangian flow on the tangent bundle described by $(g, \varsigma) \in \mathsf{TG}$.

8.6.4 Legendre Transformation and Hamilton's Equations

We identify the tangent bundle TG with $\mathsf{G} \times \mathfrak{g}$ using the left trivialization. Using this, the cotangent bundle $\mathsf{T}^*\mathsf{G}$ can be identified with $\mathsf{G} \times \mathfrak{g}^*$. For the given Lagrangian, the Legendre transformation $\mathbb{F}L : \mathsf{G} \times \mathfrak{g} \to \mathsf{G} \times \mathfrak{g}^*$ is defined as

$$\mathbb{F}L(g, \xi) = (g, \mu), \tag{8.49}$$

where $\mu \in \mathfrak{g}^*$ is given by

$$\mu = \frac{\partial L(g, \xi)}{\partial \xi}. \tag{8.50}$$

The Legendre transformation is assumed to be a diffeomorphism, that is the corresponding Lagrangian is hyperregular, which induces a Hamiltonian system on $\mathsf{G} \times \mathfrak{g}^*$.

Consider a system evolving on a Lie group G. We identify the tangent bundle TG with $\mathsf{G} \times \mathfrak{g}$ by left trivialization. Since the Lagrangian given by

$L(g,\xi) : \mathsf{G} \times \mathfrak{g} \to \mathbb{R}$ is hyperregular, the Legendre transformation yields Hamilton's equations that are equivalent to the Euler–Lagrange equations. These can be written as

$$\dot{\mu} = \mathrm{ad}^*_\xi \mu + \mathsf{T}^*_e \mathsf{L}_g \cdot \frac{\partial L(g,\xi)}{\partial g}.$$

It is convenient to introduce the Hamiltonian function $H : \mathsf{G} \times \mathfrak{g}^* \to \mathbb{R}^1$ as

$$H(g,\mu) = \mu \cdot \xi - L(g,\xi),$$

using the Legendre transformation (8.50). Thus, Hamilton's equations can be written as

$$\dot{g} = g \frac{\partial H(g,\mu)}{\partial \mu}, \tag{8.51}$$

$$\dot{\mu} = \mathrm{ad}^*_\xi \mu - \mathsf{T}^*_e \mathsf{L}_g \cdot \frac{\partial H(g,\mu)}{\partial g}. \tag{8.52}$$

Hamilton's equations (8.51) and (8.52), using the Legendre transformation (8.50), define the Hamiltonian flow on the cotangent bundle described by $(g,\mu) \in \mathsf{T}^*\mathsf{G}$.

8.6.5 Hamilton's Phase Space Variational Principle

An alternative derivation of Hamilton's equations is now provided. The Hamilton's phase space variational principle states that the infinitesimal variation of the action integral,

$$\mathfrak{G}^\epsilon = \int_{t_0}^{t_f} \mu \cdot \xi - H(g,\mu)\, dt,$$

subject to fixed endpoints for $g(t)$, along any motion is zero. The infinitesimal variation of the action integral can be written as

$$\delta\mathfrak{G} = \int_{t_0}^{t_f} \left\{ \mu \cdot \delta\xi - \frac{\partial H(g,\mu)}{\partial g} \cdot \delta g + \left(\xi - \frac{\partial H(g,\mu)}{\partial \mu} \right) \cdot \delta\mu \right\} dt.$$

Because $\xi = g^{-1}\dot{g}$, the variations of ξ are related to the variations of g, and in Section 8.6.2, we obtained expressions (8.41) and (8.42) for the variation of g and ξ, respectively, expressed in terms of a curve $\eta : [t_0, t_f] \to \mathfrak{g}$ that vanishes at the endpoints. Using these expressions, we obtain

$$\delta\mathfrak{G} = \int_{t_0}^{t_f} \left\{ \langle \mu, \dot{\eta} + \mathrm{ad}_\xi \eta \rangle - \left\langle \frac{\partial H(g,\mu)}{\partial g}, \mathsf{T}_e \mathsf{L}_g \cdot \eta \right\rangle \right. $$
$$\left. + \left(\xi - \frac{\partial H(g,\mu)}{\partial \mu} \right) \cdot \delta\mu \right\} dt$$

$$= \int_{t_0}^{t_f} \left\{ \langle \mu, \, \dot{\eta} \rangle + \left\langle \mathrm{ad}_\xi^* \mu - \mathsf{T}_e^* \mathsf{L}_g \cdot \frac{\partial H(g,\mu)}{\partial g}, \eta \right\rangle \right.$$
$$\left. + \left(\xi - \frac{\partial H(g,\mu)}{\partial \mu} \right) \cdot \delta \mu \right\} \, dt.$$

Integrating the first term on the right by parts, the infinitesimal variation of the action integral is given by

$$\delta \mathfrak{G} = \langle \mu, \, \eta \rangle \bigg|_{t_0}^{t_f} + \int_{t_0}^{t_f} \left\{ \left\langle -\dot{\mu} + \mathrm{ad}_\xi^* \mu - \mathsf{T}_e^* \mathsf{L}_g \cdot \frac{\partial H(g,\mu)}{\partial g}, \eta \right\rangle \right.$$
$$\left. + \left(\xi - \frac{\partial H(g,\mu)}{\partial \mu} \right) \cdot \delta \mu \right\} \, dt.$$

Since $\eta(t_0) = 0$ and $\eta(t_f) = 0$, the boundary terms vanish. Thus, we obtain,

$$\delta \mathfrak{G} = \int_{t_0}^{t_f} \left\{ \left\langle -\dot{\mu} + \mathrm{ad}_\xi^* \mu - \mathsf{T}_e^* \mathsf{L}_g \cdot \frac{\partial H(g,\mu)}{\partial g}, \eta \right\rangle + \left(\xi - \frac{\partial H(g,\mu)}{\partial \mu} \right) \cdot \delta \mu \right\} \, dt.$$

Then, by Hamilton's phase space variational principle and the fundamental lemma of the calculus of variations, we recover Hamilton's equations (8.51) and (8.52) on the cotangent bundle of the Lie group G.

The following property follows directly from Hamilton's equations (8.51) and (8.52),

$$\frac{dH(g,\mu)}{dt} = \frac{\partial H(g,\mu)}{\partial g} \cdot \dot{g} + \frac{\partial H(g,\mu)}{\partial \mu} \cdot \dot{\mu}$$
$$= \frac{\partial H(g,\mu)}{\partial g} \cdot \mathsf{T}_e \mathsf{L}_g \frac{\partial H(g,\mu)}{\partial \mu} + \frac{\partial H(g,\mu)}{\partial \mu} \cdot \mathrm{ad}_\xi^* \mu$$
$$- \frac{\partial H(g,\mu)}{\partial \mu} \cdot \mathsf{T}_e^* \mathsf{L}_g \frac{\partial H(g,\mu)}{\partial g},$$

where the shorthand notation for the left-trivialized derivative described by (8.40) has been applied. Therefore, the first and the third terms cancel. From (8.46) and (8.51), it is clear that $\xi = \frac{\partial H(g,\mu)}{\partial \mu}$. Substituting this,

$$\frac{dH(g,\mu)}{dt} = \xi \cdot \mathrm{ad}_\xi^* \mu = \mathrm{ad}_\xi \xi \cdot \mu = [\xi, \xi] \cdot \mu = 0.$$

This formulation exposes an important property of the Hamiltonian flow on the cotangent bundle of the Lie group G: the Hamiltonian function is constant along each solution of Hamilton's equation. It should be emphasized that this property does not hold if the Hamiltonian function has a nontrivial explicit dependence on time.

8.6.6 Reassessment of Results in the Prior Chapters

In this section, we provide a reassessment of the formulation of Euler–Lagrange equations on the Lie group configuration manifolds $SO(3)$ and $SE(3)$ by rederiving those results using the abstract results in this section. This reassessment provides a confirmation of the validity of the prior results, while also demonstrating the power of the abstract results.

8.6.6.1 Lagrangian and Hamiltonian Dynamics on $SO(3)$

The Euler–Lagrange equations derived in Chapter 6 for the configuration manifold $SO(3)$ can be obtained using the abstract Lie group formulation of this section. In particular, the configuration manifold $SO(3)$ is a Lie group with Lie algebra $\mathfrak{so}(3)$. Further, the Lagrangian $L : TSO(3) \rightarrow \mathbb{R}^1$ can be expressed as $L(R, \xi)$ where $R \in SO(3)$, $\xi \in \mathfrak{so}(3)$. Equation (8.45) can be rewritten in this case as an equation with values in $\mathfrak{so}(3)^*$. Using the identification $\xi = S(\omega)$, we obtain the Euler–Lagrange equations on $SO(3)$ given in Chapter 6, namely

$$\frac{d}{dt}\left(\frac{\partial L(R, \omega)}{\partial \omega}\right) + \omega \times \frac{\partial L(R, \omega)}{\partial \omega} + \sum_{i=1}^{3} r_i \times \frac{\partial L(R, \omega)}{\partial r_i} = 0. \qquad (8.53)$$

Hamilton's equations on $SO(3)$ were derived in Chapter 6. They also follow from the abstract results above using the Legendre transformation

$$\Pi = \frac{\partial L(R, \omega)}{\partial \omega},$$

and the Hamiltonian function

$$H(q, p) = \Pi \cdot \omega - L(R, \omega),$$

to obtain Hamilton's equations on $SO(3)$

$$\dot{r}_i = r_i \times \frac{\partial H(R, \Pi)}{\partial \Pi}, \quad i = 1, 2, 3, \qquad (8.54)$$

$$\dot{\Pi} = \Pi \times \frac{\partial H(R, \Pi)}{\partial \Pi} + \sum_{i=1}^{3} r_i \times \frac{\partial H(R, \Pi)}{\partial r_i}. \qquad (8.55)$$

Here, $r_i \in S^2$ are the columns of $R^T \in SO(3)$.

8.6.6.2 Lagrangian and Hamiltonian Dynamics on SE(3)

The Euler–Lagrange equations derived in Chapter 7 for the configuration manifold SE(3) can be obtained using the abstract formulation of this section. In particular, the configuration manifold SE(3) is a Lie group with Lie algebra $\mathfrak{se}(3)$. Further, the Lagrangian $L : \mathsf{TSE}(3) \to \mathbb{R}^1$ can be expressed as $L(R, x, \xi, \dot{x})$ where $R \in \mathsf{SO}(3)$, $x \in \mathbb{R}^3$, $\xi \in \mathfrak{so}(3)$, $\dot{x} \in \mathbb{R}^3$. Equation (8.45) can be rewritten in this case as an equation with values in $\mathfrak{se}(3)^*$.

Using the identification $\xi = S(\omega)$, we obtain the Euler–Lagrange equations on SE(3) given in Chapter 7:

$$\frac{d}{dt}\left(\frac{\partial L(R, x, \omega, \dot{x})}{\partial \omega}\right) + \omega \times \frac{\partial L(R, x, \omega, \dot{x})}{\partial \omega} + \sum_{i=1}^{3} r_i \times \frac{\partial L(R, x, \omega, \dot{x})}{\partial r_i} = 0,$$

(8.56)

$$\frac{d}{dt}\left(\frac{\partial L(R, x, \omega, \dot{x})}{\partial \dot{x}}\right) - \frac{\partial L(R, x, \omega, \dot{x})}{\partial x} = 0.$$

(8.57)

Hamilton's equations on SE(3) were derived in Chapter 7. They also follow from the abstract results above using the Legendre transformation

$$\Pi = \frac{\partial L(R, x, \omega, \dot{x})}{\partial \omega},$$

$$p = \frac{\partial L(R, x, \omega, \dot{x})}{\partial \dot{x}},$$

where we assume the Lagrangian has the property that the map $(\omega, \dot{x}) \to (\Pi, p)$ is invertible. The Hamiltonian function is defined as

$$H(R, x, \Pi, p) = \Pi \cdot \omega + p \cdot \dot{x} - L(R, x, \omega, \dot{x}),$$

using the Legendre transformation. Thus, Hamilton's equations on SE(3) are

$$\dot{r}_i = r_i \times \frac{\partial H(R, x, \Pi, p)}{\partial \Pi}, \quad i = 1, 2, 3,$$

(8.58)

$$\dot{x} = \frac{\partial H(R, x, \Pi, p)}{\partial p},$$

(8.59)

$$\dot{\Pi} = \Pi \times \frac{\partial H(R, x, \Pi, p)}{\partial \Pi} + \sum_{i=1}^{3} r_i \times \frac{\partial H(R, x, \Pi, p)}{\partial r_i},$$

(8.60)

$$\dot{p} = -\frac{\partial H(R, x, \Pi, p)}{\partial x}.$$

(8.61)

Here, $r_i \in \mathsf{S}^2$, $i = 1, 2, 3$, are the columns of $R^T \in \mathsf{SO}(3)$.

8.7 Lagrangian and Hamiltonian Dynamics on a Homogeneous Manifold

We now formulate Lagrangian dynamics on an abstract homogeneous manifold. This extension is shown to arise in a natural way from the prior results when the configuration manifold is a Lie group. In addition, these results apply to many important Lagrangian systems with configuration manifolds that have the homogeneity property.

The development is constructed around a configuration manifold Q that is a homogenous manifold and a Lagrangian function $L : TQ \to \mathbb{R}^1$ defined on the tangent bundle of the configuration manifold.

We present an abstract geometric formulation for Lagrangian systems where the configuration manifold is a homogeneous manifold Q associated with a Lie group G and its Lie algebra $\mathfrak{g}$. The action integral is defined in terms of a Lagrangian and we show that the action integral can also be expressed in terms of a Lagrangian function $\bar{L} : TG \to \mathbb{R}^1$ lifted to TG. Hamilton's principle and the results of the prior chapter are used to derive an abstract version of the Euler–Lagrange equations on TG. This leads to the same abstract form of the Euler–Lagrange equations in the prior section for Lagrangian systems that evolve on a Lie group. Specific properties of the homogeneous manifold can be used to obtain Euler–Lagrange equations on TQ.

8.7.1 Additional Material on Homogeneous Manifolds

A *left action* $\Phi : G \times Q \to Q$ of a Lie group G on a manifold Q is a smooth map satisfying $\Phi(e, q) = q$ and $\Phi(gh, q) = \Phi(g, \Phi(h, q))$ for any $g, h \in G$, the identity element $e \in G$, and $q \in Q$. It is also written as $\Phi(g, q) = gq$ for convenience. A *right action* can be defined similarly.

An action is called *transitive* if any two points on Q are connected by the group action, i.e., for any $q_1, q_2 \in Q$, there exists $g \in G$ such that $gq_1 = q_2$. A homogeneous manifold Q is a manifold with a transitive action of a Lie group G.

Any homogeneous manifold can be regarded as a set of cosets as follows. Given $q_0 \in Q$, define $H = \{g \in G : gq_0 = q_0\}$, i.e., H is the isotropy subgroup of G at q_0. Then, we can show that Q is diffeomorphic to $G/H = \{gH : g \in G\}$, that is all left cosets of H in G. Let $\pi : G \to G/H$ be the projection given by $\pi(g) = gH$ and let the origin $o = \pi(e)$.

The diffeomorphism between G/H and Q can be made explicit once we pick an element $q_0 \in Q$ to identify with the origin $o \in G/H$. By an abuse of notation, let us denote the projection from G to Q by $\pi : G \to Q$, and this is given by $\pi(g) = gq_0$.

8.7.2 A Lifting Process

Let $q : [t_0, t_f] \to Q$ define a differentiable curve on Q. Then, its tangent vector $\dot{q}(t) \in T_{q(t)}Q$ at each t can be represented in terms of an element in the Lie algebra $\mathfrak{g}$ as follows. Since the action is transitive, there is a curve $g : [t_0, t_f] \to G$ (defined up to isotropy) such that $q(t) = g(t)q(t_0)$. Suppose that the kinematics equation for the lifted curve is given by

$$\dot{g}(t) = T_e R_{g(t)} \varsigma(t) = \varsigma(t) g(t), \tag{8.62}$$

for $\varsigma(t) \in \mathfrak{g}$. Consequently, we obtain the kinematics equation on Q as

$$
\begin{aligned}
\dot{q}(t) &= \frac{d}{d\epsilon} \Phi(\exp(\epsilon\varsigma(t))g(t), q(t_0)) \bigg|_{\epsilon=0} \\
&= \frac{d}{d\epsilon} \Phi(\exp(\epsilon\varsigma(t)), g(t)q(t_0)) \bigg|_{\epsilon=0} \\
&= \frac{d}{d\epsilon} \Phi(\exp(\epsilon\varsigma(t)), q(t)) \bigg|_{\epsilon=0}.
\end{aligned}
$$

The above equation is written as

$$\dot{q} = \varsigma q. \tag{8.63}$$

Conversely, given a differential curve $g : [t_0, t_f] \to G$ such that $g(t_0) = e$, and the initial point $q(t_0) \in Q$, we obtain a differentiable curve $q : [t_0, t_f] \to Q$ on Q, given by $q(t) = g(t)q(t_0)$, i.e., $q(\cdot) = \pi \circ g(\cdot)$, where $q_0 = q(t_0)$. This yields a projection from the space of differentiable curves on G with initial point e to the space of differentiable curves Q with a prescribed initial point $q(t_0)$, and this allows us to lift an action integral defined on curves on Q to an action integral defined on curves on G.

8.7.3 Euler–Lagrange Equations

We assume that a Lagrangian function, the kinetic energy minus the potential energy, can be expressed as a function on the tangent bundle of the configuration manifold. That is the Lagrangian function $L : TQ \to \mathbb{R}^1$ is the difference of the kinetic energy $T : TQ \to \mathbb{R}^1$ and the potential energy $U : Q \to \mathbb{R}^1$:

$$L(q, \dot{q}) = T(q, \dot{q}) - U(q). \tag{8.64}$$

Define a Lagrangian lifted to (or pulled back to) TG as follows:

$$\bar{L}(g,\varsigma) = (L \circ \mathsf{T}\pi)(g,\varsigma) = (\mathsf{T}\pi)^* L(g,\varsigma). \tag{8.65}$$

The resulting Lagrangian on $\mathsf{T}\mathsf{G}$ is degenerate in the isotropy direction, and as such, there will not be a unique curve in G which satisfies Hamilton's principle. However, any curve g in G that satisfies Hamilton's principle (or equivalently, the Euler–Lagrange equations) is equally valid for our purpose, since we are only interested in the induced curve q on the homogeneous manifold Q, given by $q = \pi \circ g$, which is well defined.

The solution q of the Euler–Lagrange equation with $L(q,\dot{q})$ corresponds to the projection of the curve g in G satisfying the Euler–Lagrange equation with $\bar{L}(g,\varsigma)$, since

$$\delta \int_{t_0}^{t_f} L(q,\dot{q})\,dt = \delta \int_{t_0}^{t_f} \bar{L}(g,\dot{g}\,g^{-1})\,dt.$$

The resulting Euler–Lagrange equations for the lifted curve are equivalent to (8.48) and it is repeated below for ease of reference:

$$\frac{d}{dt}\left(\frac{\partial \bar{L}(g,\varsigma)}{\partial \varsigma}\right) + \mathrm{ad}_\varsigma^* \cdot \frac{\partial \bar{L}(g,\varsigma)}{\partial \varsigma} - \mathsf{T}_e^* \mathsf{R}_g \cdot \frac{\partial \bar{L}(g,\varsigma)}{\partial g} = 0. \tag{8.66}$$

However, as noted earlier, the curve that satisfies the resulting Euler–Lagrange equations on G are only unique up to isotropy. In practice, the issue of uniqueness of the curve in G can be addressed by constraining $\varsigma(t) \in \mathfrak{g}$ that we use in (8.62) to lie in a horizontal subspace $\mathfrak{l}$ that is transverse to the Lie subalgebra $\mathfrak{h}$ corresponding to the isotropy subgroup H. Then, we obtain a decomposition of the Lie algebra $\mathfrak{g} = \mathfrak{h} \oplus \mathfrak{l}$. This is equivalent to introducing a principle bundle connection on $\pi : \mathsf{G} \to \mathsf{G}/\mathsf{H}$, and requiring that the curves in G are horizontal with respect to that connection. While the resulting curve on G depends on the choice of connection, the induced curve on Q is independent of the choice of connection.

More concretely, the equations for the curve on the Lie group G are given by the Euler–Lagrange equations (8.48) and the kinematics equation (8.47) on G, together with the horizontal space constraint

$$\dot{g} = \varsigma g, \tag{8.67}$$

$$\frac{d}{dt}\frac{\partial \bar{L}(g,\varsigma)}{\partial \varsigma} + \mathrm{ad}_\varsigma^* \cdot \frac{\partial \bar{L}(g,\varsigma)}{\partial \varsigma} - \mathsf{T}_e^* \mathsf{R}_g \cdot \frac{\partial \bar{L}(g,\varsigma)}{\partial g} = 0, \tag{8.68}$$

$$\varsigma(t) \in \mathfrak{l}. \tag{8.69}$$

The induced curve on the homogeneous space Q is obtained from the solution $g(t)$ of the above equations by projection $q(t) = g(t)q_0$.

8.7.4 Reassessment of Results in the Prior Chapters

In this section, we provide a reassessment of the formulation of Euler–Lagrange equations on the homogeneous manifolds S^1 and S^2 by deriving these results using the abstract results in this chapter.

8.7.4.1 Lagrangian Dynamics on S^1

The Euler–Lagrange equations derived in the variational setting of Chapter 4 on the configuration manifold S^1 can be obtained using the abstract formulation of this chapter. The configuration manifold S^1 is not a Lie group but we can characterize the variations using exponential maps. This allows the development of the results in Chapter 4 for the case $n = 1$ following the procedures described in this chapter. We use the notation of Chapter 4.

We first view S^1 as a homogeneous manifold. The group $G = SO(2)$ acts on S^1 transitively. The isotropy subgroup of $e_1 = [1, 0] \in S^1$ consists of all elements in $SO(2)$ of the form:

$$\begin{bmatrix} 1 & 0 \\ 0 & A \end{bmatrix},$$

where $A \in SO(1)$. It is easy to check that in this case, the only element of the isotropy subgroup is the identity matrix. But it is still helpful to take this point of view since it generalizes to higher-dimensional spheres. Hence, S^1 can be considered as diffeomorphic to the quotient $SO(2)/SO(1)$. For $R \in SO(2)$, the projection is given by $\pi(R) = Re_1$.

The variation of a curve on $SO(2)$ can be written as

$$\delta R = \frac{d}{d\epsilon} \exp(\epsilon \eta S)R \bigg|_{\epsilon=0} = \eta SR = RS\eta,$$

where $\eta \in \mathbb{R}^1$ and defines a basis of $\mathfrak{so}(2)$; $S \in \mathbb{R}^{2\times 2}$ is the skew-symmetric matrix introduced in Chapter 4:

$$S = \begin{bmatrix} 0 & -1 \\ 1 & 0 \end{bmatrix}.$$

Suppose that the configuration manifold is given by $Q = S^1$, and the Lagrangian is given by $L(q, \dot{q})$. The lifted Lagrangian on $T(SO(2) \times \mathfrak{so}(2))$ is

$$\bar{L}(R, \omega) = L(Re_1, \omega SRe_1).$$

The derivatives of the lifted Lagrangian are given by

$$\frac{\partial \bar{L}}{\partial \omega} \cdot \delta \omega = \frac{\partial L(q, \dot{q})}{\partial \dot{q}} \cdot \delta \omega S R e_1$$

$$= -q^T S \frac{\partial L(q, \dot{q})}{\partial \dot{q}} \cdot \delta \omega,$$

$$(T_e^* R_R \cdot \frac{\partial \bar{L}}{\partial R}) \cdot \eta = \frac{\partial L(q, \dot{q})}{\partial q} \cdot \eta S R e_i + \frac{\partial L(q, \dot{q})}{\partial \dot{q}} \cdot \omega \eta S^2 R e_i$$

$$= (-q^T S \frac{\partial L(q, \dot{q})}{\partial q} + \omega q^T S^2 \frac{\partial L(q, \dot{q})}{\partial \dot{q}}) \cdot \eta.$$

Substituting these into (8.66), we obtain

$$\omega q^T S^2 \frac{\partial L(q, \dot{q})}{\partial \dot{q}} - q^T S \frac{d}{dt} \left(\frac{\partial L(q, \dot{q})}{\partial \dot{q}} \right) + q^T S \frac{\partial L(q, \dot{q})}{\partial q} - \omega q^T S^2 \frac{\partial L(q, \dot{q})}{\partial \dot{q}} = 0.$$

This reduces to

$$-q^T S \left\{ \frac{d}{dt} \left(\frac{\partial L(q, \dot{q})}{\partial \dot{q}} \right) - \frac{\partial L(q, \dot{q})}{\partial q} \right\} = 0, \qquad (8.70)$$

which is equivalent to (4.5) in Chapter 4. As in Chapter 4, this can be written as

$$(I_{2 \times 2} - q q^T) \left\{ \frac{d}{dt} \left(\frac{\partial L(q, \dot{q})}{\partial \dot{q}} \right) - \frac{\partial L(q, \dot{q})}{\partial q} \right\} = 0. \qquad (8.71)$$

8.7.4.2 Lagrangian Dynamics on S^2

The Euler–Lagrange equations derived in the variational setting of Chapter 5 on the configuration manifold S^2 can be obtained using the abstract formulation of this chapter. The configuration manifold S^2 is not a Lie group but variations can be characterized using matrix exponential maps. This allows the development of the results in Chapter 5 following the procedures described in this chapter.

We first view S^2 as a homogeneous manifold. The group $G = SO(3)$ acts on S^2 transitively. The isotropy subgroup of $e_1 = [1, 0, 0] \in S^2$ consists of all elements in $SO(3)$ of the form:

$$\begin{bmatrix} 1 & 0 \\ 0 & A \end{bmatrix},$$

where $A \in SO(2)$. Hence, S^2 is diffeomorphic to the quotient $SO(3)/SO(2)$. The projection is given by $\pi(R) = R e_1$.

Suppose that the configuration manifold is given by $Q = S^2$, and the Lagrangian is given by $L(q, \dot{q})$. The lifted Lagrangian on $T(SO(3) \times \mathfrak{so}(3))$ is

$$\bar{L}(R, \omega) = L(R e_1, \hat{\omega} R e_1).$$

The derivatives of the lifted Lagrangian are given by

$$\frac{\partial \bar{L}}{\partial \omega} \cdot \delta\omega = \frac{\partial L(q,\dot{q})}{\partial \dot{q}} \cdot S(\delta\omega)Re_1$$

$$= \left(S(q)\frac{\partial L(q,\dot{q})}{\partial \dot{q}} \right) \cdot \delta\omega,$$

$$(\mathrm{T}_e^* \mathrm{R}_R \cdot \frac{\partial \bar{L}}{\partial R}) \cdot \eta = \frac{\partial L(q,\dot{q})}{\partial q} \cdot S(\eta)Re_1 + \frac{\partial L(q,\dot{q})}{\partial \dot{q}} \cdot S(\omega)S(\eta)Re_1$$

$$= \left(S(q)\frac{\partial L(q,\dot{q})}{\partial q} - S(q)\left(S(\omega)\frac{\partial L(q,\dot{q})}{\partial \dot{q}} \right) \right) \cdot \eta.$$

Substituting these into (8.66), we obtain

$$S(q)\frac{d}{dt}\left(\frac{\partial L(q,\dot{q})}{\partial \dot{q}} \right) + S(\dot{q})\frac{\partial L(q,\dot{q})}{\partial \dot{q}} - S(\omega)\left(S(q)\frac{\partial L(q,\dot{q})}{\partial \dot{q}} \right)$$

$$- S(q)\frac{\partial L(q,\dot{q})}{\partial q} + S(q)\left(S(\omega)\frac{\partial L(q,\dot{q})}{\partial \dot{q}} \right) = 0.$$

But, we have

$$- S(\omega)\left(S(q)\frac{\partial L(q,\dot{q})}{\partial \dot{q}} \right) + S(q)\left(S(\omega)\frac{\partial L(q,\dot{q})}{\partial \dot{q}} \right)$$

$$= -(S(\omega)S(q) - S(q)S(\omega))\frac{\partial L(q,\dot{q})}{\partial \dot{q}}$$

$$= -S(S(\omega)q)\frac{\partial L(q,\dot{q})}{\partial \dot{q}}$$

$$= -S(\dot{q})\frac{\partial L(q,\dot{q})}{\partial \dot{q}}.$$

Therefore, we obtain

$$S(q)\frac{d}{dt}\left(\frac{\partial L(q,\dot{q})}{\partial \dot{q}} \right) - S(q)\frac{\partial L(q,\dot{q})}{\partial q} = 0, \qquad (8.72)$$

which recovers (5.6) in Chapter 5. As in Chapter 5 this can also be written as

$$(I_{3\times 3} - qq^T)\left\{ \frac{d}{dt}\left(\frac{\partial L(q,\dot{q})}{\partial \dot{q}} \right) - \frac{\partial L(q,\dot{q})}{\partial q} \right\} = 0. \qquad (8.73)$$

8.8 Lagrange–d'Alembert Principle

We describe a modification of Hamilton's principle to incorporate the effects of external forces that need not be derivable from a potential. An external moment is the cross product of a moment arm vector and an external force. For appropriate configuration manifolds where the physical work done is expressed as the integrated dot product of a moment vector and a configuration vector, it is convenient to express external moments in terms of equivalent external forces.

We assume the dynamics evolves on a configuration manifold M embedded in $\mathbb{R}^n$. As in Chapter 3, this modification is referred to as the Lagrange–d'Alembert's principle. It states that the infinitesimal variation of the action integral over a fixed time period equals the virtual work done by the external forces along an infinitesimal variation in the configuration during the same time period. This reduces to Hamilton's principle when there are no external forces. This version of the variational principle requires determination of the virtual work corresponding to an infinitesimal variation of the configuration.

Let $x : [t_0, t_f] \to M$ denote a differentiable curve. The external forces are described by a vertical mapping on the cotangent bundle T^*M satisfying $Q : [t_0, t_f] \to \mathsf{T}_x^*M$. The vertical assumption implies that the kinematics are unchanged, while the external forces affect only the dynamics. Thus, the virtual work along an infinitesimal variation of the configuration is given by

$$\int_{t_0}^{t_f} Q^T \delta x \, dt.$$

The Lagrange–d'Alembert principle states that

$$\delta \int_{t_0}^{t_f} L(x, \dot{x}) dt = - \int_{t_0}^{t_f} Q^T \delta x \, dt,$$

holds for all possible differentiable infinitesimal variations $\delta x(t) : [t_0, t_f] \to \mathbb{R}^n$ satisfying $\delta x(t) \in \mathsf{T}_{x(t)}M$ and $\delta x(t_0) = \delta x(t_f) = 0$.

Following the prior development, this leads to

$$\left\{ \frac{d}{dt} \left(\frac{\partial L(x, \dot{x})}{\partial \dot{x}} \right) - \frac{\partial L(x, \dot{x})}{\partial x} - Q \right\} \cdot \delta x = 0, \tag{8.74}$$

for all $\delta x \in \mathsf{T}_x M$. This is a form of the Euler–Lagrange equation, including external forces, that characterizes the Lagrangian flow that evolves on the tangent bundle of the configuration manifold.

In specifying values of the external forces in the cotangent space, it is clear that the component of the force orthogonal to the tangent space at each point is irrelevant. That is, there is no loss in generality in assuming that the external force always lies in the appropriate cotangent space.

Many, but not all, of the results described for the autonomous case, without inclusion of external forces, hold for this case. But it is important to be careful. For instance, if there are external forces it is not necessarily true that the Hamiltonian is conserved, unless the forces are always perpendicular to the motion and therefore do no work.

In practice, this version of the Euler–Lagrange equations that include external forces is important. Physical effects such as friction and other forms of energy dissipation can be incorporated into the equations of motion. Finally, external forces and moments can be used to model the effects of control actions and external disturbances. Such modifications significantly broaden the application of the Lagrangian and Hamiltonian approaches to dynamics.

8.9 Problems

8.1. Consider the following matrix that appears on the left-hand side of (8.6),

$$\begin{bmatrix} P^T M \\ I_{n \times n} - P \end{bmatrix},$$

where the matrix $P \in \mathbb{R}^{n \times n}$ corresponds to the projection operator, and the matrix $M \in \mathbb{R}^{n \times n}$ is symmetric and positive-definite. Show that the above matrix has full column rank by verifying the following two conditions.

(a) Suppose

$$\begin{bmatrix} P^T M \\ I_{n \times n} - P \end{bmatrix} v = 0,$$

for some $v \in \mathbb{R}^n$. Show that this implies:

$$v^T M v = 0.$$

(b) Using the fact that M is symmetric, positive-definite, show $v = 0$.

8.2. Let $0 \le m \le n$ and let $\{a_1, \ldots, a_m, \ldots, a_n\}$ denote an orthonormal basis for $\mathbb{R}^n$. Define a linear manifold $M = \text{span}\{a_1, \ldots, a_m\}$. Assume M is the configuration manifold for a Lagrangian function $L : \mathsf{T}M \to \mathbb{R}^1$ given by $L(x, \dot{x}) = \frac{1}{2} m \|\dot{x}\|^2 - U(x)$, where $(x, \dot{x}) \in \mathsf{T}M$.

(a) Determine the Euler–Lagrange equations, expressed in terms of differential-algebraic equations with Lagrange multipliers, that describe the evolution of the dynamical flow on the tangent bundle $\mathsf{T}M$. Show that the index is two.

(b) Determine Hamilton's equations, expressed as differential-algebraic equations with Lagrange multipliers, that describe the evolution of the dynamical flow on the cotangent bundle T^*M. Show that the index is two.

(c) Determine the Euler–Lagrange equations, as differential equations, that describe the evolution of the dynamical flow on the tangent bundle$\mathsf{T}M$.

(d) Determine Hamilton's equations, as differential equations, that describe the evolution of the dynamical flow on the cotangent bundle T^*M.

(e) Show that the Lagrangian vector field on $\mathsf{T}M$ in (a) and the Lagrangian vector field on $\mathsf{T}M$ in (c) are identical.

(f) Show that the Hamiltonian vector field on T^*M in (b) and the Hamiltonian vector field on T^*M in (d) are identical.

8.3. A particle of mass m translates on a parabolic curve embedded in $\mathbb{R}^2$ described by the manifold $M = \left\{ x \in \mathbb{R}^2 : x_1 - (x_2)^2 = 0 \right\}$. The particle is influenced by a potential function $U : \mathbb{R}^2 \to \mathbb{R}^1$.

(a) Describe the translational kinematics of the particle on the curve in terms of a single kinematics parameter.

(b) What is the Lagrangian function $L : \mathsf{T}M \to \mathbb{R}^1$ defined on the tangent bundle of the configuration manifold?

(c) What are the Euler–Lagrange equations that describe the dynamical flow on $\mathsf{T}M$?

(d) What is the Hamiltonian function $H : \mathsf{T}^*M \to \mathbb{R}^1$?

(e) What are Hamilton's equations that describe the dynamical flow on T^*M?

8.4. A particle of mass m translates on a plane embedded in $\mathbb{R}^3$ described by the manifold $M = \left\{ x \in \mathbb{R}^3 : x_1 + x_2 + x_3 - 1 = 0 \right\}$. The particle is influenced by a potential function $U : \mathbb{R}^3 \to \mathbb{R}^1$.

(a) Describe the translational kinematics of the particle on the plane in terms of two kinematics parameters.

(b) What is the Lagrangian function $L : \mathsf{T}M \to \mathbb{R}^1$ defined on the tangent bundle of the configuration manifold?

(c) What are the Euler–Lagrange equations that describe the dynamical flow on $\mathsf{T}M$?

(d) What is the Hamiltonian function $H : \mathsf{T}^*M \to \mathbb{R}^1$?

(e) What are Hamilton's equations that describe the dynamical flow on T^*M?

8.5. A particle of mass m translates on a plane embedded in $\mathbb{R}^3$ described by the manifold $M = \left\{ x \in \mathbb{R}^3 : x_1 - x_2 + x_3 - 1 = 0 \right\}$. The particle is influenced by a potential function $U : \mathbb{R}^3 \to \mathbb{R}^1$.

(a) Describe the translational kinematics of the particle on the plane in terms of two kinematics parameters.

(b) What is the Lagrangian function $L : \mathsf{T}M \to \mathbb{R}^1$ defined on the tangent bundle of the configuration manifold?

(c) What are the Euler–Lagrange equations that describe the dynamical flow on $\mathsf{T}M$?

(d) What is the Hamiltonian function $H : \mathsf{T}^*M \to \mathbb{R}^1$?

(e) What are Hamilton's equations that describe the dynamical flow on T^*M?

8.6. A particle of mass m translates on a parabolic surface embedded in $\mathbb{R}^3$ described by the manifold $M = \left\{ x \in \mathbb{R}^3 : (x_1)^2 - x_2 + x_3 = 0 \right\}$. The particle is influenced by a potential function $U : \mathbb{R}^3 \to \mathbb{R}^1$.

(a) Describe the translational kinematics of the particle on the surface in terms of two kinematics parameters.
(b) What is the Lagrangian function $L : \mathsf{T}M \to \mathbb{R}^1$ defined on the tangent bundle of the configuration manifold?
(c) What are the Euler–Lagrange equations that describe the dynamical flow on $\mathsf{T}M$?
(d) What is the Hamiltonian function $H : \mathsf{T}^*M \to \mathbb{R}^1$?
(e) What are Hamilton's equations that describe the dynamical flow on T^*M?

8.7. A particle of mass m translates on a parabolic surface embedded in $\mathbb{R}^3$ described by the manifold $M = \left\{ x \in \mathbb{R}^3 : (x_1)^2 + (x_2)^2 - x_3 = 0 \right\}$. The particle is influenced by a potential function $U : \mathbb{R}^3 \to \mathbb{R}^1$.

(a) Describe the translational kinematics of the particle on the surface in terms of two kinematics parameters.
(b) What is the Lagrangian function $L : \mathsf{T}M \to \mathbb{R}^1$ defined on the tangent bundle of the configuration manifold?
(c) What are the Euler–Lagrange equations that describe the dynamical flow on $\mathsf{T}M$?
(d) What is the Hamiltonian function $H : \mathsf{T}^*M \to \mathbb{R}^1$?
(e) What are Hamilton's equations that describe the dynamical flow on T^*M?

8.8. A particle of mass m is constrained to translate on the surface of a hyperbolic paraboloid embedded in $\mathbb{R}^3$ that is described by the manifold $M = \left\{ x \in \mathbb{R}^3 : -(x_1)^2 + (x_2)^2 - x_3 = 0 \right\}$. The particle is influenced by a potential function $U : \mathbb{R}^3 \to \mathbb{R}^1$.

(a) Describe the translational kinematics of the particle on the surface in terms of two kinematics parameters.
(b) What is the Lagrangian function $L : \mathsf{T}M \to \mathbb{R}^1$ defined on the tangent bundle of the configuration manifold?
(c) What are the Euler–Lagrange equations that describe the dynamical flow on $\mathsf{T}M$?
(d) What is the Hamiltonian function $H : \mathsf{T}^*M \to \mathbb{R}^1$?
(e) What are Hamilton's equations that describe the dynamical flow on T^*M?

8.9. A particle of mass m is constrained to translate on a line embedded in $\mathbb{R}^3$ described by the manifold $M = \left\{ x \in \mathbb{R}^3 : x_1 - x_2 = 0,\ x_1 + x_2 - x_3 = 0 \right\}$. The particle is influenced by a potential function $U : \mathbb{R}^3 \to \mathbb{R}^1$.

(a) Describe the translational kinematics of the particle on the line in terms of one kinematics parameter.
(b) What is the Lagrangian function $L : \mathsf{T}M \to \mathbb{R}^1$ defined on the tangent bundle of the configuration manifold?

(c) What are the Euler–Lagrange equations that describe the dynamical flow on $\mathsf{T}M$?

(d) What is the Hamiltonian function $H : \mathsf{T}^*M \to \mathbb{R}^1$?

(e) What are Hamilton's equations that describe the dynamical flow on T^*M?

8.10. A particle of mass m is constrained to translate on a parabolic curve embedded in $\mathbb{R}^3$ that is described by the manifold $M = \{x \in \mathbb{R}^3 : x_1 - x_2 = 0,\ (x_1)^2 + (x_2)^2 - x_3 = 0\}$. The particle is influenced by a potential function $U : \mathbb{R}^3 \to \mathbb{R}^1$.

(a) Describe the translational kinematics of the particle on the curve in terms of one kinematics parameter.

(b) What is the Lagrangian function $L : \mathsf{T}M \to \mathbb{R}^1$ defined on the tangent bundle of the configuration manifold?

(c) What are the Euler–Lagrange equations that describe the dynamical flow on $\mathsf{T}M$?

(d) What is the Hamiltonian function $H : \mathsf{T}^*M \to \mathbb{R}^1$?

(e) What are Hamilton's equations that describe the dynamical flow on T^*M?

8.11. Let $R > 0$ and $L > 0$. A particle of mass m is constrained to translate on a helical curve embedded in $\mathbb{R}^3$ described by the manifold $M = \{x \in \mathbb{R}^3 : x_1 = R\cos\left(\frac{2\pi x_3}{L}\right),\ x_2 = R\sin\left(\frac{2\pi x_3}{L}\right)\}$. The particle is influenced by a potential function $U : \mathbb{R}^3 \to \mathbb{R}^1$.

(a) Describe the translational kinematics of the particle on the curve in terms of one kinematics parameter.

(b) What is the Lagrangian function $L : \mathsf{T}M \to \mathbb{R}^1$ defined on the tangent bundle of the configuration manifold?

(c) What are the Euler–Lagrange equations that describe the dynamical flow on $\mathsf{T}M$?

(d) What is the Hamiltonian function $H : \mathsf{T}^*M \to \mathbb{R}^1$?

(e) What are Hamilton's equations that describe the dynamical flow on T^*M?

8.12. Let $a > 0$, $b > 0$. A particle of mass m is constrained to translate on an elliptical curve embedded in $\mathbb{R}^2$ that is given by $M = \{q \in \mathbb{R}^2 : \{\frac{q_1}{a}\}^2 + \{\frac{q_2}{b}\}^2 - 1 = 0\}$. The particle is influenced by a potential function $U : \mathbb{R}^2 \to \mathbb{R}^1$.

(a) Describe the translational kinematics of the particle on the curve in terms of one kinematics parameter.

(b) What is the Lagrangian function $L : \mathsf{T}M \to \mathbb{R}^1$ defined on the tangent bundle of the configuration manifold?

(c) What are the Euler–Lagrange equations that describe the dynamical flow on $\mathsf{T}M$?

(d) What is the Hamiltonian function $H : \mathsf{T}^*M \to \mathbb{R}^1$?

(e) What are Hamilton's equations that describe the dynamical flow on T^*M?

8.13. Let $a > 0$, $b > 0$, $c > 0$. A particle of mass m is constrained to translate on an ellipsoidal surface given by $M = \{q \in \mathbb{R}^3 : \{\frac{q_1}{a}\}^2 + \{\frac{q_2}{b}\}^2 + \{\frac{q_3}{c}\}^2 - 1 = 0\}$. The particle is influenced by a potential function $U : \mathbb{R}^3 \to \mathbb{R}^1$.

(a) Describe the translational kinematics of the particle on the surface in terms of two kinematics parameters.
(b) What is the Lagrangian function $L : \mathsf{T}M \to \mathbb{R}^1$ defined on the tangent bundle of the configuration manifold?
(c) What are the Euler–Lagrange equations that describe the dynamical flow on $\mathsf{T}M$?
(d) What is the Hamiltonian function $H : \mathsf{T}^*M \to \mathbb{R}^1$?
(e) What are Hamilton's equations that describe the dynamical flow on T^*M?

8.14. Consider the kinematics of a rigid rod that is constrained to translate and rotate within a fixed plane described by a two-dimensional Euclidean frame. Let A, B, and C denote three fixed points on the rigid rod: point A of the rigid rod is constrained to translate, without friction, along one axis of the Euclidean frame while point B of the rigid rod is constrained to translate, without friction, along the other axis of the Euclidean frame. The distance between points A and B is denoted by L, while the distance between point B and C is denoted by D. As previously, this mechanism is referred to as the *Trammel of Archimedes* [4] . Assume point C on the rigid rod is the center of mass of the rod. The mass of the rigid rod is m and the scalar inertia of the rigid rod, defined with respect to a body-fixed frame whose origin is located at the center of mass of the rod, is J.

(a) Let $(x_1, x_2) \in \mathbb{R}^2$ denote the position vector of the center of mass of the rigid rod in the Euclidean frame. What is the holonomic constraint that this position vector must satisfy? What is the configuration manifold M embedded in $\mathbb{R}^2$? Show that this mechanism has one degree of freedom. Describe the geometry of the configuration manifold M.
(b) Describe the kinematics relationship of the center of mass of the rod, by expressing the time derivative of the configuration in terms of the scalar angular velocity of the rigid rod and the configuration.
(c) Ignoring potential energy, what is the modified Lagrangian function $L : \mathsf{T}M \to \mathbb{R}^1$ of the translating and rotating rigid rod taking into account the constraints?
(d) What are the Euler–Lagrange equations that describe the dynamical flow on $\mathsf{T}M$?
(e) What is the Hamiltonian function $H : \mathsf{T}^*M \to \mathbb{R}^1$?
(f) What are Hamilton's equations that describe the dynamical flow on T^*M?
(g) What are conserved quantities for the dynamical flow on the tangent bundle $\mathsf{T}M$?

8.15. Consider the dynamics of a rigid body that is constrained to planar rotational motion in $\mathbb{R}^2$. Assume $R \in \mathsf{SO}(2)$, where the configuration manifold is taken as the Lie group $\mathsf{SO}(2)$. Thus, the rigid body attitude is $R \in \mathsf{SO}(2)$ and the rotational kinematics are given by

$$\dot{R} = RS\omega,$$

for some scalar angular velocity $\omega \in \mathbb{R}^1$. Assume the modified Lagrangian $\tilde{L}(R,\omega) = \frac{1}{2}J\omega^2 - U(R)$, where J is the scalar moment of inertia of the rigid body about its axis of rotation.

(a) Use the abstract results of this chapter to obtain an explicit form for the Euler–Lagrange equations that describe the dynamical flow on the tangent bundle $\mathsf{TSO}(2)$.

(b) Use the abstract results of this chapter to obtain an explicit form for Hamilton's equations that describe the dynamical flow on the cotangent bundle $\mathsf{T}^*\mathsf{SO}(2)$.

8.16. Consider the dynamics of a rigid body that is constrained to planar translation and rotational motion in $\mathbb{R}^2$. Assume $(R, x) \in \mathsf{SE}(2)$, where the configuration manifold is taken as the Lie group $\mathsf{SE}(2)$. Thus, the rigid body attitude is $R \in \mathsf{SO}(2)$ and the position vector of the center of mass of the rigid body in the inertial frame is $x \in \mathbb{R}^2$. The rotational kinematics are given by

$$\dot{R} = RS\omega,$$

for some scalar angular velocity $\omega \in \mathbb{R}^1$. Assume the modified Lagrangian $\tilde{L}(R, x, \omega, \dot{x}) = \frac{1}{2}J\omega^2 + \frac{1}{2}m\|\dot{x}\|^2 - U(R, x)$, where J is the scalar moment of inertia of the rigid body about its axis of rotation and m is the mass of the rigid body.

(a) Use the abstract results of this chapter to obtain an explicit form for the Euler–Lagrange equations that describe the dynamical flow on the tangent bundle $\mathsf{TSE}(2)$.

(b) Use the abstract results of this chapter to obtain an explicit form for Hamilton's equations that describe the dynamical flow on the cotangent bundle $\mathsf{T}^*\mathsf{SE}(2)$.

8.17. Consider n particles, of mass m_i, $i = 1,\ldots,n$, each constrained to translate on manifolds M_i, $i = 1,\ldots,n$, embedded in $\mathbb{R}^3$. For $i = 1,\ldots,n$, let $x_i \in \mathbb{R}^3$ denote the position vector of the i-th particle in an inertial Euclidean frame. Assume the existence of an orthogonal projection map $P_i(x_i) : \mathbb{R}^3 \to \mathsf{T}_{x_i}M_i$ for each $x_i \in M_i$, $i = 1,\ldots,n$. The configuration manifold is the product manifold $M = M_1 \times \ldots M_n$ and the Lagrangian function $L : \mathsf{T}M \to \mathbb{R}^1$ is given by $L(x, \dot{x}) = \sum_{i=1}^{n} m_i \|\dot{x}_i\|^2 - U(x)$.

(a) Show that the Euler–Lagrange equations are given by

$$m_i\ddot{x}_i - m_i\dot{P}_i(x_i)\dot{x}_i + P_i(x_i)\frac{\partial U(x)}{\partial x_i} = 0, \quad i = 1,\ldots,n,$$

and they define a dynamical flow on the tangent bundle $\mathsf{T}M$.

394   8 Lagrangian and Hamiltonian Dynamics on Manifolds

(b) What is the Hamiltonian function $H : T^*M \to \mathbb{R}^1$?
(c) Determine Hamilton's equations that define a dynamical flow on the cotangent bundle T^*M.
(d) Suppose the above Lagrangian function is globally defined on $T\mathbb{R}^{3n}$. Show that the Euler–Lagrange equations above can be extended to define a Lagrangian vector field on $T\mathbb{R}^{3n}$. Show that Hamilton's equations above can be extended to define a Hamiltonian vector field on $T^*\mathbb{R}^{3n}$.

8.18. In each of the following, assume that the vector $b \in S^2$ is nonzero.

(a) Show that $Q = \{q \in \mathbb{R}^3 : b^T q = 0, \|q\|^2 = 1\}$ is diffeomorphic to S^1 and it is a homogeneous manifold where the associated Lie group is $G = \{R \in SO(3) : Rb = b\}$.
(b) Show that $Q = \{q \in \mathbb{R}^3 : q = R^T b, R \in SO(3)\}$ is diffeomorphic to S^2 and it is a homogeneous manifold where the associated Lie group is $G = SO(3)$.
(c) Show that $Q = \{(q, x) \in \mathbb{R}^3 \times \mathbb{R}^3 : b^T q = 0, \|q\|^2 = 1\}$ is diffeomorphic to $S^1 \times \mathbb{R}^3$ and it is a homogeneous manifold where the associated Lie group is $G = \{(R, x) \in SE(3) : Rb = b\}$.
(d) Show that $Q = \{(q, x) \in \mathbb{R}^3 \times \mathbb{R}^3 : \|q\|^2 = 1\}$ is diffeomorphic to $S^2 \times \mathbb{R}^3$ and it is a homogeneous manifold where the associated Lie group is $SE(3)$.
(e) Show that $Q = \{(q, x) \in \mathbb{R}^3 \times \mathbb{R}^3 : x^T q = 0, \|q\|^2 = 1, \|x\|^2 = 1\}$ is diffeomorphic to $S^1 \times S^2$ and it is a homogeneous manifold where the associated Lie group is $G = \{(R, x) \in SE(3) : x \in S^2, Rx = x\}$.

8.19. Consider a Lagrangian function $L : TQ \to \mathbb{R}^1$ where $Q = \{q \in \mathbb{R}^3 : b^T q = 0, \|q\|^2 = 1\}$ is a homogeneous manifold and $L(q, \dot{q}) = \frac{1}{2} J \|\dot{q}\|^2 - U(q)$. Assume $J > 0$.

(a) What are the Euler–Lagrange equations that describe the dynamical flow on the tangent bundle of this homogeneous manifold?
(b) What are Hamilton's equations that describe the dynamical flow on the cotangent bundle of this homogeneous manifold?

8.20. Consider a Lagrangian function $L : TQ \to \mathbb{R}^1$ where $Q = \{q \in \mathbb{R}^3 : q = R^T b, R \in SO(3)\}$ is a homogeneous manifold and $L(q, \dot{q}) = \frac{1}{2} \dot{q}^T J \dot{q} - U(q)$. Assume $J \in \mathbb{R}^{3 \times 3}$ is symmetric and positive-definite.

(a) What are the Euler–Lagrange equations that describe the dynamical flow on the tangent bundle of this homogeneous manifold?
(b) What are Hamilton's equations that describe the dynamical flow on the cotangent bundle of this homogeneous manifold?

8.21. Consider a Lagrangian function $L : TQ \to \mathbb{R}^1$ where $Q = \{(q, x) \in \mathbb{R}^3 \times \mathbb{R}^3 : b^T q = 0, \|q\|^2 = 1\}$ is a homogeneous manifold and $L(q, x, \dot{q}, \dot{x}) = \frac{1}{2} J \|\dot{q}\|^2 + \frac{1}{2} m \|\dot{x}\|^2 - U(q, x)$. Assume $J > 0$, $m > 0$.

(a) What are the Euler–Lagrange equations that describe the dynamical flow on the tangent bundle of this homogeneous manifold?

(b) What are Hamilton's equations that describe the dynamical flow on the cotangent bundle of this homogeneous manifold?

8.22. Consider a Lagrangian function $L : \mathsf{T}Q \to \mathbb{R}^1$ where $Q = \{(q, x) \in \mathbb{R}^3 \times \mathbb{R}^3 : \|q\|^2 = 1\}$ is a homogeneous manifold and $L(q, x, \dot{q}, \dot{x}) = \frac{1}{2}\dot{q}^T J \dot{q} + \frac{1}{2}m \|\dot{x}\|^2 - U(q, x)$. Assume $J \in \mathbb{R}^{3 \times 3}$ is symmetric and positive-definite, and $m > 0$.

(a) What are the Euler–Lagrange equations that describe the dynamical flow on the tangent bundle of this homogeneous manifold?

(b) What are Hamilton's equations that describe the dynamical flow on the cotangent bundle of this homogeneous manifold?

8.23. Consider a Lagrangian function $L : \mathsf{T}Q \to \mathbb{R}^1$ where $Q = \{(q, x) \in \mathbb{R}^3 \times \mathbb{R}^3 : b^T q = 0, \|q\|^2 = 1, \|x\|^2 = 1\}$ is a homogeneous manifold and $L(q, x, \dot{q}, \dot{x}) = \frac{1}{2}J \|\dot{q}\|^2 + \frac{1}{2}m \|\dot{x}\|^2 - U(q, x)$. Assume $J > 0$, $m > 0$.

(a) What are the Euler–Lagrange equations that describe the dynamical flow on the tangent bundle of this homogeneous manifold?

(b) What are Hamilton's equations that describe the dynamical flow on the cotangent bundle of this homogeneous manifold?

8.24. Consider n particles, of mass m_i, $i = 1, \ldots, n$, where each particle is constrained to translate so that its inertial position vector $x_i \in \mathbb{R}^3$ lies in a two-dimensional linear manifold $M_i = \{x_i \in \mathbb{R}^3 : a_i \cdot x_i = d_i\}$, embedded in $\mathbb{R}^3$ for $i = 1, \ldots, n$. We assume $a_i \in \mathbb{R}^3$ satisfies $\|a_i\| = 1$ for $i = 1, \ldots, n$. The particles are influenced by a potential function $U : (\mathbb{R}^3)^n \to \mathbb{R}^1$.

(a) What are the orthogonal projection maps $P_i(x_i) : \mathbb{R}^3 \to \mathsf{T}_{x_i} M_i$, $i = 1, \ldots, n$?

(b) What is the Lagrangian function $L : \mathsf{T}M \to \mathbb{R}^1$ defined on the tangent bundle of the configuration manifold?

(c) What are the Euler–Lagrange equations that describe the dynamical flow on $\mathsf{T}M$?

(d) What is the Hamiltonian function $H : \mathsf{T}^*M \to \mathbb{R}^1$?

(e) What are Hamilton's equations that describe the dynamical flow on T^*M?

(f) Suppose the above Lagrangian function is globally defined on $\mathsf{T}\mathbb{R}^{3n}$. Show that the Euler–Lagrange equations above can be extended to define a Lagrangian vector field on $\mathsf{T}\mathbb{R}^{3n}$. Show that Hamilton's equations above can be extended to define a Hamiltonian vector field on $\mathsf{T}^*\mathbb{R}^{3n}$.

8.25. Consider n particles, of mass m_i, $i = 1, \ldots, n$, where each particle is constrained to translate so that its inertial position vector $x_i \in \mathbb{R}^3$ lies in a one-dimensional linear manifold $M_i = \{x_i \in \mathbb{R}^3 : a_i \cdot x_i = d_i, b_i \cdot x_i = e_i\}$, embedded in $\mathbb{R}^3$ for $i = 1, \ldots, n$. We assume $a_i, b_i \in \mathbb{R}^3$ are linearly independent and $d_i, e_i \in \mathbb{R}^1$ for $i = 1, \ldots, n$. The particles are influenced by a potential function $U : (\mathbb{R}^3)^n \to \mathbb{R}^1$.

(a) What are the orthogonal projection maps $P_i(x_i) : \mathbb{R}^3 \to \mathsf{T}_{x_i} M_i$, $i = 1, \ldots, n$?

(b) What is the Lagrangian function $L : \mathsf{T}M \to \mathbb{R}^1$ defined on the tangent bundle of the configuration manifold?

(c) What are the Euler–Lagrange equations that describe the dynamical flow on $\mathsf{T}M$?

(d) What is the Hamiltonian function $H : \mathsf{T}^*M \to \mathbb{R}^1$?

(e) What are Hamilton's equations that describe the dynamical flow on T^*M?

(f) Suppose the above Lagrangian function is globally defined on $\mathsf{T}\mathbb{R}^{3n}$. Show that the Euler–Lagrange equations above can be extended to define a Lagrangian vector field on $\mathsf{T}\mathbb{R}^{3n}$. Show that Hamilton's equations above can be extended to define a Hamiltonian vector field on $\mathsf{T}^*\mathbb{R}^{3n}$.

8.26. Consider n particles, of mass m_i, $i = 1, \ldots, n$, where each particle is constrained to translate in a circle in a fixed plane in $\mathbb{R}^3$. The inertial position vector of the i-th particle, in an inertial Euclidean frame, is $x_i \in M_i = \left\{ x_i \in \mathbb{R}^3 : a_i^T x_i = b_i, \ \|x_i\|^2 = r_i^2 \right\}$, $i = 1, \ldots, n$; here the vectors $a_1, \ldots, a_n \in \mathbb{R}^3$ are distinct, nonzero vectors, and $b_1, \ldots, b_n, r_1, \ldots, r_n$ are real scalars. The vector $x = (x_1, \ldots, x_n) \in M = M_1 \times \cdots \times M_n$ so that M is the configuration manifold. The particles are influenced by a potential function $U : M \to \mathbb{R}^1$.

(a) What are the orthogonal projection maps $P_i(x_i) : \mathbb{R}^3 \to \mathsf{T}_{x_i} M_i$, $i = 1, \ldots, n$?

(b) What is the Lagrangian function $L : \mathsf{T}M \to \mathbb{R}^1$ defined on the tangent bundle of the configuration manifold?

(c) What are the Euler–Lagrange equations that describe the dynamical flow on $\mathsf{T}M$?

(d) What is the Hamiltonian function $H : \mathsf{T}^*M \to \mathbb{R}^1$?

(e) What are the Hamilton's equations that define a dynamical flow on T^*M?

(f) Suppose the above Lagrangian function is globally defined on $\mathsf{T}\mathbb{R}^{3n}$. Show that the Euler–Lagrange equations above can be extended to define a Lagrangian vector field on $\mathsf{T}\mathbb{R}^{3n}$. Show that Hamilton's equations above can be extended to define a Hamiltonian vector field on $\mathsf{T}^*\mathbb{R}^{3n}$.

8.27. Consider n rigid bodies, each with standard moment of inertia $J_i \in \mathbb{R}^{3\times3}$ defined with respect to a body-fixed frame whose origin is located at the center of mass of the body for $i = 1, \ldots, n$. The rigid bodies can only rotate in $\mathbb{R}^3$, with their motion influenced by a potential function $U : (\mathsf{SO}(3))^n \to \mathbb{R}^1$.

(a) What is the Lagrangian function $L : \mathsf{T}(\mathsf{SO}(3))^n \to \mathbb{R}^1$ defined on the tangent bundle of the configuration manifold?

(b) What are the Euler–Lagrange equations that define a dynamical flow on $\mathsf{T}(\mathsf{SO}(3))^n$?

(c) What is the Hamiltonian function $H : \mathsf{T}^*(\mathsf{SO}(3))^n \to \mathbb{R}^1$?

(d) What are the Hamilton's equations that define a dynamical flow on $\mathsf{T}^*(\mathsf{SO}(3))^n$?

(e) Suppose the above Lagrangian function is globally defined on $T(\mathbb{R}^{3\times3})^n$. Show that the Euler–Lagrange equations above can be extended to define a Lagrangian vector field on $T(\mathbb{R}^{3\times3})^n$. Show that Hamilton's equations above can be extended to define a Hamiltonian vector field on $T^*(\mathbb{R}^{3\times3})^n$.

8.28. Consider n rigid bodies, each with mass m_i and standard moment of inertia $J_i \in \mathbb{R}^{3\times3}$ defined with respect to a body-fixed frame whose origin is located at the center of mass of the body for $i = 1,\ldots,n$. The rigid bodies can translate and rotate in $\mathbb{R}^3$, with their motion influenced by a potential function $U : (\mathsf{SE}(3))^n \to \mathbb{R}^1$.

(a) What is the Lagrangian function $L : \mathsf{T}(\mathsf{SE}(3))^n \to \mathbb{R}^1$ defined on the tangent bundle of the configuration manifold?
(b) What are the Euler–Lagrange equations that define a dynamical flow on $\mathsf{T}(\mathsf{SE}(3))^n$?
(c) What is the Hamiltonian function $H : \mathsf{T}^*(\mathsf{SE}(3))^n \to \mathbb{R}^1$?
(d) What are the Hamilton's equations that define a dynamical flow on $\mathsf{T}^*(\mathsf{SE}(3))^n$?
(e) Suppose the above Lagrangian function is globally defined on $\mathsf{T}(\mathbb{R}^{3\times3} \times \mathbb{R}^3)^n$. Show that the Euler–Lagrange equations above can be extended to define a Lagrangian vector field on $\mathsf{T}(\mathbb{R}^{3\times3} \times \mathbb{R}^3)^n$. Show that Hamilton's equations above can be extended to define a Hamiltonian vector field on $\mathsf{T}^*(\mathbb{R}^{3\times3} \times \mathbb{R}^3)^n$

8.29. Let G be a Lie group. Consider a Lagrangian system defined on the configuration manifold given by the product $\mathsf{G} \times \mathsf{G}$ with hyperregular Lagrangian function $L : \mathsf{T}(\mathsf{G} \times \mathsf{G}) \to \mathbb{R}^1$.

(a) What are the Euler–Lagrange equations for the dynamical flow of this abstract Lagrangian system on $\mathsf{T}(\mathsf{G} \times \mathsf{G})$?
(b) What is the Hamiltonian function $H : \mathsf{T}^*(\mathsf{G} \times \mathsf{G}) \to \mathbb{R}^1$?
(c) What are the Hamilton's equations for the dynamical flow of this abstract Lagrangian system on $\mathsf{T}^*(\mathsf{G} \times \mathsf{G})$?
(d) Show that the Hamiltonian is a conserved quantity of the dynamical flow on $\mathsf{T}(\mathsf{G} \times \mathsf{G})$.
(e) What conditions do the equilibrium solutions of the dynamical flow on $\mathsf{T}(\mathsf{G} \times \mathsf{G})$ satisfy?

8.30. Let G be a Lie group. Consider a Lagrangian system defined on the configuration manifold given by the product $\mathsf{G} \times \mathsf{G}$ with separable Lagrangian function $L : \mathsf{T}(\mathsf{G} \times \mathsf{G}) \to \mathbb{R}$ given by

$$L(g_1, g_2, \xi_1, \xi_2) = L_1(g_1, g_2, \xi_1) + L_2(g_1, g_2, \xi_2),$$

where each $L_i : \mathsf{T}\mathsf{G} \to \mathbb{R}^1$, $i = 1, 2$, is hyperregular.

(a) What are the Euler–Lagrange equations for the dynamical flow of this abstract Lagrangian system on $\mathsf{T}(\mathsf{G} \times \mathsf{G})$?

(b) What are the Hamilton's equations for the dynamical flow of this abstract Lagrangian system on $T^*(G \times G)$?
(c) What is the Hamiltonian function $H : T^*(G \times G) \to \mathbb{R}^1$?
(d) Show that the Hamiltonian is a conserved quantity of the dynamical flow on $T(G \times G)$.
(e) What conditions do the equilibrium solutions of the dynamical flow on $T(G \times G)$ satisfy?

8.31. Let Q be a homogeneous manifold associated with the Lie group G. Consider a Lagrangian system defined on the configuration manifold given by the product $Q \times Q$ with hyperregular Lagrangian function $L : T(Q \times Q) \to \mathbb{R}^1$.

(a) What are the Euler–Lagrange equations for this abstract Lagrangian system on $T(Q \times Q)$?
(b) What is the Hamiltonian function $H : T^*(Q \times Q) \to \mathbb{R}^1$?
(c) What are the Hamilton's equations for this abstract system on $T^*(Q \times Q)$?
(d) Show that the Hamiltonian is a conserved quantity of the dynamical flow on $T(Q \times Q)$.
(e) What conditions do the equilibrium solutions of the dynamical flow on $T(Q \times Q)$ satisfy?

Chapter 9
Rigid and Multi-Body Systems

Although the prior chapters have illustrated the theoretical and analytical benefits of using the geometric formulation that has been introduced, the benefits of this approach are even more apparent when used to study the dynamics of an interconnection of multiple bodies, that is multi-body systems. This chapter considers the detailed dynamics of several examples of rigid body and multi-body systems. In each case, a physical mechanical system is described and viewed from the perspective of geometric mechanics; the configuration manifold is identified and equations of motion are obtained. Euler–Lagrange equations, defined on the tangent bundle of the configuration manifold, and Hamilton's equations, defined on the cotangent bundle of the configuration manifold, are obtained. Conserved quantities are identified for the dynamical flow. Where appropriate, equilibrium solutions are determined. Linear vector fields are described that approximate the dynamical flow in a neighborhood of an equilibrium solution.

As in prior chapters, multi-body systems can consist of connections of lumped mass components and rigid body components. Lumped mass components are massless rigid links and concentrated mass elements; rigid body components are rigid but with spatially distributed mass. Typically, lumped mass components are used to simplify the resulting physics. The examples considered in the prior chapters, and the examples subsequently considered in this chapter, include different categories of assumptions about multi-body systems.

There is much published literature on rigid and multi-body systems. The references [3, 86] provide a traditional treatment of multi-body dynamics and demonstrate the extreme complexity that arises in both describing and analyzing multi-body dynamics. In contrast to much of the existing literature on rigid and multi-body systems, the following examples, developed from the perspective of geometric mechanics, illustrate an alternative approach that makes important use of the geometry of the identified configuration manifold

© Springer International Publishing AG 2018
T. Lee et al., *Global Formulations of Lagrangian and Hamiltonian
Dynamics on Manifolds*, Interaction of Mechanics and Mathematics,
DOI 10.1007/978-3-319-56953-6_9

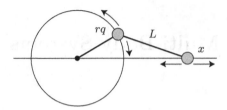

Fig. 9.1 Two masses connected by a rigid link with ends constrained to slide along a straight line and a circle

as an embedded manifold. This approach leads to novel formulations of the equations of motion that provide a basis for analyzing the dynamics globally on the configuration manifold.

9.1 Dynamics of a Planar Mechanism

A planar mechanism consists of two masses connected by a massless rigid link. One mass element is constrained to slide, without friction, along a straight line; the other mass element is constrained to slide, without friction, along a circle. For simplicity, assume that the straight line and the circle lie in a common vertical plane, with the straight line horizontal and passing through the center of the circle. The two mass elements are connected by a rigid, massless link. If the mass element constrained to the circle is made to rotate at a constant rate, this is referred to as a slider-crank mechanism since the end of the other mass element undergoes periodic motion along the straight line; the mechanism converts rotational motion of the first mass into linear or translational motion of the other mass. Here, we study the dynamics of the mechanism, under the action of uniform, constant gravity; an external force acts on the mass constrained to move on the circle and an external force acts on the mass contained to slide along a straight line. This mechanism is illustrated in Figure 9.1.

A two-dimensional inertial frame in the fixed vertical plane is constructed with the first axis along the horizontal straight line and the second axis vertical; the origin of the frame is located at the center of the circle. As usual, e_1, e_2 denote the standard basis for $\mathbb{R}^2$.

Let m_1 denote the mass of the element that slides on the circle; let m_2 denote the mass of the element that slides on the straight line. Also, L denotes the length of the link and the radius of the circle is r. We assume that $L > r$.

Let $q \in \mathsf{S}^1$ denote the direction vector of the mass element that moves on the circle in the inertial frame, and let $x \in \mathbb{R}^1$ denote the position of the mass element that moves on the straight line in the inertial frame. The geometry of the mechanism implies that the position of the sliding element always satisfies $L - r \leq x \leq L + r$ or $-L - r \leq x \leq -L + r$. Since these regions are disjoint, there are two distinct regions of operation. In the subsequent analysis, we study only the region where $L - r \leq x \leq L + r$.

In addition, the two mass elements satisfy the constraint defined by the length of the rigid link, namely

$$\|rq - xe_1\|^2 - L^2 = 0,$$

where $e_1 = [1, 0]^T \in \mathbb{R}^2$.

The configuration manifold is given by

$$M = \{(q, x) \in S^1 \times \mathbb{R}^1 : \|rq - xe_1\|^2 - L^2 = 0, \ L - r \leq x \leq L + r\},$$

and the tangent plane at $(q, x) \in M$ is

$$\mathsf{T}_{(q,x)}M = \left\{(\dot{q}, \dot{x}) \in \mathsf{T}_{(q,x)}(S^1 \times \mathbb{R}^1) : (rq - xe_1)^T(r\dot{q} - \dot{x}e_1) = 0\right\}.$$

Thus, the dimension of the configuration manifold is one and the mechanism has one degree of freedom. The dynamics of the connected mass elements constrained to slide along a straight line and circle evolve on the tangent bundle of the configuration manifold $\mathsf{T}M$.

It can be shown that the manifolds M and S^1 are diffeomorphic, so that we could also choose the latter as the configuration manifold. We use M as the configuration manifold since this choice leads to slightly less complicated expressions. In any event, the dynamics of the one degree of freedom mechanism are surprising complicated.

As shown from the kinematics analysis of this mechanism in Chapter 2, there exists an $\omega \in \mathbb{R}^1$ such that the kinematics of the mechanism are given by

$$\begin{bmatrix} \dot{q} \\ \dot{x} \end{bmatrix} = \omega \begin{bmatrix} I_{2 \times 2} \\ \frac{-rxe_1^T}{(re_1^T q - x)} \end{bmatrix} Sq. \tag{9.1}$$

We also assume that external forces act on the two mass elements. An external force with magnitude denoted by $F_1 \in \mathbb{R}^1$ acts on the mass element constrained to move on the fixed circle. The direction of this external force is assumed to lie in the tangent plane $\mathsf{T}_q S^1$; consequently this force is $F_1 Sq \in \mathbb{R}^2$. An external force with magnitude denoted by $F_2 \in \mathbb{R}^1$ acts on the mass element constrained to move on the fixed straight line. This external force is $F_2 e_1 \in \mathbb{R}^2$.

9.1.1 Euler–Lagrange Equations

The modified Lagrangian function $\tilde{L} : \mathsf{T}M \to \mathbb{R}^1$ is the sum of the kinetic energy of the mass elements minus the gravitational potential energy. It can be expressed as

$$\tilde{L}(q, x, \omega, \dot{x}) = \frac{1}{2}m_1 r^2 \omega^2 + \frac{1}{2}m_2 \dot{x}^2 - m_1 rg e_2^T q,$$

where $e_2 = [0, 1]^T \in \mathbb{R}^2$. Using the kinematic expression for $\dot{x}$, the modified Lagrangian function can be written as

$$\tilde{L}(q, x, \omega) = \frac{1}{2}\left\{m_1 + m_2 N(q, x)\right\} r^2 \omega^2 - m_1 rg e_2^T q,$$

where

$$N(q, x) = \frac{(x e_1^T S q)^2}{(r e_1^T q - x)^2}.$$

We obtain the Euler–Lagrange equations from the Lagrange–d'Alembert principle and the expressions for the Lagrangian function and the virtual work done by the external forces. The infinitesimal variation of the action integral is

$$\delta\mathfrak{G} = \int_{t_0}^{t_f} \left[\left\{m_1 + m_2 N(q, x)\right\} r^2 \omega \delta\omega \right.$$
$$\left. + \frac{1}{2}m_2 r^2 \omega^2 \left\{ \frac{\partial N(q, x)}{\partial q} + \frac{\partial N(q, x)}{\partial x}\frac{\partial x}{\partial q} \right\}^T \delta q - m_1 rg e_2^T \delta q \right] dt.$$

The virtual work done by the two external forces is

$$\int_{t_0}^{t_f} \{F_1 q^T S^T r \delta q + F_2 \delta x\}\, dt = \int_{t_0}^{t_f} \left\{ F_1 rSq + F_2 \frac{\partial x}{\partial q} \right\}^T \delta q\, dt.$$

In the above expressions, it is easy to show that

$$\frac{\partial x}{\partial q} = \frac{r x e_1}{(x - r e_1^T q)}.$$

The partial derivatives of $N(q, x)$ can also be determined to obtain

$$\frac{\partial N(q, x)}{\partial q} + \frac{\partial N(q, x)}{\partial x}\frac{\partial x}{\partial q} = \frac{1}{(x - r e_1^T q)^4} \begin{bmatrix} 2r(x e_1^T Sq)^2(x - 2r e_1^T q) \\ -2x^2 e_1^T Sq(x - r e_1^T q)^2 \end{bmatrix}.$$

Thus, the Lagrange–d'Alembert principle gives

$$\int_{t_0}^{t_f}\left[\{m_1+m_2N(q,x)\}r^2\omega\delta\omega+\frac{1}{2}m_2r^2\omega^2\Big\{\frac{\partial N(q,x)}{\partial q}+\frac{\partial N(q,x)}{\partial x}\frac{\partial x}{\partial q}\Big\}^T\delta q\right.$$
$$\left.-m_1rge_2^T\delta q\right]dt=-\int_{t_0}^{t_f}\Big\{F_1rSq+F_2\frac{\partial x}{\partial q}\Big\}^T\delta q\,dt,$$

for all infinitesimal variations $\delta q:[t_0,t_f]\to\mathsf{T}_q\mathsf{S}^1$ that vanish at the endpoints, $\delta q(t_0)=\delta q(t_f)=0$.

We use the expressions for the infinitesimal variations in S^1,

$$\delta q=\gamma Sq,$$
$$\delta\omega=\dot\gamma,$$

for differentiable curves $\gamma:[t_0,t_f]\to\mathbb{R}^2$ satisfying $\gamma(t_0)=\gamma(t_f)=0$. Substitute these into the expression for the variational principle. Integrating the result by parts and using the fundamental lemma of the calculus of variations, described in Appendix A, leads to the dynamical equations, which consist of the rotational kinematics given by (9.1), and the Euler–Lagrange equation

$$\{m_1+m_2N(q,x)\}\,r^2\dot\omega+\frac{1}{2}m_2r^2\omega^2\left(\frac{\partial N(q,x)}{\partial q}+\frac{\partial N(q,x)}{\partial x}\frac{\partial x}{\partial q}\right)^T Sq$$
$$+m_1rge_2^T Sq=F_1r+F_2\left(\frac{\partial x}{\partial q}\right)^T Sq. \tag{9.2}$$

The dynamics of the mechanism are described by the differential equations (9.1) and (9.2); these equations describe the dynamics of the mechanism expressed in terms of $(q,x,\dot q,\dot x)\in\mathsf{T}M$ on the tangent bundle of the configuration manifold M.

9.1.2 Hamilton's Equations

Hamilton's equations are determined by introducing the Legendre transformation

$$\pi=\frac{\partial\tilde L(q,x,\omega)}{\partial\omega}=(m_1+m_2N(q,x))r^2\omega.$$

Here, $\pi\in\mathsf{T}^*_{(q,x)}M$ is conjugate to $\omega\in\mathsf{T}_{(q,x)}M$. The modified Hamiltonian function $\tilde H:\mathsf{T}^*M\to\mathbb{R}^1$ is

$$\tilde H(q,x,\pi)=\frac{1}{2}\frac{\pi^2}{(m_1+m_2N(q,x))r^2}+m_1rge_2^T q.$$

Thus, from (4.24) and (4.25), Hamilton's equations are given by

$$\dot{q} = \frac{\pi S q}{(m_1 + m_2 N(q, x)) r^2},$$ (9.3)

$$\dot{\pi} = -\frac{1}{2} \frac{m_2 \pi^2}{(m_1 + m_2 N(q, x))^2 r^2} q^T S \left(\frac{\partial N(q, x)}{\partial q} + \frac{\partial N(q, x)}{\partial x} \frac{\partial x}{\partial q} \right) + m_1 r g q^T S e_2$$

$$+ F_1 r + F_2 \left(\frac{\partial x}{\partial q} \right)^T S q.$$ (9.4)

These differential equations, together with (9.1), describe the Hamiltonian flow of the mechanism in terms of $(q, x, \dot{q}, \dot{x}) \in T^* M$ on the cotangent bundle of the configuration manifold M.

9.1.3 Conservation Properties

The Hamiltonian given by

$$H = \frac{1}{2} m_1 r^2 \omega^2 + \frac{1}{2} m_2 \dot{x}^2 + m_1 r g e_2^T q$$

which coincides with the total energy E in this case is constant along each solution of the dynamical flow.

9.1.4 Equilibrium Properties

We consider the case where the external forces vanish, $F_1 = F_2 = 0$. The set of equilibrium solutions of the system occur when the mass elements are stationary, and the moment due to gravity vanishes. The condition for an equilibrium configuration is given by

$$e_2^T S q = 0.$$

Thus, there are two equilibrium configurations in M, namely $(-e_2, \sqrt{L^2 - r^2})$ and $(e_2, \sqrt{L^2 - r^2})$. Furthermore, the two equilibrium solutions in TM are $(-e_2, \sqrt{L^2 - r^2}, 0, 0)$ and $(e_2, \sqrt{L^2 - r^2}, 0, 0)$.

We examine the equilibrium solution $(-e_2, \sqrt{L^2 - r^2}, 0, 0) \in TM$, following the development in Appendix B. Equation (9.2) can be linearized to obtain the linear differential equation

$$(m_1 + m_2) r^2 \ddot{\xi}_1 + m_1 g \xi_1 = 0,$$

which is defined on the two-dimensional tangent space of TM at the equilibrium point $(-e_2, \sqrt{L^2 - r^2}, 0, 0) \in TM$. This linear vector field approximates the Lagrangian vector field of the planar mechanism in a neighborhood of this equilibrium solution. The eigenvalues are purely imaginary. This equilibrium solution can be shown to be stable since the total energy has a strict local minimum at the equilibrium with zero time derivative, thereby guaranteeing stability of the equilibrium solution.

A similar development can be carried out for the second equilibrium solution given by $(e_2, \sqrt{L^2 - r^2}, 0, 0) \in \mathsf{T}M$. Equation(9.2) can be linearized to obtain the linear differential equation

$$(m_1 + m_2)r^2\ddot{\xi}_1 - m_1 g \xi_1 = 0,$$

which is defined on the two-dimensional tangent space of $\mathsf{T}M$ at the equilibrium point $(e_2, \sqrt{L^2 - r^2}, 0, 0) \in \mathsf{T}M$. This linear vector field approximates the Lagrangian vector field of the planar mechanism in a neighborhood of this equilibrium solution. This equilibrium solution can be shown to be unstable since the linearized differential equation has a positive eigenvalue.

9.2 Dynamics of a Horizontally Rotating Pendulum on a Cart

A planar pendulum can rotate in a fixed horizontal plane about a fixed frictionless joint whose vertical axis goes through the center of mass of a cart, viewed as a flat plate, that can translate without friction in a horizontal plane. The planar pendulum is assumed to be rigid with its mass concentrated at the end of the pendulum. Since the axis of the pendulum goes through the center of mass of the cart, the pendulum exerts no moment on the cart; hence we assume the cart translates without rotating.

A two-dimensional inertial Euclidean frame lies in the horizontal plane of motion of the cart. The position vector of the center of mass of the cart in the horizontal inertial frame is given by $x \in \mathbb{R}^2$ and the attitude vector of the planar pendulum in this inertial frame is given by $q \in \mathsf{S}^1$. Thus, the configuration is described by $(x, q) \in \mathbb{R}^2 \times \mathsf{S}^1$ and the configuration manifold is $\mathbb{R}^2 \times \mathsf{S}^1$. There are three degrees of freedom. A schematic of a horizontally rotating planar pendulum on a cart is shown in Figure 9.2.

The length of the planar pendulum is L; the mass of the pendulum, concentrated at its end, is m; the mass of the cart is M.

9.2.1 Euler–Lagrange Equations

The position vector of the pendulum mass and the center of mass of the cart are $x + Lq \in \mathbb{R}^2$ and $x \in \mathbb{R}^2$, respectively, in the inertial frame. Thus, the

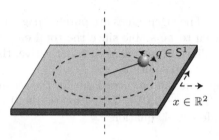

Fig. 9.2 Horizontally rotating pendulum on a cart

velocity vector of the pendulum mass is $\dot{x} + L\dot{q} \in \mathbb{R}^2$ and the velocity vector of the center of mass of the cart is $\dot{x} \in \mathbb{R}^2$.

The Lagrangian function $L : \mathsf{T}(\mathbb{R}^2 \times \mathsf{S}^1) \to \mathbb{R}^1$ is the sum of the kinetic energy of the cart and the kinetic energy of the pendulum mass; no forces act on the pendulum or the cart. Thus, the Lagrangian function $L : \mathsf{T}(\mathbb{R}^2 \times \mathsf{S}^1) \to \mathbb{R}^1$ is

$$L(x, q, \dot{x}, \dot{q}) = \frac{1}{2}M \left\|\dot{x}\right\|^2 + \frac{1}{2}m \left\|\dot{x} + L\dot{q}\right\|^2$$
$$= \frac{1}{2}(M + m) \left\|\dot{x}\right\|^2 + mL\dot{x}^T \dot{q} + \frac{1}{2}mL^2 \left\|\dot{q}\right\|^2 .$$

It is convenient to introduce the angular velocity of the horizontal pendulum, $\omega \in \mathbb{R}^1$. The rotational kinematics are

$$\dot{q} = \omega S q, \tag{9.5}$$

where S is the 2×2 skew-symmetric matrix used in prior chapters. Thus, the modified Lagrangian can be expressed in terms of the angular velocity of the pendulum as

$$\tilde{L}(x, q, \dot{x}, \omega) = \frac{1}{2}(M + m) \left\|\dot{x}\right\|^2 + mL\dot{x}^T S q \omega + \frac{1}{2}mL^2\omega^2.$$

The Euler–Lagrange equations can be obtained using Hamilton's principle:

$$\delta \mathfrak{G} = \int_{t_0}^{t_f} \Big[\{(M + m)\dot{x} + mLSq\omega\}^T \delta\dot{x}$$
$$+ \left\{ mL\dot{x}^T S q + mL^2\omega \right\} \delta\omega + \left\{ mL\dot{x}^T S\omega \right\} \delta q \Big] dt = 0,$$

for all infinitesimal variations $\delta x : [t_0, t_f] \to \mathsf{T}_x\mathbb{R}^2$ and $\delta q : [t_0, t_f] \to \mathsf{T}_q\mathsf{S}^1$ that vanish at the endpoints, $\delta x(t_0) = \delta x(t_f) = 0$ and $\delta q(t_0) = \delta q(t_f) = 0$. We use the expressions for the infinitesimal variations in S^1 in Chapter 4,

$$\delta q = \gamma S q,$$
$$\delta \omega = \dot{\gamma},$$

for differentiable curves $\gamma : [t_0, t_f] \to \mathbb{R}^2$ that vanish at the endpoints, $\gamma(t_0) = \gamma(t_f) = 0$. The infinitesimal variations in $\mathbb{R}^2$ are differentiable functions $\delta x : [t_0, t_f] \to \mathbb{R}^2$ that vanish at the endpoints, $\delta x(t_0) = \delta x(t_f) = 0$.

Integrating by parts and using the fundamental lemma of the calculus of variations in Appendix A leads to the Euler–Lagrange equations

$$\frac{d}{dt}\left\{ (M+m)\dot{x} + mLSq\omega \right\} = 0,$$

$$\frac{d}{dt}\left\{ mL\dot{x}^T Sq + mL^2\omega \right\} + mL\dot{x}^T q\omega = 0,$$

which can be expanded to obtain the equations in matrix-vector form

$$\begin{bmatrix} (M+m)I_{2\times2} & mLSq \\ mLq^T S^T & mL^2 \end{bmatrix} \begin{bmatrix} \ddot{x} \\ \dot{\omega} \end{bmatrix} + \begin{bmatrix} -mLq\omega^2 \\ 0 \end{bmatrix} = \begin{bmatrix} 0 \\ 0 \end{bmatrix}. \tag{9.6}$$

Thus, the Euler–Lagrange equations (9.6) and the kinematics (9.5) describe the dynamical flow of the horizontal pendulum on a cart in terms of $(x, q, \dot{x}, \dot{q}) \in \mathsf{T}(\mathbb{R}^2 \times \mathsf{S}^1)$ on the tangent bundle of $\mathbb{R}^2 \times \mathsf{S}^1$.

9.2.2 Hamilton's Equations

Hamilton's equations on the cotangent bundle $\mathsf{T}^*(\mathbb{R}^2 \times \mathsf{S}^1)$ are obtained by defining the momentum according to the Legendre transformation, which is given by

$$\begin{bmatrix} p \\ \pi \end{bmatrix} = \begin{bmatrix} \frac{\partial \tilde{L}(x,q,\dot{x},\omega)}{\partial \dot{x}} \\ \frac{\partial \tilde{L}(x,q,\dot{x},\omega)}{\partial \omega} \end{bmatrix} = \begin{bmatrix} (M+m)I_{2\times2} & mLSq \\ mLq^T S^T & mL^2 \end{bmatrix} \begin{bmatrix} \dot{x} \\ \omega \end{bmatrix}.$$

Here, the momenta $(p, \pi) \in \mathsf{T}^*_{(x,q)}(\mathbb{R}^2 \times \mathsf{S}^1)$ are conjugate to $(\omega, \dot{x}) \in \mathsf{T}_{(x,q)}(\mathbb{R}^2 \times \mathsf{S}^1)$. Consequently,

$$\begin{bmatrix} \dot{x} \\ \omega \end{bmatrix} = \begin{bmatrix} M_{11}^I(q) & M_{12}^I(q) \\ M_{21}^I(q) & M_{22}^I \end{bmatrix} \begin{bmatrix} p \\ \pi \end{bmatrix},$$

where

$$\begin{bmatrix} M_{11}^I(q) & M_{12}^I(q) \\ M_{21}^I(q) & M_{22}^I \end{bmatrix} = \begin{bmatrix} (M+m)I_{2\times2} & mLSq \\ mLq^T S^T & mL^2 \end{bmatrix}^{-1} = \begin{bmatrix} \frac{MI_{2\times2}+mSqq^T S^T}{M(M+m)} & -\frac{Sq}{ML} \\ \frac{-q^T S^T}{ML} & \frac{M+m}{MmL^2} \end{bmatrix}$$

is the inverse of the indicated 3×3 partitioned matrix. The modified Hamiltonian function $\tilde{H} : T^*(\mathbb{R}^2 \times S^1) \to \mathbb{R}^1$ is

$$\tilde{H}(x, q, p, \pi) = \frac{1}{2} \begin{bmatrix} p \\ \pi \end{bmatrix}^T \begin{bmatrix} M_{11}^I(q) & M_{12}^I(q) \\ M_{21}^I(q) & M_{22}^I \end{bmatrix} \begin{bmatrix} p \\ \pi \end{bmatrix}.$$

Thus, Hamilton's equations for the horizontal pendulum on a cart are

$$\begin{bmatrix} \dot{x} \\ \dot{q} \end{bmatrix} = \begin{bmatrix} M_{11}^I(q) & M_{12}^I(q) \\ SqM_{21}^I(q) & SqM_{22}^I \end{bmatrix} \begin{bmatrix} p \\ \pi \end{bmatrix}, \tag{9.7}$$

and

$$\dot{p} = 0, \tag{9.8}$$

$$\dot{\pi} = \left(\frac{m}{M(M+m)} q^T S^T p - \frac{1}{ML} \pi \right) q^T p. \tag{9.9}$$

The Hamiltonian dynamical flow for the horizontal pendulum on a cart is described by equations (9.7), (9.8), and (9.9), in terms of $(x, q, p, \pi) \in T^*(\mathbb{R}^2 \times S^1)$ on the cotangent bundle of $(\mathbb{R}^2 \times S^1)$.

9.2.3 Conservation Properties

The Hamiltonian of the horizontal pendulum and the cart, which coincides with the total energy E in this case, is given by

$$H = \frac{1}{2}(M+m) \|\dot{x}\|^2 + mL\dot{x}^T Sq\omega + \frac{1}{2}mL^2\omega^2,$$

and it is constant along each solution of the dynamical flow.

Hamilton's equations demonstrate that the translational momentum of the system is conserved. That is, the translational momentum

$$p = (M+m)\dot{x} + mLSq\omega$$

is constant along each solution of the dynamical flow. This can be viewed as a consequence of Noether's theorem applied to the translational invariance of the Lagrangian function.

9.2.4 Equilibrium Properties

The equilibrium solutions of the horizontal pendulum on a cart are easily determined. It is clear that any configuration of the system is an equilibrium solution so long as the velocity is zero. That is, the horizontal pendulum is in equilibrium for any constant horizontal position of the cart and for any constant attitude of the horizontal pendulum.

We do not study the linearized dynamics for an equilibrium solution. Since there is no potential, the eigenvalues of the linearized dynamics are necessarily all zero.

9.3 Dynamics of a Connection of a Planar Pendulum and Spherical Pendulum

A planar pendulum can rotate in a fixed vertical plane about a fixed frictionless joint whose axis is perpendicular to the vertical plane. One end of a spherical pendulum is connected to the end of the planar pendulum by a frictionless joint so that it can rotate in three dimensions. Each of the two pendulum links is assumed to be rigid with its mass concentrated at the end of the link.

An inertial Euclidean frame is constructed so that the first two axes lie in the horizontal plane containing the joint connection of the planar pendulum, and the third axis is vertical. The plane of rotation of the planar pendulum is defined by the second and third axes of the inertially fixed frame.

The configuration of the planar pendulum $q_1 \in \mathsf{S}^1$ is defined by the attitude vector of the planar pendulum with respect to the second and third axes of the inertial frame; thus S^1 is the configuration manifold of the planar pendulum. The configuration of the spherical pendulum $q_2 \in \mathsf{S}^2$ is defined by the attitude vector of the spherical pendulum with respect to the inertial frame so that S^2 is the configuration manifold of the spherical pendulum. Thus, the configuration is described by $q = (q_1, q_2) \in \mathsf{S}^1 \times \mathsf{S}^2$ and the configuration manifold is $\mathsf{S}^1 \times \mathsf{S}^2$. The connection of a planar pendulum and a spherical pendulum has three degrees of freedom. A schematic of a serial connection of a planar pendulum and a spherical pendulum is shown in Figure 9.3.

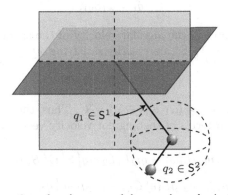

Fig. 9.3 Serial connection of a planar pendulum and a spherical pendulum

The length of the planar pendulum is L_1 and its mass is m_1; the length of the spherical pendulum is L_2 and its mass is m_2. Here, g denotes the constant acceleration of gravity.

9.3.1 Euler–Lagrange Equations

The position vector of the first mass element in the inertial frame is given by

$$x_1 = L_1 Q q_1 \in \mathbb{R}^3,$$

where the matrix $Q \in \mathbb{R}^{3 \times 2}$ is defined as

$$Q = \begin{bmatrix} 0 & 0 \\ 1 & 0 \\ 0 & 1 \end{bmatrix},$$

which defines an embedding of $\mathbb{R}^2$ into $\mathbb{R}^3$, and it satisfies $Q^T Q = I_{2 \times 2}$. The position vector of the second mass element in the inertial frame is given by

$$x_2 = x_1 + L_2 q_2 \in \mathbb{R}^3.$$

The Lagrangian function $L : \mathsf{T}(\mathsf{S}^1 \times \mathsf{S}^2) \to \mathbb{R}^1$ is given by

$$
\begin{aligned}
L(q, \dot{q}) &= \frac{1}{2} m_1 \|\dot{x}_1\|^2 + \frac{1}{2} m_2 \|\dot{x}_2\|^2 - m_1 g e_3^T x_1 - m_2 g e_3^T x_2 \\
&= \frac{1}{2}(m_1 + m_2) L_1^2 \|\dot{q}_1\|^2 + m_2 L_1 L_2 \dot{q}_1^T Q^T \dot{q}_2 + \frac{1}{2} m_2 L_2^2 \|\dot{q}_2\|^2 \\
&\quad - (m_1 + m_2) g L_1 e_3^T Q q_1 - m_2 g L_2 e_3^T q_2.
\end{aligned}
$$

It is convenient to introduce the angular velocities of the pendulums. As shown in Chapter 4, the scalar angular velocity of the planar pendulum about its joint axis is a scalar-valued function $\omega_1 \in \mathsf{T}_{q_1} \mathsf{S}^1$, such that

$$\dot{q}_1 = \omega_1 S q_1. \tag{9.10}$$

As shown in Chapter 5, the angular velocity of the spherical pendulum is a vector function $\omega_2 \in \mathsf{T}_{q_2} \mathsf{S}^2$, necessarily satisfying $\omega_2^T q_2 = 0$, such that

$$\dot{q}_2 = S(\omega_2) q_2. \tag{9.11}$$

We use the notation $\omega = (\omega_1, \omega_2) \in \mathbb{R}^1 \times \mathbb{R}^3$. Thus, the modified Lagrangian can be expressed in terms of the angular velocity vector as

$$
\begin{aligned}
\tilde{L}(q, \omega) &= \frac{1}{2}(m_1 + m_2) L_1^2 \omega_1^2 - m_2 L_1 L_2 \omega_1 q_1^T S^T Q^T S(q_2) \omega_2 + \frac{1}{2} m_2 L_2^2 \|\omega_2\|^2 \\
&\quad - (m_1 + m_2) g L_1 e_3^T Q q_1 - m_2 g L_2 e_3^T q_2.
\end{aligned}
$$

The Euler–Lagrange equations can be obtained using Hamilton's principle where the infinitesimal variation of the action integral is given by

$$\delta\mathfrak{G} = \int_{t_0}^{t_f} \Big[\{(m_1 + m_2)L_1^2\omega_1 - m_2L_1L_2q_1^T S^T Q^T S(q_2)\omega_2\} \, \delta\omega_1$$
$$- \{m_2L_1L_2\omega_2^T S^T(q_2)QS\omega_1 + (m_1+m_2)gL_1e_3^T Q\} \, \delta q_1$$
$$+ \{-m_2L_1L_2\omega_1 S^T(q_2)QSq_1 + m_2L_2^2\omega_2\}^T \, \delta\omega_2$$
$$+ \{m_2L_1L_2\omega_1 S^T(\omega_2)QSq_1 - m_2gL_2e_3\}^T \, \delta q_2 \Big] dt = 0, \quad (9.12)$$

for all infinitesimal variations $\delta q_1 : [t_0, t_f] \to \mathsf{T}_{q_1}\mathsf{S}^1$ and $\delta q_2 : [t_0, t_f] \to \mathsf{T}_{q_2}\mathsf{S}^2$ that vanish at the endpoints, $\delta q_1(t_0) = \delta q_1(t_f) = 0$ and $\delta q_2(t_0) = \delta q_2(t_f) = 0$. The infinitesimal variations in S^1 and in S^2 can be written as

$$\delta q_1 = \gamma_1 S q_1,$$
$$\delta\omega_1 = \dot\gamma_1,$$
$$\delta q_2 = S(\gamma_2)q_2,$$
$$\delta\omega_2 = S(\gamma_2)\omega_2 + (I - q_2q_2^T)\dot\gamma_2,$$

for differentiable curves $\gamma_1 : [t_0, t_f] \to \mathbb{R}^1$ satisfying $\gamma_1(t_0) = \gamma_1(t_f) = 0$, and for differentiable curves $\gamma_2 : [t_0, t_f] \to \mathbb{R}^3$ satisfying $\gamma_2(t_0) = \gamma_2(t_f) = 0$. Substitute these into the above variational principle. Integrate the result by parts and use the fundamental lemma of the calculus of variations, as in Appendix A, to obtain the Euler–Lagrange equations

$$\frac{d}{dt}\{(m_1+m_2)L_1^2\omega_1 - m_2L_1L_2q_1^T S^T Q^T(q_2 \times \omega_2)\}$$
$$+m_2L_1L_2\omega_2^T(q_2 \times Q\omega_1q_1) + (m_1+m_2)gL_1e_3^T QSq_1 = 0,$$
$$(I_{2\times2} - q_2q_2^T)\frac{d}{dt}\{-m_2L_1L_2\omega_1 S^T(q_2)QSq_1 + m_2L_2^2\omega_2\}$$
$$-q_2 \times \{m_2L_1L_2\omega_1 S^T(\omega_2)QSq_1 - m_2gL_2e_3\} = 0.$$

These can be further expanded and rearranged to obtain the following expression

$$\begin{bmatrix} (m_1+m_2)L_1^2 & -m_2L_1L_2q_1^T S^T Q^T S(q_2) \\ -m_2L_1L_2 S^T(q_2)QSq_1 & m_2L_2^2 I_{3\times3} \end{bmatrix} \begin{bmatrix} \dot\omega_1 \\ \dot\omega_2 \end{bmatrix}$$
$$+ \begin{bmatrix} -m_2L_1L_2q_1^T SQ^T S(\omega_2)^2 q_2 \\ m_2L_1L_2q_2 \times \{-Qq_1\omega_1^2 + \omega_1 S(q_2)S(QSq_1)S(\omega_2)q_2 + \omega_1 S(\omega_2)QSq_1\} \end{bmatrix}$$
$$+ \begin{bmatrix} (m_1+m_2)gL_1e_3^T QSq_1 \\ m_2gL_2q_2 \times e_3 \end{bmatrix} = 0. \quad (9.13)$$

Thus, the Euler–Lagrange equations (9.13) and the kinematics (9.10) and (9.11) describe the dynamical flow for the connection of a planar pendulum and a spherical pendulum in terms of $(q_1, q_2, \dot{q}_1, \dot{q}_2) \in \mathsf{T}(\mathsf{S}^1 \times \mathsf{S}^2)$ on the tangent bundle of $\mathsf{S}^1 \times \mathsf{S}^2$.

9.3.2 Hamilton's Equations

Hamilton's equations can be determined by introducing the Legendre transformation

$$\begin{bmatrix} \pi_1 \\ \pi_2 \end{bmatrix} = \begin{bmatrix} \dfrac{\partial \tilde{L}(q,\omega)}{\partial \omega_1} \\ (I_{3\times 3} - q_2 q_2^T)\dfrac{\partial \tilde{L}(q,\omega)}{\partial \omega_2} \end{bmatrix}$$

$$= \begin{bmatrix} (m_1 + m_2)L_1^2 & -m_2 L_1 L_2 q_1^T S^T Q^T S(q_2) \\ -m_2 L_1 L_2 S^T(q_2) Q S q_1 & m_2 L_2^2 I_{3\times 3} \end{bmatrix} \begin{bmatrix} \omega_1 \\ \omega_2 \end{bmatrix},$$

where $\pi_1 \in \mathsf{T}_{q_1}^* \mathsf{S}^1$ and $\pi_2 \in \mathsf{T}_{q_2}^* \mathsf{S}^2$ are momenta that are conjugate to $\omega_1 \in \mathsf{T}_{q_1}\mathsf{S}^1$ and $\omega_2 \in \mathsf{T}_{q_2}\mathsf{S}^2$. Consequently,

$$\begin{bmatrix} \omega_1 \\ \omega_2 \end{bmatrix} = \begin{bmatrix} M_{11}^I(q) & M_{12}^I(q) \\ M_{21}^I(q) & M_{22}^I(q) \end{bmatrix} \begin{bmatrix} \pi_1 \\ \pi_2 \end{bmatrix},$$

where

$$\begin{bmatrix} M_{11}^I(q) & M_{12}^I(q) \\ M_{21}^I(q) & M_{22}^I(q) \end{bmatrix} = \begin{bmatrix} (m_1 + m_2)L_1^2 & -m_2 L_1 L_2 q_1^T S^T Q^T S(q_2) \\ -m_2 L_1 L_2 S^T(q_2) Q S q_1 & m_2 L_2^2 I_{3\times 3} \end{bmatrix}^{-1}$$

is the inverse of the indicated 4×4 partitioned matrix. The modified Hamiltonian function $\tilde{H} : \mathsf{T}^*(\mathsf{S}^1 \times \mathsf{S}^2) \to \mathbb{R}^1$ is

$$\tilde{H}(q, \pi) = \frac{1}{2} \begin{bmatrix} \pi_1 \\ \pi_2 \end{bmatrix} \begin{bmatrix} M_{11}^I(q) & M_{12}^I(q) \\ M_{21}^I(q) & M_{22}^I(q) \end{bmatrix} \begin{bmatrix} \pi_1 \\ \pi_2 \end{bmatrix}$$
$$+ (m_1 + m_2)gL_1 e_3^T Q q_1 + m_2 g L_2 e_3^T q_2.$$

Thus, from (4.24) and (5.28), we obtain

$$\begin{bmatrix} \dot{q}_1 \\ \dot{q}_2 \end{bmatrix} = \begin{bmatrix} S q_1 M_{11}^I(q) & S q_1 M_{12}^I(q) \\ -S(q_2) M_{21}^I(q) & -S(q_2) M_{22}^I(q) \end{bmatrix} \begin{bmatrix} \pi_1 \\ \pi_2 \end{bmatrix}, \qquad (9.14)$$

and also from (4.25) and (5.29), we obtain

$$\begin{bmatrix} \dot{\pi}_1 \\ \dot{\pi}_2 \end{bmatrix} = \begin{bmatrix} q_1^T S \dfrac{\partial \tilde{H}}{\partial q_1} \\ -q_2 \times \dfrac{\partial \tilde{H}}{\partial q_2} + \dfrac{\partial \tilde{H}}{\partial \pi_2} \times \pi_2 \end{bmatrix}.$$

From (9.12), the above equation can be rewritten as

$$
\begin{bmatrix} \dot{\pi}_1 \\ \dot{\pi}_2 \end{bmatrix} = \begin{bmatrix} -m_2 L_1 L_2 \omega_2^T S(q_2) Q \omega_1 q_1 \\ m_2 L_1 L_2 \left\{ S(\omega_2) S(q_2) - S(q_2) S(\omega_2) \right\} Q S q_1 \omega_1 \end{bmatrix}
$$
$$
+ \begin{bmatrix} -(m_1 + m_2) g L_1 e_3^T Q S q_1 \\ -m_2 g L_2 q_2 \times e_3 \end{bmatrix}. \tag{9.15}
$$

These equations (9.14) and (9.15) describe the dynamical flow for a planar pendulum connected to a spherical pendulum in terms of the evolution of $(q_1, q_2, \pi_1, \pi_2) \in \mathsf{T}^*(\mathsf{S}^1 \times \mathsf{S}^2)$ on the cotangent bundle of $\mathsf{S}^1 \times \mathsf{S}^2$.

9.3.3 Conservation Properties

The Hamiltonian given by

$$
H = \frac{1}{2}(m_1 + m_2) L_1^2 \omega_1^2 - m_2 L_1 L_2 \omega_1 q_1^T S^T Q^T S(q_2) \omega_2 + \frac{1}{2} m_2 L_2^2 \|\omega_2\|^2
$$
$$
+ (m_1 + m_2) g L_1 e_3^T Q q_1 + m_2 g L_2 e_3^T q_2,
$$

which coincides with the total energy E in this case, is constant along each solution of the dynamical flow.

9.3.4 Equilibrium Properties

It is clear that the equilibrium solutions correspond to the derivative of the configuration vector, or equivalently the angular velocity vector, vanishing, and the attitudes of the two pendulums satisfying

$$
q_1^T S Q^T e_3 = 0,
$$
$$
e_3 \times q_2 = 0,
$$

which implies that the moments due to gravity vanish. This happens when the directions of the pendulums are aligned with or opposite to the direction of gravity. The hanging equilibrium solution is given by $(-Q^T e_3, -e_3, 0, 0) \in \mathsf{T}(\mathsf{S}^1 \times \mathsf{S}^2)$. The other three equilibrium solutions are $(-Q^T e_3, e_3, 0, 0)$, $(Q^T e_3, -e_3, 0, 0)$, $(Q^T e_3, e_3, 0, 0) \in \mathsf{T}(\mathsf{S}^1 \times \mathsf{S}^2)$.

We examine the first equilibrium solution, following the development in Appendix B. Equations (9.13) and the kinematics (9.10) and (9.11) can be linearized at this equilibrium to obtain

$$
\begin{bmatrix} (m_1 + m_2)L_1^2 & m_2L_1L_2 & 0 \\ m_2L_1L_2 & m_2gL_2 & 0 \\ 0 & 0 & m_2gL_2 \end{bmatrix} \begin{bmatrix} \ddot{\xi}_1 \\ \ddot{\xi}_{21} \\ \ddot{\xi}_{22} \end{bmatrix}
$$
$$
+ \begin{bmatrix} (m_1 + m_2)gL_1 & 0 & 0 \\ 0 & m_2gL_2 & 0 \\ 0 & 0 & m_2gL_2 \end{bmatrix} \begin{bmatrix} \xi_1 \\ \xi_{21} \\ \xi_{22} \end{bmatrix} = \begin{bmatrix} 0 \\ 0 \\ 0 \end{bmatrix}.
$$

This describes the linear vector field that is defined on the six-dimensional tangent space of $\mathsf{T}(\mathsf{S}^1 \times \mathsf{S}^2)$ at $(-Q^T e_3, -e_3, 0, 0) \in \mathsf{T}(\mathsf{S}^1 \times \mathsf{S}^2)$. This linear vector field approximates the Lagrangian vector field for the two pendulum in a neighborhood of $(-Q^T e_3, -e_3, 0, 0) \in \mathsf{T}(\mathsf{S}^1 \times \mathsf{S}^2)$. This equilibrium solution can be shown to be stable since the total energy has a strict local minimum at the equilibrium with zero time derivative, thereby guaranteeing stability of the equilibrium solution.

Following a similar development for the second equilibrium solution given by $(-Q^T e_3, e_3, 0, 0) \in \mathsf{T}(\mathsf{S}^1 \times \mathsf{S}^2)$, equations (9.13) and the kinematics (9.10) and (9.11) can be linearized to obtain the linear differential equation

$$
\begin{bmatrix} (m_1 + m_2)L_1^2 & m_2L_1L_2 & 0 \\ m_2L_1L_2 & m_2gL_2 & 0 \\ 0 & 0 & m_2gL_2 \end{bmatrix} \begin{bmatrix} \ddot{\xi}_1 \\ \ddot{\xi}_{21} \\ \ddot{\xi}_{22} \end{bmatrix}
$$
$$
+ \begin{bmatrix} -(m_1 + m_2)gL_1 & 0 & 0 \\ 0 & m_2gL_2 & 0 \\ 0 & 0 & m_2gL_2 \end{bmatrix} \begin{bmatrix} \xi_1 \\ \xi_{21} \\ \xi_{22} \end{bmatrix} = \begin{bmatrix} 0 \\ 0 \\ 0 \end{bmatrix}.
$$

This describes the linear vector field that is defined on the six-dimensional tangent space of $\mathsf{T}(\mathsf{S}^1 \times \mathsf{S}^2)$ at $(-Q^T e_3, e_3, 0, 0) \in \mathsf{T}(\mathsf{S}^1 \times \mathsf{S}^2)$. This linear vector field approximates the Lagrangian vector field for the two pendulum in a neighborhood of $(-Q^T e_3, e_3, 0, 0) \in \mathsf{T}(\mathsf{S}^1 \times \mathsf{S}^2)$. As a consequence, this equilibrium solution can be shown to be unstable since the linearized differential equation has a positive eigenvalue. Similarly, the equilibrium solutions $(-Q^T e_3, e_3, 0, 0)$, $(Q^T e_3, e_3, 0, 0) \in \mathsf{T}(\mathsf{S}^1 \times \mathsf{S}^2)$ can each be shown to be unstable.

9.4 Dynamics of a Spherical Pendulum on a Cart

Consider the dynamics of a spherical pendulum on a cart that moves on a horizontal plane. We assume that the spherical pendulum is a thin rigid rod or link with mass concentrated at the outboard end of the link. One end of the spherical pendulum is attached to a spherical joint or pivot that is connected to a cart that can translate, without friction, on a horizontal plane. A constant gravitational force acts on the spherical pendulum.

We demonstrate that globally valid Euler–Lagrange equations can be derived for the spherical pendulum on a cart system, and they can be expressed in a compact form without local parameterization or constraints. The results provide an intrinsic framework to study the global dynamics of a spherical pendulum on a cart system.

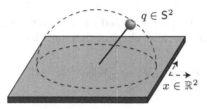

Fig. 9.4 Spherical pendulum on a cart

The cart of mass M can translate, without friction, on a horizontal plane. A spherical pendulum is attached to a pivot located at the center of mass of the cart, where the mass of the pendulum link is m and the link length is L. For simplicity, we assume that the mass of the link is concentrated at the outboard end of the link.

An inertial frame is chosen such that the first two axes are horizontal and the third axis is vertical, opposite to the direction of gravity. The direction vector of the link in the inertial frame is given by $q \in \mathsf{S}^2$. The vector $x \in \mathbb{R}^2$ denotes the position vector of the pivot of the pendulum in the horizontal two-dimensional plane. The configuration of the spherical pendulum on a cart is described by $(q, x) \in \mathsf{S}^2 \times \mathbb{R}^2$, so that the configuration manifold is $\mathsf{S}^2 \times \mathbb{R}^2$. Thus, this system has four degrees of freedom. Collisions of the spherical pendulum and the cart are ignored. A schematic of a spherical pendulum on a cart is shown in Figure 9.4.

The rotational kinematics equation for the attitude vector of the pendulum link is given by

$$\dot{q} = S(\omega)q, \tag{9.16}$$

where $\omega \in \mathsf{T}_q\mathsf{S}^2$ is the angular velocity vector of the pendulum link satisfying $\omega^T q = 0$.

9.4.1 Euler–Lagrange Equations

Consider the 3×2 matrix

$$C = \begin{bmatrix} 1 & 0 \\ 0 & 1 \\ 0 & 0 \end{bmatrix}.$$

As usual, $\{e_1, e_2, e_3\}$ denote the standard basis elements in $\mathbb{R}^3$.

The location of the center of mass of the cart, which is also the pendulum pivot, is given by $Cx \in \mathbb{R}^3$ in the inertial frame. Let $x_L \in \mathbb{R}^3$ be the position of the mass element of the spherical pendulum in the inertial frame. It can be written as

$$x_L = Cx + Lq.$$

The total kinetic energy is composed of the kinetic energy of the cart and the kinetic energy of the mass element defining the spherical pendulum:

$$T(q, x, \dot{q}, \dot{x}) = \frac{1}{2} M \|\dot{x}\|^2 + \frac{1}{2} m \|\dot{x}_L\|^2.$$

This can be written as

$$T(q, x, \dot{q}, \dot{x}) = \frac{1}{2}(M + m)\|\dot{x}\|^2 + mL\dot{x}^T C^T \dot{q} + \frac{1}{2} mL^2 \|\dot{q}\|^2.$$

The potential energy is the gravitational potential energy of the mass of the spherical pendulum link:

$$U(q, x) = mgLe_3^T q.$$

The Lagrangian function for the spherical pendulum on a cart $L : \mathsf{T}(\mathsf{S}^2 \times \mathbb{R}^2) \to \mathbb{R}^1$ is

$$L(q, x, \dot{q}, \dot{x}) = \frac{1}{2}(M + m)\|\dot{x}\|^2 + mL\dot{x}^T C^T \dot{q} + \frac{1}{2} mL^2 \|\dot{q}^T\|^2 - mgLe_3^T q.$$

The modified Lagrangian function can be expressed in terms of the angular velocity vector of the pendulum link as

$$\tilde{L}(q, x, \omega, \dot{x}) = \frac{1}{2}(M + m)\|\dot{x}\|^2 - mL\dot{x}^T C^T S(q)\omega + \frac{1}{2} mL^2 \|\omega\|^2 - mgLe_3^T q.$$

Note that the Lagrangian does not depend on the position of the cart; this property reflects a translational symmetry in the dynamics.

The infinitesimal variation of the action integral satisfies

$$\delta \mathfrak{G} = \int_{t_0}^{t_f} \left[\{(M+m)\ddot{x} - mLC^T S(q)\omega\}^T \delta \dot{x} + \{mLS(q)C\dot{x} + mL^2\omega\}^T \delta\omega \right.$$
$$\left. + \{-mLS(\omega)C\dot{x} - mgLe_3\}^T \delta q \right] dt = 0,$$

for all infinitesimal variations $\delta q : [t_0, t_f] \to \mathsf{T}_q \mathsf{S}^2$ and $\delta x : [t_0, t_f] \to \mathbb{R}^2$ that vanish at the endpoints, $\delta q(t_0) = \delta q(t_f) = 0$ and $\delta x(t_0) = \delta x(t_f) = 0$. The infinitesimal variations in S^2 can be written as

$$\delta\omega = (I_{3\times3} - qq^T)\dot{\gamma} + S(\gamma)\omega,$$
$$\delta q = S(\gamma)q,$$

for differentiable curves $\gamma : [t_0, t_f] \to \mathbb{R}^3$ satisfying $\gamma(t_0) = \gamma(t_f) = 0$. Substituting these into the infinitesimal variation of the action integral yields

$$\delta \mathfrak{G} = \int_{t_0}^{t_f} \left[\{(M+m)\ddot{x} - mLC^T S(q)\omega\}^T \delta\dot{x} \right.$$
$$+ \{mLS(q)C\dot{x} + mL^2\omega\}^T \{(I_{3\times3} - qq^T)\dot{\gamma} + S(\gamma)\omega\}$$
$$\left. + \{-mLS(\omega)C\dot{x} - mgLe_3\}^T S(\gamma)q \right] dt.$$

Since $(I_{3\times3} - qq^T)$ is the orthogonal projection onto the tangent manifold of S^2 at $q \in \mathsf{S}^2$ and the expression $\{mLS(q)C\dot{x} + mL^2\omega\}$ is necessarily orthogonal to $\mathsf{T}_q\mathsf{S}^2$, we have

$$(I_{3\times3} - qq^T)\{mLS(q)C\dot{x} + mL^2\omega\} = \{mLS(q)C\dot{x} + mL^2\omega\}.$$

Therefore,

$$\delta \mathfrak{G} = \int_{t_0}^{t_f} \left[\{(M+m)\ddot{x} - mLC^T S(q)\omega\}^T \delta\dot{x} \right.$$
$$+ \{mLS(q)C\dot{x} + mL^2\omega\}^T \dot{\gamma} + \{mLS(\omega)S(q)C\dot{x}\}^T \gamma$$
$$\left. + \{mLS(\omega)C\dot{x} + mgLe_3\}^T S(q)\gamma \right] dt.$$

Integrating by parts and using the boundary conditions, we obtain

$$\delta \mathfrak{G} = \int_{t_0}^{t_f} \left[-\frac{d}{dt}\{(M+m)\ddot{x} - mLC^T S(q)\omega\}^T \delta x \right.$$
$$- \frac{d}{dt}\{mLS(q)C\dot{x} + mL^2\omega\}^T \gamma$$
$$\left. + \{mLS(\omega)S(q)C\dot{x} - S(q)(mLS(\omega)C\dot{x} + mgLe_3)\}^T \gamma \right] dt.$$

According to Hamilton's principle, the infinitesimal variation of the action integral is zero for all differentiable curves $\gamma : [t_0, t_f] \to \mathbb{R}^3$ and $\delta x : [t_0, t_f] \to \mathbb{R}^2$ which vanish at t_0 and t_f. The fundamental lemma of variational calculus, as in Appendix A, leads to the dynamical equations for the spherical pendulum on a cart, consisting of the rotational kinematics of the spherical pendulum (9.16) and the Euler–Lagrange equations

$$\frac{d}{dt}\{mLS(q)C\dot{x} + mL^2\omega\} + mLS(C\dot{x})S(\omega)q + mgLq \times e_3 = 0,$$

$$\frac{d}{dt}\{(M+m)\dot{x} - mLC^T S(q)\omega\} = 0.$$

These can be rearranged in matrix-vector form as

$$\begin{bmatrix} mL^2 I_{3\times3} & -mLS^T(q)C \\ -mLC^T S(q) & (M+m)I_{2\times2} \end{bmatrix} \begin{bmatrix} \dot{\omega} \\ \ddot{x} \end{bmatrix} + \begin{bmatrix} mgLq \times e_3 \\ mLC^T S(\omega)^2 q \end{bmatrix} = \begin{bmatrix} 0 \\ 0 \end{bmatrix}. \qquad (9.17)$$

The kinematic equations are given by (9.16) and the Euler–Lagrange equations are given by (9.17). Together, they describe the dynamical flow of the spherical pendulum on a cart in terms of the evolution of $(q, x, \dot{q}, \dot{x}) \in T(S^2 \times \mathbb{R}^2)$ on the tangent bundle of $S^2 \times \mathbb{R}^2$.

9.4.2 Hamilton's Equations

The Legendre transformation is used to define the momenta as follows

$$\begin{bmatrix} \pi \\ p \end{bmatrix} = \begin{bmatrix} (I_{3\times3} - qq^T)\frac{\partial \tilde{L}(q,x,\omega,\dot{x})}{\partial \omega} \\ \frac{\partial \tilde{L}(q,x,\omega,\dot{x})}{\partial \dot{x}} \end{bmatrix} = \begin{bmatrix} mL^2 I_{3\times3} & -mLS^T(q)C \\ -mLC^T S(q) & (M+m)I_{2\times2} \end{bmatrix} \begin{bmatrix} \omega \\ \dot{x} \end{bmatrix}.$$

Thus, the momenta $(\pi, p) \in T^*_{(q,x)}(S^2 \times \mathbb{R}^2)$ are conjugate to the velocity vectors $(\omega, \dot{x}) \in T_{(q,x)}(S^2 \times \mathbb{R}^2)$. This can be inverted to yield

$$\begin{bmatrix} \omega \\ \dot{x} \end{bmatrix} = \begin{bmatrix} M_{11}^I(q) & M_{12}^I(q) \\ M_{21}^I(q) & M_{22}^I(q) \end{bmatrix} \begin{bmatrix} \pi \\ p \end{bmatrix},$$

where

$$\begin{bmatrix} M_{11}^I(q) & M_{12}^I(q) \\ M_{21}^I(q) & M_{22}^I(q) \end{bmatrix} = \begin{bmatrix} mL^2 I_{3\times3} & -mLS^T(q)C \\ -mLC^T S(q) & (M+m)I_{2\times2} \end{bmatrix}^{-1}$$

is the inverse of the indicated 5×5 partitioned matrix. The modified Hamiltonian function $\tilde{H} : T^*(S^2 \times \mathbb{R}^2) \to \mathbb{R}^1$ is

$$\tilde{H}(q, x, \pi, p) = \frac{1}{2} \begin{bmatrix} \pi \\ p \end{bmatrix}^T \begin{bmatrix} M_{11}^I(q) & M_{12}^I(q) \\ M_{21}^I(q) & M_{22}^I(q) \end{bmatrix} \begin{bmatrix} \pi \\ p \end{bmatrix} + mgLe_3^T q.$$

Thus, from (5.28) and (3.12), Hamilton's equations can be written in matrix-vector form as

$$\begin{bmatrix} \dot{q} \\ \dot{x} \end{bmatrix} = \begin{bmatrix} -S(q)M_{11}^I(q) & -S(q)M_{11}^I(q) \\ M_{21}^I(q) & M_{22}^I(q) \end{bmatrix} \begin{bmatrix} \pi \\ p \end{bmatrix}, \tag{9.18}$$

and from (5.29) and (3.13),

$$\begin{bmatrix} \dot{\pi} \\ \dot{p} \end{bmatrix} = \begin{bmatrix} -mLS(C\dot{x})S(\omega)q - mgLq \times e_3 \\ 0 \end{bmatrix}, \tag{9.19}$$

where the first term is expressed in terms of the momenta using the Legendre transformation. Hamilton's equations (9.18) and (9.19) describe the Hamiltonian flow of the spherical pendulum on a cart in terms of the evolution of $(q, x, \pi, p) \in T^*(S^2 \times \mathbb{R}^2)$ on the cotangent bundle of the configuration manifold $S^2 \times \mathbb{R}^2$.

9.4.3 Conservation Properties

The Hamiltonian of the spherical pendulum on a cart is given by

$$H = \frac{1}{2}(M + m)\|\dot{x}\|^2 - mL\dot{x}^T C^T S(q)\omega + \frac{1}{2}mL^2\|\omega\|^2 + mgLe_3^T q,$$

which coincides with the total energy E in this case, and it is constant along each solution of the dynamical flow of the spherical pendulum on a cart.

As seen from Hamilton's equations, the translational momentum given by

$$p = (M + m)\dot{x} - mLC^T S(q)\omega$$

is also constant along each solution of the dynamical flow. This can be viewed as a consequence of Noether's theorem and the translational invariance of the Lagrangian.

9.4.4 Equilibrium Properties

We can determine the equilibria of the spherical pendulum on a cart. Assuming the configuration is constant, the equilibrium configurations in $S^2 \times \mathbb{R}^2$ correspond to the cart in an arbitrary location with the attitude of the spherical pendulum satisfying

$$q \times e_3 = 0,$$

which implies that the moment due to gravity vanishes.

An equilibrium for which the direction of the pendulum link is in the gravity direction, $-e_3$, is referred to as a hanging equilibrium; an equilibrium for which the direction of the pendulum link is e_3, that is opposite to the gravity direction, is referred to as an inverted equilibrium.

We first examine the inverted equilibrium $(e_3, 0, 0, 0) \in T(S^2 \times \mathbb{R}^2)$, following the development in Appendix B. Equations (9.17) and the kinematics (9.16) can be linearized at this equilibrium to obtain

$$\begin{bmatrix} mL^2 & 0 & mL & 0 \\ 0 & mL^2 & 0 & mL \\ mL & 0 & (m+M) & 0 \\ 0 & mL & 0 & (m+M) \end{bmatrix} \begin{bmatrix} \ddot{\xi}_1 \\ \ddot{\xi}_2 \\ \ddot{\xi}_3 \\ \ddot{\xi}_4 \end{bmatrix} + \begin{bmatrix} -mgL & 0 & 0 & 0 \\ 0 & -mgL & 0 & 0 \\ 0 & 0 & 0 & 0 \\ 0 & 0 & 0 & 0 \end{bmatrix} \begin{bmatrix} \xi_1 \\ \xi_2 \\ \xi_3 \\ \xi_4 \end{bmatrix} = \begin{bmatrix} 0 \\ 0 \\ 0 \\ 0 \end{bmatrix}.$$

This describes the linear vector field defined on the eight-dimensional tangent space of $T(S^2 \times \mathbb{R}^2)$ at $(e_3, 0, 0, 0) \in T(S^2 \times \mathbb{R}^2)$. This linear vector field approximates the Lagrangian vector field of the spherical pendulum on a cart in a neighborhood of the inverted equilibrium solution. This equilibrium solution can be shown to be unstable since the linearized differential equations have a positive eigenvalue. This conclusion applies to any inverted equilibrium solution.

Following a similar development for the hanging equilibrium $(-e_3, 0, 0, 0) \in T(S^2 \times \mathbb{R}^2)$, equations (9.17) and the kinematics (9.16) can be linearized at this equilibrium to obtain the linear differential equations

$$\begin{bmatrix} mL^2 & 0 & mL & 0 \\ 0 & mL^2 & 0 & mL \\ mL & 0 & (m+M) & 0 \\ 0 & mL & 0 & (m+M) \end{bmatrix} \begin{bmatrix} \ddot{\xi}_1 \\ \ddot{\xi}_2 \\ \ddot{\xi}_3 \\ \ddot{\xi}_4 \end{bmatrix} + \begin{bmatrix} mgL & 0 & 0 & 0 \\ 0 & mgL & 0 & 0 \\ 0 & 0 & 0 & 0 \\ 0 & 0 & 0 & 0 \end{bmatrix} \begin{bmatrix} \xi_1 \\ \xi_2 \\ \xi_3 \\ \xi_4 \end{bmatrix} = \begin{bmatrix} 0 \\ 0 \\ 0 \\ 0 \end{bmatrix}.$$

This describes the linear vector field defined on the eight-dimensional tangent space of $T(S^2 \times \mathbb{R}^2)$ at $(-e_3, 0, 0, 0) \in T(S^2 \times \mathbb{R}^2)$. This linear vector field approximates the Lagrangian vector field of the spherical pendulum on a cart in a neighborhood of the hanging equilibrium solution. These linear differential equations have two imaginary eigenvalues and two zero eigenvalues. Since all the eigenvalues of the linearization have zero real part, stability of the hanging equilibrium cannot be determined based on a spectral analysis of the linearization.

9.5 Dynamics of a Rotating Rigid Body with Appendage

We consider the attitude dynamics of a rigid body rotating in three dimensions about its center of mass. A proof mass element, viewed as an ideal particle with mass, is constrained to translate within a frictionless slot fixed in the rigid body; a linear elastic spring connects the proof mass element and the rigid body in the slot. The elastically constrained proof mass represents a flexible appendage of the rigid body.

We select two Euclidean frames in $\mathbb{R}^3$: an inertially fixed frame and a frame fixed to the rigid body; the origin of the body-fixed frame is located at the center of mass of the rigid body, ignoring the proof mass element. The slot is assumed to lie along the first axis of the body-fixed frame. If the distance between the center of mass and the proof mass element is L, the spring exerts no restoring force on the proof mass. Let m denote the mass of the particle in the slot and let J be the 3×3 inertia matrix of the rigid body without the particle.

We let $R \in \mathsf{SO}(3)$ be the attitude of the rigid body and let $x \in \mathbb{R}^1$ be the displacement of the proof mass element along the slot from its location when the spring exerts no force; the position vector of the proof mass element in the body-fixed frame is $(L + x)e_1 \in \mathbb{R}^3$. Thus, the configuration of the rotating rigid body with an appendage is described by the pair $(R, x) \in \mathsf{SO}(3) \times \mathbb{R}^1$ and the configuration manifold is $\mathsf{SO}(3) \times \mathbb{R}^1$. There are four degrees of freedom. A schematic of a rotating rigid body with an elastic appendage is shown in Figure 9.5.

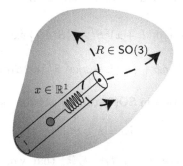

Fig. 9.5 Rotating rigid body with an elastic appendage

9.5.1 Euler–Lagrange Equations

The rotational kinematics of the rigid body are

$$\dot{R} = RS(\omega),$$

(9.20)

where $\omega \in T_R SO(3)$ is the angular velocity vector of the rigid body with respect to the body-fixed frame.

The kinetic energy of the rigid body is $\frac{1}{2}\omega^T J\omega$. Since the location of the particle in the inertially fixed frame is given by $(L+x)Re_1 \in \mathbb{R}^3$, the velocity vector of the particle in the inertial frame is given by $\dot{x}Re_1 + (L+x)RS(\omega)e_1 \in \mathbb{R}^3$. The rotational kinetic energy is

$$
\begin{aligned}
T(R, x, \omega, \dot{x}) &= \frac{1}{2}\omega^T J\omega + \frac{1}{2}m\|\dot{x}Re_1 + (L+x)RS(\omega)e_1\|^2 \\
&= \frac{1}{2}\omega^T J\omega + \frac{1}{2}m\dot{x}^2 - m\dot{x}(L+x)e_1^T S(e_1)\omega \\
&\quad + \frac{1}{2}m(L+x)^2\omega^T S^T(e_1)S(e_1)\omega \\
&= \frac{1}{2}\omega^T \left(J + m(L+x)^2 S^T(e_1)S(e_1) \right) \omega + \frac{1}{2}m\dot{x}^2,
\end{aligned}
$$

where the third step follows since $e_1^T S(e_1) = 0$. The potential energy is the energy stored in the elastic spring and is given by $\frac{1}{2}\kappa x^2$, where the positive constant κ is the elastic coefficient of the restoring spring. The resulting modified Lagrangian function is given by

$$
\tilde{L}(R, x, \omega, \dot{x}) = \frac{1}{2}\omega^T \left(J + m(L+x)^2 S^T(e_1)S(e_1) \right) \omega + \frac{1}{2}m\dot{x}^2 - \frac{1}{2}\kappa x^2.
$$

Note that the Lagrangian function is independent of the rigid body attitude, indicating a rotational symmetry of the dynamics.

The infinitesimal variation of the action integral is given by

$$
\begin{aligned}
\delta \mathfrak{G} = \int_{t_0}^{t_f} \Big[&\omega^T \left\{ J + m(L+x)^2 S^T(e_1)S(e_1) \right\} \delta\omega \\
&+ m\dot{x}^T \delta\dot{x} + \left\{ m(L+x)\omega^T S^T(e_1)S(e_1)\omega - \kappa x \right\}^T \delta x \Big] dt.
\end{aligned}
$$

The infinitesimal variation in $SO(3)$ can be written as

$$
\delta\omega = \dot{\eta} + S(\omega)\eta,
$$

for a differentiable curve $\eta : [t_0, t_f] \to \mathbb{R}^3$ that vanishes at the endpoints, $\eta(t_0) = \eta(t_f) = 0$, and the infinitesimal variation in $\mathbb{R}^1$ is described by $\delta x : [t_0, t_f] \to \mathbb{R}^1$ that vanishes at the endpoints, $\delta x(t_0) = \delta x(t_f) = 0$. Substituting these into the above and integrating by parts, we obtain

$$
\begin{aligned}
\delta \mathfrak{G} = \int_{t_0}^{t_f} \Big[&-\{(J + m(L+x)^2 S^T(e_1)S(e_1))\dot{\omega} + 2m(L+x)S^T(e_1)S(e_1)\omega\dot{x} \\
&+ S(\omega)((J + m(L+x)^2 S^T(e_1)S(e_1))\omega)\}^T \eta \\
&- \{m\ddot{x} - m(L+x)\omega^T S^T(e_1)S(e_1)\omega + \kappa x\}\delta x \Big] dt.
\end{aligned}
$$

According to Hamilton's principle, the infinitesimal variation of the action integral is zero for all differentiable curves $\eta : [t_0, t_f] \to \mathbb{R}^3$ and $\delta x : [t_0, t_f] \to \mathbb{R}^1$ that vanish at t_0 and t_f. Using the fundamental lemma of variational calculus, as in Appendix A, the dynamics is described by the rotational kinematics of the rigid body (9.20) and the Euler–Lagrange equations

$$\{J + m(L + x)^2 S^T(e_1)S(e_1)\}\dot{\omega} + 2m(L + x)S^T(e_1)S(e_1)\omega\dot{x}$$
$$+\omega \times \{J + m(L + x)^2 S^T(e_1)S(e_1)\}\omega = 0,$$
$$m\ddot{x} - m(L + x)\omega^T S^T(e_1)S(e_1)\omega + \kappa x = 0.$$

The Euler–Lagrange equations, in matrix-vector form, consist of the rotational kinematics (9.20) and

$$\begin{bmatrix} (J + m(L + x)^2 S^T(e_1)S(e_1)) & 0 \\ 0 & m \end{bmatrix} \begin{bmatrix} \dot{\omega} \\ \ddot{x} \end{bmatrix}$$
$$+ \begin{bmatrix} 2m(L + x)S^T(e_1)S(e_1)\omega\dot{x} + \omega \times \{J + m(L + x)^2 S^T(e_1)S(e_1)\}\omega \\ -m(L + x)\omega^T S^T(e_1)S(e_1)\omega \end{bmatrix}$$
$$+ \begin{bmatrix} 0 \\ \kappa x \end{bmatrix} = \begin{bmatrix} 0 \\ 0 \end{bmatrix}. \tag{9.21}$$

Equations (9.20) and (9.21) describe the dynamical flow of the rotating rigid body with an elastic appendage in terms of the evolution of $(R, x, \omega, \dot{x})$ on the tangent bundle $\mathsf{T}(\mathsf{SO}(3) \times \mathbb{R}^1)$.

9.5.2 Hamilton's Equations

Hamilton's equations of motion can be obtained by introducing the Legendre transformation

$$\Pi = \frac{\partial \tilde{L}(R, x, \omega, \dot{x})}{\partial \omega} = \{J + m(L + x)^2 S^T(e_1)S(e_1)\}\omega,$$
$$p = \frac{\partial \tilde{L}(R, x, \omega, \dot{x})}{\partial \dot{x}} = m\dot{x},$$

where the momenta $(\Pi, p) \in \mathsf{T}^*_{R,x}(\mathsf{SO}(3) \times \mathbb{R}^1)$ are conjugate to $(\omega, \dot{x}) \in \mathsf{T}_{R,x}(\mathsf{SO}(3) \times \mathbb{R}^1)$. The modified Hamiltonian function is

$$\tilde{H}(R, x, \Pi, p) = \frac{1}{2}\Pi^T \{J + m(L + x)^2 S^T(e_1)S(e_1)\}^{-1}\Pi + \frac{1}{2}\frac{p^2}{m} + \frac{1}{2}\kappa x^2.$$

From (6.10) and (3.12), Hamilton's equations can be shown to be given by

$$\dot{r}_i = r_i \times \left\{ J + m(L+x)^2 S^T(e_1)S(e_1) \right\}^{-1} \Pi, \quad i = 1,2,3, \qquad (9.22)$$

$$\dot{x} = \frac{p}{m}, \qquad (9.23)$$

and from (6.11) and (3.13),

$$\dot{\Pi} = S(\Pi) \left\{ J + m(L+x)^2 S^T(e_1)S(e_1) \right\}^{-1} \Pi, \qquad (9.24)$$

$$\dot{p} = m(L+x)\omega^T S^T(e_1)S(e_1)\omega - \kappa x, \qquad (9.25)$$

where $\omega = \left\{ J + m(L+x)^2 S^T(e_1)S(e_1) \right\}^{-1} \Pi$ and $r_i \in S^2$ is the i-th column of $R^T \in \mathsf{SO}(3)$. These differential equations describe the dynamical flow of a rotating rigid body with a flexible appendage in terms of the evolution of $(R, x, \Pi, p) \in \mathsf{T}^*(\mathsf{SO}(3) \times \mathbb{R}^1)$ on the cotangent bundle of $(\mathsf{SO}(3) \times \mathbb{R}^1)$.

9.5.3 Conservation Properties

The Hamiltonian

$$H = \frac{1}{2}\omega^T \left\{ J + m(L+x)^2 S^T(e_1)S(e_1) \right\} \omega + \frac{1}{2}m\dot{x}^2 + \frac{1}{2}\kappa x^2,$$

which coincides with the total energy E in this case, is conserved along each solution of the rotating rigid body with flexible appendage.

In addition, there is a rotational symmetry: the Lagrangian is invariant with respect to arbitrary rigid body rotations. This symmetry leads to conservation of the angular momentum in the inertial frame; that is

$$R\Pi = R\left\{ J + m(L+x)^2 S^T(e_1)S(e_1) \right\} \omega$$

is constant along each solution. Since $R \in \mathsf{SO}(3)$, the magnitude of the angular momentum in the body-fixed frame is also conserved, that is

$$\|\Pi\| = \left\| \left\{ J + m(L+x)^2 S^T(e_1)S(e_1) \right\} \omega \right\|$$

is constant along each solution.

9.5.4 Equilibrium Properties

The equilibrium solutions occur when the angular velocity vector of the rigid body is zero, the linear velocity of the proof mass is zero, and when the elastic restoring force is zero. The system can be in equilibrium at any fixed attitude of the rigid body.

Linearization of the dynamics in a neighborhood of the prototypical equilibrium point $(I_{3\times3}, 0, 0, 0) \in \mathsf{T}(\mathsf{SO}(3) \times \mathbb{R}^1)$ is obtained. Following the results in Appendix B, the linearized differential equations are given by

$$\{J + mL^2 S^T(e_1) S(e_1)\}\ddot{\xi}_1 = 0,$$
$$m\ddot{\xi}_2 + \kappa\xi_2 = 0.$$

These linearized differential equations, defined on the eight-dimensional tangent space of $\mathsf{T}(\mathsf{SO}(3) \times \mathbb{R}^1)$ at $(I_{3\times3}, 0, 0, 0) \in \mathsf{T}(\mathsf{SO}(3) \times \mathbb{R}^1)$, approximate the rotational dynamics of the rigid body and appendage in a neighborhood of the equilibrium. These linear dynamics are accurate only to first order in the perturbations. Higher-order coupling effects are important for larger deviations from the equilibrium. In particular, the linearized rotational dynamics of the rigid body and the linearized translational dynamics of the sliding particle are uncoupled, an apparent inconsistency with the conservation of the total energy that is resolved when considering higher-order terms in the expansion.

9.6 Dynamics of a Three-Dimensional Pendulum on a Cart

A three-dimensional pendulum on a cart consists of a rigid body that can rotate about a frictionless pivot that is fixed in a cart that can translate without friction on a horizontal plane. The pivot is located at the center of mass of the cart. Hence, we assume the cart translates but does not rotate in the horizontal plane. Uniform constant gravity acts on the three-dimensional pendulum.

We define two Euclidean frames: an inertial frame and a body-fixed frame attached to the pendulum. The first two axes of the inertial frame lie in the horizontal plane within which the cart moves, while the third axis of the inertial frame is vertical. The origin of the pendulum-fixed frame is located at the center of mass of the pendulum.

The following notation is used: $\rho_0 \in \mathbb{R}^3$ is the constant vector from the center of mass of the pendulum to the pivot in the pendulum-fixed frame; the constant matrix

$$Q = \begin{bmatrix} 1 & 0 \\ 0 & 1 \\ 0 & 0 \end{bmatrix}$$

maps vectors in $\mathbb{R}^2$ into vectors in $\mathbb{R}^3$; m is the mass of the three-dimensional pendulum, and M is the mass of the cart.

Let $R \in \mathsf{SO}(3)$ define the attitude of the three-dimensional pendulum as a rotation matrix from the pendulum-fixed frame to the inertial frame. Let $x \in \mathbb{R}^2$ be the position vector of the center of mass of the cart in the horizontal plane. The configuration is described by $(R, x) \in \mathsf{SO}(3) \times \mathbb{R}^2$ and $\mathsf{SO}(3) \times \mathbb{R}^2$ is the configuration manifold. Hence, there are five degrees of freedom for the three-dimensional pendulum on a cart. A schematic of a three-dimensional pendulum on a cart is shown in Figure 9.6.

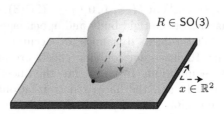

Fig. 9.6 Three-dimensional pendulum on a cart

9.6.1 Euler–Lagrange Equations

The rotational kinematics of the three-dimensional pendulum are

$$\dot{R} = RS(\omega), \tag{9.26}$$

where $\omega \in \mathsf{T}_R \mathsf{SO}(3)$ is the angular velocity vector of the three-dimensional pendulum in the pendulum-fixed frame.

The kinetic energy of the system is the sum of the kinetic energy of the cart and the kinetic energy of the three-dimensional pendulum. Since the cart is assumed to only translate in a horizontal plane, its kinetic energy is given by

$$T_{cart} = \frac{1}{2} M \|\dot{x}\|^2 .$$

To determine the kinetic energy of the three-dimensional pendulum, we need to integrate the kinetic energy of each mass element. Let $\rho \in \mathbb{R}^3$ be the vector from the center of mass of the three-dimensional pendulum to a mass element in the three-dimensional pendulum represented in the pendulum-fixed frame. The vector from the origin of the inertial frame to the mass element in the three-dimensional pendulum is given by $Qx + R(-\rho_0 + \rho)$. Then, the kinetic energy of the three-dimensional pendulum is obtained from the body integral:

$$T_{pend} = \frac{1}{2} \int_{\mathcal{B}} \left\| Q\dot{x} + \dot{R}(-\rho_0 + \rho) \right\|^2 dm(\rho)$$

$$= \frac{1}{2} \int_{\mathcal{B}} \|\dot{x}\|^2 \, dm(\rho)$$

$$+ \frac{1}{2} \int_{\mathcal{B}} \text{trace} \left\{ S(\omega)(-\rho_0 + \rho)(-\rho_0 + \rho)^T S(\omega)^T \right\} dm(\rho)$$

$$+ \int_{\mathcal{B}} \dot{x}^T Q^T R S(\omega)(-\rho_0 + \rho) \, dm(\rho).$$

From the definition of the center of mass, $\int_{\mathcal{B}} dm(\rho) = m$ and $\int_{\mathcal{B}} \rho \, dm(\rho) = 0$. This simplifies the expression for the kinetic energy of the three-dimensional pendulum to

$$T_{pend} = \frac{1}{2} m \|\dot{x}\|^2 + \frac{1}{2} \omega^T J \omega - m\dot{x}^T Q^T R S(\omega) \rho_0,$$

where $J = \int_{\mathcal{B}} S(\rho)^T S(\rho) \, dm(\rho) + m S(\rho_0)^T S(\rho_0)$ is an inertia matrix of the pendulum. The total kinetic energy is

$$T(R, x, \omega, \dot{x}) = \frac{1}{2}(M + m) \|\dot{x}\|^2 + \frac{1}{2} \omega^T J \omega - m\dot{x}^T Q^T R S(\omega) \rho_0.$$

The potential energy is the gravitational potential energy of the three-dimensional pendulum:

$$U(R) = -mg e_3^T R \rho_0.$$

Thus, the modified Lagrangian $\tilde{L} : \mathsf{T}(\mathsf{SO}(3) \times \mathbb{R}^2) \to \mathbb{R}^1$ is given by

$$\tilde{L}(R, x, \omega, \dot{x}) = \frac{1}{2}(M + m) \|\dot{x}\|^2 + \frac{1}{2} \omega^T J \omega + m\dot{x}^T Q^T R S(\rho_0) \omega + mg e_3^T R \rho_0.$$

Note that the Lagrangian function does not depend on the position vector of the center of mass of the cart. This reflects a symmetry in the dynamics of the three-dimensional pendulum on a cart.

The infinitesimal variation of the action integral is

$$\delta \mathfrak{G} = \int_{t_0}^{t_f} \Big[\left\{ J\omega - m S(\rho_0) R^T Q\dot{x} \right\}^T \delta\omega + m\dot{x}^T Q^T \delta R S(\rho_0) \omega + mg e_3^T \delta R \rho_0$$

$$- \left\{ m Q^T R S(\rho_0)^T \omega + (M + m)\dot{x} \right\}^T \delta\dot{x} \Big] dt.$$

The infinitesimal variations in $\mathsf{SO}(3)$ can be written as

$$\delta R = R S(\eta),$$
$$\delta\omega = \dot{\eta} + S(\omega)\eta,$$

for differentiable curves $\eta : [t_0, t_f] \to \mathbb{R}^3$ satisfying $\eta(t_0) = \eta(t_f) = 0$, and the infinitesimal variations in $\mathbb{R}^2$ can be expressed as $\delta x : [t_0, t_f] \to \mathbb{R}^2$ satisfying $\delta x(t_0) = \delta x(t_f) = 0$. Substituting these expressions into the above and integrating by parts, we obtain

$$
\delta \mathfrak{G} = \int_{t_0}^{t_f} \left[-\left\{ \frac{d}{dt}(J\omega - mS(\rho_0)R^T Q\dot{x}) \right\}^T \eta \right.
$$
$$
- \{S(\omega)(J\omega - mS(\rho_0)R^T Q\dot{x})\}^T \eta - m\dot{x}^T Q^T R(-\rho_0\omega^T - \omega\rho_0^T)\eta
$$
$$
\left. - mge_3^T RS(\rho_0)\eta - \left\{ \frac{d}{dt}(-mQ^T RS^T(\rho_0)\omega + (M+m)\dot{x}) \right\}^T \delta x \right] dt.
$$

According to Hamilton's principle, the infinitesimal variation of the action integral is zero for all differentiable curves $\eta(t) \in \mathbb{R}^3$ and $\delta x(t) \in \mathbb{R}^2$. Using the fundamental lemma of variational calculus, as in Appendix A, it can be shown that the Euler–Lagrange equations for the three-dimensional pendulum on a cart are

$$
J\dot{\omega} - mS(\rho_0)R^T Q\ddot{x} + \omega \times J\omega - mg\rho_0 \times R^T e_3 = 0,
$$
$$
-mQ^T RS^T(\rho_0)\dot{\omega} + (M+m)\ddot{x} - mQ^T RS(\omega)^2\rho_0 = 0.
$$

The Euler–Lagrange equations can be written in matrix-vector form

$$
\begin{bmatrix} J & -mS(\rho_0)R^T Q \\ -mQ^T RS^T(\rho_0) & (M+m)I_{2\times 2} \end{bmatrix} \begin{bmatrix} \dot{\omega} \\ \ddot{x} \end{bmatrix} + \begin{bmatrix} \omega \times J\omega \\ -mQ^T RS(\omega)^2\rho_0 \end{bmatrix}
$$
$$
+ \begin{bmatrix} -mg\rho_0 \times R^T e_3 \\ 0 \end{bmatrix} = \begin{bmatrix} 0 \\ 0 \end{bmatrix}. \quad (9.27)
$$

Equations (9.27) and the rotational kinematics equations (9.26) describe the dynamical flow of the three-dimensional pendulum on a cart in terms of the evolution of $(R, x, \omega, \dot{x}) \in \mathsf{T}(\mathsf{SO}(3) \times \mathbb{R}^2)$ on the tangent bundle of $\mathsf{SO}(3) \times \mathbb{R}^2$.

9.6.2 Hamilton's Equations

Hamilton's equations for the three-dimensional pendulum on a cart can also be obtained. The Legendre transformation is given by

$$
\begin{bmatrix} \Pi \\ p \end{bmatrix} = \begin{bmatrix} \frac{\partial \tilde{L}(R,x,\omega,\dot{x})}{\partial \omega} \\ \frac{\partial \tilde{L}(R,x,\omega,\dot{x})}{\partial \dot{x}} \end{bmatrix} = \begin{bmatrix} J & -mS(\rho_0)R^T Q \\ -mQ^T RS(\rho_0) & (M+m)I_{2\times 2} \end{bmatrix} \begin{bmatrix} \omega \\ \ddot{x} \end{bmatrix}.
$$

Thus, the angular momentum $(\Pi, p) \in \mathsf{T}^*_{(R,x)}(\mathsf{SO}(3) \times \mathbb{R}^2)$ is conjugate to $(\omega, \dot{x}) \in \mathsf{T}_{(R,x)}(\mathsf{SO}(3) \times \mathbb{R}^2)$. The above expression also gives

$$\begin{bmatrix} \omega \\ \dot{x} \end{bmatrix} = \begin{bmatrix} M_{11}^I & M_{12}^I \\ M_{21}^I & M_{22}^I \end{bmatrix} \begin{bmatrix} \Pi \\ p \end{bmatrix},$$

where

$$\begin{bmatrix} M_{11}^I & M_{12}^I \\ M_{21}^I & M_{22}^I \end{bmatrix} = \begin{bmatrix} J & -mS(\rho_0)R^TQ \\ -mQ^TRS(\rho_0) & (M+m)I_{2\times2} \end{bmatrix}^{-1}$$

is the partitioned inverse of the 5×5 matrix given above. The modified Hamiltonian function $\tilde{H} : \mathsf{T}^*(\mathsf{SO}(3) \times \mathbb{R}^2) \to \mathbb{R}^1$ is

$$\tilde{H}(R,x,\Pi,p) = \frac{1}{2} \begin{bmatrix} \Pi \\ p \end{bmatrix}^T \begin{bmatrix} M_{11}^I & M_{12}^I \\ M_{21}^I & M_{22}^I \end{bmatrix} \begin{bmatrix} \Pi \\ p \end{bmatrix} - mg\rho_0^T R^T e_3^T.$$

From (6.10) and (3.12), Hamilton's equations are given by

$$\dot{r}_i = r_i \times \left(M_{11}^I \Pi + M_{12}^I p \right), \quad i = 1,2,3, \tag{9.28}$$

$$\dot{x} = M_{21}^I \Pi + M_{22}^I p, \tag{9.29}$$

and also from (6.11) and (3.13),

$$\dot{\Pi} = \Pi \times \left(M_{11}^I \Pi + M_{12}^I p \right) + \sum_{i=1}^{3} r_i \times \frac{\partial}{\partial r_i} \begin{bmatrix} \Pi \\ p \end{bmatrix}^T \begin{bmatrix} M_{11}^I & M_{12}^I \\ M_{21}^I & M_{22}^I \end{bmatrix} \begin{bmatrix} \Pi \\ p \end{bmatrix}$$

$$+ mgS(\rho_0)R^T e_3, \tag{9.30}$$

$$\dot{p} = 0. \tag{9.31}$$

Equations (9.28), (9.29), (9.30), and (9.31) describe the Hamiltonian flow of a three-dimensional pendulum on a cart in terms of the evolution of $(R, x, \Pi, p) \in \mathsf{T}^*(\mathsf{SO}(3) \times \mathbb{R}^2)$ on the cotangent bundle of $\mathsf{SO}(3) \times \mathbb{R}^2$.

9.6.3 Conservation Properties

The Hamiltonian for the three-dimensional pendulum on a cart is

$$H = \frac{1}{2}(M+m)\|\dot{x}\|^2 + \frac{1}{2}\omega^T J\omega + m\dot{x}^T Q^T RS(\rho_0)\omega - mge_3^T R\rho_0,$$

which coincides with the total energy E in this case, and it can be shown to be constant along each solution of the dynamical flow.

In addition, the translational momentum of the three-dimensional pendulum on a cart is

$$p = -mQ^T RS(\rho_0)^T \omega + (M+m)\dot{x},$$

and it can be shown to be constant along each solution of the dynamical flow. This is a consequence of Noether's theorem, and the fact that the Lagrangian does not depend on the position vector of the center of mass of the cart, which corresponds to invariance of the Lagrangian with respect to the tangent lift of translations in $\mathbb{R}^2$.

9.6.4 Equilibrium Properties

The equilibrium solutions of the three-dimensional pendulum on a cart have the property that the horizontal position of the cart is arbitrary, and the attitude of the three-dimensional pendulum satisfies:

$$\rho_0 \times R^T e_3 = 0,$$

which implies that the moment due to gravity vanishes. Additionally, the angular velocity vector of the three-dimensional pendulum and the translation velocity of the cart must be zero.

An attitude R is an equilibrium attitude if and only if the attitude vector $R^T e_3$ is collinear with the vector ρ_0. If $R^T e_3$ is in the opposite direction to the vector ρ_0, then we obtain an inverted equilibrium of the three-dimensional pendulum; if the attitude vector $R^T e_3$ is in the same direction to the vector ρ_0, then we have a hanging equilibrium of the three-dimensional pendulum.

Note that if $R \in \mathsf{SO(3)}$ defines an equilibrium attitude for the three-dimensional pendulum, then a rotation of the three-dimensional pendulum about the vertical by an arbitrary angle is also an equilibrium. Consequently, there are two disjoint equilibrium manifolds of the three-dimensional pendulum. As in Chapter 6, the inverted equilibrium manifold is characterized by the center of mass directly above the pivot. The hanging equilibrium manifold corresponds to the center of mass directly below the pivot.

We do not develop the linearized equations at an equilibrium. However, we do state the result: the linearized equations at an inverted equilibrium can be shown to have a positive eigenvalue. Thus, each inverted equilibrium is necessarily unstable.

9.7 Dynamics of Two Rigid Bodies Constrained to Have a Common Material Point

Consider two rigid bodies that can translate and rotate in three dimensions. The rigid bodies are connected by a universal joint, a connection of the two rigid bodies at common material points, that constrains the two bodies to remain in contact; otherwise, the two bodies are unconstrained. We assume that the connection is frictionless. There are no external forces or moments on the two connected bodies.

We define three Euclidean frames; an inertial frame, and a body-fixed frame for each rigid body. The origin of each body-fixed frame is located at the point of connection in that body. Let $R_i \in \mathsf{SO}(3)$ be the rotation or attitude matrix from the i-th body-fixed frame to the inertial frame, $i = 1, 2$, and let $x \in \mathbb{R}^3$ be the position vector of the connection point in the inertial frame. Ignoring collisions of the two rigid bodies, the configuration is $(R_1, R_2, x) \in \mathsf{SO}(3) \times \mathsf{SO}(3) \times \mathbb{R}^3$ so that the configuration manifold is $\mathsf{SO}(3) \times \mathsf{SO}(3) \times \mathbb{R}^3$. This connection of two rigid bodies has nine degrees of freedom. A schematic of two rigid bodies connected by a common material point is shown in Figure 9.7.

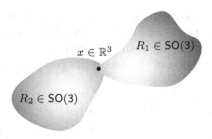

Fig. 9.7 Two rigid bodies connected by a common material point

Also, let $d_i \in \mathbb{R}^3$ be the constant vector from the connection point to the center of mass of the i-th body in the i-th body-fixed frame and let m_i be the scalar mass of the i-th body for $i = 1, 2$. The 3×3 inertia matrix of the i-th body is J_i for $i = 1, 2$; these are described subsequently.

9.7.1 Euler–Lagrange Equations

The angular velocity vector $\omega = (\omega_1, \omega_2) \in \mathsf{T}_{(R_1, R_2)}(\mathsf{SO}(3))^2$ of the two rigid bodies satisfy

$$\dot{R}_1 = R_1 S(\omega_1), \tag{9.32}$$

$$\dot{R}_2 = R_2 S(\omega_2). \tag{9.33}$$

To compute the kinetic energy of each rigid body, let $\rho_i \in \mathbb{R}^3$ denote the vector from the center of mass of the i-th body to a mass element expressed in the i-th body-fixed frame. The vector to the mass element from the origin of the inertial frame is given by $x + R_i(d_i + \rho_i)$. Thus, the kinetic energy of the i-th body is given by

$$T_i(R_i, x_i, \omega_i) = \frac{1}{2} \int_{\mathcal{B}_i} \left\| \dot{x} + \dot{R}_i(d_i + \rho_i) \right\|^2 dm(\rho_i)$$

$$= \frac{1}{2} m_i \|\dot{x}\|^2 + \frac{1}{2} \omega_i^T J_i \omega_i + m_i \dot{x}^T R_i S(\omega_i) d_i, \quad i = 1, 2,$$

where $J_i = m_i S^T(d_i) S(d_i) + \int_{\mathcal{B}_i} S^T(\rho_i) S(\rho_i) \, dm(\rho_i)$ is the 3×3 inertia matrix of the i-th body.

The modified Lagrangian of the connected two rigid bodies $\tilde{L} : \mathsf{T}(\mathsf{SO}(3) \times \mathsf{SO}(3) \times \mathbb{R}^3) \to \mathbb{R}^1$ is given by

$$\tilde{L}(R, x, \omega, \dot{x}) = \frac{1}{2}(m_1 + m_2) \|\dot{x}\|^2 + \frac{1}{2} \omega_1^T J_1 \omega_1 + \frac{1}{2} \omega_2^T J_2 \omega_2$$

$$+ \dot{x}^T \{ m_1 R_1 S(\omega_1) d_1 + m_2 R_2 S(\omega_2) d_2 \}.$$

Note that the Lagrangian does not depend on the inertial position vector of the joint that connects the two rigid bodies. This implies the existence of a translational symmetry in the resulting dynamics.

The infinitesimal variation of the action integral is given by

$$\delta \mathfrak{G} = \int_{t_0}^{t_f} \left[\frac{\partial \tilde{L}(R, x, \omega, \dot{x})}{\partial \omega_1} \cdot \delta \omega_1 + \frac{\partial \tilde{L}(R, x, \omega, \dot{x})}{\partial R_1} \cdot \delta R_1 + \frac{\partial \tilde{L}(R, x, \omega, \dot{x})}{\partial \omega_2} \cdot \delta \omega_2 \right.$$

$$\left. + \frac{\partial \tilde{L}(R, x, \omega, \dot{x})}{\partial R_2} \cdot \delta R_2 + \frac{\partial \tilde{L}(R, x, \omega, \dot{x})}{\partial \dot{x}} \cdot \delta \dot{x} \right] dt,$$

where the infinitesimal variations can be written as

$$\delta R_1 = R_1 S(\eta_1),$$

$$\delta \omega_1 = \dot{\eta}_1 + S(\omega_1) \eta_1,$$

$$\delta R_2 = R_2 S(\eta_2),$$

$$\delta \omega_2 = \dot{\eta}_2 + S(\omega_2) \eta_2,$$

for differentiable curves $\eta_1 : [t_0, t_f] \to \mathbb{R}^3$ satisfying $\eta_1(t_0) = \eta_1(t_f) = 0$ and $\eta_2 : [t_0, t_f] \to \mathbb{R}^3$ satisfying $\eta_2(t_0) = \eta_2(t_f) = 0$, and the infinitesimal variations in $\mathbb{R}^3$ can be expressed by differentiable curves $\delta x : [t_0, t_f] \to \mathbb{R}^3$ satisfying $\delta x(t_0) = \delta x(t_f) = 0$. Substituting these into the above and integrating by parts we obtain

$$\delta \mathfrak{G} = \int_{t_0}^{t_f} \left[\{ J_1 \dot{\omega}_1 + m_1 S(d_1) R_1^T \ddot{x} + S(\omega_1) J_1 \omega_1 \}^T \eta_1 \right.$$

$$+ \{ J_2 \dot{\omega}_2 + m_2 S(d_2) R_2^T \ddot{x} + S(\omega_2) J_2 \omega_2 \}^T \eta_2$$

$$+ \{ m_1 R_1 S^T(d_1) \dot{\omega}_1 + m_2 R_2 S^T(d_2) \dot{\omega}_2 + (m_1 + m_2) \ddot{x}$$

$$\left. - m_1 R_1 S^T(\omega_1) S(\omega_1) d_1 - m_2 R_2 S^T(\omega_2) S(\omega_2) d_2 \}^T \delta x \right] dt.$$

According to Hamilton's principle, the infinitesimal variation of the action integral is zero for all differentiable curves $\eta_1 : [t_0, t_f] \to \mathbb{R}^3$, $\eta_2 : [t_0, t_f] \to \mathbb{R}^3$ and $\delta x : [t_0, t_f] \to \mathbb{R}^3$ that vanish at t_0 and t_f. Using the fundamental lemma of variational calculus, as in Appendix A, it follows that the dynamics for the connection of two rigid bodies are described by the rotational kinematics of the two rigid bodies (9.32) and (9.33), and the Euler–Lagrange equations

$$J_1 \dot{\omega}_1 + m_1 S(d_1) R_1^T \ddot{x} + \omega_1 \times J_1 \omega_1 = 0,$$
$$J_2 \dot{\omega}_2 + m_2 S(d_2) R_2^T \ddot{x} + \omega_2 \times J_2 \omega_2 = 0,$$
$$m_1 R_1 S^T(d_1) \dot{\omega}_1 + m_2 R_2 S^T(d_2) \dot{\omega}_2 + (m_1 + m_2) \ddot{x}$$
$$- m_1 R_1 S^T(\omega_1) S(\omega_1) d_1 - m_2 R_2 S^T(\omega_2) S(\omega_2) d_2 = 0.$$

In a matrix-vector form, these differential equations can be written as

$$
\begin{bmatrix} J_1 & 0 & m_1 S(d_1) R_1^T \\ 0 & J_2 & m_2 S(d_2) R_2^T \\ m_1 R_1 S^T(d_1) & m_2 R_2 S^T(d_2) & (m_1 + m_2) I_{3\times 3} \end{bmatrix} \begin{bmatrix} \dot{\omega}_1 \\ \dot{\omega}_2 \\ \ddot{x} \end{bmatrix}
$$
$$
+ \begin{bmatrix} \omega_1 \times J_1 \omega_1 \\ \omega_2 \times J_2 \omega_2 \\ -m_1 R_1 S^T(\omega_1) S(\omega_1) d_1 - m_2 R_2 S^T(\omega_2) S(\omega_2) d_2 \end{bmatrix} = \begin{bmatrix} 0 \\ 0 \\ 0 \end{bmatrix}. \quad (9.34)
$$

Thus, the Euler–Lagrange equations (9.34) and the two rotational kinematics equations (9.32) and (9.33) describe the dynamical flow of the connected rigid bodies in terms of the evolution of $(R_1, R_2, x, \omega_1, \omega_2, \dot{x}) \in \mathsf{T}(\mathsf{SO}(3) \times \mathsf{SO}(3) \times \mathbb{R}^3)$ on the tangent bundle of $\mathsf{SO}(3) \times \mathsf{SO}(3) \times \mathbb{R}^3$.

9.7.2 Hamilton's Equations

Hamilton's equations can be obtained by introducing the Legendre transformation

$$
\begin{bmatrix} \Pi_1 \\ \Pi_2 \\ p \end{bmatrix} = \begin{bmatrix} \frac{\partial \tilde{L}(R,x,\omega,\dot{x})}{\partial \omega_1} \\ \frac{\partial \tilde{L}(R,x,\omega,\dot{x})}{\partial \omega_2} \\ \frac{\partial \tilde{L}(R,x,\omega,\dot{x})}{\partial \dot{x}} \end{bmatrix} = \begin{bmatrix} J_1 & 0 & m_1 S(d_1) R_1^T \\ 0 & J_2 & m_2 S(d_2) R_2^T \\ m_1 R_1 S^T(d_1) & m_2 R_2 S^T(d_2) & (m_1 + m_2) I_{3\times 3} \end{bmatrix} \begin{bmatrix} \omega_1 \\ \omega_2 \\ \dot{x} \end{bmatrix}.
$$

The angular momentum $(\Pi_1, \Pi_2, p) \in \mathsf{T}^*_{(R_1, R_2, x)}(\mathsf{SO}(3) \times \mathsf{SO}(3) \times \mathbb{R}^3)$ is conjugate to $(\omega_1, \omega_2, \dot{x}) \in \mathsf{T}_{(R_1, R_2, x)}(\mathsf{SO}(3) \times \mathsf{SO}(3) \times \mathbb{R}^3)$; thus

$$
\begin{bmatrix} \omega_1 \\ \omega_2 \\ \dot{x} \end{bmatrix} = \begin{bmatrix} M_{11}^I & M_{12}^I & M_{13}^I \\ M_{21}^I & M_{22}^I & M_{23}^I \\ M_{31}^I & M_{32}^I & M_{33}^I \end{bmatrix} \begin{bmatrix} \Pi_1 \\ \Pi_2 \\ p \end{bmatrix},
$$

where

$$
\begin{bmatrix} M_{11}^I & M_{12}^I & M_{13}^I \\ M_{21}^I & M_{22}^I & M_{23}^I \\ M_{31}^I & M_{32}^I & M_{33}^I \end{bmatrix} = \begin{bmatrix} J_1 & 0 & m_1 S(d_1) R_1^T \\ 0 & J_2 & m_2 S(d_2) R_2^T \\ m_1 R_1 S^T(d_1) & m_2 R_2 S^T(d_2) & (m_1 + m_2) I_{3\times3} \end{bmatrix}^{-1}
$$

is the inverse of the 9×9 partitioned matrix given above. The modified Hamiltonian function can be written as

$$
\tilde{H}(R, x, \Pi, p) = \frac{1}{2} \begin{bmatrix} \Pi_1 \\ \Pi_2 \\ p \end{bmatrix}^T \begin{bmatrix} M_{11}^I & M_{12}^I & M_{13}^I \\ M_{21}^I & M_{22}^I & M_{23}^I \\ M_{31}^I & M_{32}^I & M_{33}^I \end{bmatrix} \begin{bmatrix} \Pi_1 \\ \Pi_2 \\ p \end{bmatrix}.
$$

From (6.10) and (3.12), Hamilton's equations are given by

$$
\dot{r}_{1i} = r_{1i} \times \left\{ M_{11}^I \Pi_1 + M_{12}^I \Pi_2 + M_{13}^I p \right\}, \quad i = 1, 2, 3, \tag{9.35}
$$

$$
\dot{r}_{2i} = r_{2i} \times \left\{ M_{21}^I \Pi_1 + M_{22}^I \Pi_2 + M_{23}^I p \right\}, \quad i = 1, 2, 3, \tag{9.36}
$$

$$
\dot{x} = M_{31}^I \Pi_1 + M_{32}^I \Pi_2 + M_{33}^I p, \tag{9.37}
$$

and also from (6.11) and (3.13),

$$
\dot{\Pi}_1 = \Pi_1 \times \left(M_{11}^I \Pi_1 + M_{12}^I \Pi_2 + M_{13}^I p \right) + \sum_{i=1}^{3} r_{1i} \times \frac{\partial \tilde{H}}{\partial r_{1i}}, \tag{9.38}
$$

$$
\dot{\Pi}_2 = \Pi_2 \times \left(M_{21}^I \Pi_1 + M_{22}^I \Pi_2 + M_{23}^I p \right) + \sum_{i=1}^{3} r_{2i} \times \frac{\partial \tilde{H}}{\partial r_{2i}}, \tag{9.39}
$$

$$
\dot{p} = 0. \tag{9.40}
$$

The kinematic equations (9.35), (9.36), and (9.37) and Hamilton's equations (9.38), (9.39), and (9.40) describe the Hamiltonian flow of the connected rigid bodies in terms of the evolution of $(R_1, R_2, x, \Pi_1, \Pi_2, p) \in$ $\mathsf{T}^*(\mathsf{SO}(3) \times \mathsf{SO}(3) \times \mathbb{R}^3)$ on the cotangent bundle of $\mathsf{SO}(3) \times \mathsf{SO}(3) \times \mathbb{R}^3$.

9.7.3 Conservation Properties

The Hamiltonian for the connection of two rigid bodies through a universal joint, which coincides with the total energy E in this case, is given by

$$
H = \frac{1}{2}(m_1 + m_2) \|\dot{x}\|^2 + \frac{1}{2}\omega_1^T J_1 \omega_1 + \frac{1}{2}\omega_2^T J_2 \omega_2
$$
$$
+ \dot{x}^T \left\{ m_1 R_1 S(\omega_1) d_1 + m_2 R_2 S(\omega_2) d_2 \right\}.
$$

It can be shown that this Hamiltonian is constant along each solution of the dynamical flow.

It is clear from Hamilton's equations (9.40) that the translational momentum

$$p = m_1 R_1 S^T(d_1)\omega_1 + m_2 R_2 S^T(d_2)\omega_2 + (m_1 + m_2)\dot{x}$$

is constant along each solution of the dynamical flow. This can be viewed as a consequence of Noether's theorem and the translational invariance of the Lagrangian.

9.7.4 Equilibrium Properties

The potential energy is identically zero. If the angular velocity vector of the rigid body and the translational velocity vector of the connecting point are zero, it follows that the configuration will remain constant. As such, any constant attitude of each rigid body and any constant position of the rigid bodies correspond to an equilibrium solution.

Suppose that $(R_{1e}, R_{2e}, x_e, 0, 0, 0) \in \mathsf{T}(\mathsf{SO}(3) \times \mathsf{SO}(3) \times \mathbb{R}^3)$ is an equilibrium solution. It can be shown that the linearized equations at this equilibrium are given by

$$\begin{bmatrix} J_1 & 0 & m_1 S(d_1) R_{1e}^T \\ 0 & J_2 & m_2 S(d_2) R_{2e}^T \\ m_1 R_{1e} S^T(d_1) & m_2 R_{2e} S^T(d_2) & (m_1 + m_2) I_{3\times3} \end{bmatrix} \begin{bmatrix} \ddot{\xi}_1 \\ \ddot{\xi}_2 \\ \ddot{\xi}_3 \end{bmatrix} = \begin{bmatrix} 0 \\ 0 \\ 0 \end{bmatrix}.$$

These linearized dynamics are defined on the eighteen-dimensional tangent space of $\mathsf{T}(\mathsf{SO}(3) \times \mathsf{SO}(3) \times \mathbb{R}^3)$ at $(R_{1e}, R_{2e}, x_e, 0, 0, 0) \in \mathsf{T}(\mathsf{SO}(3) \times \mathsf{SO}(3) \times \mathbb{R}^3)$; the linear dynamics approximate the Lagrangian flow of the connected rigid bodies in a neighborhood of the equilibrium. The nonlinear global dynamics of the connected rigid bodies, as described previously, provide much better insight into the dynamics.

9.8 Dynamics of a Rotating and Translating Rigid Body with an Appendage

A rigid body can rotate and translate freely in a three-dimensional space. A proof mass element, viewed as an ideal particle, is constrained to move within a frictionless slot that is fixed in the rigid body. A linear elastic spring connects the proof mass element and the rigid body in the slot. The proof mass element and linear elastic spring represent a flexible appendage of the rigid body.

We select two Euclidean frames: an inertially fixed frame and a frame fixed to the rigid body; the origin of the body-fixed frame is located at the center of mass of the rigid body, ignoring the proof mass element. The slot is assumed to be along the first axis of the body-fixed frame. If the distance between the center of mass and the proof mass element is L, the spring is not stretched and it exerts no force on the proof mass. Let m denote the mass of the proof mass element in the slot, let M denote the mass of the rigid body, and let J be the 3×3 inertia matrix of the rigid body without the mass element.

Let $R \in \mathsf{SO}(3)$ be the attitude of the rigid body, let $y \in \mathbb{R}^3$ be the position vector of the center of mass of the rigid body in the inertially fixed frame, and let $x \in \mathbb{R}^1$ be the displacement of the proof mass element along the slot from the location at which the spring exerts no force; the position vector of the proof mass element in the body-fixed frame is $(L + x)e_1 \in \mathbb{R}^3$. Thus, the configuration of the rigid body and appendage is described by the triple $(R, y, x) \in \mathsf{SE}(3) \times \mathbb{R}^1$ and $\mathsf{SE}(3) \times \mathbb{R}^1$ is the configuration manifold. There are seven degrees of freedom. A schematic of a rotating and translating rigid body with an elastic appendage is shown in Figure 9.8.

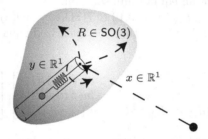

Fig. 9.8 Rotating and translating rigid body with elastic appendage

9.8.1 Euler–Lagrange Equations

The rotation matrix for the rigid body attitude satisfies the kinematics equation

$$\dot{R} = RS(\omega), \tag{9.41}$$

where $\omega \in \mathsf{T}_R\mathsf{SO}(3)$ is the angular velocity vector of the rigid body represented with respect to the body-fixed frame.

The kinetic energy of only the rigid body is $\frac{1}{2}M\|\dot{y}\|^2 + \frac{1}{2}\omega^T J\omega$. Since the location of the proof mass element in the inertially fixed frame is given by $y + (L + x)Re_1 \in \mathbb{R}^3$, the velocity vector of the proof mass element in the inertial frame is given by $\dot{y} + \dot{x}Re_1 + (L + x)RS(\omega)e_1 \in \mathbb{R}^3$. The total kinetic energy of the rigid body and proof mass is therefore given by

$$T = \frac{1}{2}M\|\dot{y}\|^2 + \frac{1}{2}\omega^T J\omega + \frac{1}{2}m\|\dot{y} + \dot{x}Re_1 + (L+x)RS(\omega)e_1\|^2$$

$$= \frac{1}{2}(M+m)\|\dot{y}\|^2 + \frac{1}{2}\omega^T J\omega + \frac{1}{2}m\dot{x}^2$$

$$- m\dot{x}(L+x)e_1^T S(e_1)\omega + m\dot{y}^T\{\dot{x}Re_1 + (L+x)RS(\omega)e_1\}$$

$$+ \frac{1}{2}m(L+x)^2\omega^T S^T(e_1)S(e_1)\omega$$

$$= \frac{1}{2}(M+m)\|\dot{y}\|^2 + \frac{1}{2}\omega^T\{J + m(L+x)^2 S^T(e_1)S(e_1)\}\omega + \frac{1}{2}m\dot{x}^2$$

$$+ m\dot{y}^T\{\dot{x}Re_1 + (L+x)RS(\omega)e_1\}.$$

The potential energy is the energy stored in the elastic spring and is given by $U = \frac{1}{2}\kappa x^2$.

The resulting modified Lagrangian function $\tilde{L} : \mathsf{T}(\mathsf{SE}(3) \times \mathbb{R}^1) \to \mathbb{R}^1$ is given by

$$\tilde{L}(R, y, x, \omega, \dot{y}, \dot{x}) = \frac{1}{2}\omega^T\{J + m(L+x)^2 S^T(e_1)S(e_1)\}\omega + \frac{1}{2}(M+m)\|\dot{y}\|^2$$

$$+ \frac{1}{2}m\dot{x}^2 + m\dot{y}^T Re_1\dot{x} - m(L+x)\dot{y}^T RS(e_1)\omega - \frac{1}{2}\kappa x^2,$$

or in matrix-vector form

$$\tilde{L}(R, y, x, \omega, \dot{y}, \dot{x})$$

$$= \frac{1}{2}\begin{bmatrix}\omega \\ \dot{y} \\ \dot{x}\end{bmatrix}^T \begin{bmatrix} J + m(L+x)^2 S^T(e_1)S(e_1) & m(L+x)S(e_1)R^T & 0 \\ m(L+x)RS^T(e_1) & (M+m)I_{3\times 3} & mRe_1 \\ 0 & me_1^T R^T & m \end{bmatrix}\begin{bmatrix}\omega \\ \dot{y} \\ \dot{x}\end{bmatrix}$$

$$- \frac{1}{2}\kappa x^2.$$

Note that the Lagrangian function is independent of the position vector of the rigid body; this implies the existence of a translational symmetry in the system. The infinitesimal variation of the action integral is given by

$$\delta\mathfrak{G} = \int_{t_0}^{t_f} \Big[\{(J + m(L+x)^2 S^T(e_1)S(e_1))\omega + m(L+x)S(e_1)R^T\dot{y}\}^T\delta\omega$$

$$+ m\dot{y}^T\{\dot{x}\delta Re_1 + (L+x)\delta RS(\omega)e_1\}$$

$$+ \{(M+m)\dot{y} + m(\dot{x}Re_1 + (L+x)RS(\omega)e_1)\}^T\delta\dot{y}$$

$$+ \{m\dot{x} + m\dot{y}^T Re_1\}^T\delta\dot{x}$$

$$+ \{m(L+x)\omega^T S^T(e_1)S(e_1)\omega + m\dot{y}^T RS(\omega)e_1 - \kappa x\}^T\delta x\Big]dt.$$

The infinitesimal variations in $\mathsf{SO}(3)$ can be written as

$$\delta R = RS(\eta),$$
$$\delta \omega = \dot{\eta} + S(\omega)\eta,$$

for differentiable curves $\eta : [t_0, t_f] \to \mathbb{R}^3$ satisfying $\eta(t_0) = \eta(t_f) = 0$, the infinitesimal variations in $\mathbb{R}^3$ can be expressed as $\delta y : [t_0, t_f] \to \mathbb{R}^3$ satisfying $\delta y(t_0) = \delta y(t_f) = 0$, and the infinitesimal variations in $\mathbb{R}^1$ can be expressed as $\delta x : [t_0, t_f] \to \mathbb{R}^1$ satisfying $\delta x(t_0) = \delta x(t_f) = 0$. Substituting these into the above and integrating by parts, we obtain

$$\delta \mathfrak{G} = \int_{t_0}^{t_f} \Big[\{(J + m(L+x)^2 S^T(e_1)S(e_1))\dot{\omega} + 2m(L+x)S(e_1)S(e_1)\omega \dot{x}$$
$$+ m(L+x)S(e_1)R^T \ddot{y} + S(\omega)((J + m(L+x)^2 S^T(e_1)S(e_1))\omega)\}^T \eta$$
$$+ \{(M+m)\ddot{y} + mRe_1 \ddot{x} + 2m\dot{x}RS(\omega)e_1$$
$$- m(L+x)(RS(\omega)^T S(\omega)e_1 - RS(e_1)\dot{\omega})\}^T \delta y$$
$$+ \{m\ddot{x} + me_1^T R^T \ddot{y} - m(L+x)\omega^T S^T(e_1)S(e_1)\omega + \kappa x\}^T \delta x \Big] dt.$$

According to Hamilton's principle, the infinitesimal variation of the action integral is zero for all differentiable curves $\eta : [t_0, t_f] \to \mathbb{R}^3$, $\delta y : [t_0, t_f] \to \mathbb{R}^3$, and $\delta x : [t_0, t_f] \to \mathbb{R}^1$ that vanish at t_0 and t_f. Using the fundamental lemma of variational calculus, we obtain the following Euler–Lagrange equations for the rotating and translating rigid body with a flexible appendage.

$$\{J + m(L+x)^2 S^T(e_1)S(e_1)\}\dot{\omega} + m(L+x)S(e_1)R^T \ddot{y}$$
$$+ 2m(L+x)S^T(e_1)S(e_1)\omega \dot{x}$$
$$+ \omega \times \{J + m(L+x)^2 S^T(e_1)S(e_1)\}\omega = 0,$$
$$m(L+x)RS^T(e_1)\dot{\omega} + (M+m)\ddot{y} + mRe_1 \ddot{x} + 2m\dot{x}RS(\omega)e_1$$
$$- m(L+x)RS(\omega)^T S(\omega)e_1 = 0,$$
$$me_1^T R^T \ddot{y} + m\ddot{x} - m(L+x)\omega^T S^T(e_1)S(e_1)\omega + \kappa x = 0.$$

These equations can be written in matrix-vector form

$$\begin{bmatrix} J + m(L+x)^2 S^T(e_1)S(e_1) & m(L+x)S(e_1)R^T & 0 \\ m(L+x)RS^T(e_1) & (M+m)I_{3\times 3} & mRe_1 \\ 0 & me_1^T R^T & m \end{bmatrix} \begin{bmatrix} \dot{\omega} \\ \ddot{y} \\ \ddot{x} \end{bmatrix}$$
$$+ \begin{bmatrix} 2m(L+x)S^T(e_1)S(e_1)\omega \dot{x} + \omega \times \{(J + m(L+x)^2 S^T(e_1)S(e_1))\} \omega \\ 2m\dot{x}RS(\omega)e_1 - m(L+x)RS(\omega)^T S(\omega)e_1 \\ -m(L+x)\omega^T S^T(e_1)S(e_1)\omega \end{bmatrix}$$
$$+ \begin{bmatrix} 0 \\ 0 \\ \kappa x \end{bmatrix} = \begin{bmatrix} 0 \\ 0 \\ 0 \end{bmatrix}. \tag{9.42}$$

The kinematic equations (9.41) and the Euler–Lagrange equations (9.42) define the dynamical flow of a rotating and translating rigid body with an elastic appendage in terms of the evolution of $(R, y, x, \omega, \dot{y}, \dot{x}) \in \mathsf{T}(\mathsf{SE}(3) \times \mathbb{R}^1)$ on the tangent bundle of $\mathsf{SE}(3) \times \mathbb{R}^1$.

9.8.2 Hamilton's Equations

We define the momenta $(\Pi, p_1, p_2) \in \mathsf{T}^*_{(R,y,x)}(\mathsf{SE}(3) \times \mathbb{R}^1)$ by using the Legendre transformation,

$$
\begin{bmatrix} \Pi \\ p_1 \\ p_2 \end{bmatrix} = \begin{bmatrix} \frac{\partial \tilde{L}(R,y,x,\omega,\dot{y},\dot{x})}{\partial \omega} \\ \frac{\partial \tilde{L}(R,y,x,\omega,\dot{y},\dot{x})}{\partial \dot{y}} \\ \frac{\partial \tilde{L}(R,y,x,\omega,\dot{y},\dot{x})}{\partial \dot{x}} \end{bmatrix}
$$

$$
= \begin{bmatrix} J + m(L+x)^2 S^T(e_1)S(e_1) & m(L+x)S(e_1)R^T & 0 \\ m(L+x)RS^T(e_1) & (M+m)I_{3\times3} & 0 \\ 0 & 0 & m \end{bmatrix} \begin{bmatrix} \omega \\ \dot{y} \\ \dot{x} \end{bmatrix}.
$$

The momentum $(\Pi, p_1, p_2) \in \mathsf{T}^*_{R,y,x}(\mathsf{SE}(3) \times \mathbb{R}^1)$ is conjugate to the velocity vector $(\omega, \dot{y}, \dot{x}) \in \mathsf{T}_{R,y,x}(\mathsf{SE}(3) \times \mathbb{R}^1)$. Thus

$$
\begin{bmatrix} \omega \\ \dot{y} \\ \dot{x} \end{bmatrix} = \begin{bmatrix} M^I_{11} & M^I_{12} & M^I_{13} \\ M^I_{21} & M^I_{22} & M^I_{23} \\ M^I_{31} & M^I_{32} & M^I_{33} \end{bmatrix} \begin{bmatrix} \Pi \\ p_1 \\ p_2 \end{bmatrix},
$$

where

$$
\begin{bmatrix} M^I_{11} & M^I_{12} & M^I_{13} \\ M^I_{21} & M^I_{22} & M^I_{23} \\ M^I_{31} & M^I_{32} & M^I_{33} \end{bmatrix} = \begin{bmatrix} J + m(L+x)^2 S^T(e_1)S(e_1) & m(L+x)S(e_1)R^T & 0 \\ m(L+x)RS^T(e_1) & (M+m)I_{3\times3} & 0 \\ 0 & 0 & m \end{bmatrix}^{-1}
$$

is the inverse of the 7×7 partitioned matrix given above. The modified Hamiltonian function $\tilde{H} : \mathsf{T}^*(\mathsf{SE}(3) \times \mathbb{R}^1) \to \mathbb{R}^1$ is

$$
\tilde{H}(R, y, x, \Pi, p_1, p_2) = \frac{1}{2} \begin{bmatrix} \Pi \\ p_1 \\ p_2 \end{bmatrix}^T \begin{bmatrix} M^I_{11} & M^I_{12} & M^I_{13} \\ M^I_{21} & M^I_{22} & M^I_{23} \\ M^I_{31} & M^I_{32} & M^I_{33} \end{bmatrix} \begin{bmatrix} \Pi \\ p_1 \\ p_2 \end{bmatrix} + \frac{1}{2} \kappa x^2.
$$

Thus, we obtain Hamilton's equations for the rotating and translating rigid body with an elastic appendage; we express the rotational kinematics of the rigid body in terms of the columns $r_i \in \mathsf{S}^2$, $i = 1, 2, 3$, of the rigid body attitude $R^T \in \mathsf{SO}(3)$.

$$\dot{r}_i = r_i \times \left\{ M_{11}^I \Pi + M_{12}^I p_1 + M_{13}^I p_2 \right\}, \quad i = 1, 2, 3, \tag{9.43}$$

$$\begin{bmatrix} \dot{y} \\ \dot{x} \end{bmatrix} = \begin{bmatrix} M_{21}^I & M_{22}^I & M_{23}^I \\ M_{31}^I & M_{32}^I & M_{33}^I \end{bmatrix} \begin{bmatrix} \Pi \\ p_1 \\ p_2 \end{bmatrix}, \tag{9.44}$$

and also from (6.11) and (3.13),

$$\dot{\Pi} = \Pi \times \left(M_{11}^I \Pi + M_{12}^I p_1 + M_{13}^I p_2 \right) + \sum_{i=1}^{3} r_i \times \frac{\partial \tilde{H}}{\partial r_i},$$

$$\dot{p}_1 = 0,$$

$$\dot{p}_2 = -\frac{\partial \tilde{H}}{\partial x}.$$

The above equations can be expressed as

$$\begin{bmatrix} \dot{\Pi} \\ \dot{p}_1 \\ \dot{p}_2 \end{bmatrix} = \begin{bmatrix} \Pi \times \omega + mS(R^T \dot{y})\{(L+x)S(e_1)\omega - e_1 \dot{x}\} \\ 0 \\ m(L+x)\omega^T S^T(e_1)S(e_1)\omega - m\dot{y}^T RS(e_1)\omega - \kappa x \end{bmatrix}, \tag{9.45}$$

where the right-hand side is evaluated using the Legendre transformation. Thus, equations (9.43), (9.44), and (9.45) define the Hamiltonian flow of a rotating and translating rigid body with an elastic appendage in terms of the evolution of $(R, y, x, \Pi, p_1, p_2) \in \mathsf{T}(\mathsf{SE}(3) \times \mathbb{R}^1)$ on the cotangent bundle of $\mathsf{SE}(3) \times \mathbb{R}^1$.

9.8.3 Conservation Properties

The Hamiltonian

$$H = \frac{1}{2}(M+m)\|\dot{y}\|^2 + \frac{1}{2}\omega^T \{J + m(L+x)^2 S^T(e_1)S(e_1)\}\omega + \frac{1}{2}m\dot{x}^2$$

$$+ m\dot{y}^T \{\dot{x}Re_1 + (L+x)RS(\omega)e_1\} + \frac{1}{2}\kappa x^2$$

which coincides with the total energy E in this case is conserved along any solution of the dynamical flow of the rotating and translating rigid body with an elastic appendage.

The total translational momentum in the inertial frame, namely

$$p_1 = m(L+x)RS^T(e_1)\omega + (M+m)\dot{y} + mRe_1\dot{x}$$

is also conserved along each solution of the dynamical flow of the rotating and translating rigid body with an elastic appendage. This is a consequence of Noether's theorem and the translational symmetry of the Lagrangian.

9.8.4 Equilibrium Properties

The equilibrium solutions occur when the angular velocity vector and the translational velocity vector of the rigid body are zero and the proof mass element is located in the slot at the point where the elastic spring is not deformed. The rigid body and appendage can be in equilibrium at any fixed attitude of the rigid body and at any fixed position vector of the rigid body.

To illustrate the linearization of the dynamics of a translating and rotating rigid body with appendage, we follow the linearization development used previously for a rotating rigid body with appendage. We examine the prototypical equilibrium solution $(I_{3\times3}, 0, 0, 0, 0, 0) \in \mathsf{T}(\mathsf{SE}(3) \times \mathbb{R}^1)$. Following the results in Appendix B, the linearized equations are given by

$$\begin{bmatrix} J + mL^2 S^T(e_1) S(e_1) & mLS(e_1) & 0 \\ mLS^T(e_1) & (M+m)I_{3\times3} & me_1 \\ 0 & me_1^T & m \end{bmatrix} \begin{bmatrix} \ddot{\xi}_1 \\ \ddot{\xi}_2 \\ \ddot{\xi}_3 \end{bmatrix} + \begin{bmatrix} 0 & 0 & 0 \\ 0 & 0 & 0 \\ 0 & 0 & \kappa \end{bmatrix} \begin{bmatrix} \xi_1 \\ \xi_2 \\ \xi_3 \end{bmatrix} = \begin{bmatrix} 0 \\ 0 \\ 0 \end{bmatrix}.$$

These linearized differential equations are defined on the fourteen-dimensional tangent space of $\mathsf{T}(\mathsf{SE}(3) \times \mathbb{R}^1)$ at $(I_{3\times3}, 0, 0, 0, 0, 0) \in \mathsf{T}(\mathsf{SE}(3) \times \mathbb{R}^1)$. These linear dynamics approximate the translational and rotational dynamics of the rigid body with appendage in a neighborhood of the equilibrium $(I_{3\times3}, 0, 0, 0, 0, 0) \in \mathsf{T}(\mathsf{SE}(3) \times \mathbb{R}^1)$. These simple linear dynamics are accurate only to first order in the perturbations. Higher-order coupling effects are important for large perturbations from the equilibrium.

9.9 Dynamics of a Full Body System

Consider the full body rotational and translational dynamics for n rigid bodies under the action of Newtonian gravitational forces between each pair of bodies. Throughout the development, we ignore the possibility of collisions between two or more bodies. Equations of motion for the full body dynamics are derived in an inertial Euclidean frame, and they are expressed in both Lagrangian and Hamiltonian form. The equations demonstrate the coupling between the translational motion and the rotational motion of the n bodies. The full body dynamics have $6n$ degrees of freedom. A schematic of a full body system with $n = 2$ is shown in Figure 9.9. The full body problem was introduced in [67]; the subsequent development follows [50, 51].

We introduce an inertial Euclidean frame in $\mathbb{R}^3$ and a body-fixed frame for each of the n bodies. The origin of the i-th body-fixed frame is located at the center of mass of body $\mathcal{B}_i$, $i = 1, \ldots, n$. The configuration space of the i-th rigid body is $\mathsf{SE}(3)$. We denote the position of the center of mass of body $\mathcal{B}_i$ in the inertial frame by $x_i \in \mathbb{R}^3$, and we denote the attitude of $\mathcal{B}_i$ by $R_i \in \mathsf{SO}(3)$,

$(R_1, x_1) \in \mathsf{SE}(3)$ $(R_2, x_2) \in \mathsf{SE}(3)$

Fig. 9.9 Full body system

which is a rotation matrix from the i-th body-fixed frame to the inertial frame, for $i = 1, \ldots, n$. Thus, the full body configuration is described by $(R_1, x_1, \ldots, R_n, x_n) \in (\mathsf{SE}(3))^n$, and the configuration manifold is $(\mathsf{SE}(3))^n$. We use the notation $R = (R_1, \ldots, R_n) \in \mathsf{SO}(3)^n$ and $x = (x_1, \ldots, x_n) \in (\mathbb{R}^3)^n$, so that $(R, x) \in \mathsf{SE}(3)^n$.

9.9.1 Euler–Lagrange Equations

To derive the equations of motion, we first construct a Lagrangian for the full body dynamics. Given $(x_i, R_i) \in \mathsf{SE}(3)$, the inertial position of a mass element of $\mathcal{B}_i$ is given by $x_i + R_i \rho_i$, where $\rho_i \in \mathbb{R}^3$ denotes the position of the mass element in the body-fixed frame. The kinetic energy of $\mathcal{B}_i$ can be written as

$$T_i(\omega_i, \dot{x}_i) = \frac{1}{2} \int_{\mathcal{B}_i} \|\dot{x}_i + \dot{R}_i \rho_i\|^2 \, dm_i(\rho_i), \quad i = 1, \ldots, n.$$

Using the fact that $\int_{\mathcal{B}_i} \rho_i dm_i(\rho_i) = 0$, and the kinematics equation

$$\dot{R}_i = R_i S(\omega_i), \quad i = 1, \ldots, n, \tag{9.46}$$

where $\omega_i \in \mathsf{T}_R \mathsf{SO}(3)$ is the angular velocity vector of body $\mathcal{B}_i$ in its body-fixed frame, the kinetic energy for each body can be rewritten as

$$T_i(\omega_i, \dot{x}_i) = \frac{1}{2} \int_{\mathcal{B}_i} \|\dot{x}_i\|^2 + \|S(\omega_i)\rho_i\|^2 \, dm_i(\rho_i)$$

$$= \frac{1}{2} m_i \|\dot{x}_i\|^2 + \frac{1}{2} \omega_i^T J \omega_i,$$

where $m_i = \int_{\mathcal{B}_i} dm_i(\rho_i)$ is the mass of $\mathcal{B}_i$, and $J_i = \int_{\mathcal{B}_i} S(\rho_i)^T S(\rho_i) dm_i(\rho_i)$ is the standard 3×3 inertia matrix of the i-th body for $i = 1, \ldots, n$. Thus, the modified Lagrangian of n full bodies $\tilde{L} : \mathsf{TSE}(3)^n \to \mathbb{R}^1$ can be written as

$$\tilde{L}(R, x, \omega, \dot{x}) = \sum_{i=1}^{n} \left\{ \frac{1}{2} \omega_i^T J \omega_i + \frac{1}{2} m_i \|\dot{x}_i\|^2 \right\} - U(R, x),$$

where $\omega = (\omega_1, \ldots, \omega_n) \in \mathsf{T}_R(\mathsf{SO}(3))^n$ and $\dot{x} = (\dot{x}_1, \ldots, \dot{x}_n) \in \mathsf{T}_x(\mathbb{R}^3)^n$.

The specific form of the Newtonian gravitational potential energy U : $\mathsf{SE}(3)^n \to \mathbb{R}^1$ is described. It is the sum of the mutual potentials over all pairs of distinct bodies given by

$$U(R,x) = -\frac{1}{2}\sum_{i=1}^{n}\sum_{\substack{j=1\\j\neq i}}^{n}\int_{\mathcal{B}_i}\int_{\mathcal{B}_j}\frac{Gdm_j(\rho_j)dm_i(\rho_i)}{\|x_i + R_i\rho_i - x_j - R_j\rho_j\|},$$

where G is the universal gravitational constant and $\rho_i \in \mathbb{R}^3$ denotes the vector from the center of mass of the i-th body to the element of mass $dm_i(\rho_i)$ in its body-fixed frame, $i = 1, \ldots, n$.

The infinitesimal variation of the action integral is given by

$$\delta\mathfrak{G} = \int_{t_0}^{t_f}\{\sum_{i=1}^{n}\{\omega_i^T J_i\delta\omega_i + m_i\dot{x}_i^T\delta\dot{x}_i\} - \delta U(R,x)\}\,dt.$$

The infinitesimal variations in $\mathsf{SO}(3)$ can be written as

$$\delta R_i = R_i S(\eta_i), \qquad i = 1, \ldots, n,$$
$$\delta\omega_i = \dot{\eta}_i + S(\omega_i)\eta_i, \quad i = 1, \ldots, n,$$

for differentiable curves $\eta_i : [t_0, t_f] \to \mathbb{R}^3$ satisfying $\eta_i(t_0) = \eta_i(t_f) = 0$, for $i = 1, \ldots, n$, and the infinitesimal variations in $\mathbb{R}^3$ can be expressed as $\delta x_i : [t_0, t_f] \to \mathbb{R}^3$ satisfying $\delta x_i(t_0) = \delta x_i(t_f) = 0$, for $i = 1, \ldots, n$. The infinitesimal variation of the potential energy is

$$\delta U(R,x) = \frac{1}{2}\sum_{i=1}^{n}\sum_{\substack{j=1\\j\neq i}}^{n}\int_{\mathcal{B}_i}\int_{\mathcal{B}_j}\frac{G(\sigma_i - \sigma_j)^T(\delta\sigma_i - \delta\sigma_j)}{\|\sigma_i - \sigma_j\|^3}\,dm_j(\rho_j)dm_i(\rho_i),$$

where

$$\sigma_i = x_i + R_i\rho_i,$$

and

$$\delta\sigma_i = \delta x_i + \delta R_i\rho_i = \delta x_i - R_i S(\rho_i)\eta_i.$$

After a careful rearrangement of terms,

$$\delta U(R,x) = \sum_{i=1}^{n}M_i^T\eta_i + \sum_{i=1}^{n}F_i^T\delta x_i,$$

where

$$M_i = \int_{\mathcal{B}_i} \sum_{\substack{j=1 \\ j \neq i}}^{n} \int_{\mathcal{B}_j} \frac{GS(\rho_i)R_i^T(x_i + R_i\rho_i - x_j - R_j\rho_j)dm_j(\rho_j)dm_i(\rho_i)}{\|x_i + R_i\rho_i - x_j + R_j\rho_j\|^3},$$

and

$$F_i = \int_{\mathcal{B}_i} \sum_{\substack{j=1 \\ j \neq i}}^{n} \int_{\mathcal{B}_j} \frac{G(x_i + R_i\rho_i - x_j - R_j\rho_j)dm_j(\rho_j)dm_i(\rho_i)}{\|x_i + R_i\rho_i - x_j + R_j\rho_j\|^3}.$$

Here, $M_i : \mathsf{SE}(3)^n \to \mathbb{R}^3$ can be interpreted as the gravitational moment on the i-th rigid body due to all the other rigid bodies and $F_i : \mathsf{SE}(3)^n \to \mathbb{R}^3$ can be interpreted as the gravitational force on the i-th rigid body due to all the other rigid bodies.

Substituting these into the above and integrating by parts we obtain

$$\delta \mathfrak{G} = \sum_{i=1}^{n} \int_{t_0}^{t_f} \{ \{J_i\dot{\omega}_i + S(\omega_i)J_i\omega_i - M_i\}^T \eta_i + \{m_i\ddot{x}_i - F_i\}^T \delta x_i \} \, dt.$$

Consequently, the Euler–Lagrange equations are given by

$$J_i\dot{\omega}_i + \omega_i \times J_i\omega_i - M_i = 0, \quad i = 1,\dots,n, \qquad (9.47)$$

$$m_i\ddot{x}_i - F_i = 0, \quad i = 1,\dots,n, \qquad (9.48)$$

where the gravitational moments and forces were given previously by integral expressions. Thus, the kinematic equations (9.46), and the Euler–Lagrange equations (9.47) and (9.48) describe the dynamical flow of the full body dynamics in terms of $(R, x, \omega, \dot{x}) \in \mathsf{TSE}(3)^n$.

9.9.2 Hamilton's Equations

Hamilton's equations can be obtained using the Legendre transformation defined by

$$\Pi_i = \frac{\partial L(R, x, \omega, \dot{x})}{\partial \omega_i} = J\omega_i, \quad i = 1,\dots,n,$$

$$p_i = \frac{\partial L(R, x, \omega, \dot{x})}{\partial \dot{x}_i} = m_i\dot{x}_i, \quad i = 1,\dots,n,$$

where $(\Pi, p) \in \mathsf{T}^*_{(R,x)}(\mathsf{SE}(3))^n$ are angular momenta and translational momenta that are conjugate to $(\omega, \dot{x}) \in \mathsf{T}_{(R,x)}(\mathsf{SE}(3))^n$. The modified Hamiltonian function is

$$\tilde{H}(R, x, \Pi, p) = \sum_{i=1}^{n} \left\{ \frac{1}{2} \Pi_i^T J_i^{-1} \Pi_i + \frac{1}{2m_i} \|p_i\|^2 \right\} + U(R, x),$$

where we use the notation $\Pi = (\Pi_1, \ldots, \Pi_n)$ and $p = (p_1, \ldots, p_n)$. Hamilton's equations of motion can be written as

$$\dot{R}_i = R_i S(J_i^{-1} \Pi_i), \qquad i = 1, \ldots, n, \tag{9.49}$$

$$\dot{x}_i = \frac{p_i}{m_i}, \qquad i = 1, \ldots, n, \tag{9.50}$$

$$\dot{\Pi}_i = \Pi_i \times J_i^{-1} \Pi_i + M_i, \quad i = 1, \ldots, n, \tag{9.51}$$

$$\dot{p}_i = F_i, \quad i = 1, \ldots, n, \tag{9.52}$$

with the gravitational moments and forces given previously. Hamilton's equations (9.49), (9.50), (9.51), and (9.52) describe the Hamiltonian flow of the full body dynamics in terms of $(R, x, \Pi, p) \in \mathsf{T}^*\mathsf{SE}(3)^n$ on the cotangent bundle of $\mathsf{SE}(3)^n$.

9.9.2.1 Comments

The equations of motion are coupled through the gravitational potential, which is described by a complicated expression that captures the most important feature of the full body dynamics. The complexity of the potential energy arises due to the presence of the body integrals and the body summations. It is challenging to evaluate the gravitational potential and to evaluate the gravitational moments and gravitational forces which entails evaluating derivatives of the potential. Various approximations can be made to simplify these expressions for the gravitational moments and gravitational forces [50, 51].

These equations of motion for the full body dynamics can have complicated solution properties that involve coupling between the rotational dynamics and the translational or orbital dynamics. In particular, this is of importance in understanding the orbital dynamics of certain binary asteroids.

9.9.3 Conservation Properties

The Hamiltonian, which coincides with the total energy E in this case, is given by

$$H = \sum_{i=1}^{n} \left\{ \frac{1}{2} m_i \left\| \dot{x}_i \right\|^2 + \frac{1}{2} \omega_i^T J \omega_i \right\}$$
$$- \frac{1}{2} \sum_{i=1}^{n} \sum_{\substack{j=1 \\ j \neq i}}^{n} \int_{\mathcal{B}_i} \int_{\mathcal{B}_j} \frac{G dm_j(\rho_j) dm_i(\rho_i)}{\left\| x_i + R_i \rho_i - x_j - R_j \rho_j \right\|}.$$

The Hamiltonian is constant along each solution of the dynamical flow of the n full bodies acting under their mutual gravity.

Depending on additional symmetry properties of the rigid bodies, there can be additional conserved quantities, but these are not considered here.

9.9.4 Equilibrium Properties

There are no meaningful equilibrium solutions. The full body dynamics can be extremely complicated but in special cases there exist important relative equilibrium solutions that correspond to solutions for which the relative motion between the bodies is constant.

9.9.5 Relative Full Body Dynamics

The motion of the full rigid bodies depends only on the relative positions and the relative attitudes of the bodies. This is a consequence of the fact that the gravitational potential only depends on these relative variables, and hence, the Lagrangian is invariant under rigid translations and rigid rotations. This implies that the equations of motion can be expressed in one of the body-fixed frames using only relative positions and relative attitudes of the bodies.

This leads to an interesting and useful alternative formulation for the equations of motion of n full rigid bodies that is developed in detail in the published literature [50, 51].

9.10 Dynamics of a Spacecraft with Reaction Wheel Assembly

Consider a spacecraft with an arbitrary number of reaction control wheels acted on by a disturbance force and a disturbance moment. The dynamics of a spacecraft with reaction wheels have been studied widely [66]. However, the dynamical models are often based on several simplifying assumptions. For example, a reaction wheel is often assumed to be inertially symmetric about its spin axis, and the spin axis corresponds to the principal axis of the spacecraft. The translational dynamics are often ignored. In some cases, the definition of the angular momentum or inertia matrices are unclear and confusing. For example, inertia matrices and reference frames for the reaction wheels are not clearly and unambiguously specified.

Here, we present the equations of motion for spacecraft reaction wheels without relying on any simplifying assumptions, according to our geometric formulation of mechanics. In this section, the *base spacecraft* denotes the spacecraft without reaction wheels, and the *(whole) spacecraft* refers to the complete spacecraft including the base spacecraft and n reaction wheels.

Define an inertial Euclidean frame and a body-fixed frame whose origin is located at the center of mass of the base spacecraft. The configuration of the base spacecraft is described by $(R, x) \in \mathsf{SE}(3)$, where $x \in \mathbb{R}^3$ is the position vector of its center of mass with respect to the inertial frame, and $R \in \mathsf{SO}(3)$ is the attitude matrix that transforms a vector from the body-fixed frame to the inertial frame (Figure 9.10).

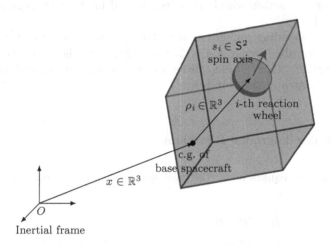

Fig. 9.10 Spacecraft with reaction wheels (only the i-th wheel is illustrated)

Next, we describe the configuration of the reaction wheels. Let $\rho_i \in \mathbb{R}^3$ be the vector from the origin of the body-fixed frame to a base point on the spin axis of the i-th reaction wheel. The spin axis of the i-th reaction wheel is denoted by the direction vector $s_i \in \mathsf{S}^2$. Since both ρ_i and s_i are represented with respect to the body-fixed frame, they are fixed, i.e., $\dot{\rho}_i = \dot{s}_i = 0$, for $i = 1, \ldots, n$.

We introduce the i-th reaction wheel frame such that its first axis corresponds to s_i, and the remaining two axes are chosen such that they constitute an orthogonal frame. Define $Q_i \in \mathsf{SO}(3)$ as the rotation matrix that transforms a vector from the i-th reaction wheel frame to the body-fixed frame for $i = 1, \ldots, n$. According to the definition of Q_i, we have $s_i = Q_i e_1$, where $e_j \in \mathbb{R}^3$ is the j-th standard basis of $\mathbb{R}^3$. It is assumed that the i-th reaction

wheel frame is fixed to the wheel, i.e., it rotates about the spin axis. Therefore, the attitude of the i-th reaction wheel can be represented by an element
of S^1. More explicitly, let $Q_i^0 \in SO(3)$ represent a fixed reference rotational
configuration of the i-th wheel for $i = 1, \ldots, n$. Since the rotation matrices
Q_i and Q_i^0 differ by a rotation about the spin axis, Q_i can be written as

$$Q_i = Q_i^0 [e_1, Cq_i, e_1 \times Cq_i], \quad i = 1, \ldots, n,$$

where the matrix $C \in \mathbb{R}^{3 \times 2}$ is given by

$$C = \begin{bmatrix} 0 & 0 \\ 1 & 0 \\ 0 & 1 \end{bmatrix},$$

and the unit vector $q_i \in S^1$ corresponds to the last two components of the
second axis of Q_i with respect to the frame defined by Q_i^0, for $i = 1, \ldots, n$.
The vector of reaction wheel attitudes, defined in this way, is given by $q = (q_1, \ldots, q_n) \in (S^1)^n$.

The corresponding configuration for the spacecraft and reaction wheel assembly is described by $(R, x, q) \in SE(3) \times (S^1)^n$ and the configuration manifold is $SE(3) \times (S^1)^n$. Thus, there are $n + 6$ degrees of freedom. This is
described in greater detail in [48].

9.10.1 Euler–Lagrange Equations

The kinematics equations for the base spacecraft and the reaction wheels are
given by

$$\dot{R} = RS(\Omega), \tag{9.53}$$

$$\dot{q}_i = q_i S \omega_i, \qquad\qquad i = 1, \ldots, n, \tag{9.54}$$

$$\dot{Q}_i = Q_i S(e_1) \omega_i = S(s_i) Q_i \omega_i, \quad i = 1, \ldots, n, \tag{9.55}$$

where $\Omega \in T_R SO(3)$ is the angular velocity vector of the base spacecraft
represented with respect to the body-fixed frame, and $\omega_i \in T_{q_i} S^1$ denotes
the spin rate of the i-th reaction wheel, for $i = 1, \ldots, n$. The corresponding
angular velocity vector of the i-th reaction wheel is $\omega_i s_i$ with respect to the
body-fixed frame, or equivalently, it is $\omega_i e_1$ with respect to the i-th reaction
wheel frame, for $i = 1, \ldots, n$. The vector of spin rates of the reaction wheels
is given by $\omega = (\omega_1, \ldots, \omega_n) \in \mathbb{R}^n$.

Consider a mass element $dm(\rho)$ of the base spacecraft, located with respect
to the body-fixed frame at $\rho \in \mathbb{R}^3$. The position vector and the velocity vector
of the mass element are $x + R\rho$ and $\dot{x} + RS(\Omega)\rho$, respectively, with respect
to the inertial frame.

Next, consider a mass element $dm_i(\xi_i)$ of the i-th reaction wheel that is located at ξ_i with respect to the i-th reaction wheel frame. Since each reaction wheel is rigid, the position vector of the mass element with respect to the inertial frame is given by $x + R(\rho_i + Q_i\xi_i)$ and its velocity vector with respect to the inertial frame is $\dot{x} + RS(\Omega)(\rho_i + Q_i\xi_i) + RQ_iS(e_1)\xi_i\omega_i$, for $i = 1, \ldots, n$.

The kinetic energy is the sum of the kinetic energy of the base spacecraft and the kinetic energy of the reaction wheels:

$$T = \int_{\mathcal{B}_b} \frac{1}{2}\|\dot{x} + RS(\Omega)\rho\|^2 \, dm(\rho)$$

$$+ \sum_{i=1}^{n} \int_{\mathcal{B}_i} \frac{1}{2}\|\dot{x} + RS(\Omega)(\rho_i + Q_i\xi_i) + RQ_iS(e_1)\xi_i\omega_i)\|^2 \, dm_i(\xi_i),$$

where $\mathcal{B}_b$ and $\mathcal{B}_i$ denote the material points of the base spacecraft and the i-th reaction wheel, respectively. Expanding the right-hand side of the above expression and rearranging, we obtain

$$T = \frac{1}{2}\left\{ \int_{\mathcal{B}_b} dm(\rho) + \sum_{i=1}^{n} \int_{\mathcal{B}_i} dm_i(\xi_i) \right\} \|\dot{x}\|^2$$

$$+ \dot{x}^T RS(\Omega) \left\{ \int_{\mathcal{B}_b} \rho \, dm(\rho) + \sum_{i=1}^{n} \int_{\mathcal{B}_i} (\rho_i + Q_i\xi_i) \, dm_i(\xi_i) \right\}$$

$$+ \dot{x}^T \sum_{i=1}^{n} RQ_iS(e_1)\omega_i \int_{\mathcal{B}_i} \xi_i \, dm_i(\xi_i)$$

$$+ \frac{1}{2}\Omega^T \left\{ \int_{\mathcal{B}_b} -S^2(\rho)dm(\rho) + \sum_{i=1}^{n} \int_{\mathcal{B}_i} -S(\rho_i + Q_i\xi_i)^2 dm_i(\xi_i) \right\} \Omega$$

$$+ \frac{1}{2}\sum_{i=1}^{n} e_1^T \left\{ \int_{\mathcal{B}_i} -S(\xi_i)^2 \, dm_i(\xi_i) \right\} e_1\omega_i^2$$

$$+ \sum_{i=1}^{n} \Omega^T \left\{ \int_{\mathcal{B}_i} -S(\rho_i + Q_i\xi_i)S(Q_i\xi_i) \, dm_i(\xi_i) \right\} s_i\omega_i.$$

Let m_b, m_i be the mass of the base spacecraft, and the mass of the i-th reaction wheel, respectively. The total mass is given by $m = m_b + \sum_{i=1}^{n} m_i$. Next, we introduce several variables that describe the mass distribution of the base spacecraft and the reaction wheels. Since the origin of the body-fixed frame is located at the center of mass of the base spacecraft, we have $\int_{\mathcal{B}_b} \rho dm(\rho) = 0$. Let $J_b \in \mathbb{R}^{3\times 3}$ be the inertia matrix of the base spacecraft with respect to the body-fixed frame:

$$J_b = \int_{\mathcal{B}_b} S(\rho)^T S(\rho)dm(\rho).$$

Note that it is defined with respect to the base spacecraft frame. Also, let $I_i \in \mathbb{R}^3$ and $J_i \in \mathbb{R}^{3 \times 3}$ be the first and the second mass moments of inertia for the i-th reaction wheel:

$$I_i = \int_{\mathcal{B}_i} \xi_i \, dm_i(\xi_i), \quad J_i = \int_{\mathcal{B}_i} S(\xi_i)^T S(\xi_i) \, dm_i(\xi_i), \quad i = 1, \ldots, n.$$

Introduce

$$J = J_b + \sum_{i=1}^{n} m_i S(\rho_i)^T S(\rho_i),$$

which corresponds to the inertia of the whole spacecraft when $I_i = 0$ and $J_i = 0$, i.e., the reaction wheel is replaced by a point mass. Substituting and using the notation $Q_i e_1 = s_i$, the kinetic energy can be reexpressed as

$$T = \frac{1}{2} m \|\dot{x}\|^2 + \dot{x}^T R \sum_{i=1}^{n} \{ S(\Omega)(m_i \rho_i + Q_i I_i) + Q_i S(e_1) I_i \omega_i \}$$

$$+ \frac{1}{2} \Omega^T \{ J + \sum_{i=1}^{n} (S(\rho_i)^T S(Q_i I_i) + S(Q_i I_i)^T S(\rho_i) + Q_i J_i Q_i^T) \} \Omega$$

$$+ \frac{1}{2} \sum_{i=1}^{n} e_1^T J_i e_1 \omega_i^2 + \sum_{i=1}^{n} \Omega^T (S(\rho_i)^T Q_i S(I_i) + Q_i J_i) e_1 \omega_i.$$

The potential function $U : \mathsf{SE}(3) \to \mathbb{R}^1$ depends on the attitude and the position of the spacecraft but not on the attitudes of the reaction wheels. The modified Lagrangian function is given by

$$\tilde{L}(R, x, q, \Omega, \dot{x}, \omega) = T(R, q, \Omega, \dot{x}, \omega) - U(R, x),$$

reflecting the assumption that the kinetic energy is independent of the inertial position of the spacecraft and the potential energy is independent of the attitudes of the reaction wheels.

The derivatives of the Lagrangian with respect to x and $\dot{x}$ are given by

$$\mathbf{D}_x L(R, x, q, \Omega, x, \omega) = -\frac{\partial U(R, x)}{\partial x} \triangleq f_u,$$

$$\mathbf{D}_{\dot{x}} L(R, x, q, \Omega, \dot{x}, \omega) = m\dot{x} + R \sum_{i=1}^{n} \{ \hat{\Omega}(m_i \rho_i + Q_i I_i) + Q_i \hat{e}_1 I_i \omega_i \}.$$

The variation of R is $\delta R = RS(\eta)$ for differentiable curves $\eta : [t_0, t_f] \to \mathbb{R}^3$. Using the notation from Chapter 8, the left-trivialized derivative of the Lagrangian with respect to R is given by

$$(\mathsf{T}_I^* \mathsf{L}_R \cdot \mathbf{D}_R L) \cdot \eta$$

$$= \mathbf{D}_R L \cdot \delta R$$

$$= \mathbf{D}_R L \cdot RS(\eta)$$

$$= \dot{x}^T RS(\eta) \sum_{i=1}^{n} \{S(\Omega)(m_i \rho_i + Q_i I_i) + Q_i S(e_1) I_i \omega_i\} + M_u \cdot \eta$$

$$= -S(R^T \dot{x}) \sum_{i=1}^{n} \{S(\Omega)(m_i \rho_i + Q_i I_i) + Q_i S(e_1) I_i \omega_i\}^T \eta + M_u^T \eta,$$

where $M_u = -\mathsf{T}_I^* \mathsf{L}_R \cdot \mathbf{D}_R U(R, x)$ denotes the moment due to the potential. The derivative of the Lagrangian with respect to Ω can be written as

$$\mathbf{D}_\Omega L(R, x, q, \Omega, \dot{x}, \omega) = -S(R^T \dot{x}) \sum_{i=1}^{n} (m_i \rho_i + Q_i I_i)$$

$$+ \{J + \sum_{i=1}^{n} (S(\rho_i)^T S(Q_i I_i) + S^T(Q_i I_i) S(\rho_i) + Q_i J_i Q_i^T)\} \Omega$$

$$+ \sum_{i=1}^{n} (S(\rho_i)^T Q_i S(I_i) + Q_i J_i) e_1 \omega_i.$$

Since the i-th reaction wheel frame has one degree of freedom corresponding to rotations about its spin axis, the infinitesimal variations of q_i, Q_i, and ω_i can be written as

$$\delta q_i = Sq_i \gamma_i,$$
$$\delta Q_i = Q_i S(e_1) \gamma_i = S(s_i) Q_i \gamma_i,$$
$$\delta \omega_i = \dot{\gamma}_i.$$

Using these expressions for the infinitesimal variations, the derivative of the Lagrangian with respect to the attitude of the i-th reaction wheel q_i is given by

$$\mathbf{D}_{q_i} L(R, x, q, \Omega, \dot{x}, \omega) \cdot \delta q_i = \Big[\dot{x}^T R\{S(\Omega) Q_i S(e_1) I_i + Q_i S(e_1)^2 I_i \omega_i\}$$

$$+ \frac{1}{2} \Omega^T \{S(\rho_i)^T S(Q_i S(e_1) I_i) + S(Q_i S(e_1) I_i)^T S(\rho_i)$$

$$+ Q_i (S(e_1) J_i - J_i S(e_1)) Q_i^T\} \Omega$$

$$+ \Omega^T (S(\rho_i)^T Q_i S(e_1) S(I_i) + Q_i S(e_1) J_i) e_1 \omega_i \Big] \gamma_i, \quad i = 1, \dots, n.$$

The derivative of the Lagrangian with respect to ω_i is

$$\mathbf{D}_{\omega_i} L(R, x, q, \Omega, \dot{x}, \omega) = \dot{x}^T R Q_i S(e_1) I_i + e_1^T J_i e_1 \omega_i$$

$$+ \Omega^T (S(\rho_i)^T Q_i S(I_i) + Q_i J_i) e_1, \quad i = 1, \dots, n.$$

Let $f_e \in \mathbb{R}^3$ be the external disturbance force applied to the center of mass of the base spacecraft, represented with respect to the inertial frame, and let $M_e \in \mathbb{R}^3$ be the external disturbance moment on the base spacecraft, represented with respect to the body-fixed frame. The control torque on the i-th reaction wheel about its spin axis s_i is denoted by $\tau_i \in \mathbb{R}$, for $i = 1\ldots, n$. The virtual work done by these force and moments is given by

$$\delta \mathcal{W} = f_e \cdot \delta x + M_e \cdot \eta + \sum_{i=1}^{n} \tau_i \gamma_i.$$

The Euler–Lagrange equations with forces are obtained using the Lagrange–d'Alembert principle. The infinitesimal variation of the action integral is

$$\delta \mathfrak{G} = \int_0^{t_f} \left\{ \mathbf{D}_\Omega L \cdot \delta \Omega + \mathbf{D}_{\dot{x}} L \cdot \delta \dot{x} + \mathbf{D}_\omega L \cdot \dot{\gamma} \right.$$
$$\left. + (\mathsf{T}_I^* \mathsf{L}_R \cdot \mathbf{D}_R L) \cdot \eta + \mathbf{D}_x L \cdot \delta\, x + \mathbf{D}_q L \cdot \gamma \right\} dt.$$

The expression for the infinitesimal variation of Ω is given by

$$\delta \Omega = \dot{\eta} + S(\Omega)\eta,$$

for differentiable curves $\eta : [0, t_f] \to \mathbb{R}^3$, $\delta x : [0, t_f] \to \mathbb{R}^3$, $\delta \theta : [0, t_f] \to \mathbb{R}^n$ satisfying $\eta(t_0) = \eta(t_f) = 0$, $\delta x(t_0) = \delta x(t_f) = 0$, and $\delta\theta(t_0) = \delta\theta(t_f) = 0$. Substituting the infinitesimal variations and integrating by parts, we obtain the Euler–Lagrange equations for the spacecraft with reaction wheels as

$$\frac{d}{dt} \mathbf{D}_{\dot{x}} L(R, x, q, \Omega, \dot{x}, \omega) - \mathbf{D}_x L(R, x, q, \Omega, \dot{x}, \omega) = f_e, \qquad (9.56)$$

$$\frac{d}{dt} \mathbf{D}_\Omega L(R, x, q, \Omega, \dot{x}, \omega) + S(\Omega) \mathbf{D}_\Omega L(R, x, q, \Omega, \dot{x}, \omega)$$
$$- \mathsf{T}_I^* \mathsf{L}_R \cdot \mathbf{D}_R L(x, R, q, \dot{x}, \Omega, \omega) = M_e, \qquad (9.57)$$

$$\frac{d}{dt} \mathbf{D}_{\omega_i} L(R, x, q, \Omega, \dot{x}, \omega) - \mathbf{D}_{q_i} L(R, x, q, \Omega, \dot{x}, \omega) = \tau_i, \quad i = 1, \ldots, n. \quad (9.58)$$

We consider a few special cases. First, suppose that the base of the spin axis corresponds to the center of mass of each reaction wheel, i.e., $I_i = 0$, and the center of mass of the whole spacecraft including reaction wheels and the center of mass of the base spacecraft are colocated, i.e., $\sum_{i=1}^{n} m_i \rho_i = 0$. Also, assume the external moments and the potential are independent of the position of the spacecraft. Then, the translational dynamics are decoupled from the rotational attitude dynamics, and the Euler–Lagrange equations can be written as

$$\left(J_b + \sum_{i=1}^{n}(m_i S(\rho)^T S(\rho_i) + Q_i J_i Q_i^T) \right) \dot{\Omega} + \sum_{i=1}^{n} Q_i J_i e_1 \dot{\omega}_i$$

$$+ \sum_{i=1}^{n} Q_i(S(e_1)J_i - J_i S(e_1))Q_i^T \Omega \omega_i + \sum_{i=1}^{n} Q_i S(e_1)J_i e_1 \omega_i^2$$

$$+ \Omega \times \{ J_b + \sum_{i=1}^{n}(m_i S(\rho)^T S(\rho_i) + Q_i J_i Q_i^T))\Omega + \sum_{i=1}^{n} Q_i J_i e_1 \omega_i \} = M, \tag{9.59}$$

$$e_1^T J_i(Q_i^T \dot{\Omega} - S(e_1)Q_i^T \Omega \omega_i + e_1 \dot{\omega}_i)$$

$$- \frac{1}{2}\Omega^T Q_i(S(e_1)J_i - J_i S(e_1))Q_i^T \Omega$$

$$- \Omega^T Q_i S(e_1)J_i e_1 \omega_i = \tau_i, \quad i = 1,\dots,n, \tag{9.60}$$

where $M = M_u + M_e \in \mathbb{R}^3$ denotes the total moment, that is the sum of the moment due to the potential and the external disturbance moment. Here, the mass distribution of each reaction wheel can be arbitrary, as long as the spin axis passes through its center of mass.

Second, we further assume that each reaction wheel is inertially symmetric about its spin axis, i.e., the inertia matrix can be written as $J_i = \text{diag}[\alpha_i, \beta_i, \beta_i]$ for some $\alpha_i, \beta_i > 0$. This implies that $S(e_1)J_i - J_i S(e_1) = 0$, $Q_i J_i e_1 = (e_1^T J_i e_1)s_i = \alpha_i s_i$, and $Q_i J_i Q_i^T = J_i$. Also, assume the external moments and the potential are independent of the spacecraft position. Then, the above equations become completely independent of Q_i and we obtain:

$$J'\dot{\Omega} + \sum_{i=1}^{n} \alpha_i s_i \dot{\omega}_i + \Omega \times (J'\Omega + \sum_{i=1}^{n} \alpha_i s_i \omega_i) = M, \tag{9.61}$$

$$\alpha_i s_i^T \dot{\Omega} + \alpha_i \dot{\omega}_i = \tau_i, \quad i = 1,\dots,n, \tag{9.62}$$

where $J' = J_b + \sum_{i=1}^{n}(m_i S(\rho)^T S(\rho_i) + J_i) \in \mathbb{R}^{3\times3}$ is the fixed inertial matrix of the whole spacecraft, represented with respect the body-fixed frame. The Euler–Lagrange equations (9.61), (9.62), describe the dynamical flow of the spacecraft with reaction wheels in the special cases indicated. Together with the kinematics equations, they are globally defined on the tangent bundle $\mathsf{T}(\mathsf{SE}(3) \times (\mathsf{S}^1)^n$ of the configuration manifold.

9.10.2 Hamilton's Equations

Hamilton's equations for the spacecraft with reaction wheels can also be obtained. Define the following conjugate momenta, which depend on the kinetic energy, according to the Legendre transformation

$$p = \mathbf{D}_{\dot{x}}L(R, x, q, \Omega, \dot{x}, \omega),$$
$$\Pi = \mathbf{D}_{\Omega}L(R, x, q, \Omega, \dot{x}, \omega),$$
$$\pi_i = \mathbf{D}_{q_i}L(R, x, q, \Omega, \dot{x}, \omega), \quad i = 1\ldots, n.$$

Then, the base spacecraft momentum $(\Pi, p) \in \mathsf{T}^*_{(R,x)}\mathsf{TSE}(3)$ is conjugate to $(\Omega, \dot{x}) \in \mathsf{T}_{(R,x)}\mathsf{TSE}(3)$ and the reaction wheel momenta $\pi_i \in \mathsf{T}^*_{q_i}\mathsf{S}^1$ are conjugate to $\omega_i \in \mathsf{T}_{q_i}\mathsf{S}^1$, for $i = 1, \ldots, n$. From this, the Euler–Lagrange equations are transformed into Hamilton's equations:

$$\dot{p} = \mathbf{D}_x L(R, x, q, \Omega, \dot{x}, \omega) + f_e, \tag{9.63}$$

$$\dot{\Pi} = -S(\Omega)\Pi + \mathsf{T}^*_I \mathsf{L}_R \cdot \mathbf{D}_R L(R, x, q, \Omega, \dot{x}, \omega) + M_e, \tag{9.64}$$

$$\dot{\pi}_i = \mathbf{D}_{q_i} L(R, x, q, \Omega, \dot{x}, \omega) + \tau_i, \quad i = 1, \ldots, n. \tag{9.65}$$

Together with the kinematics equations (9.53) and (9.55), these differential equations describe the dynamical flow of the spacecraft with reaction wheels on the cotangent bundle $\mathsf{T}^*(\mathsf{SE}(3) \times (\mathsf{S}^1)^n)$.

For the second simplified case discussed above, where the reaction wheel is axially symmetric about its spin axis, the Legendre transformation of the attitude dynamics when there is one reaction wheel, i.e., $n = 1$, is given by

$$\Pi = J'\Omega + \alpha_1 \omega_1 s_1,$$
$$\pi_1 = \alpha_1 s_1^T \Omega + \alpha_1 \omega_1.$$

The corresponding Hamilton's equations for a spacecraft with a single reaction wheel are given by

$$\dot{\Pi} = -\Omega \times \Pi + M,$$
$$\dot{\pi}_1 = \tau_1.$$

The above Hamilton's equations describe the dynamical flow of the spacecraft with reaction wheels in the special cases indicated. Together with the kinematics equations, they are globally defined on the cotangent bundle of the configuration manifold.

9.10.3 Conservation Properties

In the absence of the external force, external moment, and control torque on the reaction wheels, the Hamiltonian

$$H = \frac{1}{2}m\|\dot{x}\|^2 + \dot{x}^T R \sum_{i=1}^{n} \{S(\Omega)(m_i\rho_i + Q_i I_i) + Q_i S(e_1)I_i\omega_i\}$$

$$+ \frac{1}{2}\Omega^T\{J + \sum_{i=1}^{n}(S(\rho_i)^T S(Q_i I_i) + S(Q_i I_i)^T S(\rho_i) + Q_i J_i Q_i^T)\}\Omega$$

$$+ \frac{1}{2}\sum_{i=1}^{n} e_1^T J_i e_1 \omega_i^2 + \sum_{i=1}^{n} \Omega^T(S(\rho_i)^T Q_i S(I_i) + Q_i J_i)e_1\omega_i + U(R, x),$$

which coincides with the total energy E in this case, is conserved along each solution of the dynamical flow of the spacecraft with reaction wheel assembly.

For the second simplified case discussed above, where the reaction wheel is axially symmetric about its spin axis, this reduces to

$$E = \frac{1}{2}m\|\dot{x}\|^2 + \frac{1}{2}\Omega^T J'\Omega + \frac{1}{2}\sum_{i=1}^{n}\alpha_i\omega_i^2 + \sum_{i=1}^{n}\alpha_i\omega_i\Omega^T s_i + U(R, x).$$

Furthermore, if the potential energy is independent of the attitude of the base spacecraft, the total angular momentum expressed in the inertial frame is conserved:

$$\Pi = R(J'\Omega + \sum_{i=1}^{n}\alpha_i\omega_i s_i).$$

This vector conservation property implies that the scalar magnitude of the angular momentum in the spacecraft base frame

$$\|\Pi\|^2 = \left\| J'\Omega + \sum_{i=1}^{n}\alpha_i\omega_i s_i \right\|^2$$

is constant along each solution of the dynamical flow.

9.10.4 Equilibrium Properties

The equilibrium solutions occur when the velocity vector is zero and the position and the attitude of spacecraft are chosen such that there is no force and moment due to the potential.

9.11 Dynamics of a Rotating Spacecraft and Control Moment Gyroscope

A control moment gyroscope (CMG) is mounted on an otherwise rigid base spacecraft. The rigid spacecraft is assumed to rotate in three dimensions. The control moment gyroscope consists of a rotor that can rotate, without friction, about a symmetry axis, and this rotation axis is mounted to a gimbal that can rotate, without friction, about a gimbal support that is fixed to the rigid base spacecraft [64].

The attitude of the base spacecraft is described by an attitude matrix R in the Lie group $\mathsf{SO}(3)$. The vector $q_g \in \mathsf{S}^1$ denotes the attitude vector of the gimbal about its body-fixed axis. The vector $q_r \in \mathsf{S}^1$ denotes the attitude vector of the rotor about its gimbal-fixed axis. Thus, the configuration vector $(R, q_g, q_r) \in \mathsf{SO}(3) \times \mathsf{S}^1 \times \mathsf{S}^1$. The configuration manifold for a spacecraft and control moment gyroscope (SCCMG) is $\mathsf{SO}(3) \times \mathsf{S}^1 \times \mathsf{S}^1$. Consequently, there are five degrees of freedom. No external moments act on the spacecraft, the gimbal, or the rotor. A schematic of a control moment gyroscope is shown in Figure 9.11; the gimbal axis of this control moment gyroscope, as shown in the figure, is fixed with respect to the rigid spacecraft which is omitted from the figure.

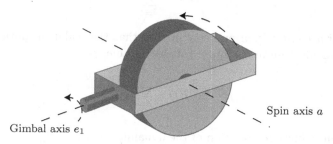

Fig. 9.11 Control moment gyroscope

Variational methods and Hamilton's principle are used to derive a geometric form of the equations of motion for the attitude dynamics of a spacecraft and control moment gyroscope. These equations are globally defined on the configuration manifold and expose important structural features of the dynamics.

9.11.1 Euler–Lagrange Equations

We consider an inertially fixed Euclidean frame and a Euclidean frame fixed to the spacecraft base, with origin located at the center of mass of the spacecraft base.

Let $R \in SO(3)$ denote the attitude matrix that transforms from the spacecraft base body-fixed frame to the inertial frame. If $\Omega \in T_R SO(3)$ is the angular velocity vector of the base body expressed in the spacecraft base body frame, then the attitude kinematics of the spacecraft base body is given by

$$\dot{R} = RS(\Omega). \tag{9.66}$$

The control moment gyroscope consists of an axially symmetric rotor rotating about its symmetry axis, which in turn is rotating about a gimbal axis fixed in the spacecraft base body frame. Without loss of generality, assume that the base spacecraft frame is selected so that the gimbal axis is aligned with the unit vector $e_1 = [1, 0, 0]^T \in S^2$ fixed in the base spacecraft frame. Let $a \in S^2$ denote the direction vector of the instantaneous symmetry axis of the rotor, expressed as a unit vector $a \in \mathbb{R}^3$ in the base spacecraft frame. The rotor axis a is always normal to the gimbal axis e_1 and the three unit vectors $e_1, a, e_1 \times a$ form a right-handed orthonormal gimbal-fixed frame. Define a rotation matrix $R_g \in SO(3)$ as

$$R_g = \begin{bmatrix} e_1, a, e_1 \times a_1 \end{bmatrix},$$

which represents the linear transformation from the gimbal-fixed frame to the base body-fixed frame. Let $\omega_g \in \mathbb{R}^1$ be the scalar angular velocity of the gimbal about the gimbal rotational axis so that the angular velocity vector of the gimbal is

$$\Omega_g = \omega_g e_1 \in \mathbb{R}^3,$$

expressed in the gimbal-fixed frame. Then the attitude kinematics equation for the gimbal-fixed frame is given by

$$\dot{R}_g = R_g S(\Omega_g) = \omega_g R_g S(e_1).$$

Let $q_g \in S^1$ denote the planar attitude of the gimbal about its base body-fixed axis e_1 in the base spacecraft frame. It consists of the last two elements of the gimbal axis a, i.e.,

$$a = Cq_g,$$

where we used an embedding of $\mathbb{R}^2$ into $\mathbb{R}^3$ defined by the 3×2 matrix

$$C = \begin{bmatrix} 0 & 0 \\ 1 & 0 \\ 0 & 1 \end{bmatrix}.$$

The planar rotational kinematics of the gimbal are given by

$$\dot{q}_g = \omega_g S q_g, \tag{9.67}$$

where the 2×2 skew-symmetric matrix

$$S = \begin{bmatrix} 0 & -1 \\ 1 & 0 \end{bmatrix}.$$

Consider an orthonormal frame that is fixed to the rotor whose first axis coincides with the gimbal axis, and its second axis corresponds to the spin axis of the rotor. Let $R_r \in SO(3)$ be the rotation matrix representing the linear transformation from the rotor-fixed frame to the body-fixed frame.

Let $\omega_r \in \mathbb{R}^1$ be the scalar angular velocity of the rotor about the rotor rotational axis $a \in S^2$. It is easy to see that the angular velocity vector of the rotor, expressed in the base body-fixed frame, is

$$\Omega_r = \omega_g e_1 + \omega_r a \in \mathbb{R}^3.$$

The kinematics equation for the attitude of the rotor-fixed frame is given by

$$\dot{R}_r = S(\omega_g e_1 + \omega_r a) R_r.$$

In contrast to other attitude kinematics equations, such as (9.66), the angular velocity term $S(\omega_g e_1 + \omega_r a)$ is placed left of R_r. This is because the rotation matrix R_r represents the linear transformation from the rotor-fixed frame to the body-fixed frame, and the angular velocity of the rotor-fixed frame, namely $\omega_g e_1 + \omega_r a$ is resolved with respect to the body-fixed frame.

Let $q_r \in S^1$ denote the planar attitude of the rotor about its gimbal-fixed axis $a \in S^3$. The planar rotational kinematics of the rotor are

$$\dot{q}_r = \omega_r S q_r. \tag{9.68}$$

The Lagrangian function for a spacecraft and control moment gyroscope, under the stated assumptions, is now derived. Denote the 3×3 inertia matrix of the spacecraft base body expressed in the base body frame by J_b. The rotational kinetic energy of the base body of the spacecraft is

$$T_b = \frac{1}{2}\Omega^T J_b \Omega. \tag{9.69}$$

Consider a mass element $dm_g(\xi_g)$ of the gimbal located at ξ_g in the gimbal-fixed frame. Let z_g be the location of the mass element with respect to the inertial frame, which can be written as

$$z_g = R(\rho_g + R_g \xi_g).$$

The time derivative of z_g is given by

$$\dot{z}_g = RS(\Omega)(\rho_g + R_g \xi_r) + \omega_g RR_g S(e_1) \xi_g$$
$$= RS(\Omega)\rho_g + RR_g S(R_g^T \Omega + \omega_g e_1) \xi_g.$$

The kinetic energy of the gimbal can be written as

$$T_g = \int_{\mathcal{G}} \frac{1}{2} \|\dot{z}_g\|^2 \, dm_g(\xi_g)$$
$$= \frac{1}{2} m_g \Omega^T S(\rho_g)^T S(\rho_g) \Omega$$
$$+ \frac{1}{2} (R_g^T \Omega + \omega_g e_1)^T \left\{ \int_{\mathcal{G}} S(\xi_g)^T S(\xi_g) \, dm_g(\xi_g) \right\} (R_g^T \Omega + \omega_g e_1),$$

where we used the fact that $\int_{\mathcal{G}} \xi_g dm_g(\xi_g) = 0$, since the origin of the gimbal-fixed frame coincides with the center of mass of the gimbal. Let $J_g \in \mathbb{R}^{3 \times 3}$ be the inertia matrix of the gimbal about the origin of the gimbal-fixed frame represented with respect to the gimbal-fixed frame:

$$J_g = \int_{\mathcal{G}} S^T(\xi_g) S(\xi_g) \, dm_g(\xi_g).$$

Note that $\dot{J}_g = 0$ since the mass distribution of the gimbal is fixed with respect to the gimbal-fixed frame. By substitution, the kinetic energy of the gimbal can be expressed as

$$T_g = \frac{1}{2} m_g \Omega^T S(\rho_g)^T S(\rho_g) \Omega + \frac{1}{2} (R_g^T \Omega + \omega_g e_1)^T J_g (R_g^T \Omega + \omega_g e_1)$$
$$= \frac{1}{2} \Omega^T (R_g J_g R_g^T + m_g S^T(\rho_g) S(\rho_g)) \Omega + \Omega^T R_g J_g e_1 \omega_g + \frac{1}{2} e_1^T J_g e_1 \omega_g^2.$$

This expression reflects the fact that the inertia matrix of the gimbal represented with respect to the base body frame depends on the relative attitude R_g of the gimbal with respect to the base spacecraft.

Consider a mass element $dm_r(\xi_r)$ of the rotor, located at ξ_r in the rotor-fixed frame. Let $z_r \in \mathbb{R}^3$ be the vector from the origin of the rotor-fixed frame to the mass element, expressed in the inertial frame. It is given by

$$z_r = R(\rho_g + R_r \xi_r).$$

The time derivative of z_r can be written as

$$\dot{z}_r = RS(\Omega)(\rho_g + R_r \xi_r) + RS(\omega_g e_1 + \omega_r a) R_r \xi_r$$
$$= RS(\Omega)\rho_g + RS(\Omega + \omega_r a + \omega_g e_1) R_r \xi_r.$$

The kinetic energy of the rotor can be written as

$$T_r = \frac{1}{2} m_r \rho_g^T S(\Omega)^T S(\Omega) \rho_g$$
$$+ \frac{1}{2} (\Omega + w_r a + w_g e_1)^T R_r \int_{\mathcal{R}} S(\xi_r)^T S(\xi_r) \, dm_r(\xi_r) R_r^T (\Omega + w_r a + w_g e_1).$$

Let $J_r \in \mathbb{R}^{3 \times 3}$ be the inertia matrix of the rotor about the origin of the rotor-fixed axis represented with respect to the rotor-fixed frame:

$$J_r = \int_{\mathcal{R}} S(\xi_r)^T S(\xi_r) \, dm_r(\xi_r).$$

The kinetic energy of the rotor is given by

$$T_r = \frac{1}{2} \Omega^T (R_r J_r R_r^T + m_r S(\rho_r)^T S(\rho_r)) \Omega + \frac{1}{2} (e_1^T R_r J_r R_r^T e_1) w_g^2$$
$$+ \frac{1}{2} (a^T R_r J_r R_r^T a) w_r^2 + \Omega^T (R_r J_r R_r^T e_1) w_g$$
$$+ \Omega^T (R_r J_r R_r^T a) w_r + w_r (a^T R_r J_r R_r^T e_1) w_g.$$

By assuming that the rotor is symmetric about its spin axis, we have

$$R_r J_r R_r^T = R_g J_r R_g^T.$$

By substitution, the kinetic energy of the rotor can be further simplified to

$$T_r = \frac{1}{2} \Omega^T (R_g J_r R_g^T + m_r S(\rho_r)^T S(\rho_r)) \Omega + \frac{1}{2} (e_1^T J_r e_1) w_g^2 + \frac{1}{2} (e_2^T J_r e_2) w_r^2$$
$$+ \Omega^T (R_g J_r e_1) w_g + \Omega^T (R_g J_r e_2) w_r + w_r (e_2 J_r e_1) w_g.$$

The total kinetic energy is the sum of the above three kinetic energies:

$$T = T_g + T_r + T_b$$
$$= \frac{1}{2} \begin{bmatrix} \Omega \\ w_g \\ w_r \end{bmatrix}^T M(q_g) \begin{bmatrix} \Omega \\ w_g \\ w_r \end{bmatrix}.$$

The inertia matrix $M(q_g) \in \mathbb{R}^{5 \times 5}$ is defined as

$$M(q_g) = \begin{bmatrix} J & R_g J_{gr} e_1 & R_g J_r e_2 \\ e_1^T J_{gr} R_g^T & e_1^T J_{gr} e_1 & e_1^T J_r e_2 \\ e_2^T J_r R_g^T & e_2^T J_r e_1 & e_2^T J_r e_2 \end{bmatrix},$$

where J_{gr}, J are matrices defined by

$$J_{gr} = J_g + J_r,$$
$$J = J_b + R_g J_g R_g^T + R_r J_r R_r^T + m_g S(\rho_g)^T S(\rho_g) + m_r S(\rho_r)^T S(\rho_r)$$
$$= J_b + R_g J_{gr} R_g^T + m_g S(\rho_g)^T S(\rho_g) + m_r S(\rho_g)^T S(\rho_g).$$

We ignore all external moments on the SCCMG so we take the potential energy to be identically zero. Hence, the modified Lagrangian function, $\tilde{L}$: $T(SO(3) \times S^1 \times S^1) \to \mathbb{R}^1$, is given by

$$\tilde{L}(R, q_g, q_r, \Omega, \omega_g, \omega_r) = T(\Omega, \omega_g, \omega_r, q_g).$$

Note that the Lagrangian function is invariant with respect to changes in $R \in SO(3)$ and $q_r \in S^1$. These symmetry properties have important implications for the attitude dynamics of the spacecraft and control moment gyroscope.

The Euler–Lagrange equations are obtained using Hamilton's principle: the infinitesimal variation of the action integral is zero for all infinitesimal variations on $SO(3) \times S^1 \times S^1$ that vanish at t_0 and t_f. The infinitesimal variation of the action integral is given by

$$\delta\mathfrak{G} = \int_0^{t_f} \left\{ \frac{\partial T}{\partial \Omega} \cdot \delta\Omega + \frac{\partial T}{\partial \omega_g} \delta\omega_g + \frac{\partial T}{\partial \omega_r} \delta\omega_r + \frac{\partial T}{\partial q_g} \cdot \delta q_g \right\} dt.$$

The expressions for the infinitesimal variations are given by

$$\delta\Omega = \dot{\eta} + S(\Omega)\eta,$$
$$\delta\omega_g = \dot{\gamma}_g,$$
$$\delta\omega_r = \dot{\gamma}_r,$$
$$\delta q_g = \gamma_g S q_g,$$

for differentiable curves $\eta : [0, t_f] \to \mathbb{R}^3$, $\gamma_g : [0, t_f] \to \mathbb{R}^1$, $\gamma_r : [0, t_f] \to \mathbb{R}^1$ satisfying $\eta(t_0) = \eta(t_f) = 0$, $\gamma_g(t_0) = \gamma_g(t_f) = 0$, and $\gamma_r(t_0) = \gamma_r(t_f) = 0$. Substituting the expressions for the infinitesimal variations, the variation of the action integral can be written as

$$\delta\mathfrak{G} = \int_0^{t_f} \left\{ \frac{\partial T}{\partial \Omega} \cdot (\dot{\eta} + S(\Omega)\eta) + \frac{\partial T}{\partial \omega_g} \dot{\gamma}_g + \frac{\partial T}{\partial \omega_r} \dot{\gamma}_r + \frac{\partial T}{\partial q_g} \cdot \gamma_g S q_g \right\} dt.$$

Integrate by parts and use the boundary conditions to obtain the infinitesimal variations for the action integral

$$\delta\mathfrak{G} = \int_0^{t_f} \left(-\frac{d}{dt}\frac{\partial T}{\partial \Omega} - S(\Omega)\frac{\partial T}{\partial \Omega} \right)^T \eta - \left(\frac{d}{dt}\frac{\partial T}{\partial \omega_g} + q_g S \frac{\partial T}{\partial q_g} \right) \gamma_g$$
$$- \frac{d}{dt}\frac{\partial T}{\partial \omega_r} \gamma_r \, dt,$$

where we have suppressed the arguments of the kinetic energy function. Using the fundamental lemma of the calculus of variations leads to

$$\frac{d}{dt}\frac{\partial T(\Omega,\omega_g,\omega_r,q_g)}{\partial\Omega} + \Omega \times \frac{\partial T(\Omega,\omega_g,\omega_r,q_g)}{\partial\Omega} = 0,$$

$$\frac{d}{dt}\frac{\partial T(\Omega,\omega_g,\omega_r,q_g)}{\partial\omega_g} + q_g S\frac{\partial T(\Omega,\omega_g,\omega_r,q_g)}{\partial q_g} = 0,$$

$$\frac{d}{dt}\frac{\partial T(\Omega,\omega_g,\omega_r,q_g)}{\partial\omega_r} = 0.$$

Thus, the Euler–Lagrange equations for the attitude dynamics of a rotating spacecraft and CMG are

$$M(q_g)\dot{V} + M_{q_g}(q_g)\omega_g V + \begin{bmatrix} \Omega \times \{J\Omega + (R_g J_{gr}e_1)\omega_g + R_g J_r e_2\omega_r\} \\ -\frac{1}{2}V^T M_{q_g}(q_g)V \\ 0 \end{bmatrix} = 0,$$

$$(9.70)$$

where the velocity vector is given by

$$V = \begin{bmatrix} \Omega \\ \omega_g \\ \omega_r \end{bmatrix},$$

and the derivative of the inertia matrix with respect to q_g has the form

$$M_{q_g}(q_g) = \begin{bmatrix} R_g(S(e_1)J_{gr} - J_{gr}S(e_1))R_g^T & R_g S(e_1)J_{gr}e_1 & R_g S(e_1)J_r e_2 \\ -e_1^T J_{gr}S(e_1)R_g^T & 0 & 0 \\ -e_2^T J_r S(e_1)R_g^T & 0 & 0 \end{bmatrix}.$$

The Euler–Lagrange equations (9.70), together with the kinematics equations (9.66), (9.67), and (9.68), describe the dynamical flow for the spacecraft with control moment gyroscope in terms of $(R, q_g, q_r, \omega, \omega_g, \omega_r) \in T(SO(3) \times S^1 \times S^1)$ on the tangent bundle of the configuration manifold.

9.11.2 Hamilton's Equations

Hamilton's equations of motion for the attitude dynamics of a rotating spacecraft and CMG can also be obtained. Using the Legendre transformation, we define the following conjugate momenta,

$$\begin{bmatrix} \Pi \\ \pi_g \\ \pi_r \end{bmatrix} = \begin{bmatrix} \frac{\partial T}{\partial \Omega} \\ \frac{\partial T}{\partial \omega_g} \\ \frac{\partial T}{\partial \omega_r} \end{bmatrix} = \begin{bmatrix} J & R_g J_{gr}e_1 & R_g J_r e_2 \\ e_1^T J_{gr}R_g^T & e_1^T J_{gr}e_1 & e_1^T J_r e_2 \\ e_2^T J_r R_g^T & e_2^T J_r e_1 & e_2^T J_r e_2 \end{bmatrix} \begin{bmatrix} \Omega \\ \omega_g \\ \omega_r \end{bmatrix}.$$

The modified Hamiltonian function is

$$\tilde{H}(q_g, \Pi, \pi_g, \pi_r) = \frac{1}{2} \begin{bmatrix} \Pi \\ \pi_g \\ \pi_r \end{bmatrix}^T \begin{bmatrix} J & R_g J_{gr} e_1 & R_g J_r e_2 \\ e_1^T J_{gr} R_g^T & e_1^T J_{gr} e_1 & e_1^T J_r e_2 \\ e_2^T J_r R_g^T & e_2^T J_r e_1 & e_2^T J_r e_2 \end{bmatrix}^{-1} \begin{bmatrix} \Pi \\ \pi_g \\ \pi_r \end{bmatrix},$$

and it is invariant with respect to changes in the configuration variables $R \in \mathsf{SO}(3)$ and $q_r \in \mathsf{S}^1$. Hamilton's equations can be written as

$$\dot{\Pi} = \Pi \times \Omega, \tag{9.71}$$

$$\dot{\pi}_g = \frac{1}{2} \begin{bmatrix} \Pi \\ \pi_g \\ \pi_r \end{bmatrix}^T M(q_g)^{-1} M_{q_g}(q_g) M(q_g)^{-1} \begin{bmatrix} \Pi \\ \pi_g \\ \pi_r \end{bmatrix}, \tag{9.72}$$

$$\dot{\pi}_r = 0, \tag{9.73}$$

where the right-hand sides of these equations are expressed in terms of the momenta using the Legendre transformation. Equations (9.71), (9.72), and (9.73) describe the rotational dynamics of the spacecraft and control moment gyroscope on the cotangent bundle $\mathsf{T}^*(\mathsf{SO}(3) \times \mathsf{S}^1 \times \mathsf{S}^1)$.

9.11.3 Conservation Properties

The Hamiltonian of the spacecraft and control moment gyroscope,

$$H = \frac{1}{2} \begin{bmatrix} \Omega \\ \omega_g \\ \omega_r \end{bmatrix}^T \begin{bmatrix} J & R_g J_{gr} e_1 & R_g J_r e_2 \\ e_1^T J_{gr} R_g^T & e_1^T J_{gr} e_1 & e_1^T J_r e_2 \\ e_2^T J_r R_g^T & e_2^T J_r e_1 & e_2^T J_r e_2 \end{bmatrix} \begin{bmatrix} \Omega \\ \omega_g \\ \omega_r \end{bmatrix},$$

which coincides with the total energy E in this case, can be shown to be constant along each solution of the dynamical flow.

The equations of motion can be used to show that the angular momentum in the inertial frame

$$R\Pi = R\{J\Omega + R_g J_{gr} e_1 \omega_g + R_g J_r e_2 \omega_r\}$$

is constant along each solution of the dynamical flow. This conservation property implies that the scalar magnitude of the angular momentum in the spacecraft base frame

$$\|\Pi\|^2 = \|J\Omega + R_g J_{gr} e_1 \omega_g + R_g J_r e_2 \omega_r\|^2 \tag{9.74}$$

is constant along each solution of the dynamical flow. This is a consequence of Noether's theorem and the invariance of the Lagrangian with respect to rigid rotations of the spacecraft and control moment gyroscope.

Further, it is easy to see that the scalar angular momentum of the rotor

$$\pi_r = e_2^T J_r R_g^T \Omega + e_2^T J_r e_1 \omega_g + e_2^T J_r e_2 \omega_r$$

is constant along each solution of the dynamical flow. This is a consequence of Noether's theorem and the invariance of the Lagrangian with respect to the rotor attitude.

These conservation laws expose important properties of the attitude dynamics of the spacecraft and control moment gyroscope.

9.12 Dynamics of Two Quad Rotors Transporting a Cable-Suspended Payload

A quad rotor aerial vehicle is a popular flight vehicle whose motion in three dimensions can be controlled by four rotors and propellers. Consider two controlled quad rotor aerial vehicles, transporting a cable-suspended payload modeled as a rigid body. The quad rotors and the payload can rotate and translate in three dimensions under the action of uniform, constant gravity. The two cables supporting the payload are each assumed to be massless and to remain straight without deformation.

The variables related to the payload are denoted by the subscript 0, and the variables for the i-th quad rotor are denoted by the subscript i, for $i = 1, 2$. We choose an inertial Euclidean frame and we attach body-fixed frames to the payload and the two quad rotors. The third axis of the inertial frame is vertical and the other axes are chosen to form an orthonormal frame. The origin of the i-th body-fixed frame is located at the center of mass of the payload for $i = 0$ and at the center of mass of each quad rotor for $i = 1, 2$. The third axis of the body-fixed Euclidean frame for each quad rotor is defined by the direction of the thrust on each quad rotor produced by the four rotors. A schematic of the quad rotor vehicles transporting a rigid body payload is shown in Figure 9.12.

Further details on the dynamics and control of quadrotors transporting a cable-suspended rigid body can be found in [31, 47, 60].

The location of the center of mass of the payload is denoted by $x_0 \in \mathbb{R}^3$ in the inertial frame, and its attitude is given by $R_0 \in \mathsf{SO}(3)$. Let $\rho_i \in \mathbb{R}^3$ be the point on the payload where the i-th cable is attached, represented in the payload body-fixed frame. The other end of the cable is attached to the center of mass of the i-th quad rotor. The direction of the cable from the center of mass of the i-th quad rotor to the point of contact with the payload is the i-th cable attitude denoted by $q_i \in \mathsf{S}^2$, and the scalar length of the i-th cable is denoted by L_i. Let $x_i \in \mathbb{R}^3$ be the location of the center of mass of the i-th quad rotor with respect to the inertial frame. As the cable is assumed to be rigid, we have $x_i = x_0 + R_0 \rho_i - L_i q_i$. The attitude of the i-th quad rotor is described by the rotation matrix $R_i \in \mathsf{SO}(3)$ that transforms a vector from the i-th body-fixed frame to the inertial frame.

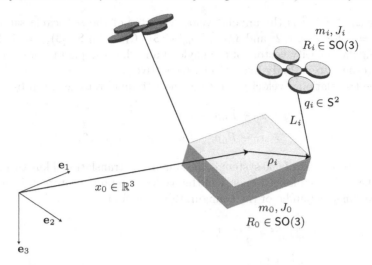

Fig. 9.12 Two quadrotors transporting a cable-suspended rigid body payload

The mass and the inertia matrix of the payload are denoted by m_0 and $J_0 \in \mathbb{R}^{3\times3}$, respectively. The dynamical model of each quad rotor is identical to the model introduced in [55]. The mass and the inertia matrix of the i-th quad rotor are denoted by m_i and $J_i \in \mathbb{R}^{3\times3}$, respectively. Each of the four rotors for the i-th quad rotor generates a thrust vector $-T_i R_i e_3 \in \mathbb{R}^3$ with respect to the inertial frame, where T_i is the total thrust magnitude and $e_3 = [0,0,1]^T \in \mathbb{R}^3$ is the direction of the thrust vector in the quad rotor-fixed frame. The four rotors for each quad rotor vehicle also generate a moment vector $M_i \in \mathbb{R}^3$ with respect to its body-fixed frame. The thrust magnitudes and moment vectors $\{T_i, M_i\}_{1=1,2}$ are viewed as control inputs.

The configuration of the quad rotor vehicles and payload is given by $(x_0, R_0, q_1, R_1, q_2, R_2) \in \mathsf{SE}(3) \times (\mathsf{S}^2 \times \mathsf{SO}(3))^2$. The corresponding configuration manifold is $\mathsf{SE}(3) \times (\mathsf{S}^2 \times \mathsf{SO}(3))^2$. Thus, there are sixteen degrees of freedom for the two quad rotor vehicles and the rigid body payload.

9.12.1 Euler–Lagrange Equations

The kinematics equations for the payload, quad rotors, and cables are given by

$$\dot{q}_i = S(\omega_i)q_i, \quad i = 1,2, \tag{9.75}$$

$$\dot{R}_0 = R_0 S(\Omega_0), \tag{9.76}$$

$$\dot{R}_i = R_i S(\Omega_i), \quad i = 1,2, \tag{9.77}$$

where $\omega_i \in \mathsf{T}_{q_i}\mathsf{S}^2$ is the angular velocity vector of the i-th cable satisfying $\omega_i^T q_i = 0$ for $i = 1, 2$, and $\Omega_0 \in \mathsf{T}_{R_0}\mathsf{SO}(3)$, $\Omega_i \in \mathsf{T}_{R_i}\mathsf{SO}(3)$, $i = 1, 2$, are the angular velocity vectors of the payload and the i-th quad rotor expressed with respect to its body-fixed frame, respectively.

The translational velocity vector of the i-th quad rotor is given by

$$
\begin{aligned}
\dot{x}_i &= \dot{x}_0 + \dot{R}_0 \rho_i - L_i \dot{q}_i \\
&= \dot{x}_0 + \dot{R}_0 \rho_i - L_i S(\omega_i) q_i, \quad i = 1, 2.
\end{aligned}
$$

The kinetic energy of the system is the sum of the translational kinetic energy and the rotational kinetic energy of the payload and quad rotors; it is defined on the tangent bundle of the configuration manifold as

$$
T = \frac{1}{2} m_0 \|\dot{x}_0\|^2 + \frac{1}{2} \Omega_0^T J_0 \Omega_0
$$

$$
+ \sum_{i=1}^{2} \left\{ \frac{1}{2} m_i \|\dot{x}_0 + \dot{R}_0 \rho_i - L_i S(\omega_i) q_i\|^2 + \frac{1}{2} \Omega_i^T J_i \Omega_i \right\}.
$$

The gravitational potential energy is given by

$$
U = m_0 g e_3^T x_0 + \sum_{i=1}^{2} m_i g e_3^T (x_0 + R_0 \rho_i - L_i q_i).
$$

The modified Lagrangian $\tilde{L} : \mathsf{T}(\mathsf{SE}(3) \times (\mathsf{S}^2 \times \mathsf{SO}(3))^2) \to \mathbb{R}^1$ is

$$
\tilde{L} = \frac{1}{2} m_0 \|\dot{x}_0\|^2 + \frac{1}{2} \Omega_0^T J_0 \Omega_0
$$

$$
+ \sum_{i=1}^{2} \left\{ \frac{1}{2} m_i \|\dot{x}_0 + R_0 S(\omega_0) \rho_i - L_i S(\omega_i) q_i\|^2 + \frac{1}{2} \Omega_i^T J_i \Omega_i \right\}
$$

$$
- m_0 g e_3^T x_0 - \sum_{i=1}^{2} m_i g e_3^T (x_0 + R_0 \rho_i - L_i q_i),
$$

which can now be used to obtain the Euler–Lagrange equations. The Lagrangian does not depend on the attitudes of the quad-rotor vehicles.

The key idea is to represent the infinitesimal variations of the cable attitudes, as we have seen previously in Chapter 5, in the form

$$
\delta q_i = \gamma_i \times q_i, \qquad\qquad i = 1, 2,
$$

$$
\delta \omega_i = -S(\omega_i)\gamma_1 + (I - q_i q_i^T)\dot{\gamma}_i, \quad i = 1, 2,
$$

for differentiable curves $\gamma_i : [t_0, t_f] \to \mathbb{R}^3$ that satisfy $\gamma_i^T q_i = 0$, for $i = 1, 2$. Similarly, the infinitesimal variations of the attitude matrices describing the attitude of the payload rigid body and the attitudes of the two quad rotor vehicles are given by

$$\delta R_i = R_i S(\eta_i), \quad i = 0, 1, 2,$$

for differentiable curves $\eta_i : [t_0, t_f] \to \mathbb{R}^3$, $i = 0, 1, 2$. The infinitesimal variation of the angular velocity vectors of the payload and the two quad rotor vehicles can be written as

$$\delta \Omega_i = \dot{\eta}_i + S(\Omega_i)\eta_i \quad i = 0, 1, 2.$$

Using the above expressions, the infinitesimal variation of the action integral can be written as

$$\delta \mathfrak{S} = \int_{t_0}^{t_f} \left[\mathbf{D}_{\dot{x}_0} L \cdot \delta \dot{x}_0 + \mathbf{D}_{x_0} L \cdot \delta x_0 \right.$$
$$+ \mathbf{D}_{\Omega_0} L \cdot (\dot{\eta}_0 + S(\Omega_0)\eta_0) + \{ \mathbf{D}_{R_0} L \cdot R_0 S(\eta_0)$$
$$+ \sum_{i=1}^{2} \mathbf{D}_{\omega_i} L \cdot (-S(\omega_i)\gamma_i + (I - q_i q_i^T)\dot{\gamma}_i) + \sum_{i=1}^{2} \mathbf{D}_{q_i} L \cdot (S(\gamma_i)q_i)$$
$$\left. + \sum_{i=1}^{2} \mathbf{D}_{\Omega_i} L \cdot (\dot{\eta}_i + S(\Omega_i)\eta_i) \right] dt.$$

The derivatives of the Lagrangian are given by

$$\mathbf{D}_{\dot{x}_0} L = m_T \dot{x}_0 + \sum_{i=1}^{2} m_i (R_0 S(\Omega_0)\rho_i - L_i \dot{q}_i),$$

$$\mathbf{D}_{\omega_i} L = \sum_{i=1}^{2} m_i (L_i^2 \omega_i - L_i S(q_i)\dot{x}_0 - L_i S(q_i) R_0 S(\Omega_0)\rho_i), \quad i = 1, 2,$$

$$\mathbf{D}_{\Omega_0} L = \bar{J}_0 \Omega_0 + \sum_{i=1}^{2} m_i S(\rho_i) R_0^T (\dot{x}_0 - L_i \dot{q}_i),$$

$$\mathbf{D}_{\Omega_i} L = J_i \Omega_i, \qquad\qquad\qquad\qquad\qquad\qquad i = 1, 2$$

$$\mathbf{D}_{x_0} L = -m_T g e_3,$$

$$\mathbf{D}_{q_i} L = +m_i L_i g e_3 + m_i L_i S(\omega_i)(\dot{x}_0 + \dot{R}_0 \rho_i), \qquad i = 1, 2,$$

where $\bar{J}_0 = J_0 + \sum_{i=1}^{2} m_i S(\rho_i)^T S(\rho_i)$. The derivative of the Lagrangian with respect to the attitude of the rigid body payload can be written as

$$\mathbf{D}_{R_0} L \cdot R_0 S(\eta_0) = \sum_{i=1}^{2} m_i R_0 S(\eta_0) S(\Omega_0) \rho_i \cdot (\dot{x}_i - L_i \dot{q}_i) - m_i g e_3 \cdot R_0 S(\eta_0) \rho_i,$$

$$= \sum_{i=1}^{2} m_i \{ S(S(\Omega_0) \rho_i) R_0^T (\dot{x}_0 - L_i \dot{q}_i) - g S(\rho_i) R_0^T e_3 \} \cdot \eta_0,$$

$$\triangleq \mathbf{d}_{R_0} L \cdot \eta_0,$$

where $\mathbf{d}_{R_0} L \in \mathbb{R}^3$ is referred to as the left-trivialized derivative.

The thrust force on the i-th quad rotor, with respect to the inertial frame, is $T_i R_i e_3 \in \mathbb{R}^3$, $i = 1, 2$. The moment vector, produced by the four rotors, on the i-th quad rotor is $M_i \in \mathbb{R}^3$, $i = 1, 2$. The corresponding virtual work done by the external forces and moments can be written as

$$\delta \mathcal{W} = \int_{t_0}^{t_f} \sum_{i=1}^{2} \{ T_i e_3^T R_i^T \delta x_i + M_i^T \eta_i \} \, dt,$$

which can be expressed as

$$\delta \mathcal{W} = \int_{t_0}^{t_f} \sum_{i=1}^{2} \{ T_i R_i e_3 \cdot (\delta x_0 + R_0 S(\eta_0) \rho_i - L_i S(\gamma_i) q_i) + M_i \cdot \eta_i \} \, dt.$$

According to the Lagrange–d'Alembert principle, we have $\delta \mathfrak{G} = -\delta \mathcal{W}$ for variations with fixed end points. Integrate by parts and rearrange to obtain the following Euler–Lagrange equations:

$$\frac{d}{dt} \mathbf{D}_{\dot{x}_0} L - \mathbf{D}_{x_0} L = \sum_{i=1}^{2} R_i e_3 T_i,$$

$$\frac{d}{dt} \mathbf{D}_{\Omega_0} L + S(\Omega_0) \mathbf{D}_{\Omega_0} L - \mathbf{d}_{R_0} L = \sum_{i=1}^{2} S(\rho_0) R_0^T R_i^T e_3 T_i,$$

$$(I - q_i q_i^T) \frac{d}{dt} \mathbf{D}_{\omega_i} L - S(q_i) \mathbf{D}_{q_i} L = -L_i S(q_i) R_i e_3 T_i, \quad i = 1, 2,$$

$$\frac{d}{dt} \mathbf{D}_{\Omega_i} L + S(\Omega_i) \mathbf{D}_{\Omega_i} L = M_i, \quad i = 1, 2.$$

Substituting the above expressions for the derivatives and rearranging using the facts that $I_{3 \times 3} - q_i q_i^T = -S(q_i)^2$ and $S(q_i)^2 S(\dot{q}_i) = S(q_i) S(\omega_i)$, for $i = 1, 2$, the Euler–Lagrange equations are given by

$$m_T \ddot{x}_0 + \sum_{i=1}^{2} m_i(-R_0 S(\rho_i)\dot{\Omega}_0 + L_i S(q_i)\dot{\omega}_i) + \sum_{i=1}^{2} m_i R_0 S(\Omega_0)^2 \rho_i$$

$$+m_i L_i \|\omega_i\|^2 q_i + m_T g e_3 = \sum_{i=1}^{2} R_i e_3 T_i, \tag{9.78}$$

$$\bar{J}_0 \dot{\Omega}_0 + \sum_{i=1}^{2} m_i S(\rho_i) R_0^T (\ddot{x}_0 + L_i S(q_i)\dot{\omega}_i + L_i \|\omega_i\|^2 q_i) + \Omega_0 \times \bar{J}_0 \Omega_0$$

$$+ \sum_{i=1}^{2} \rho_i \times R_0^T m_i g e_3 = \sum_{i=1}^{2} \rho_i \times R_0^T R_i e_3 T_i, \tag{9.79}$$

$$m_i L_i \dot{\omega}_i - m_i S(q_i)\ddot{x}_0 + m_i S(q_i)R_0 S(\rho_i)\dot{\Omega}_0 - m_i S(q_i)R_0 S(\Omega_0)^2 \rho_i$$

$$-S(q_i)m_i g e_3 = -S(q_i)R_i e_3 T_i, \quad i = 1, 2, \tag{9.80}$$

$$J_i \dot{\Omega}_i + \Omega_i \times J_i \Omega_i = M_i, \quad i = 1, 2, \tag{9.81}$$

where $m_T = m_0 + \sum_{i=1}^{2} m_i$ and $\bar{J}_0 = J_0 + \sum_{i=1}^{2} m_i S(\rho_i)^T S(\rho_i) \in \mathbb{R}^{3 \times 3}$.

Next, we substitute (9.80) into (9.78) and (9.79) to eliminate the dependency of $\dot{\omega}_i$ in the expressions for $\ddot{x}_0$ and $\dot{\Omega}_0$. Using the fact that $I_{3 \times 3} + S(q_i)^2 = q_i q_i^T$ for any $q_i \in \mathsf{S}^2$ and $S(\Omega_0)S(\rho_i)\Omega_0 = -S(\rho_i)S(\Omega_0)^2 \rho_i$ for any $\Omega_0, \rho_i \in \mathbb{R}^3$, we obtain the following form for the Euler–Lagrange equations

$$\bar{M}\ddot{x}_0 - \sum_{i=1}^{2} m_i q_i q_i^T R_0 S(\rho_i)\dot{\Omega}_0 + \sum_{i=1}^{2} m_i \left\{ L_i \|\omega_i\|^2 q_i - q_i q_i^T R_0 S(\omega_0)^2 \rho_i \right\}$$

$$+\bar{M}g e_3 = \sum_{i=1}^{2} q_i q_i^T R_i e_3 T_i, \tag{9.82}$$

$$(J_0 + \sum_{i=1}^{2} m_i S(\rho_i)^T R_0^T q_i q_i^T R_0 S(\rho_i))\dot{\Omega}_0 + \sum_{i=1}^{2} m_i S(\rho_i)R_0^T q_i q_i^T \ddot{x}_0$$

$$+\Omega_0 \times J_0 \Omega_0 + \sum_{i=1}^{2} S(\rho_i)R_0^T (m_i L_i \|\omega_i\|^2 q_i + m_i q_i q_i^T R_0 S(\Omega_0)^2 \rho_i)$$

$$- \sum_{i=1}^{2} m_i g S(\rho_i)R_0^T q_i q_i^T e_3 = \sum_{i=1}^{2} S(\rho_i)R_0^T q_i q_i^T R_i e_3 T_i, \tag{9.83}$$

$$m_i L_i \dot{\omega}_i - m_i S(q_i)(\ddot{x}_0 - R_0 S(\rho_i)\dot{\Omega}_0 + R_0 S(\Omega_0)^2 \rho_i) - m_i g S(q_i)e_3$$

$$= -q_i \times R_i e_3 T_i, \quad i = 1, 2, \tag{9.84}$$

$$J_i \dot{\Omega}_i + \Omega_i \times J_i \Omega_i = M_i, \quad i = 1, 2, \tag{9.85}$$

where $\bar{M} = m_0 I_{3\times3} + \sum_{i=1}^{2} m_i q_i q_i^T \in \mathbb{R}^{3\times3}$, which is symmetric, positive-definite for any q_i.

The dynamical equations for the two quad rotor vehicles supporting a rigid body payload consist of the kinematics equations (9.75), (9.76), and (9.77) and the Euler–Lagrange equations (9.82), (9.83), (9.84), and (9.85), which describe the dynamical flow in terms of the evolution on the tangent bundle $T(SE(3) \times (S^2 \times SO(3))^2)$ of the configuration manifold.

9.12.2 Hamilton's Equations

Due to the length of the calculations, we do not derive the Hamilton's equations of motion for the quad rotor vehicles transporting a rigid body payload.

9.12.3 Conservation Properties

Since there are external forces and moments that act on the two quad rotor vehicles, the total energy is not conserved; rather the change in total energy of the system is equal to the work done by the external thrust and moment on the quad rotors.

9.12.4 Equilibrium Properties

Assume that the thrust magnitude and moment for each quad rotor vehicle are constant. The quad rotor vehicles and payload are in equilibrium if the translational velocity vector and the angular velocity vectors of the quad rotor vehicles and the payload are zero and the following equilibrium conditions are satisfied:

$$\bar{M} g e_3 = \sum_{i=1}^{2} q_i q_i^T R_i e_3 T_i,$$

$$-\sum_{i=1}^{2} m_i g \rho_i \times R_0^T q_i q_i^T e_3 = \sum_{i=1}^{2} \rho_i \times R_0^T q_i q_i^T R_i e_3 T_i,$$

$$m_i g q_i \times e_3 = q_i \times R_i e_3 T_i, \quad i = 1, 2,$$

$$M_i = 0, \quad i = 1, 2.$$

It is possible to construct linearized dynamics that approximate the Lagrangian dynamics of the quad rotor in a neighborhood of any equilibrium solution, but we do not derive them here.

9.13 Problems

9.1. Consider the dynamics of a modified Furuta pendulum; the first link is constrained to rotate, as a planar pendulum, in a fixed horizontal plane while the rotation of the second link, as a spherical pendulum, is not constrained. Include the gravitational forces. The pendulum masses are m_1 and m_2 and the pendulum lengths are L_1 and L_2. The configuration manifold is $S^1 \times S^2$.

(a) Determine the Lagrangian function $L : T(S^1 \times S^2) \to \mathbb{R}^1$ on the tangent bundle of the configuration manifold.
(b) What are the Euler–Lagrange equations for the modified Furuta pendulum?
(c) Determine the Hamiltonian function $H : T^*(S^1 \times S^2) \to \mathbb{R}^1$ on the cotangent bundle of the configuration manifold.
(d) What are Hamilton's equations for the modified Furuta pendulum?
(e) What are conserved quantities for the dynamical flow on $T(S^1 \times S^2)$?
(f) What are the equilibrium solutions of the dynamical flow on $T(S^1 \times S^2)$? Describe the linearized equations that approximate the dynamical flow in a neighborhood of each equilibrium solution.

9.2. Assuming there are no external or potential forces and the reaction wheel is axially symmetric about its spin axis, prove the following for a spacecraft with a reaction wheel assembly:

(a) the Hamiltonian

$$ H = \frac{1}{2} m \|\dot{x}\|^2 + \frac{1}{2} \Omega^T J' \Omega + \frac{1}{2} \sum_{i=1}^{n} \alpha_i \omega_i^2 + \sum_{i=1}^{n} \alpha_i \omega_i \Omega^T s_i $$

is conserved along each solution of the dynamical flow.
(b) the total angular momentum expressed in the inertial frame

$$ \pi = R \Big(J' \Omega + \sum_{i=1}^{n} \alpha_i \omega_i s_i \Big) $$

is conserved along each solution of the dynamical flow.

9.3. Consider the full body dynamics with n bodies.

(a) Determine an expression for the position vector of the center of mass vector of all n bodies, expressed in the inertial frame.
(b) Describe the dynamics of the center of mass vector.
(c) What conservation properties are associated with the center of mass vector?

9.4. The full body dynamics of two dumbbell rigid bodies evolve due to Newtonian gravity. Assume the origin of each body-fixed frame is located at the center of mass of the body and each body-fixed frame is aligned with the principle axes of the body. The mass of each rigid body is m_i and the inertia matrix of each rigid body is J_i, for $i = 1, 2$. The gravity forces and moments on the bodies are determined by two concentrated mass elements for each dumbbell body.

(a) Determine the Lagrangian function $L : T(SE(3))^2 \to \mathbb{R}^1$ defined on the tangent bundle of the configuration manifold.
(b) What are the Euler–Lagrange equations for the full body dynamics of the two dumbbell bodies?
(c) Determine the Hamiltonian function $H : T^*(SE(3))^2 \to \mathbb{R}^1$ defined on the cotangent bundle of the configuration manifold.
(d) What are Hamilton's equations for the full body dynamics of the two dumbbell bodies?
(e) What are conserved quantities for the full body dynamics?

9.5. Consider a modification of the planar rigid body pendulum, where the rigid body is connected to an inertially fixed cylindrical joint. The cylindrical joint has a fixed axis that allows rotation of the rigid body about this axis and translation of the rigid body along this axis. The axis of the cylindrical joint, denoted by the direction vector $a \in S^2$, is inertially fixed. Let $R \in SO(3)$ be the attitude of the rigid body, which must satisfy $Ra = a$. Let $x = ya$ be the displacement of the rigid body along the axis of the cylindrical joint, where $y \in \mathbb{R}^1$. The configuration manifold is $M = \{(R, x) \in SE(3) : Ra = a,\ x = ya,\ y \in \mathbb{R}^1\}$. This modification of the planar rigid pendulum acts under uniform, constant gravity.

(a) Determine the Lagrangian function $L : TM \to \mathbb{R}^1$ defined on the tangent bundle of the configuration manifold.
(b) What are the Euler–Lagrange equations for the planar rigid body pendulum?
(c) Determine the Hamiltonian function $H : T^*M \to \mathbb{R}^1$ defined on the cotangent bundle of the configuration manifold.
(d) What are Hamilton's equations for the planar rigid body pendulum?
(e) Suppose that the axis of the cylindrical joint $a \in S^2$ is horizontal, that is $a^T e_3 = 0$. What are the equilibrium solutions of the dynamical flow on TM? Describe the linearized equations that approximate the dynamical flow in a neighborhood of each equilibrium solution.

9.6. An elastic spherical pendulum is a spherical pendulum for which a massless link connects a concentrated mass element, viewed as an ideal particle, to an inertially fixed pivot, but where the link is elastic rather than rigid. An inertial Euclidean frame is selected with origin located at the fixed pivot of the elastic spherical pendulum. Let $x \in \mathbb{R}^1$ denote the length of the massless link and let $q \in S^2$ denote the attitude vector of the massless link, which is assumed to remain straight. Thus, $(x, q) \in \mathbb{R}^1 \times S^2$ so that $\mathbb{R}^1 \times S^2$ is the configuration manifold. The mass element acts under uniform and constant gravity. The elastic link is assumed to have elastic stiffness κ and has length L when the elastic force in the link vanishes. Assume that the inequality $mg < kL$ holds.

(a) Determine the Lagrangian function $L : \mathsf{T}(\mathbb{R}^1 \times \mathsf{S}^2) \to \mathbb{R}^1$ defined on the tangent bundle of the configuration manifold.
(b) What are the Euler–Lagrange equations for the elastic spherical pendulum?
(c) Determine the Hamiltonian function $H : \mathsf{T}^*(\mathbb{R}^1 \times \mathsf{S}^2) \to \mathbb{R}^1$ defined on the cotangent bundle of the configuration manifold.
(d) What are Hamilton's equations for the elastic spherical pendulum?
(e) What are conserved quantities for the elastic spherical pendulum?
(f) What are the equilibrium solutions of the elastic spherical pendulum?
(g) For each equilibrium solution, determine linearized equations that approximate the dynamics of the elastic spherical pendulum in a neighborhood of that equilibrium.

9.7. Two concentrated mass elements, viewed as ideal particles of mass m_1 and m_2, are connected by a massless link; the link is assumed to remain straight and it can deform elastically along its longitudinal axis. An inertial Euclidean frame is selected. Let $x \in \mathbb{R}^3$ denote the position vector of the first particle; let $y \in \mathbb{R}^1$ denote the length of the massless link and let $q \in \mathsf{S}^2$ denote the attitude vector of the massless link. The vector $x + yq \in \mathbb{R}^3$ is the position vector of the second particle. Thus, $(x, y, q) \in \mathbb{R}^3 \times \mathbb{R}^1 \times \mathsf{S}^2$ so that $\mathbb{R}^3 \times \mathbb{R}^1 \times \mathsf{S}^2$ is the configuration manifold. The particles acts under uniform and constant gravity. The elastic link is assumed to have elastic stiffness κ and has length L when the elastic force in the link vanishes.

(a) Determine the Lagrangian function $L : \mathsf{T}(\mathbb{R}^3 \times \mathbb{R}^1 \times \mathsf{S}^2) \to \mathbb{R}^1$ defined on the tangent bundle of the configuration manifold.
(b) What are the Euler–Lagrange equations for the two particles connected by a massless elastic link?
(c) Determine the Hamiltonian function $H : \mathsf{T}^*(\mathbb{R}^3 \times \mathbb{R}^1 \times \mathsf{S}^2) \to \mathbb{R}^1$ defined on the cotangent bundle of the configuration manifold.
(d) What are Hamilton's equations for the two particles connected by a massless elastic link?
(e) What are conserved quantities for the two particles connected by a massless elastic link?
(f) Describe the dynamics of the position vector of the center of mass of the two particles connected by a massless elastic link.

9.8. Consider the dynamics of a rigid link, with mass concentrated at its midpoint, that can translate and rotate in three dimensions under the action of uniform gravity. Each end of the rigid link is connected to two elastic strings; the opposite ends of the elastic strings are connected to fixed inertial supports. These supports for the elastic strings are located at d_1, d_2, d_3, d_4 in $\mathbb{R}^3$ with respect to the inertial frame. The elastic strings are assumed to be in tension. Let $q \in S^2$ denote the attitude vector of the rigid link in the inertial frame; let $x \in \mathbb{R}^3$ denote the position vector of the concentrated mass, in the inertial frame. Thus, $(q, x) \in S^2 \times \mathbb{R}^3$ and the configuration manifold is $S^2 \times \mathbb{R}^3$. This can be viewed as another example of a tensegrity structure.

(a) Determine the Lagrangian function $L : T(S^2 \times \mathbb{R}^3) \to \mathbb{R}^1$ defined on the tangent bundle of the configuration manifold. Determine expressions for the tension forces in the four elastic strings.
(b) What are the Euler–Lagrange equations for the rigid link connected to four elastic strings?
(c) Determine the Hamiltonian function $H : T^*(S^2 \times \mathbb{R}^3) \to \mathbb{R}^1$ defined on the cotangent bundle of the configuration manifold.
(d) What are Hamilton's equations for the rigid link connected to four elastic strings?
(e) What are the equilibrium solutions of the rigid link connected to four elastic strings?
(f) For each equilibrium solution, determine linearized equations that approximate the dynamics of the rigid link connected to four elastic strings in a neighborhood of that equilibrium.

9.9. Consider the dynamics of a planar mechanism consisting of two ideal particles, with mass, connected by a massless, rigid link; the ends of the link are constrained to slide along a straight line and a circle. Assume the mechanism lies in a fixed horizontal plane so that there are no gravitational forces. The configuration manifold is $S^1 \times \mathbb{R}^1$.

(a) Determine the Lagrangian function $L : T(S^1 \times \mathbb{R}^1) \to \mathbb{R}^1$ defined on the tangent bundle of the configuration manifold.
(b) What are the Euler–Lagrange equations for the planar mechanism in a horizontal plane?
(c) Determine the Hamiltonian function $H : T^*(S^1 \times \mathbb{R}^1) \to \mathbb{R}^1$ defined on the cotangent bundle of the configuration manifold.
(d) What are Hamilton's equations for the planar mechanism in a horizontal plane?
(e) What are conserved quantities for the planar mechanism in a horizontal plane?
(f) What are the equilibrium solutions of the planar mechanism in a horizontal plane?
(g) For each equilibrium solution, determine linearized equations that approximate the dynamics of the planar mechanism in a neighborhood of that equilibrium.

9.10. Consider the dynamics of an ideal particle constrained to move on the surface of a right cylinder in $\mathbb{R}^3$ and assume there are no gravitational forces. The mass of the particle is m and the cylinder has radius $R > 0$. An inertial frame is selected so that the third axis of the frame has a direction along the center line of the cylinder and the origin of the frame is located on the center line of the cylinder. The configuration manifold is $S^1 \times R^1$.

(a) Show that the position vector of the particle in the inertial frame can be expressed as $x = (Rq, z)$, where $q \in S^1$ and $z \in R^1$.
(b) Determine the Lagrangian function $L : T(S^1 \times R^1) \to R^1$ on the tangent bundle of the configuration manifold.
(c) What are the Euler–Lagrange equations for the particle on a cylinder, without gravity?
(d) Determine the Hamiltonian function $H : T^*(S^1 \times R^1) \to R^1$ on the cotangent bundle of the configuration manifold.
(e) What are Hamilton's equations for the particle on a cylinder, without gravity?
(f) What are conserved quantities for the particle on a cylinder, without gravity?
(g) What are the equilibrium solutions of the dynamics of a particle on a cylinder, without gravity?

9.11. Consider the dynamics of an ideal particle constrained to move on the surface of a right cone in R^3 and assume there are no gravitational forces. The mass of the particle is m. An inertial frame is selected so that the third axis of the frame has a direction along the center line of the cone and the origin of the frame is located on the vertex of the cone. The configuration manifold is $S^1 \times R^1$.

(a) Show that the position vector of the particle in the inertial frame can be expressed as $x = \begin{bmatrix} zq \\ z \end{bmatrix}$, where $q \in S^1$ and $z \in R^1$.
(b) Determine the Lagrangian function $L : T(S^1 \times R^1) \to R^1$ on the tangent bundle of the configuration manifold.
(c) What are the Euler–Lagrange equations for the particle on a right cone, without gravity?
(d) Determine the Hamiltonian function $H : T^*(S^1 \times R^1) \to R^1$ on the cotangent bundle of the configuration manifold.
(e) What are Hamilton's equations for the particle on a right cone, without gravity?
(f) Discuss the validity of the equations of motion near the vertex of the cone, where $z = 0$.
(g) What are conserved quantities for the particle on a right cone, without gravity?
(h) What are the equilibrium solutions of the dynamics of a particle on a right cone, without gravity?

9.12. Consider the rotational and translational dynamics of two elastically connected identical rigid links with concentrated masses that are constrained to rotate and translate in a fixed horizontal two-dimensional plane. Let $q_i \in S^1$ denote the attitude vector of the i-th link and let $x_i \in R^2$ denote the position vector of the concentrated mass of the i-th link for $i = 1, 2$. The elastic potential depends only on the relative attitude and the relative position of the two links. Let J and m denote the scalar inertia and the mass of each link; κ is the rotational stiffness constant and K is the translational stiffness constant. The attitude vectors $q = (q_1, q_2) \in (S^1)^2$ and the position vectors $x = (x_1, x_2) \in (\mathbb{R}^2)^2$, so that the configuration manifold is $(S^1)^2 \times (\mathbb{R}^2)^2$. The Lagrangian function is given by

$$
L(q, x, \dot{q}, \dot{x}) = \frac{1}{2} J \|\dot{q}_1\|^2 + \frac{1}{2} J \|\dot{q}_2\|^2 + \frac{1}{2} m \|\dot{x}_1\|^2 + \frac{1}{2} m \|\dot{x}_2\|^2
$$
$$
- \kappa(1 - q_1^T q_2) - \frac{1}{2} K \|x_1 - x_2\|^2 ,
$$

on the tangent bundle of the configuration manifold.

(a) What are the Euler–Lagrange equations for the rotational and translational dynamics of the two elastically connected links in a horizontal plane?
(b) Determine the Hamiltonian function $H : T^*((S^1)^2 \times (\mathbb{R}^2)^2) \to \mathbb{R}^1$ on the cotangent bundle of the configuration manifold.
(c) What are Hamilton's equations for the rotational and translational dynamics of two elastically connected links in a horizontal plane?
(d) What are conserved quantities for the dynamical flow?
(e) What are the equilibrium solutions of the dynamical flow?
(f) Select a specific equilibrium solution and determine the linearized dynamics that approximate the Lagrangian flow in a neighborhood of that equilibrium.

9.13. Consider the dynamics of a spherical pendulum on a cart. Assume there are no gravitational forces. The configuration manifold is $S^2 \times \mathbb{R}^2$.

(a) Determine the Lagrangian function $L : T(S^2 \times \mathbb{R}^2) \to \mathbb{R}^1$ defined on the tangent bundle of the configuration manifold.
(b) Determine the Euler–Lagrange equations for the spherical pendulum on a cart with no gravity.
(c) Determine the Hamiltonian function $H : T^*(S^2 \times \mathbb{R}^2) \to \mathbb{R}^1$ on the cotangent bundle of the configuration manifold.
(d) What are Hamilton's equations for the spherical pendulum on a cart with no gravity?
(e) What are conserved quantities for the dynamical flow?
(f) What are the equilibrium solutions of the dynamical flow?
(g) Select a specific equilibrium solution and determine the linearized dynamics that approximate the Lagrangian flow in a neighborhood of that equilibrium.

9.14. Consider the dynamics of a three-dimensional pendulum on a cart. Assume there are no gravitational forces. The inertia matrix of the pendulum as a rigid body is the matrix $J = \text{diag}(J_1, J_2, J_3)$. The configuration manifold is $\text{SO}(3) \times \mathbb{R}^2$.

(a) Determine the Lagrangian function $L : \mathsf{T}(\text{SO}(3) \times \mathbb{R}^2) \to \mathbb{R}^1$ defined on the tangent bundle of the configuration manifold.
(b) What are the Euler–Lagrange equations for the three-dimensional pendulum on a cart, with no gravity?
(c) Determine the Hamiltonian function $H : \mathsf{T}^*(\text{SO}(3) \times \mathbb{R}^2) \to \mathbb{R}^1$ on the cotangent bundle of the configuration manifold.
(d) What are Hamilton's equations for the three-dimensional pendulum on a cart, with no gravity?
(e) What are conserved quantities for the dynamical flow?
(f) What are the equilibrium solutions of the dynamical flow?
(g) Select a specific equilibrium solution and determine the linearized dynamics that approximate the Lagrangian flow in a neighborhood of that equilibrium.

9.15. Consider the full body dynamics for two identical dumbbell-shaped rigid bodies; that is, each dumbbell body consists of two identical, uniform spherical bodies connected by a rigid, massless link. The inertia matrix for each rigid body is $J = \text{diag}(J_1, J_2, J_2)$, where two of the entries are equal due to the rotational symmetry about each link, and the mass of each rigid body is m. The configuration manifold is $\text{SE}(3)^2$.

(a) Obtain an expression for the gravitation potential.
(b) Determine the Lagrangian function $L : \mathsf{TSE}(3)^2 \to \mathbb{R}^1$ defined on the tangent bundle of the configuration manifold.
(c) What are the Euler–Lagrange equations for the full body dynamics?
(d) Determine the Hamiltonian function $H : \mathsf{T}^*\mathsf{SE}(3)^2 \to \mathbb{R}^1$ defined on the cotangent bundle of the configuration manifold.
(e) What are Hamilton's equations for the full body dynamics?
(f) What are conserved quantities for the dynamical flow?
(g) What are conditions for relative equilibrium solutions of the dynamical flow?

9.16. Consider the dynamics of a rigid body and an ideal mass particle. An inertial frame and a body-fixed frame are introduced and used to define the translational and rotational motion of the rigid body; the origin of the body-fixed frame is assumed to be located at the center of mass of the rigid body. The rigid body can translate and rotate in three dimensions. The mass particle can translate, without constraint in three dimensions. Let $(R, x) \in \text{SE}(3)$ denote the attitude of the rigid body and the position vector of the center of mass of the rigid body in the inertial frame and let $y \in \mathbb{R}^3$ denote the position vector of the particle in the body-fixed frame; this is interpreted as the position vector of the particle relative to the rigid body as expressed in the rotating frame defined by the rigid body. Thus, the configuration manifold is $\text{SE}(3) \times \mathbb{R}^3$. The mass of the rigid body is M and J is the standard 3×3 inertia matrix of the rigid body. The particle has mass m. There is a gravitational potential energy function $V(y)$.

(a) Determine the Lagrangian function $L : \mathsf{T}(\mathsf{SE}(3) \times \mathbb{R}^3) \to \mathbb{R}^1$ defined on the tangent bundle of the configuration manifold.
(b) What are the Euler–Lagrange equations for the rigid body and mass particle?
(c) Determine the Hamiltonian function $H : \mathsf{T}^*(\mathsf{SE}(3) \times \mathbb{R}^3) \to \mathbb{R}^1$ on the cotangent bundle of the configuration manifold.
(d) What are Hamilton's equations for the rigid body and mass particle?
(e) What are conserved quantities for the dynamical flow?

9.17. Consider the dynamics of a rigid, uniform sphere and an ideal mass particle constrained to move on the surface of the sphere. The sphere can translate and rotate in three dimensions. Assume there are no gravitational forces on the sphere or the mass particle. Let $(R, x) \in \mathsf{SE}(3)$ denote the attitude of the sphere and the position vector of the center of the sphere in an inertial frame and let $q \in \mathsf{S}^2$ denote the direction of the position vector of the particle in a sphere-fixed frame whose origin is located at the center of the sphere. Thus, the configuration manifold is $\mathsf{SE}(3) \times \mathsf{S}^2$. The mass of the sphere is M and J is the standard 3×3 inertia matrix of the rigid sphere. The particle, of mass m, moves without friction on the surface of the sphere of radius r.

(a) Determine the Lagrangian function $L : \mathsf{T}(\mathsf{SE}(3) \times \mathsf{S}^2) \to \mathbb{R}^1$ defined on the tangent bundle of the configuration manifold.
(b) What are the Euler–Lagrange equations for the sphere and mass particle?
(c) Determine the Hamiltonian function $H : \mathsf{T}^*(\mathsf{SE}(3) \times \mathsf{S}^2) \to \mathbb{R}^1$ on the cotangent bundle of the configuration manifold.
(d) What are Hamilton's equations for the sphere and mass particle?
(e) What are conserved quantities for the dynamical flow?
(f) What are the equilibrium solutions of the dynamical flow?
(g) Select a specific equilibrium solution and determine the linearized dynamics that approximate the Lagrangian flow in a neighborhood of that equilibrium.

9.18. Consider the dynamics of a single quad rotor vehicle transporting a cable-suspended rigid body. Include a thrust force and a moment on the quad rotor vehicle as control inputs. Make appropriate assumptions and suppose that the configuration manifold is $\mathsf{SE}(3) \times \mathsf{S}^2 \times \mathsf{SO}(3)$.

(a) Determine the Lagrangian function $L : \mathsf{T}(\mathsf{SE}(3) \times \mathsf{S}^2 \times \mathsf{SO}(3))$ defined on the tangent bundle of the configuration manifold.
(b) What are the Euler–Lagrange equations for the quad rotor vehicle transporting a rigid body?
(c) Determine the Hamiltonian function $H : \mathsf{T}^*(\mathsf{SE}(3) \times \mathsf{S}^2 \times \mathsf{SO}(3)) \to \mathbb{R}^1$ defined on the cotangent bundle of the configuration manifold.

(d) What are Hamilton's equations for the quad rotor vehicle transporting a rigid body?

(e) For what values of the thrust magnitude provided by the rotors can the quad rotor vehicle be in equilibrium? What are these equilibrium solutions?

(f) Select a specific equilibrium solution and determine the linearized dynamics that approximate the Lagrangian flow in a neighborhood of that equilibrium.

9.19. A rigid body consists of two particles, each of mass m, connected by a massless link of length L. Assume uniform, constant gravity. Let $x \in \mathbb{R}^3$ denotes the position vector of the first particle with respect to a fixed Euclidean frame and $q \in S^2$ denotes the attitude vector of the link. Thus, the configuration can be expressed in terms of $(x, q) \in \mathbb{R}^3 \times S^2$ and the configuration manifold is $\mathbb{R}^3 \times S^2$.

(a) Determine the Lagrangian function $L : \mathsf{T}(\mathbb{R}^3 \times S^2) \to \mathbb{R}^1$ defined on the tangent bundle of the configuration manifold.

(b) What are the Euler–Lagrange equations for the rigid body?

(c) Determine the Hamiltonian function $H : \mathsf{T}^*(\mathsf{SE}(3) \times S^2 \times \mathsf{SO}(3)) \to \mathbb{R}^1$ defined on the cotangent bundle of the configuration manifold.

(d) What are Hamilton's equations for the rigid body?

(e) What are the conservation properties of the dynamical flow?

(f) Describe the dynamics of the center of mass of the rigid body and the motion of the center of mass of the rigid body.

9.20. Consider the dynamics of a bead and rod in a fixed vertical plane (sometimes referred to as a planar ball and beam) under uniform, constant gravity. The rod is idealized as a rigid straight link that rotates, without friction, in the vertical plane about a pivot located at its center of mass. The bead, idealized as a concentrated mass element, is assumed to slide, without friction, along the axis of the rod. Let J denote the scalar inertia of the link about its center of mass and let m denote the mass of the bead. Let $q \in S^1$ denote the attitude vector of the rod in its plane of rotation and let $x \in \mathbb{R}^1$ denote the position of the bead on the rod with respect to the center of mass of the link. The configuration manifold is $S^1 \times \mathbb{R}^1$.

(a) What is the Lagrangian function $L : \mathsf{T}(S^1 \times \mathbb{R}^1) \to \mathbb{R}^1$ for the bead and rod defined on the tangent bundle of the configuration manifold?

(b) What are the Euler–Lagrange equations for the bead and rod?

(c) Determine the Hamiltonian function $H : \mathsf{T}^*(S^1 \times \mathbb{R}^1) \to \mathbb{R}^1$ defined on the cotangent bundle of the configuration manifold.

(d) What are Hamilton's equations for the bead and rod?

(e) What are conserved quantities of the dynamical flow?

(f) What are the equilibrium solutions of the dynamical flow?

(g) Select a specific equilibrium solution and determine the linearized dynamics that approximate the Lagrangian flow in a neighborhood of that equilibrium.

9.21. Consider the dynamics of a bead and rod in three dimensions (a three-dimensional version of the ball and beam) under uniform, constant gravity. The rod is idealized as a rigid straight link that rotates in three dimensions, without friction, about a spherical pivot located at its center of mass. The bead, idealized as a concentrated mass element, is assumed to slide, without friction, along the axis of the rod. Let J denote the scalar inertia of the rod about its center of mass and let m denote the mass of the bead. Let $q \in \mathsf{S}^2$ denote the attitude vector of the rod with respect to an inertial frame and let $x \in \mathbb{R}^1$ denote the axial position of the bead on the rod with respect to the center of mass of the rod. The configuration manifold is $\mathsf{S}^2 \times \mathbb{R}^1$.

(a) What is the Lagrangian function $L : \mathsf{T}(\mathsf{S}^2 \times \mathbb{R}^1) \to \mathbb{R}^1$ for the bead and rod defined on the tangent bundle of the configuration manifold?
(b) What are the Euler–Lagrange equations for the bead and rod?
(c) Determine the Hamiltonian function $H : \mathsf{T}^*(\mathsf{S}^2 \times \mathbb{R}^1) \to \mathbb{R}^1$ defined on the cotangent bundle of the configuration manifold.
(d) What are Hamilton's equations for the bead and rod?
(e) What are conserved quantities of the dynamical flow?
(f) What are the equilibrium solutions of the dynamical flow?
(g) Select a specific equilibrium solution and determine the linearized dynamics that approximate the Lagrangian flow in a neighborhood of that equilibrium.

9.22. Two ideal particles of concentrated mass m_1 and m_2 are connected by a massless, rigid link of length L. The first particle is constrained to translate along a horizontal frictionless rail. The first axis of the inertial Euclidean frame is aligned along the rail and lies in a horizontal plane; the third axis is vertical. A linear elastic restoring force acts on the first mass along the rail; the elastic stiffness is κ and the restoring force is zero when the particle is located at the origin of the inertial frame. The second particle is free to move in three dimensions, under uniform, constant gravity, subject to its connection to the first particle. Let $x \in \mathbb{R}^1$ denote the position vector of the first particle on the rail and let $q \in \mathsf{S}^2$ denote the attitude vector of the link connecting the two particles, expressed in the inertial Euclidean frame. Thus, $\mathbb{R}^1 \times \mathsf{S}^2$ is the configuration manifold. The Lagrangian function is given by

$$L(x, q, \dot{x}, \dot{q}) = \frac{1}{2}m_1\dot{x}^2 + \frac{1}{2}m_2\left\|\dot{x}e_1 + L\dot{q}\right\|^2 - \frac{1}{2}\kappa x^2 - m_2 g e_3^T q.$$

(a) What are the Euler–Lagrange equations for the two particles?
(b) Determine the Hamiltonian function $H : \mathsf{T}^*(\mathbb{R}^1 \times \mathsf{S}^2) \to \mathbb{R}^1$ defined on the cotangent bundle of the configuration manifold.
(c) What are Hamilton's equations for the two particles?

(d) What are conserved quantities for the dynamical flow?
(e) What are the equilibrium solutions of the dynamical flow?
(f) Select a specific equilibrium solution and determine the linearized dynamics that approximate the Lagrangian flow in a neighborhood of that equilibrium.

9.23. Consider the dynamics of an ideal particle of mass m that moves on the surface of a right cylinder in $\mathbb{R}^3$ under the influence of uniform, constant gravity. Assume the axis of the cylinder is in the horizontal plane along the positive first axis of an inertial Euclidean frame. The radius of the cylinder is $r > 0$. Let $q \in S^1$ denote the projection of the position vector of the particle onto the circular cross section of the cylinder; let $x \in \mathbb{R}^1$ denote the projection of the position vector of the particle onto the first axis of the Euclidean frame (the axis of the cylinder). The configuration is $(q, x) \in S^1 \times \mathbb{R}^1$ and the configuration manifold is $S^1 \times \mathbb{R}^1$.

(a) What is the position vector of the particle, expressed in the inertial Euclidean frame, in terms of the configuration?
(b) What is the Lagrangian function $L : T(S^1 \times \mathbb{R}^1) \to \mathbb{R}^1$ for the particle on the cylinder defined on the tangent bundle of the configuration manifold?
(c) Determine the Euler–Lagrange equations for the particle on the cylinder.
(d) Determine Hamilton's equations for the particle on the cylinder.
(e) What are conserved quantities of the dynamical flow?
(f) What are the equilibrium solutions of the dynamical flow?
(g) Select a specific equilibrium solution and determine the linearized dynamics that approximate the Lagrangian flow in a neighborhood of that equilibrium.

9.24. Consider the dynamics of an ideal massless, rigid link with length L and concentrated masses m, viewed as ideal particles, located at each end of the link. The link is constrained so that each end of the link moves on the surface of a right cylinder in $\mathbb{R}^3$ under the influence of uniform, constant gravity. Assume the axis of the cylinder is in the horizontal plane along the positive first axis of an inertial Euclidean frame. The radius of the cylinder is $r > 0$. Let $q_i \in S^1$ denote the direction of the projection of the position vector of the i-th particle onto the circular cross section of the cylinder; let $x_i \in \mathbb{R}^1$ denote the projection of the position vector of the i-th particle onto the first axis of the Euclidean frame (the axis of the cylinder). The configuration vector is $(q, x) = (q_1, q_2, x_1, x_2) \in M \subset (S^1 \times \mathbb{R}^1)^2$, where M is the constraint manifold that includes the link length constraint.

(a) What is the position vector of each particle, expressed in the inertial Euclidean frame, in terms of the configuration?
(b) Describe the constraint manifold M as an embedded manifold in $(S^1 \times \mathbb{R}^1)^2$ corresponding to the holonomic constraint imposed by the link length.

(c) Determine the augmented Lagrangian function $L^a : \mathsf{T}(M \times \mathbb{R}^1) \to \mathbb{R}^1$ defined on the tangent bundle of the constraint manifold.
(d) What are the Euler–Lagrange equations for the rigid link constrained to the cylinder?
(e) Determine the augmented Hamiltonian function $H^a : \mathsf{T}^*(M \times \mathbb{R}^1) \to \mathbb{R}^1$ defined on the cotangent of the configuration manifold.
(f) What are Hamilton's equations for the rigid link constrained to the cylinder?
(g) What are conserved quantities of the dynamical flow?
(h) What are the equilibrium solutions of the dynamical flow?

9.25. A frictionless pulley can rotate about an inertially fixed horizontal axis. A massless, inextensible, flexible cable is partially wound around the pulley; one end of the cable is fixed to the pulley while the other end of the cable is attached to a mass element, viewed as an ideal particle. The cable is free to rotate as a spherical pendulum with respect to the point where the cable makes contact with the pulley; the cable is assumed to remain straight except at the point of contact. The mass element attached to the end of the cable acts under uniform, constant gravity. The pulley has scalar moment of inertia J about its fixed axis of rotation, the pulley radius is $r > 0$, and the mass of the pendulum element is m. Let $x \in \mathbb{R}^1$ be the distance of the mass element from the instantaneous point of contact of the cable with the pulley; this is the instantaneous length of the spherical pendulum. Let $q \in \mathsf{S}^2$ denote the attitude vector of the spherical pendulum in the inertial frame. Thus, the configuration is $(q, x) \in \mathsf{S}^2 \times (0, \infty)$ and the configuration manifold is $\mathsf{S}^2 \times (0, \infty)$. The Lagrangian function is

$$L(q, x, \dot{q}, \dot{x}) = \frac{1}{2}\frac{J}{r^2}(\dot{x})^2 + \frac{1}{2}m\|\dot{x}q + x\dot{q}\|^2 - mgx.$$

(a) What are the Euler–Lagrange equations for the pulley and spherical pendulum?
(b) Determine the Hamiltonian function $H : \mathsf{T}^*(\mathsf{S}^2 \times (0, \infty)) \to \mathbb{R}^1$ defined on the cotangent bundle of the configuration manifold.
(c) What are Hamilton's equations for the pulley and spherical pendulum?
(d) What are conserved quantities for the dynamical flow?

9.26. A rigid, massless cable is stretched over two frictionless rollers. Each end of the cable is attached to a mass element, viewed as an ideal particle of mass m. Assume the cable segment between the two rollers is always horizontal, and the cable and mass elements hanging off of each roller can rotate in three dimensions under uniform gravity as a spherical pendulum. The inertia of the rollers is ignored; the total length of the cable is L and the horizontal distance between the rollers is $\frac{L}{2}$. Let $x \in \mathbb{R}^1$ denote the hanging length of the first element from the first roller so that $\frac{L}{2} - x \in \mathbb{R}^1$ denotes the hanging length of the second mass element from the second roller. Assume that $0 < x < \frac{L}{2}$. An inertial frame is selected so that the third axis is vertical. The attitudes of the two spherical pendulums, with variable lengths, are $q_1 \in S^2$ and $q_2 \in S^2$. Thus, the configuration vector is $(q, x) = (q_1, q_2, x) \in (S^2)^2 \times (0, \frac{L}{2})$ and the configuration manifold is $(S^2)^2 \times (0, \frac{L}{2})$. The Lagrangian function is

$$L(q, x, \dot{q}, \dot{x}) = \frac{1}{2} m \{ (\dot{x})^2 + (x)^2 \| \dot{q}_1 \|^2 \}$$

$$+ \frac{1}{2} m \left\{ (\dot{x})^2 + \left(\frac{L}{2} - x \right)^2 \| \dot{q}_2 \|^2 \right\} - 2mgx.$$

(a) What are the Euler–Lagrange equations for the two mass elements connected by a cable?
(b) Determine the Hamiltonian function $H : T^*((S^2)^2 \times (0, \frac{L}{2})) \to \mathbb{R}^1$ on the cotangent bundle of the configuration manifold.
(c) What are Hamilton's equations for the two mass elements connected by a cable?
(d) What are conserved quantities for the dynamical flow?
(e) What are the equilibrium solutions of the dynamical flow?

9.27. The frictionless pivots of two identical spherical pendulums are constrained to translate along a common horizontal, frictionless rail. The massless pivots are separated by a linear elastic spring, with stiffness constant κ, which is not stretched when the pivots are a distance D apart. The concentrated mass of each spherical pendulum is m, located at distance L from its pivot. Constant, uniform gravity acts on the two pendulums. A three-dimensional inertial frame is selected so that the horizontal rail lies along its first axis, while the third axis is vertical. Ignore any collision of the two spherical pendulums. Let $x_1, x_2 \in \mathbb{R}^1$ denote the position vectors of the two pivots along the first axis of the inertial frame; assume that $x_2 > x_1$. The attitude vector of the first spherical pendulum is denoted by $q_1 \in S^2$ and the attitude vector of the second spherical pendulum is denoted by $q_2 \in S^2$. Thus, the configuration vector is $(q, x) = (q_1, q_2, x_1, x_2) \in (S^2)^2 \times (\mathbb{R}^1)^2$ and the configuration manifold is $(S^2)^2 \times (\mathbb{R}^1)^2$. The Lagrangian function is

$$L(q, x, \dot{q}, \dot{x}) = \frac{1}{2} m \| \dot{x}_1 e_1 + L \dot{q}_1 \|^2$$

$$+ \frac{1}{2} m \| \dot{x}_2 e_1 + L \dot{q}_2 \|^2 + \kappa (x_2 - x_1 - D)^2.$$

(a) What are the Euler—Lagrange equations for the two spherical pendulums whose pivots are connected by an elastic spring?
(b) Determine the Hamiltonian function $H : T^*((S^2)^2 \times (\mathbb{R}^1)^2) \to \mathbb{R}^1$ on the cotangent bundle of the configuration manifold.
(c) What are Hamilton's equations for the two spherical pendulums whose pivots are connected by an elastic spring?

(d) What are conserved quantities for the dynamical flow?
(e) What are the equilibrium solutions of the dynamical flow?
(f) Select a specific equilibrium solution and determine the linearized dynamics that approximate the Lagrangian flow in a neighborhood of that equilibrium.

9.28. This is an extension of a problem at the end of Chapter 2. A knife-edge can slide on a horizontal plane without friction; the knife-edge is assumed to have a single point of contact with the plane. The motion of the point of contact of the knife-edge is constrained so that its velocity vector is always in the direction of the axis of the knife-edge. This constraint on the direction of the velocity vector is an example of a nonholonomic or non-integrable constraint. The motion of the knife-edge is controlled by an axial force $f \in \mathbb{R}^1$ and a rotational torque $\tau \in \mathbb{R}^1$ about the vertical. A two-dimensional Euclidean frame is introduced for the horizontal plane, so that $x \in \mathbb{R}^2$ denotes the position vector of the contact point of the knife-edge; let $q \in \mathsf{S}^1$ denote the direction vector of the knife-edge in the horizontal plane. The configuration vector is $(x, q) \in \mathbb{R}^2 \times \mathsf{S}^1$ so that $\mathbb{R}^2 \times \mathsf{S}^1$ is the configuration manifold. We also introduce the scalar speed V of the knife-edge along its axis and the scalar rotation rate ω of the knife-edge about the vertical. The knife-edge has scalar mass M and scalar rotational inertia J.

(a) Describe the nonholonomic constraint by expressing $\dot{x}$ in terms of V and q.
(b) Describe the rotational kinematics of the knife-edge by expressing $\dot{q}$ in terms of ω and q.
(c) Develop the equations for the dynamics of the knife-edge. Assume the Lagrangian function $L : \mathsf{T}(\mathbb{R}^2 \times \mathsf{S}^1) \to \mathbb{R}^1$ is given by the sum of the translational kinetic energy and the rotational kinetic energy

$$ L(x, q, \dot{x}, \omega) = \frac{1}{2} M \left\| \dot{x} \right\|^2 + \frac{1}{2} J \left\| \omega \right\|^2 . $$

Develop the equations of motion for the knife-edge dynamics using the d'Alembert principle. Show that the work done by the axial force f and the torque τ for a variation in the configuration is given by $f\delta x + \tau \delta q$; show that the work done by the contact force of the horizontal plane on the knife-edge is zero. Express your equations of motion in terms of (x, q, V, ω), including the nonholonomic constraint equation.
(d) Suppose that the axial force $f = 0$ and the rotational torque $\tau = 0$. Describe the motion of the point of contact of the knife-edge in the horizontal plane as it depends on the initial conditions.

Chapter 10
Deformable Multi-Body Systems

The developments introduced in the prior chapters can be applied to deformable bodies, that is bodies that deform or change their shape. A geometric formulation of Lagrangian dynamics and Hamiltonian dynamics of deformable multi-body systems is provided, based on the assumption of a finite-dimensional configuration manifold.

This chapter explores, in detail, the dynamics of several examples of deformable multi-body systems. In each case, the physical multi-body is described and viewed from the perspective of geometric mechanics; the configuration manifold is identified. Euler–Lagrange equations and Hamilton's equations are obtained for each example. Conserved quantities are identified for the dynamical flows. Where appropriate, equilibrium solutions are determined. Linear vector fields are obtained that approximate the dynamical flow in a neighborhood of an equilibrium solution.

10.1 Infinite-Dimensional Multi-Body Systems

These results can be extended to infinite-dimensional deformable multi-body systems. Some publications that adopt a geometric perspective for dynamical systems on an infinite-dimensional configuration manifold include [52, 56, 61, 88].

It is worth noting that the finite-dimensional multi-body systems that are studied in this chapter can often be viewed as approximations of infinite-dimensional multi-body systems. For example, an inextensible string with a distributed mass can be approximated by mass lumping so that the mass of the string is concentrated at a set of discrete points along the string. To further simplify, we can approximate the string as being straight and inextensible between two mass lumped points. This leads to a finite-dimensional

© Springer International Publishing AG 2018
T. Lee et al., *Global Formulations of Lagrangian and Hamiltonian Dynamics on Manifolds*, Interaction of Mechanics and Mathematics, DOI 10.1007/978-3-319-56953-6_10

model of a string as a chain of spherical pendula, which we discuss in the next section.

More generally, given an infinite-dimensional variational problem, we can construct a finite-dimensional approximation by projecting the problem onto finite-dimensional subspaces of the infinite-dimensional configuration space. In the setting of numerical solutions of partial differential equations, this is referred to as a semi-discretization, or the method of lines, as the infinite-dimensional degrees of freedom in the spatial directions are replaced by a finite-dimensional approximation.

10.2 Dynamics of a Chain Pendulum

A chain pendulum is a connection of n rigid links, that are serially connected by two degree of freedom spherical joints. We assume that each link of the chain pendulum is a rigid link with mass concentrated at the outboard end of the link. One end of the chain pendulum is connected to a spherical joint that is supported by a fixed base. A constant gravity potential acts on each link of the chain pendulum. A schematic of a chain pendulum is shown in Figure 10.1.

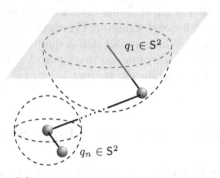

Fig. 10.1 A chain pendulum

We demonstrate that globally valid Euler–Lagrange equations can be developed for the chain pendulum, and they can be expressed in a compact form. The results provide an intrinsic and unified framework to study the dynamics of a chain pendulum, that is applicable for an arbitrary number of links, and globally valid for any configuration of the links. This problem, in a more general formulation, was first studied in [59].

The mass of the i-th link is denoted by m_i and the link length is denoted by L_i. For simplicity, assume that the mass of each link is concentrated at the outboard end of the link.

An inertial frame is chosen such that the first two axes are horizontal and the third axis is vertical. The origin of the inertial frame is located at the fixed spherical joint. The vector $q_i \in \mathsf{S}^2$ represents the attitude of the i-th link in the inertial frame, $i = 1, \ldots, n$. Thus, the configuration of the chain pendulum is the n-tuple of link attitudes $q = (q_1, \ldots, q_n) \in (\mathsf{S}^2)^n$, so that the configuration manifold is $(\mathsf{S}^2)^n$. The chain pendulum has $2n$ degrees of freedom. Collisions of the mass elements and links of the chain pendulum are assumed not to occur.

10.2.1 Euler–Lagrange Equations

Let $x_i \in \mathbb{R}^3$ be the position of the outboard end of the i-th link in the inertial frame. It can be written as

$$x_i = \sum_{j=1}^{i} L_j q_j.$$

The total kinetic energy is composed of the kinetic energy of each mass:

$$T(q, \dot{q}) = \frac{1}{2} \sum_{i=1}^{n} m_i \left\| \sum_{j=1}^{i} L_j \dot{q}_j \right\|^2.$$

This can be rewritten as

$$T(q, \dot{q}) = \frac{1}{2} \sum_{i,j=1}^{n} M_{ij} L_i L_j \dot{q}_i^T \dot{q}_j,$$

where the real inertia constants M_{ij} are given by

$$M_{ij} = \sum_{k=\max\{i,j\}}^{n} m_k, \quad i, j = 1, \ldots, n.$$

The potential energy consists of the gravitational potential energy of all the mass elements. The potential energy can be written as

$$U(q) = \sum_{i=1}^{n} m_i g e_3^T x_i = \sum_{i=1}^{n} \sum_{j=i}^{n} m_j g L_i e_3^T q_i.$$

The Lagrangian function $L : \mathsf{T}(\mathsf{S}^2)^n \to \mathbb{R}^1$ of the chain pendulum is

$$L(q, \dot{q}) = \frac{1}{2} \sum_{i=1}^{n} \sum_{j=1}^{n} M_{ij} L_i L_j \dot{q}_i^T \dot{q}_j - \sum_{i=1}^{n} \sum_{j=i}^{n} m_j g L_i e_3^T q_i.$$

The rotational kinematics equation for the attitude vector of the i-th link can be expressed in terms of the angular velocity vector of that link by

$$\dot{q}_i = S(\omega_i)q_i, \quad i = 1, \dots, n, \tag{10.1}$$

where $\omega_i \in \mathsf{T}_{q_i}\mathsf{S}^2$ is the angular velocity vector of the i-th link, satisfying $\omega_i^T q_i = 0$, for $i = 1, \dots, n$. We use the notation $\omega = (\omega_1, \dots, \omega_n) \in \mathsf{T}_q(\mathsf{S}^2)^n$. Thus, the modified Lagrangian function $\tilde{L} : \mathsf{T}(\mathsf{S}^2)^n \to \mathbb{R}^1$ of a chain pendulum can be expressed in terms of the angular velocity vector as

$$\tilde{L}(q, \omega) = \frac{1}{2}\sum_{i=1}^{n}\sum_{j=1}^{n} \omega_i^T S^T(q_i) M_{ij} L_i L_j S(q_j)\omega_j - \sum_{i=1}^{n}\sum_{j=i}^{n} m_j g L_i e_3^T q_i.$$

This can be written as

$$\tilde{L}(q, \omega) = \frac{1}{2}\sum_{i=1}^{n} \omega_i^T M_{ii} L_i^2 \omega_i + \frac{1}{2}\sum_{i=1}^{n}\sum_{\substack{j=1 \\ j \neq i}}^{n} \omega_i^T S^T(q_i) M_{ij} L_i L_j S(q_j)\omega_j$$

$$- \sum_{i=1}^{n}\sum_{j=i}^{n} m_j g L_i e_3^T q_i.$$

The infinitesimal variation of the action integral is given by

$$\delta \mathfrak{G} = \int_{t_0}^{t_f} \sum_{i=1}^{n} \left\{ \frac{\partial \tilde{L}(q, \omega)}{\partial \omega_i} \cdot \delta\omega_i + \frac{\partial \tilde{L}(q, \omega)}{\partial q_i} \cdot \delta q_i \right\} dt,$$

and the infinitesimal variations are given by

$$\delta q_i = S(\gamma_i)q_i, \qquad\qquad\qquad i = 1, \dots, n,$$
$$\delta\omega_i = -S(\omega_i)\gamma_i + (I_{3\times 3} - q_i q_i^T)\dot{\gamma}_i, \quad i = 1, \dots, n,$$

for differentiable curves $\gamma_i : [t_0, t_f] \to \mathbb{R}^3$ satisfying $\gamma_i(t_0) = \gamma_i(t_f) = 0$, for $i = 1, \dots, n$.

Following the procedure described in Chapter 5, we substitute the infinitesimal variations into the expression for the infinitesimal variation of the action integral, integrate by parts and carry out a simplification using various matrix identities. Finally, we use Hamilton's principle and the fundamental lemma of the calculus of variations, as in Appendix A, to obtain the Euler–Lagrange equations for the chain pendulum:

$$M_{ii}L_i^2\dot{\omega}_i + \sum_{\substack{j=1 \\ j\neq i}}^{n} M_{ij}L_iL_j S^T(q_i)S(q_j)\dot{\omega}_j - \sum_{\substack{j=1 \\ j\neq i}}^{n} M_{ij}L_iL_j \|\omega_j\|^2 S(q_i)q_j$$

$$-\sum_{j=i}^{n} m_j g L_i S(e_3) q_i = 0, \quad i = 1, \ldots, n,$$

or equivalently in matrix-vector form

$$
\begin{bmatrix}
M_{11} L_1^2 I_3 & M_{12} L_1 L_2 S^T(q_1) S(q_2) & \cdots & M_{1n} L_1 L_n S^T(q_1) S(q_n) \\
M_{21} L_2 L_1 S^T(q_2) S(q_1) & M_{22} L_2^2 I_3 & \cdots & M_{2n} L_2 L_n S^T(q_2) S(q_n) \\
\vdots & \vdots & \ddots & \vdots \\
M_{n1} L_n L_1 S^T(q_n) S(q_1) & M_{n2} L_n L_2 S^T(q_n) S(q_2) & \cdots & M_{nn} L_n^2 I_3
\end{bmatrix}
$$

$$
\begin{bmatrix} \dot{\omega}_1 \\ \dot{\omega}_2 \\ \vdots \\ \dot{\omega}_n \end{bmatrix}
-
\begin{bmatrix}
\sum_{j=2}^{n} M_{1j} L_1 L_j \|\omega_j\|^2 S(q_1) q_j \\
\sum_{j=1, j \neq 2}^{n} M_{2j} L_2 L_j \|\omega_j\|^2 S(q_2) q_j \\
\vdots \\
\sum_{j=1}^{n-1} M_{nj} L_n L_j \|\omega_j\|^2 S(q_n) q_j
\end{bmatrix}
-
\begin{bmatrix}
\sum_{j=1}^{n} m_j g L_1 S(e_3) q_1 \\
\sum_{j=2}^{h} m_j g L_2 S(e_3) q_2 \\
\vdots \\
m_n g l_n S(e_3) q_n
\end{bmatrix}
$$

$$
=
\begin{bmatrix} 0 \\ 0 \\ \vdots \\ 0 \end{bmatrix}.
\tag{10.2}
$$

These Euler–Lagrange equations (10.2) and the rotational kinematics (10.1) describe the Lagrangian flow of the chain pendulum dynamics in terms of the evolution of $(q, \dot{q}) \in \mathsf{T}(\mathsf{S}^2)^n$ on the tangent bundle of $(\mathsf{S}^2)^n$.

10.2.2 Hamilton's Equations

Hamilton's equations for the chain pendulum dynamics as they evolve on the cotangent bundle $\mathsf{T}^*(\mathsf{S}^1)^2$ can be obtained by considering the momentum $\pi = (\pi_1, \ldots, \pi_n) \in \mathsf{T}_q^*(\mathsf{S}^2)^n$ that is conjugate to the angular velocity vector $\omega = (\omega_1, \ldots, \omega_n) \in \mathsf{T}_q(\mathsf{S}^2)^n$. This is defined by the Legendre transformation, which is given by

$$\pi_i = (I_{3\times 3} - q_i q_i^T) \frac{\partial \tilde{L}(q, \omega)}{\partial \omega_i}, \quad i = 1, \ldots, n,$$

and which can be written in matrix form as

$$
\begin{bmatrix} \pi_1 \\ \vdots \\ \pi_n \end{bmatrix}
=
\begin{bmatrix}
M_{11} L_1^2 I_3 & \cdots & M_{1n} L_1 L_n S^T(q_1) S(q_n) \\
\vdots & \ddots & \vdots \\
M_{n1} L_n L_1 S^T(q_n) S(q_1) & \cdots & M_{nn} L_n^2 I_3
\end{bmatrix}
\begin{bmatrix} \omega_1 \\ \vdots \\ \omega_n \end{bmatrix}.
$$

Thus, we obtain

$$
\begin{bmatrix} \omega_1 \\ \vdots \\ \omega_n \end{bmatrix} = \begin{bmatrix} M_{11}^I(q) & \cdots & M_{1n}^I(q) \\ \vdots & \ddots & \vdots \\ M_{n1}^I(q) & \cdots & M_{nn}^I(q) \end{bmatrix} \begin{bmatrix} \pi_1 \\ \vdots \\ \pi_n \end{bmatrix}, \tag{10.3}
$$

where

$$
\begin{bmatrix} M_{11}^I(q) & \cdots & M_{1n}^I(q) \\ \vdots & \ddots & \vdots \\ M_{n1}^I(q) & \cdots & M_{n1}^I(q) \end{bmatrix} = \begin{bmatrix} M_{11}L_1^2 I_3 & \cdots & M_{1n}L_1 L_n S^T(q_1)S(q_n) \\ \vdots & \ddots & \vdots \\ M_{n1}L_n L_1 S^T(q_n)S(q_1) & \cdots & M_{nn}L_n^2 I_3 \end{bmatrix}^{-1}.
$$

Here, the functions $M_{ij}^I(q)$ denote the 3×3 matrices that partition the inverse matrix above. The modified Hamiltonian function $\tilde{H} : \mathsf{T}^*(\mathsf{S}^2)^n \to \mathbb{R}^1$ is

$$
\tilde{H}(q, \pi) = \frac{1}{2} \begin{bmatrix} \pi_1 \\ \vdots \\ \pi_n \end{bmatrix}^T \begin{bmatrix} M_{11}^I(q) & \cdots & M_{1n}^I(q) \\ \vdots & \ddots & \vdots \\ M_{n1}^I(q) & \cdots & M_{nn}^I(q) \end{bmatrix} \begin{bmatrix} \pi_1 \\ \vdots \\ \pi_n \end{bmatrix} + \sum_{i=1}^n \sum_{j=i}^n m_j g L_i e_3^T q_i.
$$

Thus, Hamilton's equations can be obtained from (5.28) and (5.29) and written as

$$
\dot{q}_i = -S(q_i) \sum_{j=1}^n M_{ij}^I \pi_j, \qquad\qquad i = 1, \ldots, n, \tag{10.4}
$$

$$
\dot{\pi}_i = -S(q_i) \left\{ \frac{1}{2} \frac{\partial}{\partial q_i} \sum_{j,k=1}^n \pi_j^T M_{jk}^I(q) \pi_k \right\}
$$

$$
+ \sum_{j=1}^n M_{ij}^I \pi_j \times \pi_i - S(q_i) \sum_{j=i}^n m_j g L_i e_3, \quad i = 1, \ldots, n. \tag{10.5}
$$

The nongravitational terms on the right-hand side of (10.5) are quadratic in the angular momenta. Hamilton's equations (10.4) and (10.5) describe the Hamiltonian flow of the chain pendulum dynamics in terms of the evolution of $(q, \pi) \in \mathsf{T}^*(\mathsf{S}^2)^n$ on the cotangent bundle of $(\mathsf{S}^2)^n$.

10.2.3 Comments

The Euler–Lagrange equations and Hamilton's equations for the chain pendulum dynamics are valid for arbitrary deformations of the chain so long as there are no collisions. Part or all of the chain pendulum may be above the spherical joint support point while part or all of the chain pendulum may be below the spherical joint support point.

If there is a single link, $n = 1$, the spherical pendulum is obtained; if there are two serial links, $n = 2$, the double spherical pendulum is obtained. The resulting dynamics of the chain pendulum can be complicated due to the coupling and the energy transfer between the links, and even the double spherical pendulum can exhibit interesting and nontrivial dynamics as a consequence of the coupling effects.

10.2.4 Conservation Properties

The Hamiltonian of the chain pendulum

$$H = \frac{1}{2} \sum_{i=1}^{n} \sum_{j=1}^{n} M_{ij} L_i L_j \dot{q}_i^T \dot{q}_j + \sum_{i=1}^{n} \sum_{j=i}^{n} m_j g L_i e_3^T q_i,$$

which coincides with the total energy E in this case, is constant along each solution of the Lagrangian flow.

10.2.5 Equilibrium Properties

We now determine the equilibrium solutions of the chain pendulum. These arise when the velocity vector is zero, and the equilibrium configuration vector in $(S^2)^n$ satisfies

$$S(e_3)q_i = 0, \quad i = 1, \ldots, n,$$

which implies that the moments due to gravity vanishes. Consequently, there are 2^n equilibrium configurations that correspond to all the n link configurations being aligned with either e_3 or $-e_3$.

The equilibrium configuration where all the n links are aligned with the $-e_3$ gravity direction is referred to as the hanging equilibrium. All other equilibria correspond to at least one link oriented upwards, opposite to the direction of gravity. Two interesting equilibrium solutions correspond to the case that all adjacent links are counter-aligned, that is $q_i^T q_{i+1} = -1$. These two equilibrium solutions, one with the first link pointing in the direction of e_3 and the other with the first link pointing in the direction of $-e_3$, are referred to as folded equilibria since the chain pendulum is completely folded in each case.

Consider the hanging equilibrium $(-e_3, -e_3, \ldots, -e_3, 0, 0, \ldots, 0) \in \mathsf{T}(S^2)^n$. The linearized differential equations at this hanging equilibrium can be shown to be

$$M_{ii} L_i^2 \ddot{\xi}_i + \sum_{\substack{j=1 \\ j \neq i}}^{n} M_{ij} L_i L_j \ddot{\xi}_j + \sum_{j=i}^{n} m_j g L_i \xi_i = 0, \quad i = 1, \ldots, n,$$

defined on the $4n$-dimensional tangent space of $\mathsf{T}(\mathsf{S}^2)^n$ at the equilibrium $(-e_3, -e_3, \ldots, -e_3, 0, 0, \ldots, 0) \in \mathsf{T}(\mathsf{S}^2)^n$. These linearized differential equations provide an approximation to the Lagrangian flow of the chain pendulum in a neighborhood of the hanging equilibrium. All of the eigenvalues can be shown to be purely imaginary. The hanging equilibrium can be shown to be stable since the total energy has a strict local minimum at the hanging equilibrium with compact level sets in a neighborhood and the derivative of the total energy is identically zero.

Now consider the equilibrium $(-e_3, e_3, \ldots, e_3, 0, 0, \ldots, 0) \in \mathsf{T}(\mathsf{S}^2)^n$ where the first link is in the direction of gravity while the direction of all other links are vertical. The linearized differential equations at this equilibrium can be shown to be

$$M_{11}L_1^2\ddot{\xi}_1 + \sum_{j=2}^{n} M_{1j}L_1L_j\ddot{\xi}_j + \sum_{j=1}^{n} m_j g L_1 \xi_1 = 0,$$

$$M_{ii}L_i^2\ddot{\xi}_i + \sum_{\substack{j=1 \\ j \neq i}}^{n} M_{ij}L_iL_j\ddot{\xi}_j - \sum_{j=i}^{n} m_j g L_i \xi_i = 0, \quad i = 2, \ldots, n,$$

defined on the $4n$-dimensional tangent space of $\mathsf{T}(\mathsf{S}^2)^n$ at the equilibrium $(-e_3, e_3, \ldots, e_3, 0, 0, \ldots, 0) \in \mathsf{T}(\mathsf{S}^2)^n$. These linearized differential equations provide an approximation to the Lagrangian flow of the chain pendulum in a neighborhood of this equilibrium. It can be shown that these linearized equations have a positive eigenvalue, so that this equilibrium is unstable. Similarly, it can be shown that all of the equilibrium solutions, except for the hanging equilibrium, have at least one positive eigenvalue and are therefore unstable.

10.3 Dynamics of a Chain Pendulum on a Cart

Consider the dynamics of a chain pendulum on a cart: a serial connection of n rigid links, connected by two degree of freedom spherical joints, attached to a cart that moves on a horizontal plane. We assume that each link of the chain pendulum is a thin rigid rod with mass concentrated at the outboard end of the link. One end of the chain pendulum is attached to a spherical joint that is located at the center of mass of the cart. The cart can translate, without friction, in the horizontal plane; it is assumed that the cart does not rotate in the plane. Constant uniform gravity acts on each link of the chain pendulum. A schematic of a chain pendulum on a cart is shown in Figure 10.2.

Globally valid Euler–Lagrange equations are developed for the chain pendulum on a cart, and they are expressed in a compact form. The results provide an intrinsic, unified framework to study the dynamics of a chain pen-

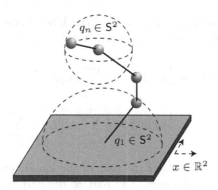

Fig. 10.2 Chain pendulum on a cart

dulum on a cart, that is applicable for an arbitrary number of links, and globally valid for any configuration of the links. The dynamics of a chain pendulum on a cart have been studied in [59].

A serial connection of n rigid links, connected by spherical joints, is attached to the cart, where the mass of the i-th link is denoted by m_i and the link length is denoted by L_i for $i = 1, \ldots, n$. The mass of each link is concentrated at the outboard end of the link.

An inertial frame is chosen with the first two axes horizontal and the third axis vertical. The cart of mass m can translate in the horizontal plane defined by the first two axes of the inertial frame. The position of the center of mass in this horizontal plane is denoted by $x \in \mathbb{R}^2$. The attitude vector of each link in the inertial frame is given by $q_i \in \mathsf{S}^2$ for $i = 1, \ldots, n$. The configuration of the chain pendulum on a cart is $(x, q) = (x, q_1, \ldots, q_n) \in \mathbb{R}^2 \times (\mathsf{S}^2)^n$, and the configuration manifold is $\mathbb{R}^2 \times (\mathsf{S}^2)^n$. The chain pendulum on a cart has $2n + 2$ degrees of freedom. Collisions of the mass elements, links, and cart of the chain pendulum on a cart are assumed not to occur.

10.3.1 Euler–Lagrange Equations

The 3×2 matrix

$$C = \begin{bmatrix} 1 & 0 \\ 0 & 1 \\ 0 & 0 \end{bmatrix},$$

defines an embedding of $\mathbb{R}^2$ into $\mathbb{R}^3$. The location of the cart is given by $Cx \in \mathbb{R}^3$ in the inertial frame. Let $x_i \in \mathbb{R}^3$ be the position vector of the outboard end of the i-th link in the inertial frame. It can be written as

$$x_i = Cx + \sum_{j=1}^{i} L_j q_j, \quad i = 1, \ldots, n.$$

The total kinetic energy is composed of the kinetic energy of the cart and the kinetic energy of the link masses; the total kinetic energy is

$$
T(x, q, \dot{x}, \dot{q}) = \frac{1}{2} m \|\dot{x}\|^2 + \frac{1}{2} \sum_{i=1}^{n} m_i \|\dot{x}_i\|^2
$$

$$
= \frac{1}{2} m \|\dot{x}\|^2 + \frac{1}{2} \sum_{i=1}^{n} m_i \left\| C\dot{x} + \sum_{j=1}^{i} L_j \dot{q}_j \right\|^2.
$$

For simplicity, we first consider the part of the kinetic energy that is dependent on the motion of the cart. The part of the cart kinetic energy that is dependent on $\dot{x}$ is given by

$$
\frac{1}{2} \left(m + \sum_{i=1}^{n} m_i \right) \|\dot{x}\|^2 + \dot{x}^T C^T \sum_{i=1}^{n} \sum_{j=i}^{n} m_j L_i \dot{q}_i,
$$

which can be written as

$$
\frac{1}{2} M_{00} \|\dot{x}\|^2 + \dot{x}^T \sum_{i=1}^{n} M_{0i} \dot{q}_i,
$$

where the inertia matrices $M_{00} \in \mathbb{R}^1$, $M_{0i} \in \mathbb{R}^{2 \times 3}$, and $M_{i0} \in \mathbb{R}^{3 \times 2}$ are given by

$$
M_{00} = m + \sum_{i=1}^{n} m_i, \quad M_{0i} = C^T \sum_{j=i}^{n} m_j, \quad M_{i0} = M_{0i}^T,
$$

for $i = 1, \ldots, n$. The part of the kinetic energy that is independent of $\dot{x}$ is given by

$$
\frac{1}{2} \sum_{i=1}^{n} m_i \left\| \sum_{j=1}^{i} L_j \dot{q}_j \right\|^2 = \frac{1}{2} \sum_{i=1}^{n} \sum_{j=1}^{n} M_{ij} L_i L_j \dot{q}_i^T \dot{q}_j,
$$

where the real-valued inertia constants M_{ij} are given by

$$
M_{ij} = \sum_{k=\max\{i,j\}}^{n} m_k,
$$

for $i, j = 1, \ldots, n$. Thus, the total kinetic energy is given by

$$
T(x, q, \dot{x}, \dot{q}) = \frac{1}{2} M_{00} \|\dot{x}\|^2 + \dot{x}^T \sum_{i=1}^{n} M_{0i} L_i \dot{q}_i + \frac{1}{2} \sum_{i=1}^{n} \sum_{j=1}^{n} M_{ij} L_i L_j \dot{q}_i^T \dot{q}_j.
$$

The potential energy consists of the gravitational potential energy of all the mass elements. The potential energy can be written as

$$U(x,q) = \sum_{i=1}^{n} m_i g e_3^T x_i = \sum_{i=1}^{n} \sum_{j=i}^{n} m_j g L_i e_3^T q_i.$$

The Lagrangian function $L : \mathsf{T}(\mathbb{R}^2 \times (\mathsf{S}^2)^n) \to \mathbb{R}^1$ of a chain pendulum on a cart is

$$L(x,q,\dot{x},\dot{q}) = \frac{1}{2} M_{00} \|\dot{x}\|^2 + \dot{x}^T \sum_{i=1}^{n} M_{0i} L_i \dot{q}_i$$

$$+ \frac{1}{2} \sum_{i=1}^{n} \sum_{j=1}^{n} M_{ij} L_i L_j \dot{q}_i^T \dot{q}_j - \sum_{i=1}^{n} \sum_{j=i}^{n} m_j g L_i e_3^T q_i.$$

Note that the Lagrangian does not depend on the position of the cart, which implies a translational symmetry in the dynamics.

The rotational kinematics for the attitude vector of the i-th link can be expressed in terms of the angular velocity vector of the i-th link as

$$\dot{q}_i = S(\omega_i) q_i, \quad i = 1, \ldots, n, \tag{10.6}$$

where $\omega_i \in \mathsf{T}_{q_i} \mathsf{S}^2$ is the angular velocity vector of the i-th link satisfying $\omega_i^T q_i = 0$. We use the notation $\omega = (\omega_1, \ldots, \omega_n) \in \mathsf{T}_q(\mathsf{S}^2)^n$. Thus, the modified Lagrangian function $\tilde{L} : \mathsf{T}(\mathbb{R}^2 \times (\mathsf{S}^2)^n) \to \mathbb{R}^1$ of the chain pendulum on a cart can be expressed in terms of the angular velocity vector as

$$\tilde{L}(x,q,\dot{x},\omega) = \frac{1}{2} M_{00} \|\dot{x}\|^2 - \dot{x}^T \sum_{i=1}^{n} M_{0i} L_i S(q_i) \omega_i$$

$$+ \frac{1}{2} \sum_{i=1}^{n} \sum_{j=1}^{n} \omega_i^T S^T(q_i) M_{ij} L_i L_j S(q_j) \omega_j - \sum_{i=1}^{n} \sum_{j=i}^{n} m_j g L_i e_3^T q_i.$$

This can be written as

$$\tilde{L}(x,q,\dot{x},\omega) = \frac{1}{2} M_{00} \|\dot{x}\|^2 - \dot{x}^T \sum_{i=1}^{n} M_{0i} L_i S(q_i) \omega_i + \frac{1}{2} \sum_{i=1}^{n} \omega_i^T M_{ii} L_i^2 \omega_i$$

$$+ \frac{1}{2} \sum_{i=1}^{n} \sum_{\substack{j=1 \\ j \neq i}}^{n} \omega_i^T S^T(q_i) M_{ij} L_i L_j S(q_j) \omega_j$$

$$- \sum_{i=1}^{n} \sum_{j=i}^{n} m_j g L_i e_3^T q_i.$$

The infinitesimal variation of the action integral is given by

$$\delta\mathfrak{G} = \int_{t_0}^{t_f} \sum_{i=1}^{n} \left\{ \frac{\partial \tilde{L}(x,q,\dot{x},\omega)}{\partial \dot{x}} \cdot \delta\dot{x} + \frac{\partial \tilde{L}(x,q,\dot{x},\omega)}{\partial \omega_i} \cdot \delta\omega_i \right.$$

$$\left. + \frac{\partial \tilde{L}(x,q,\dot{x},\omega)}{\partial x} \cdot \delta x + \frac{\partial \tilde{L}(x,q,\dot{x},\omega)}{\partial q_i} \cdot \delta q_i \right\} dt,$$

and the infinitesimal variations are given by

$$\delta q_i = S(\gamma_i)q_i, \qquad\qquad i = 1,\ldots,n,$$
$$\delta\omega_i = -S(\omega_i)\gamma_i + (I_{3\times3} - q_i q_i^T)\dot{\gamma}_i, \quad i = 1,\ldots,n,$$

with differentiable curves $\gamma_i : [t_0, t_f] \to \mathbb{R}^3$ satisfying $\gamma_i(t_0) = \gamma_i(t_f) = 0$, for $i = 1,\ldots,n$, and differentiable curves $\delta x : [t_0, t_f] \to \mathbb{R}^2$ satisfying $\delta x(t_0) = \delta x(t_f) = 0$.

Following the procedure described in Chapters 3 and 5, we substitute the infinitesimal variations into the expression for the infinitesimal variation of the action integral, integrate by parts and carry out a simplification using various matrix identities. Finally, we use Hamilton's principle and the fundamental lemma of the calculus of variations, as in Appendix A, to obtain the Euler–Lagrange equations for the chain pendulum:

$$M_{00}\ddot{x} + \sum_{i=1}^{n} M_{0i}L_i S^T(q_i)\omega_i - \sum_{i=1}^{n} M_{0i}L_i \|\omega_i\|^2 q_i = 0, \qquad (10.7)$$

$$M_{ii}L_i^2\dot{\omega}_i + \sum_{\substack{j=1 \\ j\neq i}}^{n} M_{ij}L_i L_j S^T(q_i)S(q_j)\dot{\omega}_j + M_{i0}L_i S(q_i)\ddot{x}$$

$$- \sum_{\substack{j=1 \\ j\neq i}}^{n} M_{ij}L_i L_j \|\omega_j\|^2 S(q_i)q_j + \sum_{j=i}^{n} m_j g L_i S(q_i)e_3 = 0, \quad i = 1,\ldots,n.$$

$$(10.8)$$

These Euler–Lagrange equations (10.7) and (10.8), and the rotational kinematics (10.6) describe the dynamical flow of the chain pendulum on a cart in terms of the evolution of $(x, q, \dot{x}, \dot{q}) \in \mathsf{T}(\mathbb{R}^2 \times (\mathsf{S}^2)^n)$ on the tangent bundle of $\mathbb{R}^2 \times (\mathsf{S}^2)^n$.

10.3.2 Hamilton's Equations

Hamilton's equations for the chain pendulum on a cart describe the dynamics on the cotangent bundle $\mathsf{T}^*(\mathbb{R}^2 \times (\mathsf{S}^2)^2)$, and can be obtained by considering the momentum $(p, \pi) = (p, \pi_1, \ldots, \pi_n) \in \mathsf{T}^*_{(x,q)}(\mathbb{R}^2 \times (\mathsf{S}^2)^n)$ that is conjugate to the velocity vector $(\dot{x}, \omega) = (\dot{x}, \omega_1, \ldots, \omega_n) \in \mathsf{T}_{(x,q)}(\mathbb{R}^2 \times (\mathsf{S}^2)^n)$. This is defined by the Legendre transformation

$$p = \frac{\partial \tilde{L}(x, q, \dot{x}, \omega)}{\partial \dot{x}},$$

$$\pi_i = (I_{3 \times 3} - q_i q_i^T) \frac{\partial \tilde{L}(x, q, \dot{x}, \omega)}{\partial \omega_i}, \quad i = 1, \ldots, n,$$

which can be written in matrix form as

$$
\begin{bmatrix} p \\ \pi_1 \\ \vdots \\ \pi_n \end{bmatrix}
$$

$$
= \begin{bmatrix}
M_{00} & -M_{01}L_1S(q_1) & \cdots & -M_{0n}L_nS(q_n) \\
-M_{10}L_1S^T(q_1) & M_{11}L_1^2 I_3 & \cdots & M_{1n}L_1L_nS^T(q_1)S(q_n) \\
\vdots & \vdots & \ddots & \vdots \\
-M_{n0}L_nS^T(q_n) & M_{n1}L_nL_1S^T(q_n)S(q_1) & \cdots & M_{nn}L_n^2 I_3
\end{bmatrix}
$$

$$
\begin{bmatrix} \dot{x} \\ \omega_1 \\ \vdots \\ \omega_n \end{bmatrix}.
$$

Thus, we obtain

$$
\begin{bmatrix} \dot{x} \\ \omega_1 \\ \vdots \\ \omega_n \end{bmatrix}
=
\begin{bmatrix}
M_{00}^I & M_{01}^I & \cdots & M_{0n}^I \\
M_{10}^I & M_{11}^I & \cdots & M_{1n}^I \\
\vdots & \vdots & \ddots & \vdots \\
M_{n0}^I & M_{n1}^I & \cdots & M_{nn}^I
\end{bmatrix}
\begin{bmatrix} p \\ \pi_1 \\ \vdots \\ \pi_n \end{bmatrix},
\tag{10.9}
$$

where

$$
\begin{bmatrix}
M_{00}^I & M_{01}^I & \cdots & M_{0n}^I \\
M_{10}^I & M_{11}^I & \cdots & M_{1n}^I \\
\vdots & \vdots & \ddots & \vdots \\
M_{n0}^I & M_{n1}^I & \cdots & M_{nn}^I
\end{bmatrix}
=
$$

$$
\begin{bmatrix}
M_{00} & -M_{01}L_1S(q_1) & \cdots & -M_{0n}L_nS(q_n) \\
-M_{10}L_1S^T(q_1) & M_{11}L_1^2 I_3 & \cdots & M_{1n}L_1L_nS^T(q_1)S(q_n) \\
\vdots & \vdots & \ddots & \vdots \\
-M_{n0}L_nS^T(q_n) & M_{n1}L_nL_1S^T(q_n)S(q_1) & \cdots & M_{nn}L_n^2 I_3
\end{bmatrix}^{-1}.
$$

Here, the functions M_{ij}^I denote the 3×3 matrices that partition the inverse matrix above. The modified Hamiltonian function $\tilde{H} : \mathsf{T}^*(\mathbb{R}^2 \times (\mathsf{S}^2)^n) \to \mathbb{R}^1$ can also be expressed as

$$\tilde{H}(x, q, p, \pi) = \frac{1}{2} \begin{bmatrix} p \\ \pi_1 \\ \vdots \\ \pi_n \end{bmatrix}^T \begin{bmatrix} M_{00}^I & M_{01}^I & \cdots & M_{0n}^I \\ M_{10}^I & M_{11}^I & \cdots & M_{1n}^I \\ \vdots & \vdots & \ddots & \vdots \\ M_{n0}^I & M_{n1}^I & \cdots & M_{nn}^I \end{bmatrix} \begin{bmatrix} p \\ \pi_1 \\ \vdots \\ \pi_n \end{bmatrix}$$
$$+ \sum_{i=1}^n \sum_{j=i}^n m_j g L_i e_3^T q_i.$$

Thus, Hamilton's equations can be expressed in the following form

$$\dot{x} = M_{00}^I p + \sum_{j=1}^n M_{0j}^I \pi_j, \tag{10.10}$$

$$\dot{q}_i = -S(q_i) \left\{ M_{i0}^I p + \sum_{j=1}^n M_{ij}^I \pi_j \right\}, \quad i = 1, \ldots, n, \tag{10.11}$$

$$\dot{p} = 0, \tag{10.12}$$

$$\dot{\pi}_i = -S(q_i) \left\{ \frac{1}{2} \frac{\partial}{\partial q_i} p^T M_{00}^I p + \frac{\partial}{\partial q_i} \sum_{j=1}^n p^T M_{0j}^I \pi_j + \frac{1}{2} \frac{\partial}{\partial q_i} \sum_{j,k=1}^n \pi_j^T M_{jk} \pi_j \right\}$$
$$+ \left\{ M_{i0}^I p + \sum_{j=1}^n M_{ij}^I \pi_j \right\} \times \pi_i - S(q_i) \sum_{j=1}^n m_j g L_i e_3, \quad i = 1, \ldots, n. \tag{10.13}$$

The nongravitational terms on the right-hand side of (10.13) are quadratic in the momenta. Hamilton's equations (10.10), (10.11), (10.12), and (10.13) describe the Hamiltonian flow of the chain pendulum on a cart in terms of the evolution of $(x, q, p, \pi) \in \mathsf{T}^*(\mathbb{R}^2 \times (\mathsf{S}^2)^n)$ on the cotangent bundle of $\mathbb{R}^2 \times (\mathsf{S}^2)^n$.

10.3.3 Comments

The Euler–Lagrange equations and Hamilton's equations given above for the chain pendulum on a cart are valid for arbitrary deformations of the chain with respect to the cart; part or all of the chain pendulum may be above the cart while part or all of the chain pendulum may be below the cart. Collisions of the chain pendulum with itself and of the chain pendulum with the cart are ignored.

If there is a single link, $n = 1$, the spherical pendulum on a cart is obtained; if there are two serial links, $n = 2$, the double spherical pendulum on a cart is obtained. The resulting dynamics of the chain pendulum on a cart for any number of links can be very complicated due to the coupling between the chain pendulum dynamics and the cart dynamics.

10.3.4 Conservation Properties

The Hamiltonian of the chain pendulum on a cart is given by

$$H = \frac{1}{2} M_{00} \|\dot{x}\|^2 - \dot{x}^T \sum_{i=1}^{n} M_{0i} L_i S(q_i) \omega_i$$

$$+ \frac{1}{2} \sum_{i=1}^{n} \sum_{j=1}^{n} \omega_i^T S^T(q_i) M_{ij} L_i L_j S(q_j) \omega_j + \sum_{i=1}^{n} \sum_{j=i}^{n} m_j g L_i e_3^T q_i,$$

which coincides with the total energy E in this case, and it is constant along each solution of the dynamical flow.

As seen from Hamilton's equations, the translational momentum in the inertial frame, given by

$$p = M_{00} \dot{x} + \sum_{i=1}^{n} M_{0i} L_i S^T(q_i) \omega_i$$

is also constant along each solution of the dynamical flow. As observed earlier, the Lagrangian is invariant under translations, and by Noether's theorem, this implies that the translational momentum is preserved along the flow.

10.3.5 Equilibrium Properties

The equilibrium solutions of the chain pendulum on a cart can be determined using (10.7) and (10.8). If the velocity vector is zero, the equilibrium configuration vector in $(\mathbb{R}^2 \times (S^2)^n)$ occurs for an arbitrary location of the cart and for each link of the chain pendulum aligned vertically, i.e.,

$$S(q_i)e_3 = 0, \quad i = 1, \dots, n,$$

which implies that the moments due to gravity vanish. For the typical equilibrium condition where $x = 0$, there are 2^n possible equilibrium link attitudes that correspond to the case where all n links are vertical: $q_i = \pm e_3$ for $i = 1, \dots, n$. The equilibrium for which all links are aligned with the gravity direction, $q_i = -e_3$ for $i = 1, \dots, n$, is referred to as the hanging equilibrium; the equilibrium for which all links are aligned opposite to the gravity direction $q_i = e_3$ for $i = 1, \ldots, n$, is referred to as the inverted equilibrium. All other equilibria correspond to at least one link oriented opposite to the direction of gravity.

Consider the hanging equilibrium $(0, -e_3, \ldots, -e_3, 0, 0, \ldots, 0) \in \mathsf{T}(\mathbb{R}^2 \times (\mathsf{S}^2)^n)$. The linearized differential equations at this hanging equilibrium can be shown to be

$$M_{00}\ddot{\xi}_0 + \sum_{i=1}^{n} M_{0i}L_i S C^T \ddot{\xi}_i = 0,$$

$$M_{i0}L_i C S^T \ddot{\xi}_0 + M_{ii}L_i^2 \ddot{\xi}_i + \sum_{\substack{j=1 \\ j \neq i}}^{n} M_{ij}L_i L_j \ddot{\xi}_j + \sum_{j=i}^{n} m_j g L_i \xi_i = 0, \quad i = 1, \ldots, n.$$

These linear differential equations are defined on the $(4n + 4)$-dimensional tangent space of the manifold $\mathsf{T}(\mathbb{R}^2 \times (\mathsf{S}^2)^n)$ at the hanging equilibrium $(0, -e_3, \ldots, -e_3, 0, 0, \ldots, 0) \in \mathsf{T}(\mathbb{R}^2 \times (\mathsf{S}^2)^n)$. They provide an approximation to the Lagrangian flow of the chain pendulum on a cart in a neighborhood of the hanging equilibrium. It can be shown that there are two zero eigenvalues and all other eigenvalues are purely imaginary. We cannot establish stability of the hanging equilibrium by Lyapunov techniques as the energy minimum is not isolated, since all hanging equilibria have the same energy independent of the cart position.

Now consider the equilibrium $(0, e_3, -e_3, \ldots, -e_3, 0, 0, 0, \ldots, 0) \in \mathsf{T}(\mathbb{R}^2 \times (\mathsf{S}^2)^n)$. The linearized differential equations at this equilibrium can be shown to be

$$M_{00}\ddot{\xi}_0 + \sum_{j=1}^{n} M_{0j}L_j S C^T \ddot{\xi}_j = 0,$$

$$M_{10}L_1 C S^T \ddot{\xi}_0 + M_{1i}L_i^2 \ddot{\xi}_1 + \sum_{j=2}^{n} M_{1j}L_1 L_j \ddot{\xi}_j + \sum_{j=1}^{n} m_j g L_1 \xi_1 = 0,$$

$$M_{i0}L_i C S^T \ddot{\xi}_0 + M_{ii}L_i^2 \ddot{\xi}_i + \sum_{\substack{j=1 \\ j \neq i}}^{n} M_{ij}L_i L_j \ddot{\xi}_j - \sum_{j=i}^{n} m_j g L_i \xi_i = 0, \quad i = 2, \ldots, n.$$

These linear differential equations are defined on the $(4n + 4)$-dimensional tangent space of the manifold $\mathsf{T}(\mathbb{R}^2 \times (\mathsf{S}^2)^n)$ at the equilibrium point $(0, e_3, -e_3, \ldots, -e_3, 0, 0, \ldots, 0) \in \mathsf{T}(\mathbb{R}^2 \times (\mathsf{S}^2)^n)$. They provide an approximation to the Lagrangian flow of the chain pendulum on a cart in a neighborhood of this equilibrium. It can be shown that there is a positive eigenvalue so that this equilibrium is unstable. It can also be shown that there is a positive eigenvalue for each set of linearized equations, linearized at any equilibrium other than the hanging equilibrium. Consequently, all equilibrium solutions, other than the hanging equilibrium, are unstable.

10.4 Dynamics of a Free-Free Chain

Consider the dynamics of a chain, viewed as a serial connection of n rigid links, that is in free fall under uniform, constant gravity in three dimensions. Each link of the chain is a thin rigid rod with mass concentrated at the out-board end of the link. The motion of each end of the chain is unconstrained.

We demonstrate that globally valid Euler–Lagrange equations and Hamilton's equations can be derived for the chain dynamics, and they can yield compact expressions that are more amenable to analysis. The results provide an intrinsic and unified framework to study the dynamics of a chain that is applicable for an arbitrary number of links and globally valid for any configuration of the links.

The chain is viewed as a serial connection of n rigid links, connected by $n-1$ spherical joints. The mass of the i-th link is denoted by m_i and the link length is denoted by L_i for $i = 1, \ldots, n$. For simplicity, we assume that the mass of each link is concentrated at the end of the link, with mass element m_0 located at the free end of the first link. The chain is referred to as a free-free chain since the motion of each end of the chain is not constrained. A schematic of a free-free chain is shown in Figure 10.3.

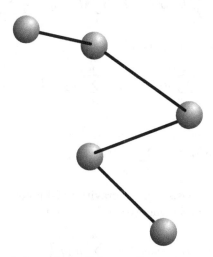

Fig. 10.3 A free-free chain under uniform gravity

An inertial frame is chosen such first two axes are horizontal and the third axis is vertical. The inertial position vector of the free end of the first link is $x \in \mathbb{R}^3$. The attitude vector of each link in the inertial frame is given by $q_i \in \mathsf{S}^2$ for $i = 1, \ldots, n$. The configuration of the chain is $(x, q) = (x, q_1, \ldots, q_n) \in \mathbb{R}^3 \times (\mathsf{S}^2)^n$, so that the configuration manifold of the chain is $\mathbb{R}^3 \times (\mathsf{S}^2)^n$. The free-free chain has $2n + 3$ degrees of freedom. Collisions of the mass elements and links of the chain are assumed not to occur.

10.4.1 Euler–Lagrange Equations

Let $x_i \in \mathbb{R}^3$ be the position of the i-th mass element in the inertial frame. It can be written as

$$x_i = x + \sum_{j=1}^{i} L_j q_j, \quad i = 1, \dots, n.$$

The total kinetic energy is the sum of the kinetic energies of the concentrated mass at the end of each link:

$$T(x, q, \dot{x}, \dot{q}) = \frac{1}{2} m_0 \|\dot{x}\|^2 + \frac{1}{2} \sum_{i=1}^{n} m_i \|\dot{x}_i\|^2$$

$$= \frac{1}{2} m_0 \|\dot{x}\|^2 + \frac{1}{2} \sum_{i=1}^{n} m_i \|\dot{x} + \sum_{j=1}^{i} L_j \dot{q}_j\|^2.$$

The kinetic energy can be written as

$$T(x, q, x, \dot{x}, \dot{q}) = \frac{1}{2} M_{00} \|\dot{x}\|^2 + \dot{x}^T \sum_{i=1}^{n} M_{0i} L_i \dot{q}_i + \frac{1}{2} \sum_{i,j=1}^{n} M_{ij} L_i L_j \dot{q}_i^T \dot{q}_j,$$

where the real inertia constants are

$$M_{00} = \sum_{i=0}^{n} m_i, \quad M_{0i} = \sum_{j=i}^{n} m_j, \quad i = 1, \dots, n,$$

and

$$M_{ij} = \sum_{k=\max\{i,j\}}^{n} m_k, \quad i, j = 1, \dots, n.$$

The potential energy consists of the gravitational potential energy of all links of the chain. The potential energy can be written as

$$U(x, q) = m_0 g e_3^T x + \sum_{i=1}^{n} m_i g e_3^T x_i$$

$$= M_{00} g e_3^T x + \sum_{i=1}^{n} \sum_{j=i}^{n} m_j g L_i e_3^T q_i.$$

The Lagrangian of the chain is

$$L(x, q, x, \dot{x}, \dot{q}) = \frac{1}{2} M_{00} \|\dot{x}\|^2 + \dot{x}^T \sum_{i=1}^{n} M_{0i} L_i \dot{q}_i + \frac{1}{2} \sum_{i=1}^{n} \sum_{i=1}^{n} M_{ij} L_i L_j \dot{q}_i^T \dot{q}_j$$

$$-M_{00}ge_3^T x - \sum_{i=1}^{n}\sum_{j=i}^{n} m_j g L_i e_3^T q_i.$$

The rotational kinematics for the attitude vector of the i-th link are expressed in terms of the angular velocity vector of the i-th link as

$$\dot{q}_i = S(\omega_i)q_i, \quad i = 1,\ldots,n, \tag{10.14}$$

where $\omega_i \in \mathsf{T}_{q_i}\mathsf{S}^2$ is the angular velocity vector of the i-th link satisfying $\omega_i^T q_i = 0$ for $i = 1,\ldots,n$. We use the notation $\omega = (\omega_1,\ldots,\omega_n) \in \mathsf{T}_q(\mathsf{S}^2)^n$. Thus, the modified Lagrangian function $\tilde{L} : \mathsf{T}(\mathbb{R}^3 \times (\mathsf{S}^2)^n) \to \mathbb{R}^1$ of the free-free chain can be expressed in terms the angular velocities as

$$\tilde{L}(x,q,\dot{x},\omega) = \frac{1}{2}M_{00}\|\dot{x}\|^2 - \dot{x}^T \sum_{i=1}^{n} M_{0i}L_i S(q_i)\omega_i$$

$$+ \frac{1}{2}\sum_{i=1}^{n}\sum_{j=1}^{n} \omega_i^T S^T(q_i) M_{ij}L_i L_j S(q_j)\omega_j$$

$$- M_{00}ge_3^T x - \sum_{i=1}^{n}\sum_{j=i}^{n} m_j g L_i e_3^T q_i.$$

This can be rewritten as

$$\tilde{L}(x,q,\dot{x},\omega) = \frac{1}{2}M_{00}\|\dot{x}\|^2 - \dot{x}^T \sum_{i=1}^{n} M_{0i}L_i S(q_i)\omega_i + \frac{1}{2}\sum_{i=1}^{n} \omega_i^T M_{ii} L_i^2 \omega_i$$

$$+ \frac{1}{2}\sum_{i=1}^{n}\sum_{\substack{j=1 \\ j\neq i}}^{n} \omega_i^T S^T(q_i) M_{ij}L_i L_j S(q_j)\omega_j$$

$$- M_{00}ge_3^T x - \sum_{i=1}^{n}\sum_{j=i}^{n} m_j g L_i e_3^T q_i.$$

The infinitesimal variation of the action integral is given by

$$\delta\mathfrak{G} = \int_{t_0}^{t_f} \sum_{i=1}^{n} \left\{ \frac{\partial \tilde{L}(x,q,\dot{x},\omega)}{\partial \dot{x}} \cdot \delta\dot{x} + \frac{\partial \tilde{L}(x,q,\dot{x},\omega)}{\partial \omega_i} \cdot \delta\omega_i \right.$$

$$\left. + \frac{\partial \tilde{L}(x,q,\dot{x},\omega)}{\partial x} \cdot \delta x + \frac{\partial \tilde{L}(x,q,\dot{x},\omega)}{\partial q_i} \cdot \delta q_i \right\} dt,$$

where

$$\delta q_i = S(\gamma_i)q_i, \qquad\qquad i = 1,\ldots,n,$$

$$\delta\omega_i = -S(\omega_i)\gamma_i + (I_{3\times3} - q_i q_i^T)\dot{\gamma}_i, \quad i = 1,\ldots,n,$$

for differentiable curves $\gamma_i : [t_0, t_f] \to \mathbb{R}^3$ satisfying $\gamma_i(t_0) = \gamma_i(t_f) = 0$, for $i = 1, \ldots, n$, and differentiable curves $\delta x : [t_0, t_f] \to \mathbb{R}^3$ satisfying $\delta x(t_0) = \delta x(t_f) = 0$.

Following the procedure described in Chapters 3 and 5, we substitute the infinitesimal variations into the expression for the infinitesimal variation of the action integral, integrate by parts and carry out a simplification using various matrix identities. Finally, we use Hamilton's principle and the fundamental lemma of the calculus of variations, as in Appendix A, to obtain the Euler–Lagrange equations for the free-free chain dynamics:

$$M_{00}\ddot{x} + \sum_{i=1}^{n} M_{0i} L_i S^T(q_i)\omega_i - \sum_{i=1}^{n} M_{0i} L_i \|\omega_i\|^2 q_i + M_{00} g e_3 = 0, \quad (10.15)$$

$$M_{ii} L_i^2 \dot{\omega}_i + \sum_{\substack{j=1 \\ j\neq i}}^{n} M_{ij} L_i L_j S^T(q_i) S(q_j)\dot{\omega}_j + M_{0i} L_i S(q_i)\ddot{x}$$

$$- \sum_{\substack{j=1 \\ j\neq i}}^{n} M_{ij} L_i L_j \|\omega_j\|^2 S(q_i) q_j + \sum_{j=i}^{n} m_j g L_i S(q_i) e_3 = 0, \quad i = 1, \ldots, n.$$

$$(10.16)$$

These Euler–Lagrange equations (10.15), (10.16), and the rotational kinematics (10.14) describe the dynamical flow of the free-free chain under uniform gravity in terms of $(x, q, \dot{x}, \omega) \in \mathsf{T}(\mathbb{R}^3 \times (\mathsf{S}^2)^n)$ on the tangent bundle of $\mathbb{R}^3 \times (\mathsf{S}^2)^n$.

10.4.2 Hamilton's Equations

Hamilton's equations for the free-free chain dynamics as they evolve on the cotangent bundle $\mathsf{T}^*(\mathbb{R}^3 \times (\mathsf{S}^2)^2)$ can be obtained by considering the momentum $(p, \pi) = (p, \pi_1, \ldots, \pi_n) \in \mathsf{T}^*_{(x,q)}(\mathbb{R}^3 \times (\mathsf{S}^2)^n)$ that is conjugate to the velocity vector $(\dot{x}, \omega) = (\dot{x}, \omega_1, \ldots, \omega_n) \in \mathsf{T}_{(x,q)}(\mathbb{R}^3 \times (\mathsf{S}^2)^n)$. This is defined by the Legendre transformation, which is given by

$$p = \frac{\partial \tilde{L}(x, q, \dot{x}, \omega)}{\partial \dot{x}},$$

$$\pi_i = (I_{3\times 3} - q_i q_i^T)\frac{\partial \tilde{L}(x, q, \dot{x}, \omega)}{\partial \omega_i}, \quad i = 1, \ldots, n,$$

which can be written as

$$
\begin{bmatrix} p \\ \pi_1 \\ \vdots \\ \pi_n \end{bmatrix} =
$$

$$
\begin{bmatrix} M_{00} & -M_{01}L_1 S(q_1) & \cdots & -M_{0n}L_n S(q_n) \\ -M_{10}L_1 S^T(q_1) & M_{11}L_1^2 I_3 & \cdots & M_{1n}L_1 L_n S^T(q_1)S(q_n) \\ \vdots & \vdots & \ddots & \vdots \\ -M_{n0}L_n S^T(q_n) & M_{n1}L_n L_1 S^T(q_n)S(q_1) & \cdots & M_{nn}L_n^2 I_3 \end{bmatrix}
$$

$$
\begin{bmatrix} \dot{x} \\ \omega_1 \\ \vdots \\ \omega_n \end{bmatrix} .
$$

Thus, we obtain

$$
\begin{bmatrix} \dot{x} \\ \omega_1 \\ \vdots \\ \omega_n \end{bmatrix} = \begin{bmatrix} M_{00}^I & M_{01}^I & \cdots & M_{0n}^I \\ M_{10}^I & M_{11}^I & \cdots & M_{1n}^I \\ \vdots & \vdots & \ddots & \vdots \\ M_{n0}^I & M_{n1}^I & \cdots & M_{nn}^I \end{bmatrix} \begin{bmatrix} p \\ \pi_1 \\ \vdots \\ \pi_n \end{bmatrix} , \tag{10.17}
$$

where

$$
\begin{bmatrix} M_{00}^I & M_{01}^I & \cdots & M_{0n}^I \\ M_{10}^I & M_{11}^I & \cdots & M_{1n}^I \\ \vdots & \vdots & \ddots & \vdots \\ M_{n0}^I & M_{n1}^I & \cdots & M_{nn}^I \end{bmatrix} =
$$

$$
\begin{bmatrix} M_{00} & -M_{01}L_1 S(q_1) & \cdots & -M_{0n}L_n S(q_n) \\ -M_{10}L_1 S^T(q_1) & M_{11}L_1^2 I_3 & \cdots & M_{1n}L_1 L_n S^T(q_1)S(q_n) \\ \vdots & \vdots & \ddots & \vdots \\ -M_{n0}L_n S^T(q_n) & M_{n1}L_n L_1 S^T(q_n)S(q_1) & \cdots & M_{nn}L_n^2 I_3 \end{bmatrix}^{-1} .
$$

Here, the functions M_{ij}^I denote 3×3 matrices that partition the inverse matrix above. The modified Hamiltonian function $\tilde{H} : T^*(\mathbb{R}^3 \times (S^2)^n) \to \mathbb{R}^1$ can also be expressed as

$$\tilde{H}(x,q,p,\pi) = \frac{1}{2} \begin{bmatrix} p \\ \pi_1 \\ \vdots \\ \pi_n \end{bmatrix}^T \begin{bmatrix} M_{00}^I & M_{01}^I & \cdots & M_{0n}^I \\ M_{10}^I & M_{11}^I & \cdots & M_{1n}^I \\ \vdots & \vdots & \ddots & \vdots \\ M_{n0}^I & M_{n1}^I & \cdots & M_{nn}^I \end{bmatrix} \begin{bmatrix} p \\ \pi_1 \\ \vdots \\ \pi_n \end{bmatrix}$$

$$+ M_{00}ge_3^T x + \sum_{i=1}^{n}\sum_{j=i}^{n} m_j g L_i e_3^T q_i.$$

Thus, Hamilton's equations can be expressed in the following form

$$\dot{x} = M_{00}^I p + \sum_{j=1}^{n} M_{0j}^I \pi_j, \tag{10.18}$$

$$\dot{q}_i = -S(q_i)\left\{ M_{i0}^I p + \sum_{j=1}^{n} M_{ij}^I \pi_j \right\}, \quad i = 1,\ldots,n, \tag{10.19}$$

$$\dot{p} = -M_{00}ge_3, \tag{10.20}$$

$$\dot{\pi}_i = -S(q_i)\left\{ \frac{1}{2}\frac{\partial}{\partial q_i} p^T M_{00}^I p + \frac{\partial}{\partial q_i} \sum_{j=1}^{n} p^T M_{0j}^I \pi_j + \frac{1}{2}\frac{\partial}{\partial q_i} \sum_{j,k=1}^{n} \pi_j^T M_{jk}\pi_k \right\}$$

$$+ \left\{ M_{i0}^I p + \sum_{j=1}^{n} M_{ij}^I \pi_j \right\} \times \pi_i - S(q_i)\sum_{j=1}^{n} m_j g L_i e_3, \quad i = 1,\ldots,n. \tag{10.21}$$

The nongravitational terms on the right-hand side of (10.21) are quadratic in the momenta. Hamilton's equations (10.18), (10.19), (10.20), and (10.21) describe the Hamiltonian flow of the free-free chain dynamics in terms of the evolution of $(x,q,p,\pi) \in \mathsf{T}^*(\mathbb{R}^3 \times (\mathsf{S}^2)^n)$ on the cotangent bundle of $\mathbb{R}^3 \times (\mathsf{S}^2)^n$.

10.4.3 Conservation Properties

The Hamiltonian of the chain under uniform gravity is given by

$$H = \frac{1}{2}M_{00}\|\dot{x}\|^2 - \dot{x}^T \sum_{i=1}^{n} M_{0i}S(q_i)\omega_i$$

$$+ \frac{1}{2}\sum_{i,j=1}^{n} M_{ij}\omega_i^T S^T(q_i)S(q_j)\omega_j + M_{00}ge_3^T x + \sum_{i=1}^{n}\sum_{j=i}^{n} m_j g L_i e_3^T q_i,$$

which coincides with the total energy E in this case, and it is constant along each solution of the dynamical flow.

Hamilton's equations can be used to show that the horizontal components of the translational momentum in the inertial frame are conserved along the flow of the free-free chain dynamics under uniform gravity. That is, the scalar components of the translational momentum

$$
e_1^T p = \left\{ M_{00}\dot{x} + \sum_{i=1}^{n} M_{0i} S^T(q_i)\omega_i \right\}^T e_1,
$$

$$
e_2^T p = \left\{ M_{00}\dot{x} + \sum_{i=1}^{n} M_{0i} S^T(q_i)\omega_i \right\}^T e_2,
$$

are each conserved along each solution of the dynamical flow. This can also be viewed to be consequence of Noether's theorem, applied to the translational invariance of the system in the directions that are orthogonal to gravity.

10.4.4 Equilibrium Properties

Since the chain is in free fall, there are no equilibrium solutions.

10.5 Dynamics of a Fixed-Free Elastic Rod

The dynamics of a fixed-free elastic rod are studied, assuming bending deformation of the elastic rod under the influence of gravity. We approximate the elastic rod by $n > 1$ slender rigid rod elements that are serially connected by spherical joints. This finite element approximation is geometrically exact in the sense that its length, even as it deforms, is constant. Each rigid rod element is modeled as a massless link with element mass concentrated at its centroid. We assume that one end of the first rod element is fixed to a rigid wall but with a spherical joint connection. Each spherical joint has two rotational degrees of freedom; the tip of the i-th rod lies on a sphere centered at the i-th connection.

A fixed inertial frame is constructed so that its first two axes lie in a horizontal plane, its third axis is vertical; the origin of the inertial frame is located at the point where the first rod element is attached to the rigid wall. Let $q_i \in S^2$ be the attitude vector of the i-th rod element in the inertial frame, $i = 1, \ldots, n$. The configuration vector is $q = (q_1, \ldots, q_n) \in (S^2)^n$, so that the configuration manifold is $(S^2)^n$. Thus, the geometrically exact elastic rod has $2n$ degrees of freedom. A schematic of a fixed-free elastic rod is shown in Figure 10.4.

Fig. 10.4 Geometrically exact fixed-free elastic rod

The mass of each rod element is m_i and the length of each rod element is L_i for $i = 1, \ldots, n$. The position vectors for the mass elements in the inertial frame are given by $\frac{L}{2}q_1$ and $\sum_{j=1}^{i-1} L_j q_j + \frac{L}{2}q_i$ for $i = 2, \ldots, n$. The total length of the fixed-free elastic rod is $L = \sum_{i=1}^{n} L_i$.

10.5.1 Euler–Lagrange Equations

We now compute the kinetic energy and the potential energy of each rod element to obtain the Lagrangian function.

Let $s_1 \in [0, L_1]$ be the distance from the fixed joint of the first link to a mass element dm_1 in the first rod. Since the mass is uniformly distributed, we have $dm_1 = \frac{m_1}{L_i} ds_1$. The kinetic energy of the first rod is

$$
\begin{aligned}
T_1(q_1, \dot{q}_1) &= \frac{1}{2} \frac{m_1}{L_1} \int_0^{L_1} \| s\dot{q}_1 \|^2 \, ds_1 \\
&= \frac{1}{6} m_1 L_1^2 \|\dot{q}_1\|^2.
\end{aligned}
$$

Let $s_i \in [0, L_i]$ be the distance from the i-th joint to a mass element dm_i in the i-th rod. Since the mass is uniformly distributed, we have $dm_i = \frac{m_i}{L_i} ds_i$. The kinetic energy of the i-th rod, $i = 2, \ldots, n$, is given by

$$
\begin{aligned}
T_i(q_i, \dot{q}_i) &= \frac{1}{2} \frac{m_i}{L_i} \int_0^{L_i} \left\| \sum_{j=1}^{i-1} L_j \dot{q}_j + s_i \dot{q}_i \right\|^2 ds_i \\
&= \frac{1}{6} m_i L_i^2 \|\dot{q}_i\|^2 + \frac{1}{2} m_i L_i \sum_{j=1}^{i-1} L_j \dot{q}_j^T \dot{q}_i + \frac{1}{2} m_i \left\| \sum_{j=1}^{i-1} L_j \dot{q}_j \right\|^2.
\end{aligned}
$$

Using this, the total kinetic energy can be written as

$$
T(q, \dot{q}) = \frac{1}{2} \sum_{i,j=1}^{n} M_{ij} \dot{q}_i^T \dot{q}_j,
$$

where the inertia constants are defined as

$$M_{ii} = \left[\frac{1}{3}m_i + \sum_{p=i+1}^{n} m_p\right] L_i^2, \qquad i = 1, \ldots, n,$$

$$M_{ij} = \left[\frac{1}{2}m_i + \sum_{p=i+1}^{n} m_p\right] L_i L_j, \quad i = 1, \ldots, n, \; j = 1, \ldots, i.$$

The remaining terms are defined by the symmetry $M_{ij} = M_{ji}$.

The potential energy is composed of a gravitational potential and a strain energy. The position vector of the center of mass of the i-th rod is given by $L_1 q_1 + \cdots + L_{i-1} q_{i-1} + \frac{1}{2} L_i q_i$. Thus, the gravitational potential is given by

$$U_g(q) = \sum_{i=1}^{n} m_i g e_3^T \left(\sum_{j=1}^{i-1} q_j L_j + \frac{1}{2} L_i q_i\right),$$

which can be expressed as

$$U_g(q) = \sum_{i=1}^{n} \left(\sum_{j=i+1}^{n} m_j + \frac{1}{2} m_i\right) g e_3^T L_i q_i.$$

The strain potential energy for pure bending of an elastic rod is given by

$$U_\epsilon(q) = \int_0^L \frac{EI}{2R^2} ds,$$

where E is Young's modulus, I is the sectional area moment, and R is the radius of curvature. We assume that the radius of curvature is constant along the i-th rod and is approximated by

$$R_i = \frac{L_i}{\sin(\frac{\theta_i}{2})},$$

where θ_i denotes the angle between the i-th rod and the $(i-1)$-th rod. The strain potential energy is approximated as

$$U_\epsilon(q) = \sum_{i=1}^{n} \frac{EI \sin^2(\frac{\theta_i}{2})}{2L_i^2} L_i$$

$$= \sum_{i=1}^{n} \frac{EI}{4L_i^2}(1 - \cos\theta_i)$$

$$= \sum_{i=1}^{n} \frac{EI}{4L_i^2}(1 - q_{i-1}^T q_i).$$

Therefore, the total potential energy is given by

$$U(q) = \sum_{i=1}^{n} \left(\sum_{j=i+1}^{n} m_j + \frac{1}{2} m_i \right) g e_3^T L_i q_i + \frac{EI}{4L_i^2} (1 - q_{i-1}^T q_i).$$

Thus, the Lagrangian function $L : \mathsf{T}(\mathsf{S}^2)^n \to \mathbb{R}^1$ is

$$L(q, \dot{q}) = \frac{1}{2} \sum_{i=1}^{n} \sum_{j=1}^{n} M_{ij} \dot{q}_i^T \dot{q}_j$$

$$- \sum_{i=1}^{n} \left(\sum_{j=i+1}^{n} m_j + \frac{1}{2} m_i \right) g e_3^T L_i q_i - \frac{EI}{4L_i^2} (1 - q_{i-1}^T q_i).$$

The rotational kinematics of each rod element is given by

$$\dot{q}_i = S(\omega_i) q_i, \quad i = 1, \ldots, n, \tag{10.22}$$

where the angular velocity vector $\omega_i \in \mathsf{T}_{q_i} \mathsf{S}^2$ satisfies $\omega_i^T q_i = 0$ for $i = 1, \ldots, n$. The Lagrangian can be expressed in terms of the angular velocity vector as follows

$$\tilde{L}(q, \omega) = \frac{1}{2} \sum_{i=1}^{n} \sum_{j=1}^{n} M_{ij} \omega_i S^T(q_i) S(q_j) \omega_j$$

$$- \sum_{i=1}^{n} \left\{ \left(\sum_{j=i+1}^{n} m_j + \frac{1}{2} m_i \right) g e_3^T L_i q_i + \frac{EI}{4L_i^2} (1 - q_{i-1}^T q_i) \right\}.$$

The results from Chapter 5 can be used to obtain the Euler–Lagrange equations of motion based on the Lagrangian for the finite-element model of the elastic rod

$$M_{ii} \dot{\omega}_i + \sum_{\substack{j=1 \\ j \neq i}}^{n} M_{ij} S^T(q_i) S(q_j) \dot{\omega}_j) - \sum_{\substack{j=1 \\ j \neq i}}^{n} M_{ij} \|\omega_j\|^2 S(q_i) q_j$$

$$+ S(q_i) \left\{ \left(\sum_{j=i+1}^{n} m_j + \frac{1}{2} m_i \right) g L_i e_3 + \frac{\partial U_\epsilon(q)}{\partial q_i} \right\} = 0, \quad i = 1, \ldots, n, \tag{10.23}$$

where

$$\frac{\partial U_\epsilon(q)}{\partial q_i} = \begin{cases} -\dfrac{EI}{4L_i^2} q_{i-1} - \dfrac{EI}{4L_{i+1}^2} q_{i+1} & i = 1, \ldots, n-1, \\ -\dfrac{EI}{4L_n^2} q_{n-1} & i = n. \end{cases}$$

These Euler–Lagrange equations (10.23), together with the rotational kinematics (10.22), describe the dynamics of the geometrically exact elastic rod in terms of $(q, \omega) \in \mathsf{T}(\mathsf{S}^2)^n$ on the tangent bundle of $(\mathsf{S}^2)^n$.

10.5.2 Hamilton's Equations

The Legendre transformation

$$\pi_i = (I_{3\times3} - q_i q_i^T) \frac{\partial \tilde{L}(q, \omega)}{\partial \omega_i}, \quad i = 1, \ldots, n,$$

is used to define the momentum $(\pi_1, \ldots, \pi_n) \in \mathsf{T}_q^*(\mathsf{S}^2)^n$ conjugate to $(\omega_1, \ldots, \omega_n) \in \mathsf{T}_q(\mathsf{S}^2)^n$ by

$$\begin{bmatrix} \pi_1 \\ \pi_2 \\ \vdots \\ \pi_n \end{bmatrix} = \begin{bmatrix} M_{11}I_{3\times3} & M_{12}S^T(q_1)S(q_2) & \cdots & M_{1n}S^T(q_1)S(q_n) \\ M_{21}S^T(q_2)S(q_1) & M_{22}I_{3\times3} & \cdots & M_{2n}S^T(q_2)S(q_n) \\ \vdots & \vdots & \ddots & \vdots \\ M_{n1}S^T(q_n)S(q_1) & M_{n2}S^T(q_n)S(q_2) & \cdots & M_{nn}I_{3\times3} \end{bmatrix} \begin{bmatrix} \omega_1 \\ \omega_2 \\ \vdots \\ \omega_n \end{bmatrix}.$$

Thus, we can write

$$\begin{bmatrix} \omega_1 \\ \omega_2 \\ \vdots \\ \omega_n \end{bmatrix} = \begin{bmatrix} M_{11}^I & M_{12}^I & \cdots & M_{1n}^I \\ M_{21}^I & M_{22}^I & \cdots & M_{2n}^I \\ \vdots & \vdots & \ddots & \vdots \\ M_{n1}^I & M_{n2}^I & \cdots & M_{nn}^I \end{bmatrix} \begin{bmatrix} \pi_1 \\ \pi_2 \\ \vdots \\ \pi_n \end{bmatrix},$$

where

$$\begin{bmatrix} M_{11}^I & M_{12}^I & \cdots & M_{1n}^I \\ M_{21}^I & M_{22}^I & \cdots & M_{2n}^I \\ \vdots & \vdots & \ddots & \vdots \\ M_{n1}^I & M_{n2}^I & \cdots & M_{nn}^I \end{bmatrix} =$$

$$\begin{bmatrix} M_{11}I_{3\times3} & M_{12}S^T(q_1)S(q_2) & \cdots & M_{1n}S^T(q_1)S(q_n) \\ M_{21}S^T(q_2)S(q_1) & M_{22}I_{3\times3} & \cdots & M_{2n}S^T(q_2)S(q_n) \\ \vdots & \vdots & \ddots & \vdots \\ M_{n1}S^T(q_n)S(q_1) & M_{n2}S^T(q_n)S(q_2) & \cdots & M_{nn}I_{3\times3} \end{bmatrix}^{-1}$$

is the inverse of the indicated $3n \times 3n$ partitioned matrix. The modified Hamiltonian function is

$$
\tilde{H}(q,\pi) = \frac{1}{2}
\begin{bmatrix} \pi_1 \\ \pi_2 \\ \vdots \\ \pi_n \end{bmatrix}
\begin{bmatrix}
M_{11}^I & M_{12}^I & \cdots & M_{1n}^I \\
M_{21}^I & M_{22}^I & \cdots & M_{2n}^I \\
\vdots & \vdots & \ddots & \vdots \\
M_{n1}^I & M_{n2}^I & \cdots & M_{nn}^I
\end{bmatrix}
\begin{bmatrix} \pi_1 \\ \pi_2 \\ \vdots \\ \pi_n \end{bmatrix}
$$

$$
+ \left\{ \sum_{i=1}^{n} \left(\sum_{j=i+1}^{n} m_j + \frac{1}{2} m_i \right) g e_3^T L_i q_i + \frac{EI}{4L_i^2} (1 - q_{i-1}^T q_i) \right\}.
$$

Thus, Hamilton's equations are

$$
\dot{q}_i = -S(q_i) \left\{ \sum_{k=1}^{n} M_{ik}^I \pi_k \right\}, \quad i = 1, \ldots, n, \tag{10.24}
$$

$$
\dot{\pi}_i = -S(q_i) \left\{ \frac{1}{2} \frac{\partial}{\partial q_i} \sum_{j,k=1}^{n} \pi_j^T M_{jk} \pi_j \right\} + \left\{ \sum_{j=1}^{n} M_{ij}^I \pi_j \right\} \times \pi_i
$$

$$
- S(q_i) \left(\sum_{j=i+1}^{n} m_j + \frac{1}{2} m_i \right) g L_i e_3 - S(q_i) \frac{\partial U_\epsilon(q)}{\partial q_i}, \quad i = 1, \ldots, n.
$$

$$\tag{10.25}$$

Hamilton's equations (10.24) and (10.25) describe the Hamiltonian dynamics of the geometrically exact elastic rod in terms of $(q,\pi) \in T^*(S^2)^n$ on the cotangent bundle of $(S^2)^n$.

10.5.3 Conservation Properties

The Hamiltonian of the elastic rod,

$$
H = \frac{1}{2} \sum_{i=1}^{n} M_{ii} \omega_i \omega_i + \frac{1}{2} \sum_{i=1}^{n} \sum_{\substack{j=1 \\ j \neq i}}^{n} M_{ij} \omega_i S^T(q_i) S(q_j) \omega_j
$$

$$
+ \sum_{i=1}^{n} \left\{ \sum_{j=i+1}^{n} (m_j + \frac{1}{2} m_i) g e_3^T L_i q_i + \frac{EI}{4L_i^2} (1 - q_{i-1}^T q_i) \right\},
$$

which coincides with the total energy E in this case, is constant along each solution of the Lagrangian flow.

10.5.4 Equilibrium Properties

For the elastic rod to be in equilibrium, the angular velocity vector of all the rod elements must be zero, and the configuration must satisfy the conditions

$$S(q_i) \left\{ \sum_{j=i+1}^{n} (m_j + \frac{1}{2} m_i) g L_i e_3 + \frac{\partial U_\epsilon(q)}{\partial q_i} \right\} = 0, \quad i = 1, \ldots, n,$$

which imply that there is a balance between the gravity forces and the bending forces on each rod element.

It is possible to determine linearized differential equations that approximate the Lagrangian flow of the bending dynamics of the geometrically exact elastic rod.

10.6 Problems

10.1. A planar version of a chain pendulum consists of a serial connection of n rigid links. Each link is assumed to move within a fixed vertical plane; a two-dimensional Euclidean frame is constructed within this vertical plane. Each link is a rigid rod with mass m_i concentrated at the outboard end of the link; the link lengths are L_i, $i = 1, \ldots, n$. One end of the chain pendulum is connected to a fixed pivot located at the origin of the Euclidean frame. Constant gravity acts on each link of the chain pendulum. There is no friction at any of the joints. Since the links are constrained to move within a fixed vertical plane, $q_i \in S^1$, $i = 1, \ldots, n$ denotes the attitude of the i-th link in the two-dimensional Euclidean frame, so that $q = (q_1, \ldots, q_n) \in (S^1)^n$ is the configuration vector and $(S^1)^n$ is the configuration manifold.

(a) Determine the Lagrangian function $L : T(S^1)^n \to \mathbb{R}^1$ on the tangent bundle of the configuration manifold.
(b) What are the Euler–Lagrange equations for the planar chain pendulum?
(c) Determine the Hamiltonian function $H : T^*(S^1)^n \to \mathbb{R}^1$ on the cotangent bundle of the configuration manifold.
(d) What are Hamilton's equations for the planar chain pendulum?
(e) What are conserved quantities for the planar chain pendulum?
(f) Describe the hanging equilibrium solution and the inverted equilibrium solution.
(g) What are the linear dynamics that approximate the planar chain pendulum dynamics in a neighborhood of the hanging equilibrium solution?
(h) What are the linear dynamics that approximate the planar chain pendulum dynamics in a neighborhood of the inverted equilibrium solution?

10.2. A planar version of a chain pendulum on a cart consists of a serial connection of n rigid links where one end of the chain is connected to a cart. Each link is assumed to move within a fixed vertical plane; a two-dimensional Euclidean frame is constructed within this vertical plane. The cart can translate, without friction, along the first axis of the two-dimensional Euclidean frame. Each link is a rigid rod with mass m_i concentrated at the outboard end of the link; the link lengths are L_i, $i = 1, \ldots, n$. One end of the chain pendulum is connected to a pivot on the cart. Constant gravity acts on each link of the chain pendulum. There is no friction at any of the joints. Since the cart and links are constrained to move within a fixed vertical plane, $x \in \mathbb{R}^1$ denotes the position of the cart on the first axis of the Euclidean frame and $q_i \in \mathsf{S}^1$, $i = 1, \ldots, n$ denotes the attitude of the i-th link in the two-dimensional Euclidean frame, so that $(x, q) = (x, q_1, \ldots, q_n) \in \mathbb{R}^1 \times (\mathsf{S}^1)^n$ is the configuration vector and $\mathbb{R}^1 \times (\mathsf{S}^1)^n$ is the configuration manifold.

(a) Determine the Lagrangian function $L : \mathsf{T}(\mathbb{R}^1 \times (\mathsf{S}^1)^n) \to \mathbb{R}^1$ on the tangent bundle of the configuration manifold.
(b) What are the Euler–Lagrange equations for the planar chain pendulum on a cart?
(c) Determine the Hamiltonian function $H : \mathsf{T}^*(\mathbb{R}^1 \times (\mathsf{S}^1)^n) \to \mathbb{R}^1$ on the cotangent bundle of the configuration manifold.
(d) What are Hamilton's equations for the planar chain pendulum on a cart?
(e) What are conserved quantities for the planar chain pendulum on a cart?
(f) Describe the hanging equilibrium solution and the inverted equilibrium solution.
(g) What are the linear dynamics that approximate the planar chain pendulum on a cart dynamics in a neighborhood of the hanging equilibrium solution?
(h) What are the linear dynamics that approximate the planar chain pendulum on a cart dynamics in a neighborhood of the inverted equilibrium solution?

10.3. A planar version of a uniform chain is constrained to deform in a vertical plane. The vertical plane rotates about an inertially fixed axis with constant rotation rate $\Omega \in \mathbb{R}^1$. The chain consists of a serial connection of n identical rigid links, each with mass m concentrated at the outboard end of the link; each link has length L. There are no gravitational forces or friction moments at the joints. The configuration vector $q = (q_1, \ldots, q_n) \in (\mathsf{S}^1)^n$ is the vector of attitudes of the n links, each expressed in the inertial frame.

(a) Determine the Lagrangian function $L : \mathsf{T}(\mathsf{S}^1)^n \to \mathbb{R}^1$ on the tangent bundle of the configuration manifold.
(b) What are the Euler–Lagrange equations for the chain rotating at a constant angular rate?
(c) Determine the Hamiltonian function $H : \mathsf{T}^*(\mathsf{S}^1)^n \to \mathbb{R}^1$ defined on the cotangent bundle of the configuration manifold.

(d) What are Hamilton's equations for the chain rotating at a constant angular rate?
(e) What are conserved quantities for the chain rotating at a constant angular rate?
(f) What are the equilibrium solutions for the dynamics of a chain rotating at a constant angular rate?
(g) What are the linear dynamics that approximate the rotating chain dynamics in a neighborhood of an equilibrium solution?

10.4. A planar version of a free-free chain consists of a serial connection of n rigid links. Each link is assumed to move within a fixed vertical plane; a two-dimensional Euclidean frame is constructed within this vertical plane. Each link is a rigid rod with mass m_i concentrated at the outboard end of the link with mass m_0 concentrated at the other end of the first link; the link lengths are L_i, $i = 1, \ldots, n$. Constant gravity acts on each link of the chain. There is no friction at any of the joints. Since the links are constrained to move within a fixed vertical plane, let $x \in \mathbb{R}^2$ denote the position vector of the free end of the first link and let $q_i \in \mathsf{S}^1$, $i = 1, \ldots, n$ denote the attitude of the i-th link in the two-dimensional Euclidean frame, so that $(x, q) = (x, q_1, \ldots, q_n) \in \mathbb{R}^2 \times (\mathsf{S}^1)^n$ is the configuration vector and $\mathbb{R}^2 \times (\mathsf{S}^1)^n$ is the configuration manifold.

(a) Determine the Lagrangian function $L : \mathsf{T}(\mathbb{R}^2 \times (\mathsf{S}^1)^n) \to \mathbb{R}^1$ on the tangent bundle of the configuration manifold.
(b) What are the Euler–Lagrange equations for the planar free-free chain?
(c) Determine the Hamiltonian function $H : \mathsf{T}^*(\mathbb{R}^2 \times (\mathsf{S}^1)^n) \to \mathbb{R}^1$ on the cotangent bundle of the configuration manifold.
(d) What are Hamilton's equations for the planar free-free chain?
(e) What are conserved quantities for the planar free-free chain?

10.5. A planar version of a fixed-free, geometrically exact elastic rod is viewed as a serial connection of n rigid links. Each link is assumed to move within a fixed vertical plane; a two-dimensional Euclidean frame is constructed within this vertical plane. Each link is a rigid rod with mass m_i concentrated at the outboard end of the link; the link lengths are L_i, $i = 1, \ldots, n$. One end of the elastic rod is connected to rigid wall, where the attachment point is located at the origin of the Euclidean frame; the other end of the elastic rod is free. Constant gravity acts on each link of the elastic rod. There is no friction at any of the joints. Since the links are constrained to move within a fixed vertical plane, $q_i \in \mathsf{S}^1$, $i = 1, \ldots, n$ denotes the attitude of the i-th link in the two-dimensional Euclidean frame, so that $q = (q_1, \ldots, q_n) \in (\mathsf{S}^1)^n$ is the configuration vector and $(\mathsf{S}^1)^n$ is the configuration manifold.

(a) Following the development in the text, obtain an approximate expression for the strain potential energy of the elastic rod, which is used in (b).
(b) Determine the Lagrangian function $L : \mathsf{T}(\mathsf{S}^1)^n \to \mathbb{R}^1$ on the tangent bundle of the configuration manifold.

(c) What are the Euler–Lagrange equations for the planar fixed-free elastic rod?

(d) Determine the Hamiltonian function $H : \mathsf{T}^*(\mathsf{S}^1)^n \to \mathbb{R}^1$ on the cotangent bundle of the configuration manifold.

(e) What are Hamilton's equations for the planar fixed-free elastic rod?

(f) What are conserved quantities for the planar fixed-free elastic rod?

(g) What algebraic equations characterize the equilibrium solutions of the planar fixed-free elastic rod?

10.6. Consider the dynamics of n ideal particles, with the mass of the i-th particle being m_i, $i = 1, \ldots, n$. The mass particles are constrained to move, without friction, on the surface of a unit sphere in three dimensions. An inertial Euclidean frame is selected with origin located at the center of the sphere. Let $q_i \in \mathsf{S}^2$ denote the direction vector of the i-th particle in the inertial frame for $i = 1, \ldots, n$; thus, $q = (q_1, \ldots, q_n) \in (\mathsf{S}^2)^n$ and $(\mathsf{S}^2)^n$ is the configuration manifold. There are forces between the i-th particle and the j-th particle that arise from the potential $U(q) = \frac{1}{2} \sum_{i=1}^{n} \sum_{j \neq i} \kappa_{ij}(1 - q_i^T q_j)$. Here $\kappa_{ij} = \kappa_{ji}$, $i \neq j$ are constants that characterize the potential forces between the i-th particle and the j-th particle. No other external forces act on the mass particles.

(a) Characterize the potential forces in terms of the signs of the constants. When are the forces attractive? When are the forces repellent?

(b) Determine the Lagrangian function $L : \mathsf{T}(\mathsf{S}^2)^n \to \mathbb{R}^1$ on the tangent bundle of the configuration manifold.

(c) What are the Euler–Lagrange equations for the n mass particles on a sphere?

(d) Determine the Hamiltonian function $H : \mathsf{T}^*(\mathsf{S}^1)^n \to \mathbb{R}^1$ on the cotangent bundle of the configuration manifold.

(e) What are Hamilton's equations for the n mass particles on a sphere?

(f) What are conserved quantities for the n mass particles on a sphere?

(g) What algebraic equations characterize the equilibrium solutions of n mass particles on a sphere?

10.7. A chain pendulum, consisting of a serial connection of n rigid links, is supported by multiple elastic strings in tension. Each link of the chain pendulum is rigid with mass m_i concentrated at the outboard end of the link; the link length is L_i. A three-dimensional inertial Euclidean frame is defined. One end of the chain pendulum is connected to a spherical joint or pivot supported at a fixed base located at the origin of the Euclidean frame. Constant gravity acts on each link of the chain pendulum in the negative direction of the third axis of the Euclidean frame. The vector $q_i \in \mathsf{S}^2$, $i = 1, \ldots, n$ is the attitude of the i-th link in the inertial frame, so that $q = (q_1, \ldots, q_n) \in (\mathsf{S}^2)^n$ is the configuration vector and $(\mathsf{S}^2)^n$ is the configuration manifold. The chain pendulum is supported in a neighborhood of its inverted equilibrium by $4n$ massless elastic strings in tension. The $4n$ elastic strings connect the outboard end of the i-th link to the inertially fixed locations $(D_i, 0, \sum_{j=1}^{i} L_j)$, $(-D_i, 0, \sum_{j=1}^{i} L_j)$, $(0, D_i, \sum_{j=1}^{i} L_j)$, $(0, -D_i, \sum_{j=1}^{i} L_j)$ for $i = 1, \ldots, n$. The elastic strings connected to the outboard end of the i-th link have elastic stiffness κ_i; the restoring force in the elastic string is zero when the string length is $d_i < D_i$, $i = 1, \ldots, n$. Assume all $4n$ elastic strings remain in tension and ignore any collisions between the strings and links. This is another example of a tensegrity structure.

(a) Determine the Lagrangian function $L : \mathsf{T}(\mathsf{S}^2)^n \to \mathbb{R}^1$ for the elastically supported chain pendulum, viewing the forces of the elastic strings as external forces.
(b) Use the Lagrange–d'Alembert principle to obtain the Euler–Lagrange equations for the elastically supported chain pendulum.
(c) What are Hamilton's equations for the elastically supported chain pendulum?
(d) What algebraic conditions guarantee that the inverted equilibrium of the chain pendulum is an equilibrium solution?

10.8. Consider a free-free elastic rod that undergoes elastic deformation by bending under uniform, constant gravity. The elastic rod is modeled as a serial connection of n links, where the connections are spherical joints with proportional elastic restoring forces. Each link is a rigid rod with mass m_i concentrated at the outboard end of the link; the link lengths are L_i, $i = 1, \ldots, n$; a mass m_0 is concentrated at the free end of the first link. A three-dimensional inertial Euclidean frame is introduced. Gravity is assumed to act in the negative direction of the third axis of the Euclidean frame. Let $x \in \mathbb{R}^3$ denote the position vector of the free end of the first link and let $q_i \in \mathsf{S}^2$, $i = 1, \ldots, n$ denote the attitude of the i-th link in the three-dimensional Euclidean frame, so that $(x, q) = (x, q_1, \ldots, q_n) \in \mathbb{R}^3 \times (\mathsf{S}^2)^n$ is the configuration vector and $\mathbb{R}^3 \times (\mathsf{S}^2)^n$ is the configuration manifold.

(a) Determine the Lagrangian function $L : \mathsf{T}(\mathbb{R}^3 \times (\mathsf{S}^2)^n) \to \mathbb{R}^1$ defined on the tangent bundle of the configuration manifold.
(b) What are the Euler–Lagrange equations for the free-free elastic rod?
(c) Determine the Hamiltonian function $H : \mathsf{T}^*(\mathbb{R}^3 \times (\mathsf{S}^2)^n) \to \mathbb{R}^1$ defined on the cotangent bundle of the configuration manifold.
(d) What are Hamilton's equations for the free-free elastic rod?
(e) What are conserved quantities for the free-free elastic rod?

10.9. Consider a serial connection of n elastic spherical pendulums. Each spherical pendulum consists of a massless elastic link and a mass element, viewed as an ideal particle. The masses of the elements are denoted by $m_1, \ldots, m_n$. The in-board end of the first massless elastic link is attached to an inertially supported spherical pivot. The out-board end of the n-th spherical pendulum is free. Each massless link is assumed to be elastic in that it can deform axially while always remaining straight. An inertial Euclidean frame is selected with origin located at the fixed pivot of the first elastic spherical pendulum. Let $x_i \in \mathbb{R}^1$ denote the length of the i-th mass-less link and let $q_i \in S^2$ denote the attitude vector of the i-th massless link for $i = 1,\ldots,n$; each massless link is assumed to remain straight. Thus, $(x,q) = (x_1,\ldots,x_n,q_1,\ldots,q_n) \in \mathbb{R}^n \times (S^2)^n$ so that $\mathbb{R}^n \times (S^2)^n$ is the configuration manifold. The mass particles act under uniform and constant gravity. The elastic links are assumed to have elastic stiffness $\kappa_i, 1 = 1,\ldots,n$, and lengths L_i, $i = 1,\ldots,n$, corresponding to zero elastic force in the links. Assume that the inequalities $m_i g < k_i L_i$, $i = 1,\ldots,n$, hold.

(a) Determine the Lagrangian function $L : \mathsf{T}(\mathbb{R}^n \times (S^2)^n) \to \mathbb{R}^1$ defined on the tangent bundle of the configuration manifold.
(b) What are the Euler–Lagrange equations for the serial connection of elastic spherical pendulums?
(c) Determine the Hamiltonian function $H : \mathsf{T}^*(\mathbb{R}^n \times (S^2)^n) \to \mathbb{R}^1$ defined on the cotangent bundle of the configuration manifold.
(d) Determine Hamilton's equations for the serial connection of elastic spherical pendulums.
(e) Determine conserved quantities for the serial connection of elastic spherical pendulums.
(f) What are the equilibrium solutions of the serial connection of elastic spherical pendulums?
(g) For a selected equilibrium solution, determine linearized equations that approximate the dynamics of the serial connection of elastic spherical pendulums in a neighborhood of that equilibrium.

10.10. Consider a free-free serial connection of n mass elements, viewed as ideal particles, with masses are denoted by $m_1,\ldots,m_n$. The mass elements are connected by $n-1$ massless elastic links; each link is assumed to remain straight and it can deform elastically along its longitudinal axis. An inertial Euclidean frame is selected. Let $x \in \mathbb{R}^3$ denote the position vector of the first mass element; let $y_i \in \mathbb{R}^1$ denote the length of the i-th massless link and let $q_i \in S^2$ denote the attitude vector of the i-th massless link for $i = 1,\ldots,n$. Thus, $x + \sum_{j=1}^{i} y_j q_j \in \mathbb{R}^3$ is the position vector of the i-th mass particle. The configuration vector is $(x,y,q) = (x, y_1,\ldots,y_n, q_1,\ldots,q_n) \in \mathbb{R}^3 \times \mathbb{R}^n \times (S^2)^n$ so that $\mathbb{R}^3 \times \mathbb{R}^n \times (S^2)^n$ is the configuration manifold. The mass elements act under uniform and constant gravity. The i-th elastic link is assumed to have elastic stiffness κ_i and length L_i corresponding to zero elastic force in the link for $i = 2,\ldots,n$. Thus, the massless link with elastic stiffness κ_i connects the mass elements denoted by indices $i-1$ and i.

(a) Determine the Lagrangian function $L : \mathsf{T}(\mathbb{R}^3 \times \mathbb{R}^n \times (S^2)^n) \to \mathbb{R}^1$ defined on the tangent bundle of the configuration manifold.
(b) What are the Euler–Lagrange equations for the free-free serial connection of mass elements?
(c) Determine the Hamiltonian function $H : \mathsf{T}^*(\mathbb{R}^3 \times \mathbb{R}^n \times (S^2)^n) \to \mathbb{R}^1$ defined on the cotangent bundle of the configuration manifold.

(d) Determine Hamilton's equations for the free-free serial connection of mass elements.
(e) Determine conserved quantities for the free-free serial connection of mass elements.
(f) Describe the dynamics of the center of mass of the free-free serial connection of mass elements.

10.11. Consider a serial connection of n identical mass elements, viewed as ideal particles, each of mass m, interlaced with $n-1$ identical linear elastic springs, each with stiffness κ. The mass elements translate along a horizontal straight line, assumed to be the first axis of an inertial Euclidean frame. The ends of the first and last elastic spring are inertially fixed. Each mass element serves as the pivot point for n identical spherical pendulums, each of length L and mass M assumed concentrated at the free end of each pendulum link. Constant gravity acts on each of the spherical pendulums in the negative direction of the third axis of the Euclidean frame. There is no friction on any of the translating mass elements or at any of the pendulum pivots. Let $x_i \in \mathbb{R}^1$, $i = 1, \ldots, n$ denote the one-dimensional positions of the translating mass elements and let $q_i \in \mathsf{S}^2$, $i = 1, \ldots, n$ denote the attitude vectors of the pendulum links in the Euclidean frame, so that the configuration vector is $(x, q) = (x_1, q_1, \ldots, x_n, q_n) \in (\mathbb{R}^1 \times \mathsf{S}^2)^n$.

(a) Determine the Lagrangian function $L : \mathsf{T}(\mathbb{R}^1 \times (\mathsf{S}^2)^n) \to \mathbb{R}^1$ defined on the tangent bundle of the configuration manifold.
(b) What are the Euler–Lagrange equations for the serial connection of mass elements and spherical pendulums?
(c) Determine the Hamiltonian function $H : \mathsf{T}^*(\mathbb{R}^3 \times (\mathsf{S}^2)^n) \to \mathbb{R}^1$ defined on the cotangent of the configuration manifold.
(d) What are Hamilton's equations for the serial connection of mass elements and spherical pendulums?
(e) What are conserved quantities for the serial connection of mass elements and spherical pendulums?
(f) Describe the equilibrium solutions of the serial connection of mass elements and spherical pendulums.
(g) Consider the equilibrium solution corresponding to hanging equilibrium of each spherical pendulum. Determine the linearized equations that approximate the dynamics of the serial connection of mass elements and spherical pendulums in a neighborhood of this equilibrium.

10.12. Consider the dynamics of a uniform chain under the influence of uniform, constant gravity. The chain consists of a serial connection of n identical rigid links. Each end of the chain is supported by an inertially fixed pivot; this is referred to as a chain with inertially fixed ends. Each link is a rigid rod with mass m concentrated at the mid-point of the link; each link length is L. Each end of the chain is connected to a fixed pivot; one pivot is located at the origin of a three-dimensional inertial Euclidean frame while the other end of the chain is connected to a fixed pivot located a distance d along the first axis of the Euclidean frame. Constant gravity acts on each link of the chain in the negative direction of the third axis of the Euclidean frame. There is no friction at any of the joints. Let $q_i \in S^2$, $i = 1, \ldots, n$ denote the attitude of the i-th link in the Euclidean frame, so that $q = (q_1, \ldots, q_n) \in (S^2)^n$. The fact that the ends of the chain are connected to fixed pivot locations defines a holonomic constraint.

(a) Describe the constraint manifold M as a sub-manifold of $(S^2)^n$.
(b) Determine the augmented Lagrangian function $L^a : TM \times \mathbb{R}^3 \to \mathbb{R}^1$ on the tangent bundle of the constraint manifold using Lagrange multipliers.
(c) What are the Euler–Lagrange equations for the chain with inertially fixed ends?
(d) Determine the augmented Hamiltonian function $H^a : T^*M \times \mathbb{R}^3 \to \mathbb{R}^1$ on the cotangent bundle of the constraint manifold using Lagrange multipliers.
(e) What are Hamilton's equations for the chain with inertially fixed ends?
(f) What algebraic equations characterize the equilibrium solutions of the chain with inertially fixed ends?

10.13. Consider the dynamics of a uniform elastic rod that bends under the influence of uniform, constant gravity. The elastic rod is viewed as a serial connection of n identical rigid links, with elastic restoring moments as described previously in this chapter. Each link is a rigid rod with mass m concentrated at the mid-point of the link; each link length is L. An inertial Euclidean frame is established, with the third axis of the frame vertical. The position of one end of the elastic rod is inertially fixed at the origin of the Euclidean frame; the other end of the elastic rod is allowed to translate, without friction, along the first axis of the Euclidean frame. Let $q_i \in S^2$, $i = 1, \ldots, n$ denote the attitude vectors of the links, expressed in the Euclidean frame, so that $q = (q_1, \ldots, q_n) \in (S^2)^n$. The fact that one end of the elastic rod is fixed while the other end of the elastic rod is constrained requires the introduction of a holonomic constraint.

(a) Describe the constraint manifold M as a sub-manifold of $(S^2)^n$.
(b) Determine the augmented Lagrangian function $L^a : TM \times \mathbb{R}^2 \to \mathbb{R}^1$ on the tangent bundle of the constraint manifold using Lagrange multipliers.
(c) What are the Euler–Lagrange equations for the constrained elastic rod?
(d) Determine the augmented Hamiltonian function $H^a : T^*M \times \mathbb{R}^2 \to \mathbb{R}^1$ on the cotangent bundle of the constraint manifold using Lagrange multipliers.
(e) What are Hamilton's equations for the constrained elastic rod?
(f) What algebraic equations characterize the equilibrium solutions of the constrained elastic rod?

Appendix A
Fundamental Lemmas of the Calculus of Variations

Several versions of the fundamental lemma in the calculus of variations are presented. The classical version of the fundamental lemma when the configuration manifold is $\mathbb{R}^n$ is summarized first, and then versions are summarized when the configuration manifold is an embedded manifold in $\mathbb{R}^n$ or a Lie group embedded in $\mathbb{R}^{n \times n}$.

A.1 Fundamental Lemma of Variational Calculus on $\mathbb{R}^n$

The fundamental lemma of the calculus of variations on $\mathbb{R}^n$ [5] is the following statement. Let $x : [t_0, t_f] \to \mathbb{R}^n$ be continuous and suppose that $F : [t_0, t_f] \to \mathbb{R}^n$ is continuous. If

$$\int_{t_0}^{t_f} F(t) \cdot \delta x(t) \, dt = 0,$$

holds for all continuous variations $\delta x : [t_0, t_f] \to \mathbb{R}^n$ satisfying $\delta x(t_0) = \delta x(t_f) = 0$, then it follows that

$$F(t) = 0, \quad t_0 \leq t \leq t_f.$$

As usual, the notation $F \cdot \delta x$ indicates the inner or dot product of a covector $F \in (\mathbb{R}^n)^*$ and a vector $\delta x \in \mathbb{R}^n$.

Suppose that the result is not true, that is $F(\bar{t}) \neq 0$ for some $t_0 \leq \bar{t} \leq t_f$. Since F is continuous, it is necessarily nonzero and of a fixed sign in some neighborhood of $\bar{t}$. Then, an admissible variation can be constructed that is zero outside of this neighborhood and leads to a violation of the assumption. This contradiction proves the validity of the statement.

© Springer International Publishing AG 2018   521
T. Lee et al., *Global Formulations of Lagrangian and Hamiltonian
Dynamics on Manifolds*, Interaction of Mechanics and Mathematics,
DOI 10.1007/978-3-319-56953-6

A.2 Fundamental Lemma of the Calculus of Variations on an Embedded Manifold

Let M be a differentiable manifold embedded in $\mathbb{R}^n$. The fundamental lemma of the calculus of variations on M [5] is the following statement. Let $x : [t_0, t_f] \to M$ be continuous and suppose that $F : [t_0, t_f] \to (\mathbb{R}^n)^*$ is continuous. If

$$\int_{t_0}^{t_f} F(t) \cdot \xi(t)\, dt = 0,$$

holds for all continuous variations $\xi : [t_0, t_f] \to \mathsf{T}_{x(t)}M$ satisfying $\xi(t_0) = \xi(t_f) = 0$, then it follows that

$$F(t) \cdot \xi = 0, \quad t_0 \leq t \leq t_f,$$

that is $F(t)$ is orthogonal to $\mathsf{T}_x M$ for each $t_0 \leq t \leq t_f$. As usual, the notation $F \cdot \xi$ indicates the inner or dot product of a covector $F \in (\mathbb{R}^n)^*$ and a vector $\xi \in \mathsf{T}_{x(t)}M$.

The proof is essentially the same as the one above.

A.3 Fundamental Lemma of Variational Calculus on a Lie Group

The fundamental lemma of the calculus of variations can also be applied to a Lie group G, since we can associate the tangent space with the corresponding Lie algebra $\mathfrak{g}$. Let $\mathfrak{g}^*$ denote the vector space of linear functionals on $\mathfrak{g}$. This leads to the following statement.

The fundamental lemma of the calculus of variations on the Lie group G is the following statement. Let $g : [t_0, t_f] \to \mathsf{G}$ be continuous and suppose that $F : [t_0, t_f] \to \mathfrak{g}^*$ is continuous. If

$$\int_{t_0}^{t_f} \langle F(t) \cdot \eta \rangle\, dt = 0,$$

holds for all continuous variations $\eta : [t_0, t_f] \to \mathfrak{g}$ satisfying $\eta(t_0) = \eta(t_f) = 0$, then it follows that

$$\langle F(t) \cdot \eta \rangle = 0, \quad t_0 \leq t \leq t_f,$$

that is $F(t)$ is orthogonal to $\mathfrak{g}$ for each $t_0 \leq t \leq t_f$. As usual, the notation $\langle F \cdot \eta \rangle$ indicates the pairing of a covector $F \in \mathfrak{g}^*$ and a vector $\eta \in \mathfrak{g}$.

Appendix B
Linearization as an Approximation to Lagrangian Dynamics on a Manifold

Linearization of a (typically) nonlinear vector field on $\mathbb{R}^n$, in a neighborhood of an equilibrium of the vector field, is a classical and widely used technique. It provides a way to approximate the local nonlinear dynamics near an equilibrium solution, it may provide information about the stability of the equilibrium according to the stable manifold theorem [33], and it provides the basis for developing mathematical equations that are widely used in control applications. This technique is so widely used that we give special attention to it in this Appendix. A summary of linearization has been given in Chapter 1. Further background on linearization of a vector field on $\mathbb{R}^n$ is given in [44, 76, 93].

Here, we apply the concept of linearization to a vector field defined on a manifold in the form developed and studied in this book. We begin with a vector field defined on an $(n-m)$-dimensional differentiable manifold $M \subset \mathbb{R}^n$, where $1 \leq m < n$; the vector field is denoted by $F : M \to \mathbb{R}^n$ and it has the property that for each $x \in M, F(x) \in \mathsf{T}_x M$.

Consistent with the assumption that M is a manifold embedded in $\mathbb{R}^n$, we consider an extension of the vector field F on M to the embedding vector space. We denote this extension vector field by $F^e : \mathbb{R}^n \to \mathbb{R}^n$. Thus, for each $x \in M$, $F^e(x) = F(x)$. From the perspective of the embedding space, M can be viewed as an invariant manifold of the vector field $F^e : \mathbb{R}^n \to \mathbb{R}^n$. If $x_e \in M$ is an equilibrium of the vector field F, that is $F(x_e) = 0$, then it is also an equilibrium of the extended vector field F^e.

Linearization of the vector field $F : M \to \mathbb{R}^n$ in the neighborhood of an equilibrium solution $x_e \in M$ leads to a linear vector field on $\mathsf{T}_{x_e} M$ that is first-order accurate in approximating $F : M \to \mathbb{R}^n$ in a neighborhood of $x_e \in M$. It is convenient to describe the linear vector field on the $(n-m)$-dimensional tangent space in terms of $n-m$ basis vectors for $\mathsf{T}_{x_e} M$.

© Springer International Publishing AG 2018   523
T. Lee et al., *Global Formulations of Lagrangian and Hamiltonian Dynamics on Manifolds*, Interaction of Mechanics and Mathematics, DOI 10.1007/978-3-319-56953-6

There are two equivalent approaches that can be followed:

- Linearize the vector field F^e on $\mathbb{R}^n$ at the equilibrium $x_e \in M$; then restrict this linear vector field to the tangent space $\mathsf{T}_{x_e} M$ by introducing a basis for the tangent space $\mathsf{T}_{x_e} M$.
- Introduce $n - m$ local coordinates on M in a neighborhood of the equilibrium $x_e \in M$; express the vector field F in terms of these local coordinates and linearize at the equilibrium.

Either of these approaches leads to a linear vector field defined on an $(n-m)$-dimensional subspace of $\mathbb{R}^n$. The two approaches lead to equivalent realizations of the linearized vector field.

This linearized vector field can be viewed as approximating the original nonlinear vector field on M, at least in a small neighborhood of the equilibrium. In this sense, the linearized equations are viewed as describing local perturbations of the dynamical flow near the equilibrium. This interpretation provides important motivation for the linearization technique.

The second approach is most common in many engineering applications, but it requires the introduction of local coordinates on the manifold M and the description of the resulting vector field on M in terms of local coordinates; this is often a challenging step. In this book we emphasize the first approach, but we use whichever approach is most convenient for a particular case.

We now illustrate the linearization process for three classes of Lagrangian vector fields with configuration manifolds that are studied in this book: S^1, S^2, and $\mathsf{SO}(3)$. These illustrations provide the necessary background for linearization on configuration manifolds that are products of these. We obtain Euler–Lagrange equations that describe the dynamics for a particle or rigid body under the action of a potential. In each case, linearized differential equations are determined that approximate the dynamics on a manifold in a neighborhood of an equilibrium solution.

B.1 Linearization on TS^1

Consider an ideal particle, of mass m, that moves on a circle of radius $r > 0$ in a fixed plane in $\mathbb{R}^3$, centered at the origin of an inertial frame, under the action of a potential. The configuration manifold is S^1, embedded in $\mathbb{R}^2$. As shown in Chapter 4, the Euler–Lagrange equation, using standard notation, is given by

$$mL^2\ddot{q} + mr^2 \left\| \dot{q} \right\|^2 q + (I_{2\times 2} - qq^T)\frac{\partial U(q)}{\partial q} = 0.$$

These differential equations define a Lagrangian vector field on the tangent bundle of the configuration manifold TS^1. Assume that $q_e \in \mathsf{S}^1$ satisfies

$\frac{\partial U(q_e)}{\partial q} = 0$ so that $(q_e, 0) \in \mathsf{TS}^1$ is an equilibrium of the Lagrangian vector field. We obtain linearized equations following the first approach described above.

We can view the above differential equations as defining an extended Lagrangian vector field on the tangent bundle $\mathsf{T}\mathbb{R}^2$, assuming the potential function is defined everywhere on $\mathbb{R}^2$. We can linearize the extended differential equations at the equilibrium to obtain the linearized vector field defined on $\mathsf{T}\mathbb{R}^2$ as

$$mr^2\ddot{\xi} + \frac{\partial^2 U(q_e)}{\partial q^2}\xi = 0.$$

This linearization of the extended vector field can be restricted to the tangent space of TS^1 at $(q_e, 0) \in \mathsf{TS}^1$, which can be described as

$$\mathsf{T}_{(q_e,0)}\mathsf{TS}^1 = \left\{ (\xi, \dot{\xi}) \in \mathbb{R}^4 : q_e^T\xi = 0, \dot{\xi} \in \mathsf{T}_{q_e}\mathsf{S}^1 \right\}.$$

The above description of the linearized vector field is in the form of differential-algebraic equations. It is convenient to describe this linearized vector field in a more accessible form by introducing a basis for the tangent space $\mathsf{T}_{(q_e,0)}\mathsf{TS}^1$. To this end, select Sq_e as a basis vector for $\mathsf{T}_{q_e}\mathsf{S}^1$ so that any $(\xi, \dot{\xi}) \in \mathsf{T}_{(q_e,0)}\mathsf{TS}^1$ can be written as

$$\xi = \sigma Sq_e,$$
$$\dot{\xi} = \dot{\sigma} Sq_e,$$

where $\sigma \in \mathbb{R}^1$ can be viewed as a local coordinate for $\mathsf{T}_{q_e}\mathsf{S}^1$. Thus, $(\xi, \dot{\xi}) \in \mathsf{T}_{(q_e,0)}\mathsf{TS}^1$ for all $\sigma \in D$, where $D \subset \mathbb{R}^1$ is an open set containing the origin. Substituting these expressions into the equation for the linearization of the extended vector field and taking the inner product with Sq_e gives

$$mr^2\ddot{\sigma} + \left\{ q_e^T S^T \frac{\partial^2 U(q_e)}{\partial q^2} Sq_e \right\}\sigma = 0.$$

This scalar second-order differential equation describes the linearized vector field of the original Lagrangian vector field on TS^1. Thus, this differential equation, expressed in terms of $(\sigma, \dot{\sigma}) \in \mathsf{T}\mathbb{R}^1$, describes the Lagrangian dynamics on the manifold TS^1 in a neighborhood of $(q_e, 0) \in \mathsf{TS}^1$ to first order in the perturbations.

This linearized equation has been described as a second-order differential equation in σ, consistent with the usual formulation of the Euler–Lagrange equations. They can also be expressed as a system of first-order differential equations by including perturbations of the angular velocity or momentum.

B.2 Linearization on $\mathbf{TS}^2$

Consider an ideal particle, of mass m, that moves on a sphere in $\mathbb{R}^3$ of radius $r > 0$, centered at the origin of an inertial frame, under the action of a potential. The configuration manifold is S^2, embedded in $\mathbb{R}^3$. As shown in Chapter 5, the Euler–Lagrange equation, using standard notation, is

$$mr^2\ddot{q} + mr^2 \left\| \dot{q} \right\|^2 q + (I_{3\times 3} - qq^T)\frac{\partial U(q)}{\partial q} = 0.$$

These differential equations define a Lagrangian vector field on the tangent bundle of the configuration manifold TS^2. Assume that $q_e \in \mathsf{S}^2$ satisfies $\frac{\partial U(q_e)}{\partial q} = 0$ so that $(q_e, 0) \in \mathsf{TS}^2$ is an equilibrium of the Lagrangian vector field. We obtain linearized equations following the first approach described above. The same approach, for different dynamics, is followed in [58].

We can view the above differential equations as defining an extended Lagrangian vector field on the tangent bundle $\mathsf{T}\mathbb{R}^3$, assuming the potential function is defined everywhere on $\mathbb{R}^3$. We can linearize the extended differential equations at the equilibrium to obtain the linearized vector field defined on $\mathsf{T}\mathbb{R}^3$ as described by

$$mr^2\ddot{\xi} + \frac{\partial^2 U(q_e)}{\partial q^2}\xi = 0.$$

This linearization of the extended vector field can be restricted to the tangent space of TS^2 at $(q_e, 0) \in \mathsf{TS}^2$, that is

$$\mathsf{T}_{(q_e,0)}\mathsf{TS}^2 = \left\{ (\xi, \dot{\xi}) \in \mathbb{R}^6 : q_e^T\xi = 0, \ \dot{\xi} \in \mathsf{T}_{q_e}\mathsf{S}^2 \right\}.$$

The above description of the linearized vector field is in the form of differential-algebraic equations. It is convenient to describe this linearized vector field in a more accessible form by introducing a basis for the tangent space $\mathsf{T}_{(q_e,0)}\mathsf{TS}^2$. To this end, select ξ_1, ξ_2 as an orthonormal basis for $\mathsf{T}_{(q_e,0)}\mathsf{TS}^2$. Thus, any $(\xi, \dot{\xi}) \in \mathsf{T}_{(q_e,0)}\mathsf{TS}^2$ can be written as

$$\xi = \sigma_1\xi_1 + \sigma_2\xi_2,$$
$$\dot{\xi} = \dot{\sigma}_1\xi_1 + \dot{\sigma}_2\xi_2,$$

where $\sigma = (\sigma_1, \sigma_2) \in \mathbb{R}^2$ can be viewed as local coordinates for $\mathsf{T}_{q_e}\mathsf{S}^2$. Thus, $(\xi, \dot{\xi}) \in \mathsf{T}_{(q_e,0)}\mathsf{TS}^2$ for all $\sigma \in D$, where $D \subset \mathbb{R}^2$ is an open set containing the origin. Substituting these expressions into the equation for the linearization of the extended vector field and taking the inner product with the basis vectors ξ_1 and ξ_2 gives

$$mr^2 \begin{bmatrix} \ddot{\sigma}_1 \\ \ddot{\sigma}_2 \end{bmatrix} + \begin{bmatrix} \xi_1^T \frac{\partial^2 U(q_e)}{\partial q^2} \xi_1 & \xi_1^T \frac{\partial^2 U(q_e)}{\partial q^2} \xi_2 \\ \xi_2^T \frac{\partial^2 U(q_e)}{\partial q^2} \xi_1 & \xi_2^T \frac{\partial^2 U(q_e)}{\partial q^2} \xi_2 \end{bmatrix} \begin{bmatrix} \sigma_1 \\ \sigma_2 \end{bmatrix} = \begin{bmatrix} 0 \\ 0 \end{bmatrix}.$$

This system of second-order linear differential equations describe the linearized vector field of the original Lagrangian vector field on TS^2. Thus, this differential equation, expressed in terms of $(\sigma, \dot{\sigma}) \in \mathsf{T}\mathbb{R}^2$, describes the Lagrangian dynamics on the manifold TS^2 in a neighborhood of $(q_e, 0) \in \mathsf{TS}^2$ to first order in the perturbations.

These linearized equations have been described as a system of second-order differential equations, consistent with the usual formulation of the Euler–Lagrange equations. They can also be expressed as a system of first-order differential equations by including perturbations of the angular velocity vector or the momentum.

B.3 Linearization on TSO(3)

Consider a rigid body with moment of inertia J that rotates under the action of a potential. The configuration manifold is $\mathsf{SO}(3)$, embedded in $\mathbb{R}^{3\times 3}$. As shown in Chapter 6, the rotational kinematics are

$$\dot{R} = RS(\omega),$$

and the Euler equations are

$$J\dot{\omega} + \omega \times J\omega - \sum_{i=1}^{3} r_i \times \frac{\partial U(R)}{\partial r_i} = 0.$$

These differential equations define a vector field on $\mathsf{TSO}(3)$. Assume $R_e \in \mathsf{SO}(3)$ satisfies $\frac{\partial U(R_e)}{\partial r_i} = 0$, $i = 1, 2, 3$, so that $(R_e, 0) \in \mathsf{TSO}(3)$ is an equilibrium of the vector field defined by the rotational kinematics and Euler's equations. We obtain linearized equations following the second approach described above. The same approach, for different dynamics, is followed in [20].

We can linearize the above differential equations by first introducing local coordinates on $\mathsf{SO}(3)$. We use the exponential representation

$$R = R_e e^{S(\theta)},$$

where $\theta = (\theta_1, \theta_2, \theta_3) \in D \subset \mathbb{R}^3$, and D is an open set containing the origin. As indicated in Chapter 1, the exponential map $\theta \in D \rightarrow R \in \mathsf{SO}(3)$ is a local diffeomorphism.

The kinematics of a rotating rigid body in $\mathsf{SO}(3)$ are now expressed in terms of local coordinates in $D \subset \mathbb{R}^3$. The angular velocity vector of the rotating rigid body can be expressed in terms of the time derivatives of the local coordinates. The following implications

$$S(\omega) = R^T \dot{R}$$
$$= e^{-S(\theta)} R_e^T R_e e^{S(\theta)} S(\dot{\theta})$$
$$= S(\dot{\theta}),$$

demonstrate that $\omega = \dot{\theta}$, where we have used the fact that the skew-symmetric function $S : \mathbb{R}^3 \to \mathfrak{so}(3)$ is locally invertible.

The Euler equations can be written in terms of local exponential coordinates as

$$J\ddot{\theta} + \dot{\theta} \times J\dot{\theta} - \sum_{i=1}^{3} R_e e^{-S(\theta)} e_i \times \frac{\partial U(R_e e^{S(\theta)})}{\partial r_i} = 0.$$

These equations define the rotational dynamics of the rigid body on a subset of $\mathsf{TSO}(3)$ in terms of local coordinates $\theta \in D \subset \mathbb{R}^3$. These are complicated equations, but they can be easily linearized in a neighborhood of $(R_e, 0) \in \mathsf{TSO}(3)$, or equivalently in terms of local coordinates in a neighborhood of the origin in D. Linearization of the differential equations in local coordinates gives

$$J\ddot{\sigma} - \sum_{i=1}^{3} r_{ei} \times \frac{\partial^2 U(R_e)}{\partial r_i^2} \sigma = 0.$$

Here, $r_{ei} = R_e^T e_i \in \mathsf{S}^2$, $i = 1, 2, 3$. This linear vector differential equation describes the linearized vector field of the Lagrangian vector field on $\mathsf{TSO}(3)$; it is expressed in terms of perturbations σ from the equilibrium $(R_e, 0) \in \mathsf{TSO}(3)$.

These linearized equations have been described as a system of second-order differential equations, consistent with the usual formulation of the Euler–Lagrange equations. They can also be expressed as a system of first-order differential equations by including perturbations of the angular velocity vector or the momentum.

References

1. R. Abraham, J.E. Marsden, *Foundations of Mechanics*, 2nd edn. (Benjamin/Cummings, Reading, 1978)
2. R. Abraham, J.E. Marsden, T.S. Ratiu, *Manifolds, Tensor Analysis, and Applications* (Springer, New York, 1988)
3. F. Amirouche, *Fundamentals of Multibody Dynamics* (Birkhäuser, Boston, 2006)
4. T.M. Apostol, M.A. Mnatsakanian, A new look at the so-called trammel of Archimedes. Am. Math. Mon. **116**, 115–133 (2009)
5. V.I. Arnold, *Mathematical Methods of Classical Mechanics* (Springer, New York, 1989)
6. G.L. Baker, J.A. Blackmun, *The Pendulum: A Case Study in Physics* (Oxford University Press, New York, 2005)
7. R.E. Bellman, *Introduction to Matrix Analysis*, 2nd edn. (Society for Industrial and Applied Mathematics, Philadelphia, 1997)
8. D.S. Bernstein, *Matrix Mathematics, Theory, Facts, and Formulas with Applications to Linear System Theory* (Princeton University Press, Princeton, 2005)
9. B. Bittner, K. Sreenath, Symbolic computation of dynamics on smooth manifolds, in *Workshop on Algorithmic Foundations of Robotics* (2016)
10. A.M. Bloch, *Nonholonomic Mechanics and Control*. Interdisciplinary Applied Mathematics, vol. 24 (Springer, New York, 2003)
11. A.M. Bloch, P.E. Crouch, A.K. Sanyal, A variational problem on Stiefel manifolds. Nonlinearity **19**, 2247–2276 (2006)
12. K.E. Brenan, S.L. Campbell, L.R. Petzold, *Numerical Solution of Initial-Value Problems in Differential-Algebraic Equations* (Society for Industrial and Applied Mathematics, Philadelphia, 1996)
13. A.J. Brizard, *An Introduction to Lagrangian Mechanics* (World Scientific, Singapore, 2008)

© Springer International Publishing AG 2018
T. Lee et al., *Global Formulations of Lagrangian and Hamiltonian Dynamics on Manifolds*, Interaction of Mechanics and Mathematics,
DOI 10.1007/978-3-319-56953-6

14. R.W. Brockett, Lie theory and control systems defined on spheres. SIAM J. Appl. Math. **25**, 213–225 (1973)
15. B. Brogliato, *Nonsmooth Mechanics* (Springer, London, 1999)
16. F. Bullo, A.D. Lewis, *Geometric Control of Mechanical Systems*. Texts in Applied Mathematics, vol. 49 (Springer, New York, 2005). Modeling, analysis, and design for simple mechanical control systems
17. H. Cendra, J.E. Marsden, T.S. Ratiu, Lagrangian reduction by stages. Mem. Am. Math. Soc. **152**(722), x+108 (2001)
18. N.A. Chaturvedi, F. Bacconi, A.K. Sanyal, D.S. Bernstein, N.H. Mc-Clamroch, Stabilization of a 3D rigid pendulum, in *Proceedings of the American Control Conference* (2005), pp. 3030–3035
19. N.A. Chaturvedi, N.H. McClamroch, D.S. Bernstein, Stabilization of a 3D axially symmetric pendulum. Automatica **44**(9), 2258–2265 (2008)
20. N.A. Chaturvedi, N.H. McClamroch, D. Bernstein, Asymptotic smooth stabilization of the inverted 3-D pendulum. IEEE Trans. Autom. Control **54**(6), 1204–1215 (2009)
21. N.A. Chaturvedi, T. Lee, M. Leok, N. Harris McClamroch, Nonlinear dynamics of the 3D pendulum. J. Nonlinear Sci. **21**(1), 3–32 (2011)
22. G.S. Chirikjian, *Stochastic Models, Information Theory, and Lie Groups*. Analytic Methods and Modern Applications, vol. 2 (Birkhauser, Basel, 2011)
23. G.S. Chirikjian, A.B. Kyatkin, *Harmonic Analysis for Engineers and Applied Scientists* (Dover Publications, Mineola, 2016)
24. D. Condurache, V. Martinusi, Foucault pendulum-like problems: a tensorial approach. Int. J. Non Linear Mech. **43**(8), 743–760 (2008)
25. J. Cortés Monforte, *Geometric, Control, and Numerical Aspects of Nonholonomic Systems*. Lecture Notes in Mathematics, vol. 1793 (Springer, New York, 2002)
26. P.E. Crouch, Spacecraft attitude control and stabilizations: applications of geometric control theory to rigid body models. IEEE Trans. Autom. Control **29**(4), 321–331 (1984)
27. M. de León, P.R. Rodrigues, *Methods of Differential Geometry in Analytical Mechanics*. Mathematics Studies (North Holland, Amsterdam, 1989)
28. J.J. Duistermaat, J.A.C. Kolk, *Lie groups*. Universitext (Springer, Berlin, 2000)
29. C. Gignoux, B. Silvestre-Brac, *Solved Problems in Lagrangian and Hamiltonian Mechanics* (Springer, Dordrecht, 2009)
30. H. Goldstein, C.P. Poole, J.L. Safko, *Classical Mechanics* (Addison Wesley, Reading, 2001)
31. F. Goodarzi, D. Lee, T. Lee, Geometric stabilization of a quadrotor UAV with a payload connected by flexible cable, in *Proceedings of the American Control Conference* (2014), pp. 4925–4930
32. D. Greenwood, *Classical Dynamics* (Prentice Hall, Englewood Cliffs, 1977)

33. J. Guckenheimer, P. Holmes, *Nonlinear Oscillations, Dynamical Systems, and Bifurcations of Vector Fields* (Springer, Berlin, 1983)
34. E. Hairer, C. Lubich, G. Wanner, *Geometric Numerical Integration.* Springer Series in Computational Mechanics, vol. 31 (Springer, Berlin, 2000)
35. S. Helgason, *Differential Geometry, Lie Groups, and Symmetric Spaces.* Graduate Studies in Mathematics, vol. 34. (American Mathematical Society, Providence, 2001). Corrected reprint of the 1978 original
36. A. Hernández-Garduño, J.K. Lawson, J.E. Marsden, Relative equilibria for the generalized rigid body. J. Geom. Phys. **53**(3), 259–274 (2005)
37. D. Holm, *Geometric Mechanics, Part I: Dynamics and Symmetry* (Imperial College Press, London, 2008)
38. D. Holm, *Geometric Mechanics, Part II: Rotating, Translating and Rolling* (Imperial College Press, London, 2011)
39. D. Holm, T. Schmah, C. Stoica, *Geometric Mechanics and Symmetry: From Finite to Infinite Dimensions.* (Oxford University Press, Oxford, 2009)
40. P.C. Hughes, *Spacecraft Attitude Dynamics* (Wiley, New York, 1986)
41. A. Iserles, H. Munthe-Kaas, S.P. Nørsett, A. Zanna, Lie-group methods, in *Acta Numerica*, vol. 9 (Cambridge University Press, Cambridge, 2000), pp. 215–365
42. A. Jones, A. Gray, R. Hutton, *Manifolds and Mechanics* (Australian Mathematical Society, 1987)
43. S. Kobayashi, K. Furuta, M. Yamakita, Swing-up control of inverted pendulum using pseudo-state feedback. J. Syst. Control Eng. **206**(6), 263–269 (1992)
44. H.K. Khalil, *Nonlinear Systems* (Prentice Hall, Upper Saddle River, 2002)
45. C. Lanczos, *The Variational Principles of Mechanics* (University of Toronto Press, Toronto, 1962)
46. T. Lee, Computational Geometric Mechanics and Control of Rigid Bodies. Ph.D Thesis, University of Michigan (2008)
47. T. Lee, Geometric control of multiple quadrotor UAVs transporting a cable-suspended rigid body, in *Proceedings of the IEEE Conference on Decision and Control* (2014), pp. 6155–6160
48. T. Lee, F. Leve, Lagrangian mechanics and Lie group variational integrators for spacecraft with imbalanced reaction wheels, in *Proceedings of the American Control Conference* (2014), pp. 3122–3127
49. T. Lee, M. Leok, N.H. McClamroch, A Lie group variational integrator for the attitude dynamics of a rigid body with application to the 3D pendulum, in *Proceedings of the IEEE Conference on Control Application* (2005), pp. 962–967
50. T. Lee, M. Leok, N.H. McClamroch, Lie group variational integrators for the full body problem. Comput. Methods Appl. Mech. Eng. **196**, 2907–2924 (2007)

51. T. Lee, M. Leok, N.H. McClamroch, Lie group variational integrators for the full body problem in orbital mechanics. Celest. Mech. Dyn. Astron. **98**(2), 121–144 (2007)

52. T. Lee, M. Leok, N.H. McClamroch, Dynamics of a 3D elastic string pendulum, in *Proceedings of IEEE Conference on Decision and Control* (2009), pp. 3447–3352

53. T. Lee, M. Leok, N.H. McClamroch, Dynamics of connected rigid bodies in a perfect fluid, in *Proceedings of the American Control Conference* (2009), pp. 408–413

54. T. Lee, M. Leok, N.H. McClamroch, Lagrangian mechanics and variational integrators on two-spheres. Int. J. Numer. Methods Eng. **79**(9), 1147–1174 (2009)

55. T. Lee, M. Leok, N.H. McClamroch, Geometric tracking control of a quadrotor UAV on SE(3), in *Proceedings of the IEEE Conference on Decision and Control* (2010), pp. 5420–5425

56. T. Lee, M. Leok, N.H. McClamroch, Computational dynamics of a 3D elastic string pendulum attached to a rigid body and an inertially fixed reel mechanism. Nonlinear Dyn. **64**(1–2), 97–115 (2011)

57. T. Lee, M. Leok, N.H. McClamroch, Geometric numerical integration for complex dynamics of tethered spacecraft, in *Proceedings of the American Control Conference* (2011), pp. 1885–1891

58. T. Lee, M. Leok, N.H. McClamroch, Stable manifolds of saddle points for pendulum dynamics on S^2 and SO(3), in *Proceedings of the IEEE Conference on Decision and Control* (2011), pp. 3915–3921

59. T. Lee, M. Leok, N.H. McClamroch, Dynamics and control of a chain pendulum on a cart, in *Proceedings of the IEEE Conference on Decision and Control* (2012), pp. 2502–2508

60. T. Lee, K. Sreenath, V. Kumar, Geometric control of cooperating multiple quadrotor UAVs with a suspended load, in *Proceedings of the IEEE Conference on Decision and Control* (2013), pp. 5510–5515

61. T. Lee, M. Leok, N.H. McClamroch, High-fidelity numerical simulation of complex dynamics of tethered spacecraft. Acta Astronaut. **99**, 215–230 (2014)

62. T. Lee, M. Leok, N.H. McClamroch, Global formulations of Lagrangian and Hamiltonian dynamics on embedded manifolds, in *Proceedings of the IMA Conference on Mathematics of Robotics* (2015)

63. T. Lee, M. Leok, N.H. McClamroch, Global formulations of Lagrangian and Hamiltonian mechanics on two-spheres, in *Proceedings of the IEEE Conference on Decision and Control* (2015), pp. 6010–6015

64. T. Lee, F. Leve, M. Leok, N.H. McClamroch, Lie group variational integrators for spacecraft with variable speed control moment gyros, in *Proceedings of the U.S. National Congress on Computational Mechanics* (2015)

65. B. Leimkuhler, S. Reich, *Simulating Hamiltonian Dynamics*, vol. 14 of *Cambridge Monographs on Applied and Computational Mathematics* (Cambridge University Press, Cambridge, 2004)
66. F. Leve, B. Hamilton, M. Peck, *Spacecraft Momentum Control Systems*. Space Technology Library (Springer, Cham, 2015)
67. A.J. Maciejewski, Reduction, relative equilibria and potential in the two rigid bodies problem. Celest. Mech. Dyn. Astron. **63**, 1–28 (1995)
68. F.L. Markley, J.L. Crassidis, *Fundamentals of Spacecraft Attitude Determination and Control* (Springer, New York, 2014)
69. J.E. Marsden, *Lectures on Mechanics*. London Mathematical Society Lecture Note Series, vol. 174 (Cambridge University Press, Cambridge, 1992)
70. J.E. Marsden, T.S. Ratiu, *Introduction to Mechanics and Symmetry*. Texts in Applied Mathematics, vol. 17, 2nd edn. (Springer, New York, 1999)
71. J.E. Marsden, M. West, Discrete mechanics and variational integrators, in *Acta Numerica*, vol. 10 (Cambridge University Press, Cambridge, 2001), pp. 317–514
72. J.E. Marsden, R. Montgomery, T.S. Ratiu, *Reduction, Symmetry and Phases in Mechanics* (American Mathematical Society, Providence, 1990)
73. J.E. Marsden, J. Scheurle, J.M. Wendlandt, Visualization of orbits and pattern evocation for the double spherical pendulum. Z. Angew. Math. Phys., **44**(1), 17–43 (1993)
74. J.E. Marsden, G. Misiołek, J.-P. Ortega, M. Perlmutter, T.S. Ratiu, *Hamiltonian Reduction by Stages*. Lecture Notes in Mathematics, vol. 1913 (Springer, Berlin, 2007)
75. L. Meirovitch, *Methods of Analytical Dynamics* (McGraw-Hill, New York, 1970)
76. A.N. Michel, L. Liu, D. Liu, *Stability of Dynamical Systems: On the Role of Monotonic and Non-monotonic Lyapunov Functions*, 2nd edn. (Birkhauser, Boston, 2015)
77. R.M. Murray, Z. Li, S.S. Sastry, *A Mathematical Introduction to Robotic Manipulation* (CRC Press, Boca Raton, 1993)
78. P.J. Olver, *Applications of Lie Groups to Differential Equations*. Graduate Texts in Mathematics, vol. 107, 2nd edn. (Springer, New York, 1993)
79. L.R. Petzold, Differential/algebraic equations are not odes. SIAM J. Sci. Stat. Comput. **3**, 367–384 (1982)
80. J.E. Prussing, B.A. Conway, *Orbital Mechanics* (Oxford University Press, Oxford, 1993)
81. P.J. Rabier, W.C. Rheinboldt, *Nonholonomic Motion of Rigid Mechanical Systems from a DAE Viewpoint* (Society for Industrial and Applied Mathematics, Philadelphia, 2000)

82. P.J. Rabier, W.C. Rheinboldt, *Theoretical and Numerical Analysis of Differential-Algebraic Equations*. (North Holland, New York, 2002)
83. W.C. Rheinboldt, Differential-algebraic systems as differential equations on manifolds. Math. Comput. **43**(168), 473–482 (1984)
84. C. Rui, I.V. Kolmanovsky, N.H. McClamroch, Nonlinear attitude and shape control of spacecraft with articulated appendages and reaction wheels. IEEE Trans. Autom. Control **45**, 1455–1469 (2000)
85. S. Sastry, *Nonlinear Systems: Analysis, Stability and Control* (Springer, New York, 1999)
86. A.A. Shabana, Flexible multibody system dynamics: review of past and recent developments. Multibody Syst. Dyn. **1**, 189–222 (1997)
87. J. Shen, A.K. Sanyal, N.A. Chaturvedi, D.S. Bernstein, N.H. McClamroch, Dynamics and control of a 3D pendulum, in *Proceedings of IEEE Conference on Decision and Control* (2004), pp. 323–328
88. J. Simo, L. Vu-Quoc, On the dynamics in space of rods undergoing large motions – a geometrically exact approach. Comput. Methods Appl. Mech. Eng. **66**, 125–161 (1988)
89. S.F. Singer, *Symmetry in Mechanics: A Gentle, Modern Introduction* (Birkhauser, Boston, 2001)
90. R.E. Skelton, M.C. Oliveira, *Tensegrity Systems* (Springer, Berlin, 2009)
91. M. Spivak, *A Comprehensive Introduction to Differential Geometry*, vol. I, 2nd edn. (Publish or Perish, Inc., Wilmington, 1979)
92. R. Talman, *Geometric Mechanics* (Wiley, New York, 2008)
93. G. Teschi, *Ordinary Differential Equations and Dynamical Systems* (American Mathematical Society, Providence, 2012)
94. F. Udwadia, Constrained motion of Hamiltonian systems. Nonlinear Dyn. **84**, 1135–1145 (2015)
95. F. Udwadia, R.E. Kalaba, Equations of motion for constrained mechanical systems and the extended d'Alembert principle. Q. Appl. Math. **LV**, 321–331 (1997)
96. A.J. van der Schaft, Equations of motion for Hamiltonian systems with constraints. J. Phys. A **20**, 3271–3277 (1987)
97. V.S. Varadarajan, *Lie Groups, Lie Algebras, and their Representations*. Graduate Texts in Mathematics, vol. 102 (Springer, New York, 1984) Reprint of the 1974 edition.
98. R.W. Weber, Hamiltonian constraints and their meaning in mechanics. Arch. Ration. Mech. Anal. **91**, 309–335 (1996)
99. E.T. Whittaker, *A Treatise on the Analytical Dynamics of Particles and Rigid Bodies* (Cambridge University Press, Cambridge, 1961)
100. S. Wiggins, *Introduction to Applied Dynamical Systems and Chaos*. (Springer, New York, 2003)

Index

action
  left, 381
  right, 381
  transitive, 381
action integral, viii, 91, 134, 211, 277,
    317, 373
  infinitesimal variation, 135, 138, 211,
    215, 278, 318, 374, 387
adjoint operator, 372
angular momentum, 282, 323
angular velocity, 132, 275
atlas, 12

calculus of variations, 521
  fundamental lemma, 136, 139, 212,
    350, 355, 375, 521
canonical transformation, vi
Cayley transformation, 8
centripetal, 121
charts, vii, 12
Christoffel, 94, 214
closed loop, 85, 86
coadjoint operator, 372
computational dynamics, ix, 42
configuration, vii, 55, 89, 131, 313
configuration manifold
  infinite-dimensional, 485
conjugate momentum, 95, 99, 143, 147,
    222, 226, 228, 453, 462
conservative, vii
conserved quantity, 29
control effects, 109
coordinate-free, vii, 59

coordinates, vii, 12, 59
Coriolis, 121
cotangent bundle, 14, 98, 99, 132, 146,
    148, 151, 208, 226, 228, 231,
    282–284, 325, 326, 368, 377
cotangent space, 14, 43, 132, 208
cotangent vector, 14, 208
covector, 14, 43, 521
covector field
  on manifold, 43
cross product, vi, 6
cyclic coordinate, 101

Darboux's theorem, 104
deformable body, x, 54, 485
derivative, 11
  directional, 11
  gradient, 11
diffeomorphic, 11, 401
diffeomorphism, 11, 13, 134, 205, 206,
    210, 270, 353, 372, 373, 376, 381
differential equations, 27, 28
differential form, 104
differential geometry, vi, x, 12, 29, 131,
    207
differential-algebraic equations, 29, 107,
    368
  index one, 30
  index two, 30, 31, 107, 367, 368
dimension, 2
discrete-time, ix, 42
dot product, vi, 2, 521
dual, 14, 132

© Springer International Publishing AG 2018   535
T. Lee et al., *Global Formulations of Lagrangian and Hamiltonian
Dynamics on Manifolds*, Interaction of Mechanics and Mathematics,
DOI 10.1007/978-3-319-56953-6

Printed in the United States
By Bookmasters